High Performance Computing Symposium 2011

(HPCS 2011)

Journal of Physics: Conference Series Volume 341

Quebec, Montreal, Canada
15-17 June 2011

ISBN: 978-1-62276-687-1
ISSN: 1742-6588

Printed from e-media with permission by:

Curran Associates, Inc.
57 Morehouse Lane
Red Hook, NY 12571

Printed by Curran Associates, Inc. (2013)

For permission requests, please contact the Institute of Physics
at the address below.

Institute of Physics
Dirac House, Temple Back
Bristol BS1 6BE UK

Phone: 44 1 17 929 7481
Fax: 44 1 17 920 0979

techtracking@iop.org

Additional copies of this publication are available from:

Curran Associates, Inc.
57 Morehouse Lane
Red Hook, NY 12571 USA
Phone: 845-758-0400
Fax: 845-758-2634
Email: curran@proceedings.com
Web: www.proceedings.com

TABLE OF CONTENTS

HPC INFRASTRUCTURE IN CANADA

HPC IN ENGINEERING, PHYSICS AND MATERIALS SCIENCE

HPC IN MEDICAL SCIENCE

HPC ENABLING TO EXPLORE OUR WORLD

NEW ALGORITHMS FOR HPC

Journal of Physics: Conference Series **341** (2012) 011001 doi:10.1088/1742-6596/341/1/011001

PREFACE

HPCS (High Performance Computing Symposium) is a multidisciplinary conference that focuses on research involving High Performance Computing and its application. Attended by Canadian and international experts and renowned researchers in the sciences, all areas of engineering, the applied sciences, medicine and life sciences, mathematics, the humanities and social sciences, it is Canada's pre-eminent forum for HPC.

The 25th edition was held in Montréal, at the Université du Québec à Montréal, from 15-17 June and focused on *HPC in Medical Science.* The conference was preceded by tutorials held at Concordia University, where 56 participants learned about HPC best practices, GPU computing, parallel computing, debugging and a number of high-level languages. 274 participants from six countries attended the main conference, which involved 11 invited and 37 contributed oral presentations, 33 posters, and an exhibit hall with 16 booths from our sponsors. The work that follows is a collection of papers presented at the conference covering HPC topics ranging from computer science to bioinformatics. They are divided here into four sections: HPC in Engineering, Physics and Materials Science, HPC in Medical Science, HPC Enabling to Explore our World and New Algorithms for HPC.

We would once more like to thank the participants and invited speakers, the members of the Scientific Committee, the referees who spent time reviewing the papers and our invaluable sponsors.

To hear the invited talks and learn about 25 years of HPC development in Canada visit the Symposium website http://2011.hpcs.ca/lang/en/conference/keynote-speakers/.

Enjoy the excellent papers that follow, and we look forward to seeing you in Vancouver for HPCS 2012!

Gilles Peslherbe
Chair of the Scientific Committee

Normand Mousseau
Co-Chair of HPCS 2011

Suzanne Talon
Chair of the Organizing Committee

Sponsors

Conference Photographs

All papers published in this volume of *Journal of Physics: Conference Series* have been peer reviewed through processes administered by the proceedings Editors. Reviews were conducted by expert referees to the professional and scientific standards expected of a proceedings journal published by IOP Publishing.

Journal of Physics: Conference Series **341** (2012) 012001 doi:10.1088/1742-6596/341/1/012001

Compute Canada: Advancing Computational Research

Susan Baldwin
Executive Director, Compute Canada, 1200 Montreal Road, Building M-50 IPF 226B
Ottawa, Ontario K1A 0R6

E-mail: susan.baldwin@computecanada.org

Abstract. High Performance Computing (HPC) is redefining the way that research is done. Compute Canada's HPC infrastructure provides a national platform that enables Canadian researchers to compete on an international scale, attracts top talent to Canadian universities and broadens the scope of research.

1. Overview

Compute Canada is leading the creation of a powerful national HPC platform for research. This national platform integrates High Performance Computing (HPC) resources at seven partner consortia across the country to create a dynamic computational resource. Compute Canada integrates high-performance computers, data resources and tools, and academic research facilities around the country. These integrated resources currently represent close to a petaflop of computing capability and online and long term storage with rapid access and retrieval over Canada's national, provincial and territorial high-performance networks.

Working in collaboration, Compute Canada and the seven university-based regional HPC consortia (WestGrid, SHARCNET, SciNet, HPCVL, RQCHP, CLUMEQ and ACEnet) provide for overall architecture and planning, software integration, operations and management, and coordination of user support for the national HPC platform. As a national organization, Compute Canada coordinates and promotes the use of HPC in Canadian research and works to ensure that Canadian researchers have the computational facilities and expert services necessary to advance scientific knowledge and innovation.

2. Resources

Compute Canada supports researchers in many different disciplines who have different requirements and varying skill levels with respect to HPC, and projects that range in scope from an individual project to multi-country projects requiring international agreements that stipulate the HPC commitment. The use of high-performance computing across many research disciplines is evidence of its necessity as a scientific tool. The national platform is composed of highly qualified personnel, computational facilities, and software and is accessible by researchers independent of system or researcher location.

3. Highly Qualified Personnel

The key "product" academia provides is Highly Qualified Personnel (HQP). The development of HQP constitutes the most effective form of knowledge transfer. Computationally trained individuals are a very valuable output of the academic environment. These individuals go into businesses and enable those businesses to take advantage of HPC for reducing the time for innovation and the time to get a product to market. We must continue to develop these talented and creative people in order to position our private sector to be early adopters of new computing power and technologies and realize the resulting economic advantages and wealth creation.

As computing architectures evolve to provide enhanced performance, the associated software, both system software and applications software, must be re-written to accommodate the new more complex and sophisticated features. This highly specialized software will continue to need to be re-written and a new generation of software tools must be developed in order to keep pace with architectural advances. Individual scientists and researchers are usually not specialists in the development of the software they need to perform their research. For this reason, HPC specialists within Compute Canada are an essential component of the scientific research that is dependent upon HPC. These individuals are part of the competitive advantage Compute Canada's provides.

4. Facilities

In order to serve our broad base of constituencies and to provide pathways for researchers to develop expertise to use more sophisticated systems as demanded by their research programs, Compute Canada provides a full ecosystem of capabilities. High Performance Computing requires the availability of many different capabilities from desktop/small lab computing, to mid-range computing to the very largest systems (Tier 1) as well as new architectures such as GPGPU and cloud computing.

Tied to the demand for computational power is the need for vast amounts of storage. In the past data volumes have been driven primarily by larger models running on larger computers; however, the volume of data from other sources is now beginning to dominate. Large experiments and instruments such as ATLAS and the SKA are generating or will generate petabytes to 100s of petabytes of data. In addition, many other activities are generating data: gene sequencing, medical imaging, sensor networks, satellite data to name just a few. The sources of digital data are growing at a rate that is greater than the rate at which hardware storage capacity is growing. The provision of data storage and the management and archiving of large datasets is a key focus of Compute Canada.

5. Towards the Future

The rapid and continuous progress in computer technology is a powerful driver of the HPC environment and community. This dramatic growth enables researchers to tackle problems today that were unimaginable a few years ago. It has also resulted in a number of strategic challenges that Compute Canada must meet in order to leverage these trends effectively. The scale of modern massively parallel systems presents extreme challenges in terms of both the infrastructure required to support them and the software tools to effectively utilize them.

The combination of the sheer scale of large systems together with their increasingly heterogeneous nature makes writing efficient software for them an extreme challenge. It is generally recognized that current architectures have far outstripped the availability of software that can take advantage of their capabilities. As an example, when an application program must take advantage not only of all cores in a node but of thousands of cores across nodes, the challenge can be extreme. These trends highlight the need for skilled technical support, including programmers, to ensure that Canadian researchers are able to take advantage of contemporary and future systems. It will be increasingly challenging for individual programmers to develop applications for such systems and the possibility of writing a one-off program for a specific short-term need for such a machine is becoming remote. Compute Canada will actively encourage and enable the use of these systems – at both the mid-range and Tier 1 levels — through training and programming support. The development of a culture of software is a long-term investment and one in which we must invest.

The success of Canadian research depends upon our researchers having access to the mandatory tools for research in the twenty-first century. In an era of massive data volumes and the need to understand complex phenomena, High Performance Computing has become an indispensable tool for

Journal of Physics: Conference Series **341** (2012) 012001 doi:10.1088/1742-6596/341/1/012001

the creation and advancement of knowledge and the resulting innovation, wealth creation and well being for Canadians. With the capabilities afforded by HPC resources, scientists and researchers are able to address issues of immediate significance for contemporary society.

6. References

[1] Compute Canada Strategic Plan
https://computecanada.org/en/Compute_Canada_Strategic_Plan_2011

High performance modeling of atmospheric re-entry vehicles

Alexandre Martin[1], Leonardo C. Scalabrin[2] and Iain D. Boyd[3]

[1] Department of Mechanical Engineering, University of Kentucky, Lexington, KY, 40506, USA
[2] Embraer S.A., São José dos Campos, 12243-000, Brazil
[3] Department of Aerospace Engineering, University of Michigan, Ann Arbor, MI, 48109, USA

E-mail: alexandre.martin@uky.edu, leonardo.scalabrin@embraer.com.br, iainboyd@umich.edu

Abstract. Re-entry vehicles designed for space exploration are usually equipped with thermal protection systems made of ablative material. In order to properly model and predict the aerothermal environment of the vehicle, it is imperative to account for the gases produced by ablation processes. In the case of charring ablators, where an inner resin is pyrolyzed at a relatively low temperature, the composition of the gas expelled into the boundary layer is complex and may lead to thermal chemical reactions that cannot be captured with simple flow chemistry models. In order to obtain better predictions, an appropriate gas flow chemistry model needs to be included in the CFD calculations. Using a recently developed chemistry model for ablating carbon-phenolic-in-air species, a CFD calculation of the Stardust re-entry at 71 km is presented. The code used for that purpose has been designed to take advantage of the nature of the problem and therefore remains very efficient when a high number of chemical species are involved. The CFD result demonstrates the need for such chemistry model when modeling the flow field around an ablative material. Modeling of the nonequilibrium radiation spectra is also presented, and compared to the experimental data obtained during Stardust re-entry by the *Echelle* instrument. The predicted emission from the CN lines compares quite well with the experimental results, demonstrating the validity of the current approach.

1. Introduction

The Thermal Protection System (TPS) of a re-entry vehicle is one of the key components of its design. The materials used for the TPS can be classified into two main categories: ablative materials, such as the ones used on the Apollo missions, and non-ablative materials, such as the ceramic tiles used on the Space Shuttle. The former can also be divided into two sub-categories: charring (also know as pyrolyzing) and non-charring ablators. The theory behind the use of ablators is quite simple: the energy absorbed by the removal of material from the surface is not used to heat the TPS, thus keeping the vehicle at a relatively "cold" temperature. In the case of charring ablators, the ablative material is a resin which fills the pores of a carbon matrix. Although the matrix may ablate, it usually does not, thus preserving the original geometry of the aerodynamic surface during re-entry.

In order to properly model the heating rates at the surface of the vehicle, the ablating boundary condition must take into account many phenomena: surface recession, wall temperature, blowing rates, gas composition, surface chemistry, etc. However, to account for the effects of the pyrolysis gas on the vehicle, the chemistry model of the flow field must include

Published under licence by IOP Publishing Ltd

the reactions associated with the presence of this gas. Because ablation coupling is becoming an increasingly important research topic [1–4], the development of an accurate, yet usable, chemistry model is of great importance. Models have been proposed in the past [5–7] but important reactions were not included, and some of the reaction rates were inappropriate or simply outdated.

Recently, a more complete model was proposed [8], which includes an extensive set of kinetic rates, taken from the combustion community. The model was reduced using 0-D sensitivity analysis over a parameter space relevant to the re-entry conditions that such a material would be exposed to. It was established that such a model was necessary to study carbon-phenolic TPS through a review of past models, which gave a wide range of results, especially when radiative heat transfer calculations were performed [9]. The reduced model was later integrated into the hypersonic Computational Fluid Dynamics (CFD) code LeMANS. Since the model is composed of 38 species and 158 reactions, the number of conservation equations to be solved significantly increases, as does the computation times. However, because LeMANS was parallelized efficiently with respect to the intrinsic characteristic of the problem, the code remains efficient, robust and fast, even in these conditions. As a test case, this paper presents results of the 71 km trajectory point of the Stardust return vehicle, and compares them with measurements of the experimental data obtained from the *Echelle* spectrometer.

2. LeMANS: an unstructured three-dimensional Navier-Stokes solver for hypersonic nonequilibrium aerothermodynamics

The hypersonic aerothermodynamic CFD code used to analyze the chemistry model is LeMANS, a finite volume Navier-Stokes solver currently being developed at The University of Michigan [10–13]. The code assumes that the translational and rotational energy modes of all species can be described by their respective temperatures T and T_r, as is the free electron temperature T_e. The vibrational energy mode and electronic energy mode of all species is however described by a single temperature, T_{ve} [14]. The viscous stresses are modeled assuming a Newtonian fluid, using Stokes' hypothesis, and the species mass diffusion fluxes are modeled using a modified version of Fick's law. Mixture transport properties are calculated using one of two models; the first uses Wilke's semi-empirical mixing rule with species viscosities calculated using Blottner's model and species thermal conductivities determined using Eucken's relation, and the other uses Gupta's mixing rule with species viscosities and thermal conductivities calculated using non-coulombic/coulombic collision cross section data. Heat fluxes are modeled according to Fourier's law for all energy modes. Finally, the source terms of the species conservation equations are modeled using a standard finite-rate chemistry model for reacting air in conjunction with Park's two-temperature model to account for thermal nonequilibrium effects on the reaction rates.

To account for the ablating TPS, a blowing boundary condition which solves the conservation of momentum at the wall and the mass balance is available in LeMANS [15]. The energy balance equation, which can be solved using the strongly coupled material response code MOPAR [16], is not required in the current analysis.

The code has the capability to handle meshes containing any mix of hexahedra, tetrahedra, prisms and pyramids in 3D, or triangles and quadrilaterals in 2D. Numerical fluxes between the cells are discretized using a modified Steger-Warming Flux Vector Splitting scheme, which has low dissipation and is appropriate to calculate boundary layers. A point or line implicit method is used to perform the time integration. The code has been extensively validated against experimental data, and has also been compared to other similar codes such as NASA Ames' DPLR [17] and NASA Langley's LAURA [18].

2.1. Implicit Line Solver

The system of equations used to numerically simulate hypersonic, chemically reacting viscous flows is very stiff. One of the options to overcome the time-step limitations in such a case is using implicit time integration. The developed code can use a point-implicit or a line-implicit method to solve the set of equations implicitly. The line implicit method has better convergence performance but it comes at the expenses of a more complicated algorithm. A line in the system of equations to be solved can be written as

$$A_i \Delta Q_i + \sum_j B_j \Delta Q_{nb} = R_i^n \tag{1}$$

where A and B contain the inviscid and viscous jacobians, R is the residual of the system of equations and ΔQ is the correction used in the time march. The summation in the last expression is over all j-th faces that form the i-th cell. Moreover, nb stands for the neighbor of the i-th cell that shares the j-th face. Details of the formulation can be obtained in Ref. [11]. When using unstructured grids, Eq. 1 is a sparse matrix without any structure. Considering the mesh shown in Fig. 1a), the corresponding matrix form of Eq. 1 is shown in Fig. 2a). There are many algorithms available for efficient solution of sparse linear systems. Such algorithms rely heavily on general mathematical properties of matrices to solve the system. These algorithms, however, do not take into account some flow characteristics that can simplify the system. For instance, flow gradients will be stronger in the direction normal to the body according to boundary layer theory. It was shown [19] that solving the system of equations exactly in the normal direction for structured grids and relaxing the system in the other directions was a good option to parallelize an implicit code.

The idea of solving the linear system originated by an implicit method using special directions in unstructured grids is not new [20, 21]. The first step in such a method is to define the special directions that will be used. In this work, such directions are called lines that are grown from the body surface by a simple algorithm that chooses the next cell such that the line stays nearly normal to the body. Cells that are left over are grouped together and handled by a point implicit algorithm. The lines formed from the mesh in Fig. 1a) are indicated in Fig. 1b). That figure also shows a renumbering of the cells that will be shown to facilitate the solution of the linear system. The renumbering of the cells is never carried out in the code. Instead, an array keeps track of the relation between the two cell numbering systems.

Using the numbering for the cells provided in Fig. 1b), the sparse system is greatly simplified as can be observed in Fig. 2b). The matrix is still sparse, but one can see a tridiagonal structure embedded in the matrix. In order to take advantage of the tridiagonal structure, an iterative process is assembled. All the terms off the tridiagonal part of the matrix are moved to the right hand side of the equation. By doing that, the iterative process is now of the form depicted in Fig. 3. The last three lines of the system represent the three cells left over from the line finding procedure. Those cells corrections are calculated using a simple point implicit method which is naturally integrated in the tridiagonal solver. Due to its diagonal characteristic, the system is very simple to solve. Assuming that $\Delta Q(n+1,0) = 0$, $\Delta Q(n+1,1)$ is obtained by solving a simple tridiagonal system of equations. The next step requires a multiplication of the sparse matrix on the right hand side by the newly calculated $\Delta Q(n+1,1)$. That multiplication is straightforward when using the face based data structure of the mesh and the mesh numbering. The result of that multiplication is mapped to the linear system numbering and a new right hand side is fed to the tridiagonal solver. Following Ref. [19], four iterations are taken in the iterative process. The code starts running with a very small CFL number which is increased to a maximum value of 10 when using a second order scheme or 1000 when using a first order scheme.

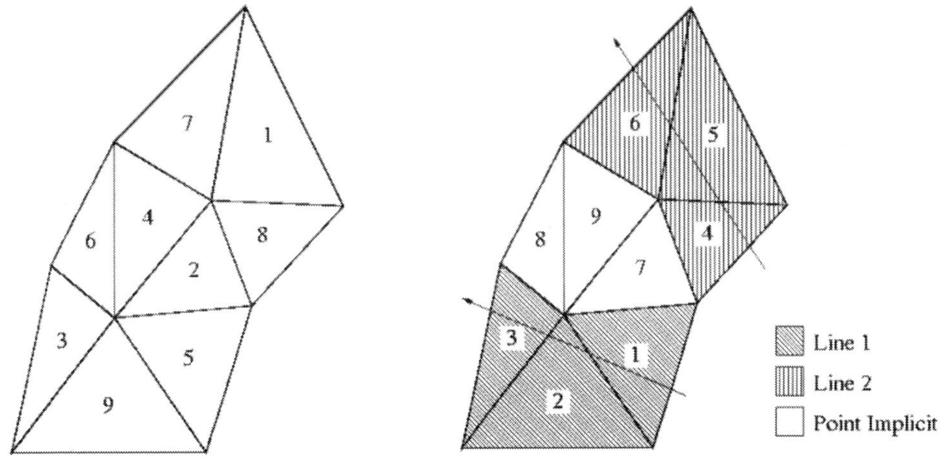

(a) Mesh (b) Linear system

Figure 1. Distinct numbering used in the code

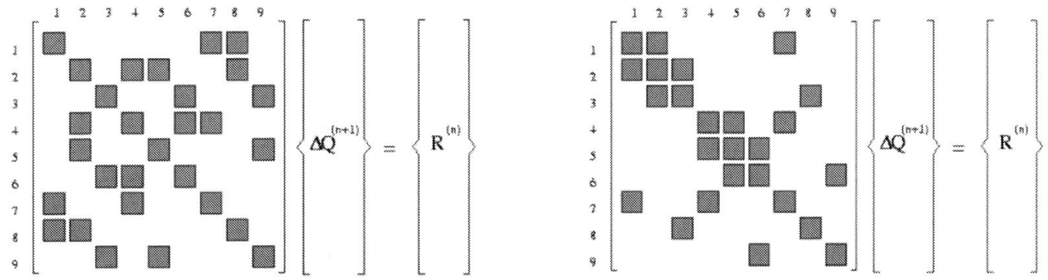

(a) Original numbering (b) Renumbered

Figure 2. Typical linear systems for the unstructured grids of Fig. 1

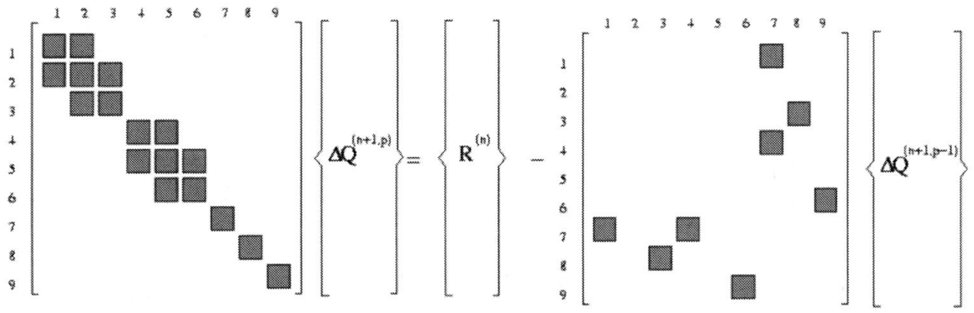

Figure 3. Iterative solution of linear system

2.2. Parallel Processing
The code developed in this work takes full advantage of modern parallel machines by dividing the computational work among many processors. This task is accomplished using METIS [22] to partition the mesh between the processors and MPI to communicate necessary information from one processor to another. METIS uses graph theory to partition unstructured grids minimizing the amount of inter-processor communication. Each node in the graph corresponds to a cell in the mesh. Each edge in the graph connecting two nodes corresponds to the face shared by the cells represented by the two nodes. A partition generated in this way usually is not optimal for a line implicit solver because most of the lines are significantly shortened. In order to avoid such a problem, the graph is modified so that each node in fact corresponds to a line and each edge corresponds to all the faces connecting one line to another. The weights of the nodes in the graph are set to the number of cells that form the line and the weight of each edge is equal to the number of faces connecting each line so that the load balancing of the partitions is not reduced. The implemented algorithm is highly effective as can be observed in Fig. 4 which shows the partitions for a hybrid unstructured grid and for a structured like grid.

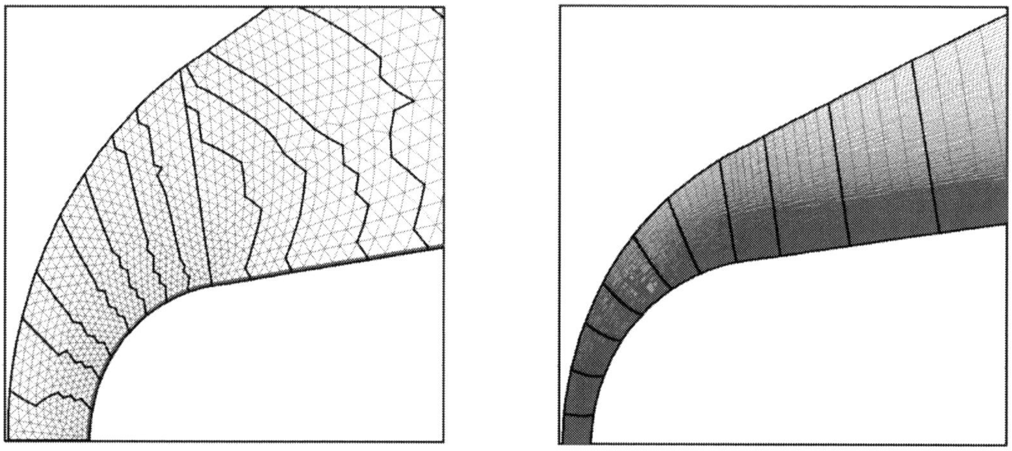

(a) Triangular grid (b) Quadrilateral grid

Figure 4. Mesh partitioning

2.3. Chemistry model
The chemistry model used in this analysis has been developed and optimized for vehicles equipped with ablative heat shields composed of PICA, re-entering in a N_2-O_2 atmosphere [8, 23]. The species and reaction mechanisms used in the model were carefully selected using a material response analysis. The corresponding kinetic chemistry rates were taken from the GRI-MECH model, and then reduced using a sensitivity analysis [8, 9, 23]. The reduced model contains 38 species and 158 reaction rates. The selected species can be grouped into three categories; the air species:

$$N_2,\ O_2,\ NO,\ N,\ O,\ N_2^+,\ O_2^+,\ NO^+,\ N^+,\ O^+,\ e$$

the surface species:

$$H_2,\ CO,\ CH_4,\ H_2O,\ CO_2,\ OH,\ C_2H_2,\ HCN,\ C_2H,\ C_3,\ CN$$

and the reacting species:

$$H, NH, HO_2, H_2O_2, HCO, C, C_2, CH, CH_2, CH_3, NCO, HNO, CO^+, CN^+, C^+, H^+$$

The reaction mechanisms and kinetics rates are listed in Ref. [9]. The model has been validated with multiple zero-dimensional simulations performed using the CHEMKIN [24] package, and compared to available experimental data that are representative of an ablative boundary layer during hypersonic re-entry [8].

3. Test-case: Stardust re-entry vehicle at 71 km

3.1. Problem description

In order to evaluate and validate the model used in LeMANS, the forebody of the Stardust return capsule is modeled at the 71 km trajectory point [25]. This re-entry point is chosen because it is well into the continuum region where the Navier-Stokes equations are valid, and spectral emission data obtained by the *Echelle* instrument is available. The geometry and mesh are presented in Fig. 5 and the flow and surface parameters are given in Tables 1 and 2, respectively. To complete the chemistry model described earlier, Gibb's Free Energy is used to calculate the equilibrium constants needed for the backward reaction rates, and the transport properties are calculated from Lennard-Jones potentials, using CHEMKIN [24], to produce temperature dependent viscosity curve fits for each species.

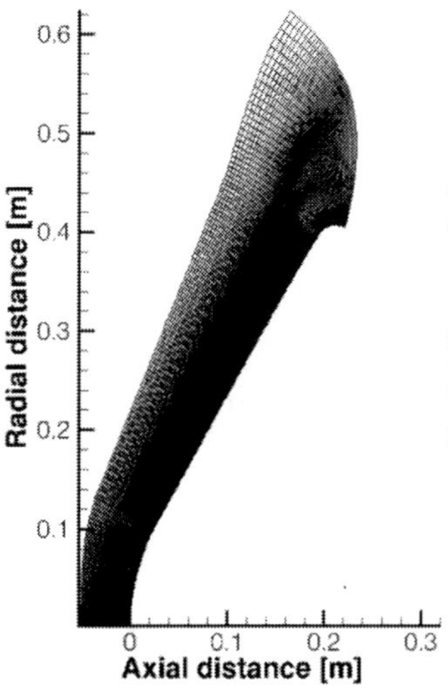

(a) Geometry and overall mesh

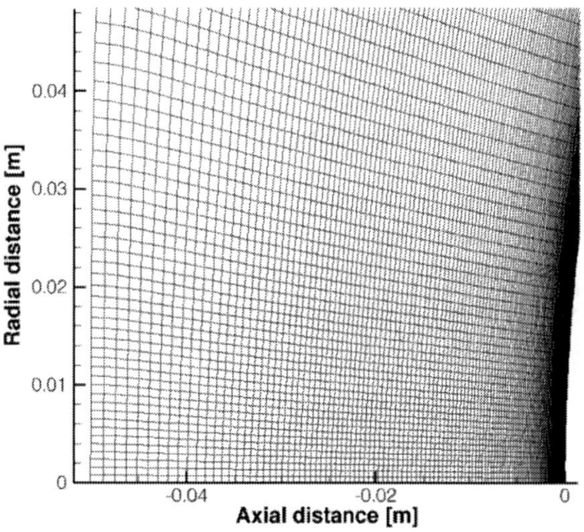

(b) Stagnation region mesh

Figure 5. Geometry and mesh of the Stardust re-entry capsule, used for the chemistry model comparison

The values used for the surface temperature, blowing rates and gas composition are obtained with an uncoupled approach, using a combination of three NASA software tools. First, the aeroheating environment is calculated using DPLR [26]. The heat transfer coefficient, the surface pressure and the freestream enthalpy are then used in the material response code, FIAT [27]. The pressure and the non-dimensionalized ablation rate, B'_g and B'_c, values predicted by FIAT are then extracted and input to the Multicomponent Ablation Thermochemistry (MAT) code [28]. When given this information, MAT calculates back from the B' tables the species mole fractions according to JANNAF information corresponding to the wall temperature predicted by

Table 1. Free stream conditions for the 71 km trajectory point of the Stardust re-entry vehicle

Altitude [km]	U_∞ [km/s]	T_∞ [K]	ρ_∞ [kg/m^3]	Y_{N_2}	Y_{O_2}	Kn
71.2	12.1	222.	5.55×10^{-4}	0.763	0.237	0.00145

Table 2. Wall condition at the stagnation point for the 71 km trajectory point of the Stardust re-entry vehicle

T_w [K]	$\dot{m}_w$ [kg/m^2/s]	Y_{N_2}	Y_{CO}	Y_{H_2}	Y_{H_2O}
3240.	0.0453	6.75×10^{-1}	1.23×10^{-1}	5.76×10^{-3}	5.23×10^{-2}
Y_{OH}	Y_O	Y_{CO_2}	Y_{NO}	Y_{O_2}	Y_N
3.30×10^{-2}	3.68×10^{-2}	2.70×10^{-2}	1.74×10^{-2}	2.99×10^{-2}	0.00

FIAT. This procedure therefore provides input values for blowing rates, wall temperature and equilibrium species composition. It is to be noted that the equilibrium condition is computed while accounting for the boundary layer edge gas (air) and the surface material (carbon), and therefore accounts for surface ablation. The downside of this procedure is that mass is introduced at the surface as the boundary layer edge gas and is considered to be injected at the surface. However, at this altitude the added mass is negligible, since the blowing rate is less than 1% of the freestream mass flux.

The values obtained through this procedure are only calculated at the stagnation point and are expected to be significantly lower elsewhere on the forebody of the vehicle. To reflect this, a temperature profile is obtained using a fully radiative equilibrium boundary condition at the wall, without ablation. The surface temperature profile is then normalized and re-applied for the ablating wall simulation, multiplied by the surface temperature at the stagnation point that is listed in Table 2. This method has been proven to give a good estimate of the surface conditions [25]. For the blowing rate, a linear relation between the temperature and the mass flux is derived using the second and third columns of Table 2, and applied proportionally using the normalized surface temperature profile.

3.2. Results – convective heat flux

The convective heat flux obtained with the ablative boundary condition, using the proposed chemistry model, is presented in Fig. 6. As expected, the heat flux is significantly reduced when compared to the radiative equilibrium heat flux, also presented in that figure. As can be seen in Fig. 7, most of the heat reduction comes in the form of the mass diffusion term. This is a different phenomenon than that observed at higher altitude [23], where the majority of the reduction in heat flux occurred through the translational-rotational term. A decrease of this latter term is still observed, of approximately the same magnitude as at the higher altitude trajectory point, but it is no longer the dominating factor. The reduction in the mass diffusion term is explained by the increased blowing rates which causes the mixing region to be thicker, thus reducing the species gradients near the wall. The reduction of the translational-rotational conduction term is caused by a slight modification of the temperature gradients, due to the fact that the shock is pushed away from the surface by the blowing at the surface. Additionally, the change in boundary layer composition causes a modification of the thermal conductivity of the gas, which also acts to reduce this term.

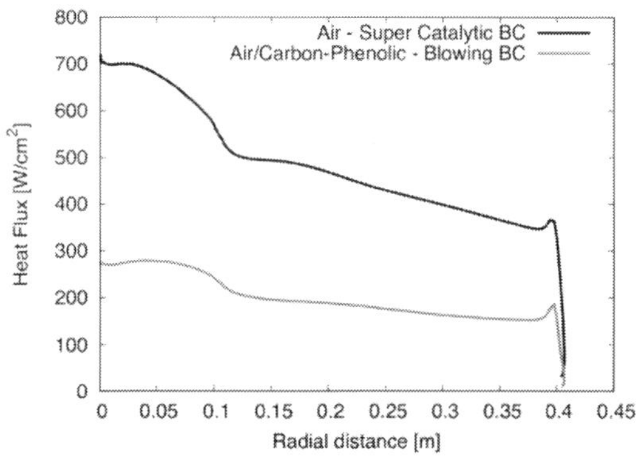

Figure 6. Heat fluxes at the surface of the Stardust re-entry vehicle at 42 s into re-entry (71 km), computed using an air-only chemistry model in the flow field with supercatalytic boundary conditions, and the carbon-phenolic-in-air chemistry model, with a blowing boundary condition

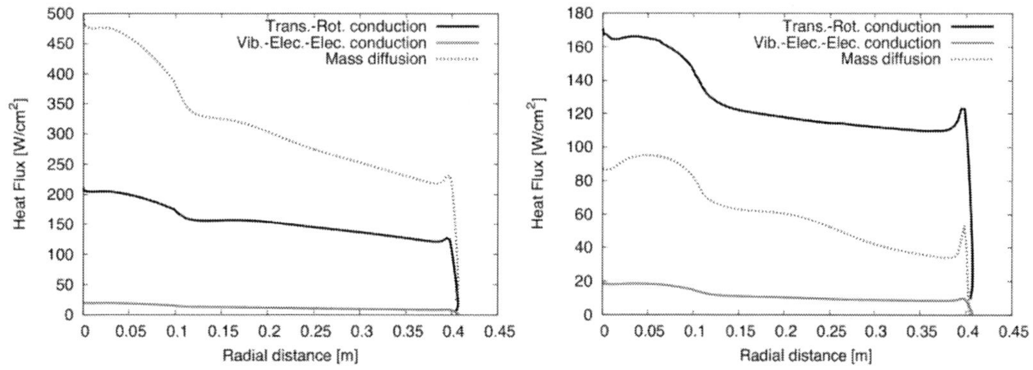

(a) Air-only chemistry; supercatalytic boundary condition (b) Carbon-phenolic-in-air chemistry; blowing boundary condition

Figure 7. Individual heat flux components

3.3. Results – species concentrations

The species composition along the stagnation line is presented in Fig. 8. It is interesting to note in Fig. 8b) that most of the C-H blowing species (CO_2, H_2, OH and H_2O) are destroyed almost immediately once they enter the flow, although CO remains in high concentration. This species is important since it is itself a strong radiator, but more importantly, it is the main source of CN through the CO/CN exchange reaction, and CN is one of the strongest radiators at re-entry conditions. In Fig. 8c), it can be seen that most blowing species are transformed into atomic species (H and C), as they are present in high concentrations near the boundary. This figure also shows that CN is also present in great quantities. It is also interesting to note, in that same figure, that the concentration of HNO is relatively high near the shock; this species was neglected in past models [5–7]. Finally, Fig. 8d) presents the species which are not present in a significant concentration (species that had a number density less than the machine precision are not shown). Although it might be tempting to discard them, they remain important since they start to appear in greater concentration at other re-entry conditions (i.e. other altitudes). The same calculations were performed for the other trajectory points, as presented in Ref. [8], and although some differences are noticeable, the general conclusions remain the same.

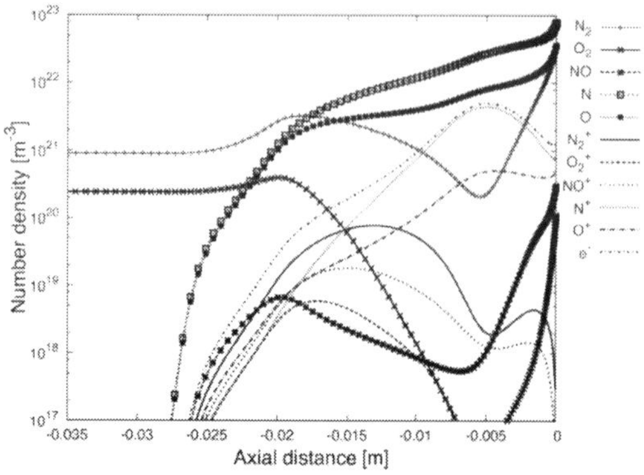

(a) Air species

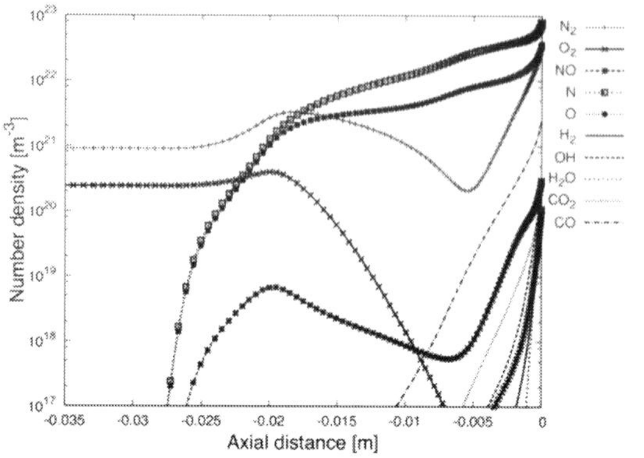

(b) Surface blowing species

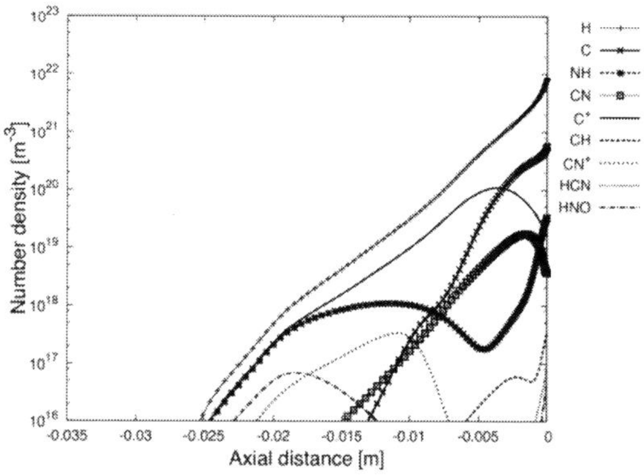

(c) High concentration species

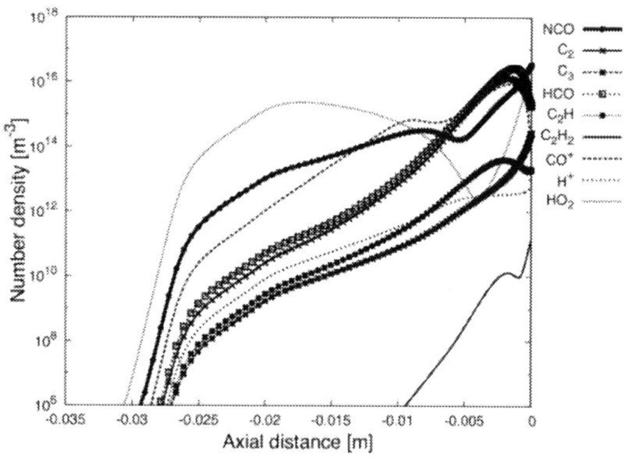

(d) Low concentration species

Figure 8. Species concentrations along the stagnation line for the Stardust vehicle 42 s into re-entry (altitude of 71 km)

3.4. Results – radiative heat flux

Certain species present in the boundary layer are strong radiative emitters, and are expected to significantly contribute to the overall radiative heat flux at the surface. For the Stardust re-entry vehicle, the contribution of the radiative heat flux to the overall heat flux has been evaluated, at maximum intensity, to be approximately 10% [25].

A radiative emission calculation is performed using the flow field solution of the 42 s trajectory point (71 km), with the NEQAIR (version 99d) [29] non-equilibrium radiation code. Apart from the usual air species, radiative emission from the CN violet and CN red systems is included.

The stagnation line temperatures and species concentrations used as inputs to NEQAIR are presented in Fig. 9.Since the observation took place far away from Stardust, and the radiation is mostly emitted from the volumetric plasma in front of the vehicle and not from the surface, there is no need to account for the observation angle from which the data was acquired.

From Fig. 9b), it is interesting to note that CN^+ plays a more significant role in the determination of the species concentrations than was previously thought. It is created at the cost of CN, and therefore reduces the CN concentration by up to two orders of magnitude in a region where the vibrational-electron-electronic temperature is high. The same cannot be said about CO^+, as nowhere in the flow is that molecule present in quantities that approach the concentration of CO.

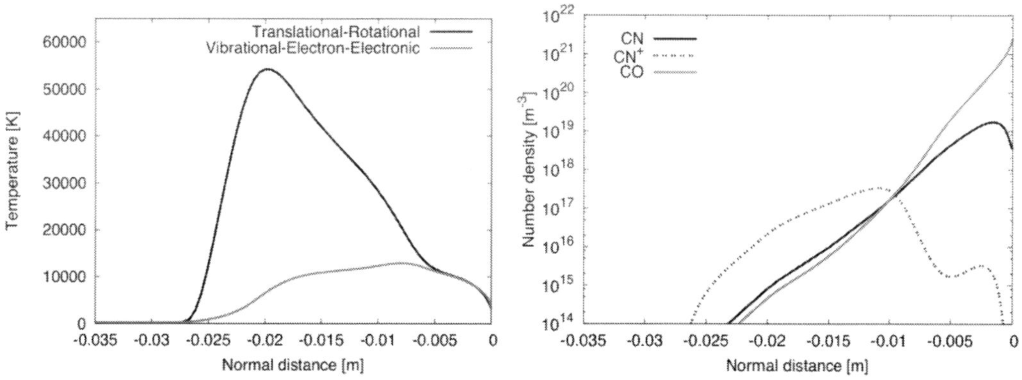

(a) Temperatures (b) Radiating species from ablation

Figure 9. Temperatures and CN/CO species concentrations along the stagnation line for the Stardust vehicle at 42 s into re-entry (71 km).

A comparison of the spectral emission from CN is presented in Fig. 10. It can be seen in that figure that the spectral computations match the *Echelle* data remarkably well, as the numerical results are within the same order of magnitude.

Finally, the radiative heat flux has also been computed for the test-case. The total radiative flux is approximately 1% of the convective heat flux, making it negligible at this trajectory point. What is more significant, however, is that a large portion (30-40%) of the radiative heat flux is generated by CN over the wavelength range of the *Echelle* instrument (about 360 to 890 nm), which further confirms the importance of properly modeling that particular species.

4. Conclusion

In an attempt to assess the importance of accounting for ablation species in the flow field, a new 38 species carbon/phenolic-in-air chemistry set was implemented in the hypersonic non-equilibrium CFD code LeMANS. Because of the way LeMANS has been parallelized, the code remains efficient, robust and fast, even with the increased number of equations to solve. As a test case, the Stardust vehicle at the 71 km trajectory point of its re-entry was used. This point was chosen because it is well into the continuum regime, and experimental data is available in the form of spectral radiative emission gathered by the *Echelle* instrument. As expected, the convective heat flux predicted using the carbon-phenolic-in-air chemistry model was significantly reduced relative to the prediction obtained using the equilibrium radiative boundary conditions. The species concentrations along the stagnation line were also presented and it was shown that at chemical equilibrium conditions, most species blown from the surface immediately react in the flow field and are transformed. These results clearly indicate the need to use an appropriate chemistry model in the flow field, and that the chemistry model should be significantly different than that used to model pyrolysis gas behavior inside the TPS. The flow field solution was also used to perform analysis of the CN radiative spectral emission using NEQAIR. The result compared remarkably well to the experimental data obtained by the *Echelle* instrument, and therefore provides increased confidence in the current approach.

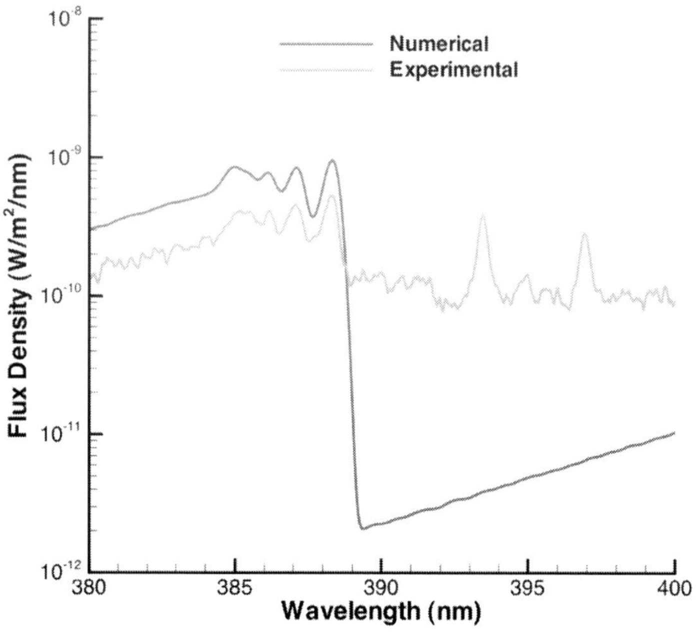

Figure 10. Comparison of the CN spectral line for the Stardust re-entry vehicle at 71 km

Acknowledgments

Financial support for this work was provided by NASA SBIR Phase-1 Award NNX10CC53P, NASA Prime Contract NNA04BC25C to ELORET Corporation and NASA Kentucky EPSCoR RIDG program. The authors would also like to thank Dr. Erin D. Farbar at the University of Michigan as well as Dr. Mike Wright, Dr. Ioana Cozmuta, Dr. David Hash, Dr. Nagi Mansour, Dr. Mike Barnhardt, Dr. Jean Lachaud and Dr. Tahir Gökçen from NASA Ames Research Center for numerous insightful discussions.

References

[1] Suzuki T, Sawada K, Yamada T and Inatani Y 2005 *Journal of Thermophysics and Heat Transfer* **19** 266–272

[2] Amar A J, Blackwell B F and Edward J R 2007 *39th AIAA Thermophysics Conference* AIAA-2007-4535 (Miami, FL)

[3] Milos F and Chen Y K 2008 *46th AIAA Aerospace Sciences Meeting and Exhibit* AIAA-2008-1223 (Reno, NV)

[4] Martin A and Boyd I D 2008 *40th AIAA Thermophysics Conference* AIAA-2008-3805 (Seattle, WA) p 20

[5] Park C, Jaffe R L and Partridge H 2001 *Journal of Thermophysics and Heat Transfer* **15** 76–90

[6] Suzuki K, Kubota H, Fujita K and Abe T 1997 *32nd AIAA Thermophysics Conference* AIAA-1997-2481

[7] Olynick D, Chen Y K and Tauber M E 1999 *Journal of Spacecraft and Rockets* **36** 442–462

[8] Martin A, Boyd I D, Cozmuta I and Wright M J 2010 *48th AIAA Aerospace Sciences Meeting and Exhibit* AIAA-2010-1175 (Orlando, FL)

[9] Martin A and Boyd I D 2010 *10th AIAA/ASME Joint Thermophysics and Heat Transfer Conference* AIAA-2010-4656 (Chicago, IL)

[10] Scalabrin L C and Boyd I D 2007 *39th AIAA Thermophysics Conference* AIAA-2007-4044 (Miami, FL)

[11] Scalabrin L C 2007 *Numerical Simulation of Weakly Ionized Hypersonic Flow Over Reentry Capsules.* Ph.D. thesis The University of Michigan Ann Arbor, MI

[12] Scalabrin L C and Boyd I D 2005 *38th AIAA Thermophysics Conference* AIAA-2005-5203 (Toronto, Ontario)

[13] Scalabrin L C and Boyd I D 2006 *9th AIAA/ASME Joint Thermophysics and Heat Transfer Conference* AIAA-2006-3773 (San Francisco, CA)

[14] Farbar E D, Boyd I D, Kim M and Martin A 2011 *42nd AIAA Thermophysics Conference* (Honolulu, HI)

[15] Martin A and Boyd I D 2010 *Journal of Thermophysics and Heat Transfer* **24**

[16] Martin A and Boyd I D 2009 *41th AIAA Thermophysics Conference* AIAA-2009-3597 (San Antonio, TX)

[17] Wright M J, Candler G V and Bose D 1998 *AIAA Journal* **36** 1603–1609

[18] Gnoffo P A 1989 *9th AIAA Computational Fluid Dynamics Conference* AIAA-1989-1972-CP (Buffalo, NY) pp 415–425

[19] Wright M J 1997 *A Family of Data-Parallel Relaxation Methods for the Navier-Stokes equations* Ph.D. thesis University of Minnesota Minneapolis, MN

[20] Venkatakrishnan V 1995 *VKI Lecture Series* VKI–LS–1995–02

[21] Nompelis I, Drayna T W and Candler G V 2005 *17th AIAA Computational Fluid Dynamics Conference* AIAA 2005-4867 (Toronto, ON)

[22] Karypis G and Kumar V 1998 *METIS: A Software Package for Partitioning Unstructured Graphs, Partitioning Meshes, and Computing Fill-Reducing Orderings of Sparse Matrices* Univeristy of Minnesota Minneapolis, MN

[23] Martin A and Boyd I D 2011 *49th AIAA Aerospace Sciences Meeting and Exhibit* AIAA 2011-143 (Orlando, FL)

[24] Kee R J, Rupley F M, Miller J A, Coltrin M E, Grcar J F, Meeks E, Moffat H K, Lutz A E, Dixon-Lewis G, Smooke M D, Warnatz J, Evans G H, Larson R S, Mitchell R E, Petzold L R, Reynolds W C, Caracotsios M, Stewart W E, Glarborg P, Wang C, McLellan C L, Adigun O, Houf W, Chou C P, Miller S F, Ho P, Young P D, Young D J, Hodgson D W, Petrova M V and Puduppakkam K V 2006 CHEMKIN, release 4.1

[25] Trumble K A, Cozmuta I, Sepka S and Jenniskens P 2008 *46th AIAA Aerospace Sciences Meeting and Exhibit* AIAA-2008-1201 (Reno, NV)

[26] NASA Ames Research Center *DPLR Package Users Guide* 3rd ed

[27] Chen Y K and Milos F S 1999 *Journal of Spacecraft and Rockets* **36** 475–483

[28] Milos F S and Chen Y K 1997 *35th Aerospace Sciences Meeting and Exhibit* AIAA-1997-141 (Reno, NV)

[29] Whiting E E, Park C, Liu Y, Arnold J O and Paterson J A 1996 NEQAIR96, nonequilibrium and equilibrium radiative transport and spectra program: User's manual Reference Publication 1389 NASA

Data intensive high energy physics analysis in a distributed cloud

A. Charbonneau[1], A. Agarwal[2], M. Anderson[2], P. Armstrong[2], K. Fransham[2], I. Gable[2], D. Harris[2], R. Impey[1], C. Leavett-Brown[2], M. Paterson[2], W. Podaima[1], R.J. Sobie[2] and M. Vliet[2]

[1] National Research Council Canada
100 Sussex Drive, Ottawa, Canada
[2] Department of Physics and Astronomy
University of Victoria, Victoria, Canada V8W 3P6

E-mail: `Andre.Charbonneau@nrc-cnrc.gc.ca`

Abstract. We show that distributed Infrastructure-as-a-Service (IaaS) compute clouds can be effectively used for the analysis of high energy physics data. We have designed a distributed cloud system that works with any application using large input data sets requiring a high throughput computing environment. The system uses IaaS-enabled science and commercial clusters in Canada and the United States. We describe the process in which a user prepares an analysis virtual machine (VM) and submits batch jobs to a central scheduler. The system boots the user-specific VM on one of the IaaS clouds, runs the jobs and returns the output to the user. The user application accesses a central database for calibration data during the execution of the application. Similarly, the data is located in a central location and streamed by the running application. The system can easily run one hundred simultaneous jobs in an efficient manner and should scale to many hundreds and possibly thousands of user jobs.

1. Introduction

Infrastructure as a Service (IaaS) cloud computing is emerging as a new and efficient way to provide computing to the research community. The growing interest in clouds can be attributed, in part, to the ease of encapsulating complex research applications in Virtual Machines (VMs) with little or no performance degradation [1]. Studies have shown, for example, that high energy physics application code runs equally well in a VM or on the native system [2]. Virtualization technologies not only offers several advantages such as abstraction from the underlying hardware and simplified application deployment, but in some situations where traditional computing clusters have hardware and software configurations which are incompatible with the scientific application's requirements, virtualization is the only option available. A key question is how to manage large data sets in a cloud or distributed cloud environment.

We have developed a system for running high throughput batch processing applications using any number of IaaS clouds [3]. This system uses a number of off-the-shelf software such as Condor [4] and Nimbus [5], in addition to custom components such as a cloud scheduling element and a VM image repository. The results presented in this work use the IaaS clouds at the National Research Council (NRC) in Ottawa and two clouds (elephant and hermes) at the University of Victoria (UVIC). The Amazon EC2 cloud was not included in these tests because of network

Published under licence by IOP Publishing Ltd

performance issues between the NRC and UVIC clouds and Amazon. The total amount of memory and CPU of each computational cluster in our clouds are divided evenly into what we call *VM slots*, where each of these slots can be assigned to run a VM. When a VM has finished running, that slot's resources are then released and available to run another VM. In other words, the number of VM slots in our system determines the maximum number of simultaneously running VMs at any given time. A total of 84 VM slots on UVIC and NRC clouds were used in generating the results presented later in this paper. Two additional clouds have since been added to the system raising the capacity to over 200 VM slots. The input data and analysis software are located on one of the UVIC clouds and the VM images are stored in a repository on the NRC cloud. The Canadian sites are connected by a research network provided by CANARIE while the commodity network is used to connect the NRC and UVIC clouds to Amazon EC2.

We describe the system architecture in the next section, however, here we highlight the main features. Users are provided with a set of VMs that are configured with the application software. The user adds code to run their specific analysis and saves the VM to a repository. The user submits their jobs to a Condor scheduler where the job script contains a link to the required VM. A cloud scheduling component we implemented (called Cloud Scheduler [3]) searches the job queue, identifies the VM required for each queued jobs, and sends out a request to one of the clouds to boot the user specific VM. Once the VM is booted, the scheduler submits the user job to the running VM. The job runs and returns any output to a user specified location. If there are no further jobs requiring that specific VM, then Cloud Scheduler shuts it down.

The system has been demonstrated to work well for applications with modest I/O requirements such as the production of simulated data [6]. The input files for this type of application are small and the rate of production of the output data is modest (though the files can be large). The system has also been used for the analysis of astronomical survey data where the images are pre-staged to the storage on the cloud [6].

In this work, we focus on data intensive high energy physics applications where the job reads large sets of input data at higher rates. In particular, we use the analysis application of the BaBar experiment [7] that recorded electron-positron collisions at the SLAC National Accelerator Laboratory from 2000-2008. We show that the data can be quickly and efficiently streamed from a single data storage location to each of the clouds. We will describe the issues that have arisen and the potential for scaling the system to many hundreds or thousands of simultaneous user jobs.

2. System architecture

The architecture of the distributed IaaS cloud system is shown in fig. 1 (see ref. [3] for a detailed description). The UVIC and NRC clouds use the Nimbus software to manage VMs while Amazon EC2 uses its own proprietary software. The system is designed so that the user submits their jobs to a single Condor scheduler. The system then boots the user-specific VMs on any one of the available clouds.

Users submit jobs using X.509 Proxy Certificates [8] to authenticate. The certificates are also used to authenticate with Nimbus clusters when starting, shutting down, or polling VMs. Authentication with EC2 is done by using a standard shared access key and secret key.

To simplify the management of user VMs we have developed a system called Repoman [9]. Repoman implements an online VM repository where the user can retrieve copies of a variety of standard VM images preloaded with the application software. The user clones the standard VM, adds and tests their software, saves it in the VM repository and includes a reference to their VM image in their job submission script.

The user submits their jobs to the Condor job scheduler. We selected Condor as the job schedule since it was designed to utilize heterogeneous idle workstations and is an ideal job scheduler for a dynamic VM environment where worker nodes become available on demand.

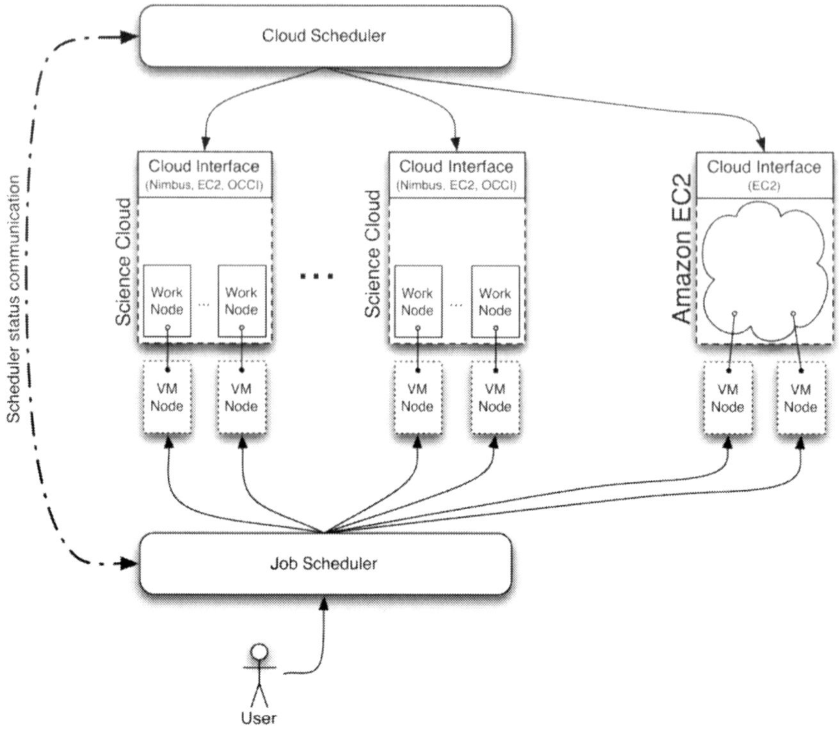

Figure 1. An overview of the architecture used for the system. A user prepares a VM image and a job script. The job script is submitted to the job scheduler. The Cloud Scheduler reads the job queue and makes a request to boot the user VM on one of the available clouds. Once there are no more user jobs requiring that VM type, the Cloud Scheduler makes a request to the proper cloud to shutdown the user VM.

Users submit jobs by issuing the `condor_submit` command. The user must add a number of additional parameters specifying the location of the image and the properties of the required VM.

The management of the VMs on the clouds is done by Cloud Scheduler [3]. Cloud Scheduler monitors the queue of jobs and if one of the jobs needs a VM that is not already booted on one of the clouds, then it sends a request to the next available cloud. Cloud Scheduler will also shut down a VM if there are no jobs needing that particular type of VM anymore. A complete description of Cloud Scheduler can be found in ref. [3].

3. Database and data management

Analysis jobs in high energy physics typically require two inputs: event data and configuration data. The configuration data also includes a BaBar conditions database [10], which contains time-dependant information about the conditions under which the events where taken. The event data can be the real data recorded by the detector or simulated data. Each event contains information about the particles seen in detector such as their trajectories and energies. The real and simulated data are nearly identical in format; the simulated data contains additional information describing how it was generated. The user analysis code analyzes one event at a time. In the BaBar experiment the total size of the real and simulated data is approximately 2

PB but users typically read a small fraction of this sample. In this work we use a subset of the data containing approximately 8 TB of simulated and real data. The configuration data that describes the state of the detector and the conditions database totals 24 GB.

The event data for this analysis was stored in a distributed Lustre file system at UVIC. The Lustre file system is hosted on a cluster of six nodes, consisting of a Management/Metadata server (MGS/MDS), and five Object Storage servers (OSS). Each node has dual quad core Intel Nehalem processors and 24 GB of memory. All servers have six disk drives: two 450 GB, 15K RPM, Serial-Attached SCSI (SAS) system drives in a RAID 1 configuration; and four metadata/object storage drives in a RAID 5 configuration. The drives for the metadata storage on the MGS/MDS are 450 GB, 15K RPM, SAS drives. The drives for the object storage on each OSS are 1 TB, 7200 RPM, Serial ATA (SATA) drives. Originally configured to allow remote access via mounting over the public network, the Lustre file system uses a single gigabit interface/VLAN to communicate both internally and externally. This is an important consideration for the test results presented, because these same nodes also host the IaaS front-end (MGS/MDT server) and Virtual Machine Monitors (OSS servers) for the UVIC cloud.

The jobs use Xrootd [11] to read the data. Xrootd is a file server providing byte level access and is used by many high energy physics experiments. Xrootd provides read only access to the Lustre data (read/write access is also possible) and the capability of implementing Grid Security Infrastructure (GSI) authentication. Though the implementation of Xrootd is fairly trivial, some optimization was necessary to achieve good performance across the network: a read-ahead value of 1 MB and a read-ahead cache size of 10 MB was set on each Xrootd client.

The VM images are stored at NRC in a Repoman server and propagated to the worker nodes by http. For analysis runs that includes the Amazon EC2 cloud, we store another copy of the VM images on Amazon EC2 in an Amazon S3 bucket using the `ec2_bundle_vol` and `ec2_upload_bundle` tools provided by the Amazon EC2 API Tools package.

In addition to transferring the input data on demand using Xrootd, the BaBar software is also staged to the VMs on demand using a specialized network file system called CernVM File System (CernVM-FS) [12]. Using CernVM-FS can considerably reduce the size of the VM images since the software files are stored in a remote location and read by the application at runtime. The rationale of using CernVM-FS for this project was to reduce the amount of data initially transferred to the clouds when the VM starts by reducing the size of the VM images transferred from the image repository to each cloud site. This not only makes the VM start faster, but also helps mitigate the network saturation after job submission by postponing some of the data transfer to happen later after the job has started.

4. Results

A Xen VM image based on Scientific Linux SL release 5.5 (Boron) was created and the latest release of BaBar software was installed. The BaBar software, which contains C++ and FORTRAN code, requires 1 GB of RAM and is completely contained in a 16 GB VM. Using CernVM-FS to load the BaBar software on demand, we were able to bring down the size of the VM to approximately 4GB. In this study, a member of the team and also a researcher in the BaBar Collaboration added one of their analysis codes to this VM. After compiling, linking and testing the code, the modified VM was saved for access by the batch system. A server was setup as a login node which contained the tools to allow users to submit and manage their jobs. A separate VM contained both the Condor job scheduler and Cloud Scheduler for the tests presented in this paper.

A typical user job in high energy physics reads one event at a time where the event contains the information of a single particle collision. Electrons and positrons circulate in opposite directions in a storage ring and are made to collide millions of times per second in the centre of the BaBar detector. The BaBar detector is a cylindrical detector with a size of approximately 5 meters in

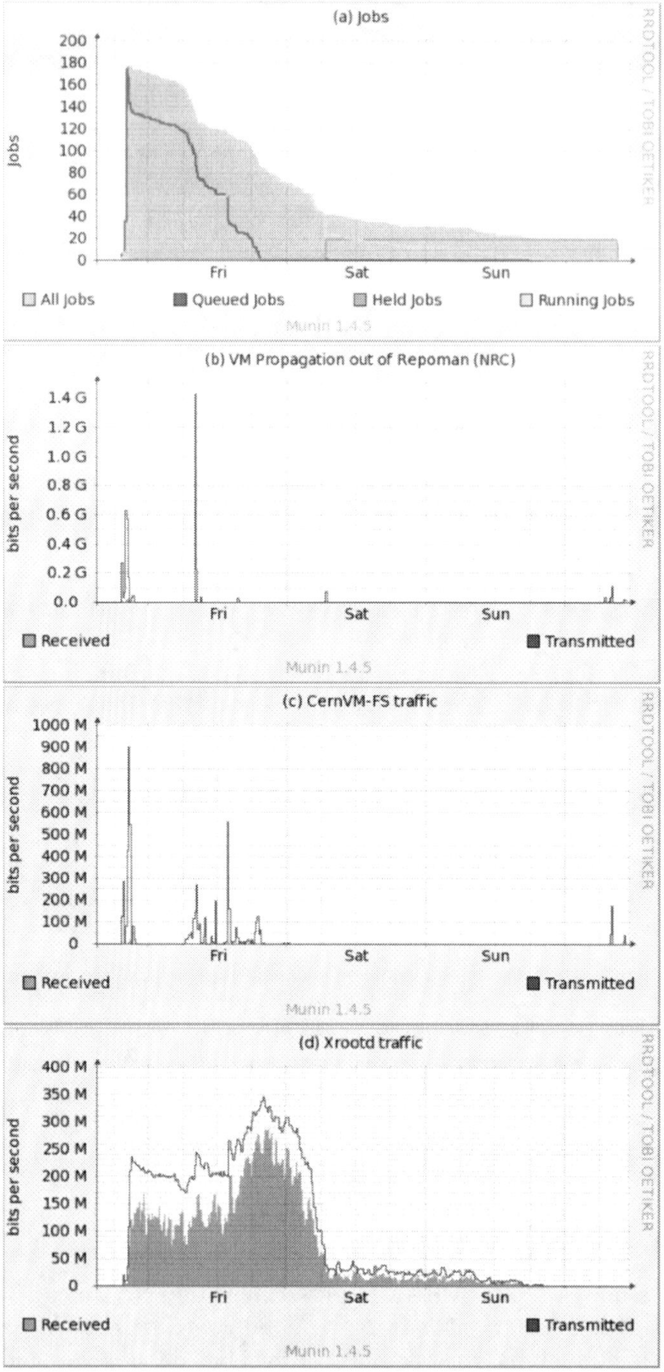

Figure 2. The number of running, queued and failed jobs (a), the network traffic for the transfer of VM images from NRC (b), the traffic for the loading of the required BaBar software files at runtime using CernVM-FS (c), and the traffic from the Lustre file system at UVIC which is used to store the event data (d).

each dimension. The detector measures the trajectories of charged particles and the energy of both neutral and charged particles. A fraction of those events are considering interesting from a scientific standpoint and the information in the detector is written to a storage medium. The size of the events in BaBar are a few kilobytes depending on the number of particles produced in the collision.

The user code analyzes each event independently and writes out a subset of the information into an output file for subsequent analysis. A typical job reads millions of events (the size of each event is approximately 3 KB) and the user submits tens to hundreds of these jobs in parallel. The output files are a few percent of the original input data set size, making it easier to study the events of particular interest to the researcher.

In this work, the analysis is focused on a particular subset of the BaBar data set where there are a small number of particles produced in each collisions. We use two separate samples for our tests which we label Tau11-data and Tau11-MC. The Tau11-data and Tau11-MC samples are identical except that the first is real data and the second is simulated data.

We submitted approximately 170 analysis jobs (80 jobs analyzed the real data sample and 90 analyzed the simulation data sample) where a single job runs for approximately 12 hours. The input rate ranges up to a few megabits/second depending on whether the jobs is running on a site local to the data repository or at a remote location. In addition, the processing time is different for each sample as the fraction of interesting events in each sample is different. For example, background or non-interesting events are quickly identified and discarded so that the job does not waste processing time.

Plot (a) of fig. 2 shows the number of VMs on the system over a 3 day period. The yellow curve is the number of running VMs, the blue curve is the number of queued jobs and the orange curve is the number of jobs in an error state. There were 44 VM slots at NRC Ottawa and 40 VM slots at the two University of Victoria clusters. Initially, only the 2 clusters at UVIC were online; the NRC cluster was offline for maintenance. This explains why the number of running VMs plateaus at 40 until approximately 22:00 on Thursday, at which point the NRC cluster came online, adding 44 extra resources to the cloud. The Cloud Scheduler then started to submit jobs to the NRC resources and at approximately 3:00 on Friday all the 84 slots were allocated and running jobs. The 40 VMs required at UVIC corresponds to nearly 160 GB and took approximately 1 hour to transfer from NRC (3000 km away). We are working on a method so that only a single copy of a VM needs to be transferred. It is interesting to note that without the use of CernVM-FS, this data transfer would have been in the range of 640 GB. At around 7:00 on Saturday, we can see that there are approximately 20 jobs in error state. Further investigation revealed that these jobs failed because of a communication failure between the Condor processes, most probably caused by an overly congested network (we are investigating solutions to this problem after discussions with Condor developers). The system kept the failed jobs in the queue and these jobs were eventually resubmitted to the cloud at a later time and ran to completion.

Plot (b) of fig. 2 shows the network traffic at the NRC cloud generated by the transfer of VMs from the Repoman image repository to the clouds. The 40 VMs required at UVIC were obtained from NRC and the blue curve shows the outbound network traffic of approximately 500 Mb/s over 1 hour period (the NRC site has a 1 Gb/s connection to the CANARIE research network in Canada). Once all the VMs were booted, the outbound network traffic at NRC fell to zero until the NRC cluster came online late Thursday, at which point 44 additional VMs were transferred to the available NRC resources. Since this later transfer was done internally on the NRC network, it peaked at a much higher rate of approximately 1.4 Gb/s.

In plot (c) of fig. 2 we observe that the network traffic generated by the VM retrieving the required sotware files at runtime using CernVM-FS. This figure shows a high throughput transfer which peaked at 900 Mb/s for the VMs running at UVIC and, as expected, a lower transfer rate for the files sent to the VMs which started later at NRC. Because CernVM-FS implements some client-side caching, we see that most of the files are transferred soon after the VMs start, and then the network traffic goes down considerably, even though these files are still used by the running application.

Plot (d) of fig. 2 shows the network bandwidth for the data transfers from the UVIC cluster that hosts all the BaBar data. The total bandwidth plateaus around 330 Mb/s for the 84 running jobs. We see an unusual amount of input traffic to the Xrootd server (shown by green histogram). This is abnormal because these type of jobs typically read data from the Xrootd server but write the results at another location. After further investigation, it was found that this anomaly was caused by a faulty network configuration, which was later rectified.

Within a given sample, we find that the time to process the data is affected by the location of the data repository. Jobs that ran at UVIC, which had access to the Xrootd server locally on the network, ran 2 to 3 times faster than jobs that ran on the NRC cloud in Ottawa. Since the data rate out of the Xrootd server at UVIC was not the limiting factor (see plot (d) of fig. 2), this suggests that the overall Xrootd performance is somewhat sensitive to network latency. Investigation is underway on how to further improve Xrootd performance when transfering data over long distances.

One of the features of the system is its ability to recover from faults arising either from local system problems at each of the clouds or network issues. We list some of the problems we identified in the processing of the jobs. For example, we find that:

- A number of VMs at NRC went into an error state (see the orange line in plot (a) of fig. 2 at Friday 22:00) and this was attributed to some error with the Condor Connection Brokering (CCB). We believe that this problem was caused by connection timeouts that were set to values too low to accommodate for the high load on the network at that time. In any case, Cloud Scheduler killed those VMs and held these job in the queue for further analysis by the user.
- Cloud resources can be brought down for maintenance and back up again. In our test, the NRC cloud resources were added to the pool of resources after the set of jobs was submitted. The Cloud Scheduler automatically detected the new resources available and successfully scheduled jobs to these newly available resources without affecting already running jobs.

The overall system has performed remarkably well. We are planning to upgrade the network of the data storage system to 10 Gb/s and this would allows us to scale the number of jobs by an order of magnitude. There are also plans in 2011 to connect Amazon EC2 to the Canadian research network (provided by CANARIE).

A significant fraction of the data transfer in this test was the movement of VMs. It is true that as the number of available VM slots increases, the amount of time taken to transfer the VMs for larger sets of jobs also grows and as a consequence can lead to under-utilization of compute resources. This was alleviated by using CernVM-FS to reduce the VM size, but there is still room for further improvement. We will soon implement a VM caching system so that only a single copy of the VM image would need to be transferred to each cloud, significantly reducing the start-up time of the VMs at remote sites.

5. Conclusion

We have presented the results for running a data intensive particle physics applications in a distributed compute cloud. The results show that multi-TB data sets can be easily accessed from remote locations over a high-speed research network. Further improvements can also be achieved by loading software on demand using a specialized network file system such as CernVM-FS, which not only makes the VMs start faster, but also alleviates network saturation by averaging out the overall bandwidth requirement over a longer period. There appears to be no limitations to scaling the number of jobs by an order of magnitude once we increase the network bandwidth from the event storage servers and implement a VM image caching mechanism. From the users perspective, the system is robust and is able to handle intermittent network issues gracefully. We have shown that the use of distributed compute clouds can be an effective way of analyzing large research data sets.

Acknowledgments

The support of CANARIE, the Natural Sciences and Engineering Research Council, the National Research Council of Canada and Amazon are acknowledged.

References

[1] Alef M and Gable I 2010 Hep specific benchmarks of virtual machines on multi-core cpu architectures *Journal of Physics: Conference Series* **219** 052015 URL `http://stacks.iop.org/1742-6596/219/i=5/a=052015`

[2] Agarwal A, Desmarais R, Gable I, Grundy D, P-Brown D, Seuster R, Vanderster D C, Charbonneau A, Enge R and Sobie R 2008 Deploying hep applications using xen and globus virtual workspaces *Journal of Physics: Conference Series* **119** 062002 URL `http://stacks.iop.org/1742-6596/119/i=6/a=062002`

[3] Armstrong P, Agarwal A, Bishop A, Charbonneau A, Desmarais R J, Fransham K, Hill N, Gable I, Gaudet S, Goliath S, Impey R, Leavett-Brown C, Ouellete J, Paterson M, Pritchet C, Penfold-Brown D, Podaima W, Schade D and Sobie R J 2010 Cloud scheduler: a resource manager for distributed compute clouds *CoRR* **abs/1007.0050**

[4] Thain D, Tannenbaum T and Livny M 2005 Distributed computing in practice: the condor experience *Concurrency - Practice and Experience* **17** 323–356

[5] Keahey K, Foster I, Freeman T and Zhang X 2005 Virtual workspaces: Achieving quality of service and quality of life in the grid *Sci. Program.* **13**(4) 265–275 ISSN 1058-9244 URL `http://portal.acm.org/citation.cfm?id=1239655.1239657`

[6] Fransham K, Agarwal A, Armstrong P, Bishop A, Charbonneau A, Desmarais R, Hill N, Gable I, Gaudet S, Goliath S, Impey R, Leavett-Brown C, Ouellete J, Paterson M, Pritchet C, Penfold-Brown D, Podaima W, Schade D and Sobie R J 2010 Research computing in a distributed cloud environment *Journal of Physics: Conference Series* **256** 012003 URL `http://stacks.iop.org/1742-6596/256/i=1/a=012003`

[7] Boutigny D *et al.* 1995 Babar technical design report Tech. Rep. SLAC-R-95-457

[8] Tuecke S, Welch V, Engert D, Pearlman L and Thompson M 2004 Internet X.509 Public Key Infrastructure (PKI) Proxy Certificate Profile RFC 3820 (Proposed Standard) URL `http://www.ietf.org/rfc/rfc3820.txt`

[9] Vliet M, Agarwal A, Anderson M, Armstrong P, Charbonneau A, Fransham K, Gable I, Harris D, Impey R, Leavett-Brown C, Paterson M and Podaima W 2011 Repoman: A simple restful x.509 virtual machine image repository *to appear in the Proc. of the Int. Symp. on Grids & Clouds* (Taipei)

[10] Khan A, Gaponenko I and Brown D 2008 Distributed online conditions database of the babar experiment *Nuclear Science, IEEE Transactions on* **55** 2579–2583 ISSN 0018-9499

[11] Dorigo A, Elmer P, Furano F and Hanushevsky A 2005 Xrootd/txnetfile: a highly scalable architecture for data access in the root environment *Proceedings of the 4th WSEAS International Conference on Telecommunications and Informatics* (Stevens Point, Wisconsin, USA: World Scientific and Engineering Academy and Society (WSEAS)) pp 46:1–46:6 ISBN 960-8457-11-4 URL `http://portal.acm.org/citation.cfm?id=1391157.1391203`

[12] Buncic P, Sanchez C A, Blomer J, Franco L, Harutyunian A, Mato P and Yao Y 2010 Cernvm a virtual software appliance for lhc applications *Journal of Physics: Conference Series* **219** 042003 URL `http://stacks.iop.org/1742-6596/219/i=4/a=042003`

Journal of Physics: Conference Series **341** (2012) 012004 doi:10.1088/1742-6596/341/1/012004

Numerical method for dynamic imaging of nonlinear polyatomic molecules

Catherine Lefebvre, HuiZhong Lu and André D. Bandrauk

Département de chimie, Université de Sherbrooke, Sherbrooke, QC, J1K 2R1 Canada

E-mail: `Andre.Dieter.Bandrauk@USherbrooke.ca`

Abstract. We present electron and nuclear dynamic imaging of a polyatomic molecule in an ultrashort intense laser pulse. A numerical solution of the time-dependent Schrödinger equation is described to study the non-perturbative nonlinear dynamic of a nonlinear polyatomic molecule in interaction with ultrashort intense laser pulse. In a quantum description of the molecular system, beyond the Born-Oppenheimer approximation, we investigate the full influence of the nuclear motion on the photoionization and harmonic generation processes.

1. Introduction

Quantum dynamic imaging of molecules is currently studied with the help of the recent advances in laser technology [1]. To perform dynamic imaging of molecules one has to take into account the different time scale in the motion of the particles within a molecule. Nuclei evolve on the time scale of the femtosecond ($1\ fs = 10^{-15}\ s$). For instance, by applying a femtosecond laser pulse of a few optical cycles, in the near infrared spectral region, the ultrafast vibrational wavepacket dynamics in the dissociative ionization of H_2 molecule has been probed [2]. On the other hand, electrons evolve on the time scale of the attosecond ($1\ as = 10^{-18}\ s$). Since only a decate, such ultrashort laser pulses are produced from the nonlinear interaction of matter with light through the process of high-order harmonics generation (HHG) in atoms [3] or molecules [4]. In this process an atom or a molecule is ionized by a femtosecond infrared laser pulse. The photoelectron is subsequently accelerated by the laser field. As the electric field changes phase within an optical cycle, the photoelectron changes its direction and recollide with the parent ion [3]. Alternatively, the electron may recollide with the neighbors at higher energies than recollision with the parent ion [4],[5]. It emits energy (photon) that it has accumulated under the laser field in the form of high-order harmonics of the carrier-wave frequency. To such high frequencies correspond ultrashort pulses of duration of few attoseconds, which can be used to image the electron dynamics in atoms and molecules [6].

Numerical simulations are playing an important role in the understanding of laser-matter interaction, in particular in quantum molecular dynamic imaging. The exact description of molecular dynamics should include the correlated electron-nuclei motion and electron-electron motion. Large-scale numerical solutions of the multidimensional time-dependent Schrödinger equation (TDSE) is used in this paper to simulate the nonlinear nonperturbative interaction of a nonlinear polyatomic molecule with ultrashort laser pulses on time scales of nuclear and electron motion. These simulations are possible, thanks to the constant advances of highly parallel computational ressources.

2. Numerical method

We study the one electron molecular system H_3^{++} in a nonlinear (triangular) configuration in a two-dimension plane. The nuclear coordoninates are defined with Jacobi coordinates, r, being the relative coordinate between two lower nuclei (forming the basis of the triangle), and R, the relative coordinate between the upper nucleus and the center of mass of the two lower nuclei as shown in Fig. 1. The angle between these two Jacobi coordinates is fixed to $\theta = 90°$ to ensure a C_{2v} symmetry, i.e. an isosceles triangle. The nuclear coordinates are related to cartesian coordinates as follow: $x_{N_1} = 0, y_{N_1} = \frac{2}{3}R, x_{N_2} = -\frac{1}{2}r, y_{N_2} = -\frac{1}{3}R, x_{N_3} = \frac{1}{2}r, y_{N_3} = -\frac{1}{3}R$. The electron coordinates are defined with polar coordinates ϕ and ρ, with correspondence to cartesian coordinates as $\rho = \sqrt{x_e^2 + y_e^2}$ and $\phi = \arccos(x_e/\rho)$ [7].

The complete equation of motion of the molecule is prescribed by the time-dependent Schrödinger equation (TDSE) (in atomic units, $e = \hbar = m_e = 1$) beyond the Born-Oppenheimer approximation,

$$i\hbar\frac{\partial}{\partial t}\Psi(\rho,\phi,r,R,t) = \Big(T_e + T_N + V(\rho,\phi,r,R,t)\Big)\Psi(\rho,\phi,r,R,t), \tag{1}$$

with the kinetic energy operators

$$T_e = -\frac{1}{2\mu_e}\Big[\frac{1}{\rho}\frac{\partial}{\partial\rho}\rho\frac{\partial}{\partial\rho} + \frac{1}{\rho^2}\frac{\partial^2}{\partial\phi^2}\Big], \tag{2}$$

$$T_N = -\frac{1}{2\mu_r}\frac{\partial^2}{\partial r^2} - \frac{1}{2\mu_R}\frac{\partial^2}{\partial R^2}, \tag{3}$$

and the total potential

$$V(\rho,\phi,r,R,t) = V_{eN}(\rho,\phi,r,R) + V_{NN}(r,R) + V_{int}(\rho,\phi,t), \tag{4}$$

where the ecletron-nuclei attraction, nuclear repulsion and interaction potentials are given respectively by

$$V_{eN}(\rho,\phi,R,r) = -\sum_{k=1}^{3}\frac{1}{\sqrt{\rho^2 + x_{N_k}(r)^2 + y_{N_k}(R)^2 - 2\rho(\cos\phi\, x_{N_k}(r) + \sin\phi\, y_{N_k}(R)) + c_{en}}} \tag{5}$$

$$V_{NN}(r,R) = \sum_{k,l=1,..,3}^{k<l}\frac{1}{\sqrt{(x_{N_k}(r) - x_{N_l}(r))^2 + (y_{N_k}(R) - y_{N_l}(R))^2 + c_{NN}}}, \tag{6}$$

$$V_{int}(\phi,\rho,t) = -\rho\cos(\phi - \theta)E(t). \tag{7}$$

The reduced masses in the kinetic operators T_e and T_N, in Eqs. (2) and (3), are defined as

$$\mu_e = \frac{m_{N_1} + m_{N_2} + m_{N_3}}{1 + m_{N_1} + m_{N_2} + m_{N_3}};\ \mu_r = \frac{m_{N_2}m_{N_3}}{m_{N_2} + m_{N_3}};\ \mu_R = \frac{m_{N_1}(m_{N_2} + m_{N_3})}{m_{N_1} + m_{N_2} + m_{N_3}}, \tag{8}$$

with $m_{N_i}, i = 1,..,3$ denoting the mass of the three nucleus.

In the attraction and repulsion potentials in Eqs (5) and (6), we use Coulomb softening parameters $c_{eN} = 0.35$ and $c_{NN} = 1$ which gives potentials for H_3^+ in good agreement with experimental geometries. The electric laser pulse in the interaction potential, Eq. (7), is linearly polarized along $(\cos\theta, \sin\theta)$ direction with electric strength $E(t) = \epsilon(t)\cos(\omega t + \varphi)$. Here, ω (a.u.) is the carrier-wave frequency which is linked to wavelength λ(nm) by $\omega = 2\pi c/\lambda$, φ is the carrier-envelope phase and $\epsilon(t)$ is the laser pulse envelope of trapezoidal shape with two optical

Journal of Physics: Conference Series **341** (2012) 012004 doi:10.1088/1742-6596/341/1/012004

cycles ($T = 2\pi/\omega$) ramp on, two constant optical cycles at intensity $E_0 = \sqrt{I}$, and two optical cycles ramp off, as depicted in Fig. 2.

In order to solve numerically the TDSE, Eq. (1), we adopt the second order split-operator method [8] for the temporal propagation of the wavefunction

$$\begin{aligned} \Psi(t+\delta t) &= e^{[-iV(t+\delta t/2)*\delta t/2]} \\ &* \; e^{[-iT_N*\delta t/2]} e^{[-iT_e*\delta t]} e^{[-iT_N*\delta t/2]} \quad (9) \\ &* \; e^{[-iV(t+\delta t/2)*\delta t/2]} \Psi(t) + \mathcal{O}(\delta t^3), \quad (10) \end{aligned}$$

where the exponential operator of the electron kinetic energy is approximated by the Crank-Nicholson method [9]

$$e^{[-iT_e*\delta t]} \Psi(t) = \frac{1 - i(\delta t/2)T_e}{1 + i(\delta t/2)T_e} \Psi(t) + \mathcal{O}(\delta t^3), \quad (11)$$

implied by the choice of spatial discretization of the wavefunction for the electron variable.

The wavefunction is periodic in the r, R and ϕ directions where we apply an absober at the r_{max} and R_{max} boundaries, so we choose momentum Fourier space via Fast Fourier Transform (FFT) for the approximation in these variables. Furthermore, as it is well known, the exponential of the Laplacian (kinetic energy T_N here) operator becomes a multiplication operator in the Fourier spectral space, so the Fourier representation of wavefunction in r and R is very efficient. The finite difference (FD) method is chosen to discretize ρ. The combined FD and Fourier series is applied to Eq. (11): The T_e operator is written in the Fourier space of ϕ, and the partial derivate operator for each Fourier coefficient of ϕ is discretized with 4th order finite difference scheme. At origin, we profit from the symmetry (anti-symmetry) of these Fourier coefficient on ρ in order to avoid numerical boundary condition. More details can be found in [10].

The final algorithm requires FFT and diagonal band limited linear system solver. It is implemented by calling multithreads version subroutines of FFTW and OpenMP (multithreads) parallelized from Lapack linear solver.

3. Numerical results

3.1. Initial conditions and grid specifications

We apply the numerical method described above to the problem of electron and nuclear dynamics in the H_3^{++} molecular ion in interaction with an intense and ultrashort laser pulse. The initial molecular wavefunction is defined by $\Psi(\rho, \phi, r, R, t = 0) = f(r, R)\Phi(\rho, \phi; r, R)$, where $\Phi(\rho, \phi; r, R)$ is the electronic eigenfunction at fixed nuclear coordinates r and R obtained by imaginary-time propagation [11],[12]. $f(r, R)$ is the initial proton function in terms of the nuclear coordinates centered at $r = R = 2$ $a.u.$ and describes a Franck-Condon transition from H_3^+ to H_3^{++} at the equilibrium distance $R_{eq}(H_3^+) = 2$ $a.u.$ [13], [14], [15].

The electric laser pulse, linearly polarized along y axis, with the carrier-wave frequency $\omega = 2\pi c/(\lambda = 800\ nm) = 0.057$ $a.u.$ of period $T = 2\pi/\omega$=2.7 fs, carrier-envelope phase $\varphi = 0$ and intensity $I = E_0^2 = 2\times10^{14} W/cm^2 (7.4\times10^{-2}\ a.u.)$, is depicted in Fig. 2. Recent experiments have shown that ionization probability is most favorable with a linearly polarized laser pulse oriented at 0° and 90°, which justifies the choice of nuclear coordinates (with fixed angle $\theta = 90°$) with the laser orientation along the R nuclear coordinate, (y axis), (see Fig. 1) [16].

The TDSE is solved numerically, with time step $\delta t = 0.05$ $a.u.$, and we use an electron grid of size $\rho < \rho_{max} = 128$ $a.u.$ with 512 points and $0 < \phi < 2\pi$ with 128 points and a proton grid of size $r < r_{max}$ and $R < R_{max} = 19.2$ $a.u.$ with 192 points in both r and R directions. For very short pulses, as considered here, it has been verified that the nuclear wavefunction is contained within the grid at all times, and that the electron grid size ρ_{max} is always longer than the maximum free electron trajectory defined by $\alpha = E/\omega^2$ in a laser field of amplitude E and frequency ω. Thus, at $I = 7.4 \times 10^{-2}$ $a.u.$ and $\omega = 0.057$ $a.u.$, $\alpha = 23$ $a.u.$ $<< \rho_{max}$. Absorbing boundaries are used and we ensure that less that 20% of the population leaks to the absorbing boundaries during the propagation.

3.2. Observables

During the time propagation, we calculate the electron density,

$$P_e(t) = \int_0^\infty dr \int_0^\infty dR |\Psi(\rho, \phi, r, R, t)|^2, \tag{12}$$

and the nuclear density

$$P_N(t) = \int_0^{+\rho_{max}} d\rho \int_0^{2\pi} d\phi |\Psi(\rho, \phi, r, R, t)|^2 \rho. \tag{13}$$

We are also interested in the laser induced electron dipole moment in the acceleration form [17],

$$\begin{aligned} a(t) &= \int d\rho \int d\phi \int dR \int dr \Psi(\rho, \phi, R, r, t)^* \left(\frac{\partial H}{\partial y_e}\right) \Psi(\rho, \phi, R, r, t)\rho \\ &= \int d\rho \int d\phi \int dR \int dr \Psi(\rho, \phi, R, r, t)^* \left(-\nabla V_{eN} + E(t)\right) \Psi(\rho, \phi, R, r, t)\rho. \end{aligned} \tag{14}$$

The Fourier transform of the dipole gives the molecular high-order harmonic generation (MHOHG) power spectrum,

$$P(\omega) \propto |\int_0^{6T} a(t) \exp(-i\omega t) dt|^2. \tag{15}$$

3.3. Discussion

In this paper, we study the electron-nuclei correlation in the dissociating molecule H_3^{++}. The electron and nuclear motion is clearly illustrated in Fig. 2 which describes the electron density as the molecule is dissociating under the electric laser pulse. Initially, the electron density is uniformly delocalized on the three protons, which form an equilateral triangle. Once the laser field is turned on, it oscillates at frequency ω and period $T = 2\pi/\omega$ and the electron density is pushed towards the upper proton when the field is negative (e.g. $t = T, 2T, 3T, ...$), and towards the two lower protons when the field is positive (e.g. $t = 0.5T, 1.5T, 2.5T, ...$). The electron, being a negative particle, moves in the opposite direction of the electric laser field. Moreover, after only few optical cycles, the electron density spreads out because the nuclei are moving apart, i.e. the molecule is dissociating. At $t = 5T$, the electron density is delocalized on the three distinct protons, while at the end of the pulse, it is mainly localized on the upper proton, forming the product $H + 2H^+$. Also, as the nuclei move to larger distance due to nuclear repulsion, the field interaction becomes stronger compared to the Coulomb field and the molecule starts to ionize. The ionization probability is the most important at each peak of the laser pulse.

Once the molecule is ionized, there is a certain probability that the photoelectron recombines with the ion and emits high order harmonics. The MHOHG spectrum is shown in Fig. 3 where we compare the MHOHG generated from the case of the moving molecular ion with the case of the static molecular ion at fixed geometry $R = r = 5$ $a.u.$. From a time-frequency analysis (Gabor transform [18], not shown here) of the acceleration dipole, we observed that the main recollision event occurs at around four optical cycles ($t = 4T$). At this time, the molecular geometry is described by $r = R = 5$ $a.u.$ (see Fig. 2). We observe similarities in the MHOHG spectra for the dynamic and static cases: there is a well-defined plateau, followed by a rapid drop in the harmonic signal. This cutoff is predicted by the semi-classical electron recollision model [3] and defines the maximum harmonic N_{max} with energy $N_{max} = 3.17U_p + I_p = 43$, where the ponderomotive energy is $U_p = E_0^2/4\omega^2 = 0.4387$ $a.u.$ and the ionization potential at $r = R = 5$ $a.u.$ is $I_p = 1.029$ $a.u.$. The predicted value of N_{max} is very close to the one calculated $N_{max} = 46$ [19]. Even and odd harmonics are observed in both cases [7], due to the non-linearity of the molecule (absence of center of inversion) [20]. The odd harmonics disapear at vey high frequencies and the global amplitude of the harmonics profile is lower in the moving nuclear case due to the non-Born-Oppenheimer dynamics. This shows that electron dynamics is strongly influenced by the nuclear motion.

4. Conclusion

We have demonstrated with our multidimensional non-Born-Oppenheimer model the effect of nuclear motion on the electron dynamics for a nonlinear polyatomic molecule. This is illustrated in the MHOHG spectrum, which is a phenomenon of great importance in attosecond science [6]. Work is in progress to extend this model and method to the stable two-electron molecular system H_3^+. To include nuclear and electron motion, this is at least a five body problem. With respect to memory resource, we have estimated about $60TB$ of fast accessible disk space necessary to perform calculations on relatively small grids ($n\rho = 512, n\phi = 128, nR = 192$). Also, because of the very large number of spatial discretization points, one typical calculation may need ≈ 5 days on a 12000 core parallel computer. With constant advances in computational ressources, such intensive calculations will make possible the study of electron-electron correlation, influenced by the nuclear motion.

5. Aknowledgement

We thank RQCHP, SciNet and Compute Canada for access to massively parallel computer clusters, and the Natural Sciences and Engineering Research Council of Canada (NSERC) and CIPI (Canadian Institute for Photonic Innovations) for financial support of this research in its ultrafast science program. C.L. aknowledge the FQRNT (Fonds québécois de la recherche sur la nature et les technologies) for financial support.

References

[1] A. D. Bandrauk and M. Ivanov. *Quantum Dynamic Imaging*. Springer, N.Y., 2011.
[2] F. Kelkensberg, C. Lefebvre, W. Siu, O. Ghafur, T. T. Nguyen-Dang, O. Atabek, A. Keller, V. Serov, P. Johnsson, M. Swoboda, T. Remetter, A. L'Huillier, S. Zherebstov, G. Sansone, E. Benedetti, F. Ferrari, M. Nisoli, F. Lépine, M. F. Kling, and M. J. J. Vrakking. Molecular dissociative ionization and wavepacket dynamics studies using two-colour $XUV + IR$ pump-probe spectroscopy. *Phys. Rev. Lett.*, 103:123005, 2009.
[3] P. B. Corkum. Plasma perspective on strong field multiphoton ionization. *Phys. Rev. Lett.*, 71:1994, 1993.
[4] A. D. Bandrauk, S. Barmaki, and G. Lagmago Kamta. Laser phase control of high-order harmonic generation at large internuclear distance: the H^+-H_2^+ system. *Phys. Rev. Lett.*, 98:013001, 2007.
[5] S. Barmaki and A. D. Bandrauk. Laser control of electron transfer with ultrashort intense laser pulses. *J. Mod. Opt.*, 54:1047, 2007.
[6] F. Krausz and M. Ivanov. Attosecond physics. *Rev. Mod. Phys.*, 81:163, 2009.
[7] D. S. Tchitchekova, H. Lu, S. Chelkowski, and A. D. Bandrauk. Molecular high-order harmonic generation in a nonlinear two-electron molecule: The equilateral H_3^+. *J. Phys. B*, 44:065601, 2011.
[8] A. D. Bandrauk and H. Shen. Exponential split operator methods for solving couple time-dependent Shcrödinger equations. *J. Chem. Phys.*, 99:1185, 1993.
[9] D. Zwillinger. *CRC Standard Mathematical Tables and Formulae*, pages 711–714. CRC Press, N.Y., 1996.
[10] A. D. Bandrauk and H. Lu. Efficient numrical methods for the time-dependent Schrödinger equation for molecules in intense laser fields. In Douglas J. K. Mewhort, Natalie M. Cann, Gary W. Slater, and Thomas J. Naughton, editors, *HPCS*, volume 5976 of *Lecture Notes in Computer Science*, pages 99–108. Spinger, 2009.

[11] F. He, C. Ruiz, and A. Becker. Absorbing boundaries in numerical solutions of the time-dependent Schrödinger equation on a grid using exterior complex scaling. *Phys. Rev. A*, 75:053407, 2007.
[12] R. Kosloff and H. Tal-Ezer. A direct relaxation method for calculating eigenfunctions and eigenvalues of the Schrödinger equation on a grid. *Chem. Phys. Lett.*, 127:223, 1986.
[13] L. P. Viegas, A. Alijah, and A. J. Varandas. Accurate ab initio multisheeted double many-body expansion potential energy surface for the three lowest electronic singlet states of H_3^+. *J. Chem. Phys.*, 126:074309, 2006.
[14] P. Barragán, L. F. Errea, A. Macía, L. Méndez, I. Rabadán, and A. Riera. A study of conical intersection for the H_3^+ system. *J. Chem. Phys.*, 124:184303, 2006.
[15] A. Ichihara and K. Yokoyama. Ab initio potential energy surfaces for the two lowest $1A'$ states of H_3^+. *J. Chem. Phys.*, 103:2110, 1995.
[16] J. McKenna, A. M. Sayler, B. Gaire, N. G. Johnson, K. D. Carnes, B. D. Esry, and I. Ben-Itzhak. Benchmark measurments of H_3^+ nonlinear dynamics in intense ultrashort laser pulses. *Phys. Rev. Lett.*, 103:103004, 2009.
[17] A. D. Bandrauk, S. Chelkowski, D. J. Diestler, J. Manz, and K.-J. Yuan. Quantum simulation of high-order harmonic spectra of the hydrogen atom. *Phys. Rev. A*, 79:023403, 2009.
[18] A. D. Bandrauk, S. Chelkowski, and H. Lu. Correlated electron-nuclear motion visualized using a wavelet time-frequency analysis. In *Quantum Dynamic Imaging*, pages 23–35, N.Y., 2011. Springer.
[19] A. D. Bandrauk, S. Chelkowski, and H. Lu. Signature of nuclear motion in molecular high-order harmonics and in the generation of attosecond pulse trains by ultrashort intense laser pulese. *J. Phys. B*, 42:075602, 2009.
[20] T. Kreibich, M. Lein, V. Engel, and E. K. U. Gross. Even-harmonic generation due to beyond-*B*orn-*O*ppenheimer dynamics. *Phys. Rev. Lett.*, 87(10):103901, 2001.

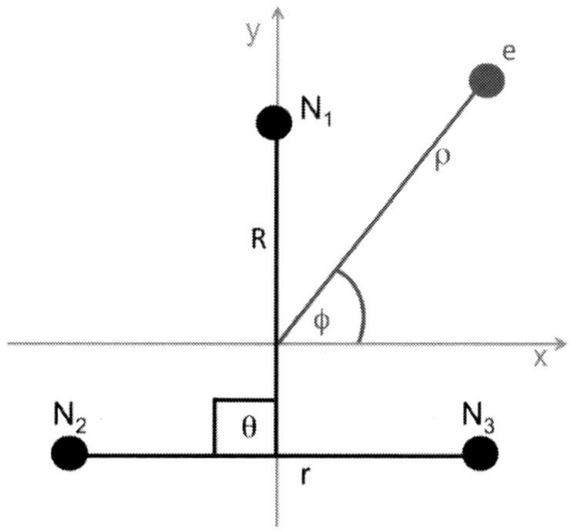

Figure 1. Geometry of H_3^{++} molecular ion in the xy plane with the origin at the center of mass of the nuclei. Nuclear coordinates are labeled r, R and θ (in black) and electron coordinates are labeled ρ and ϕ (in blue).

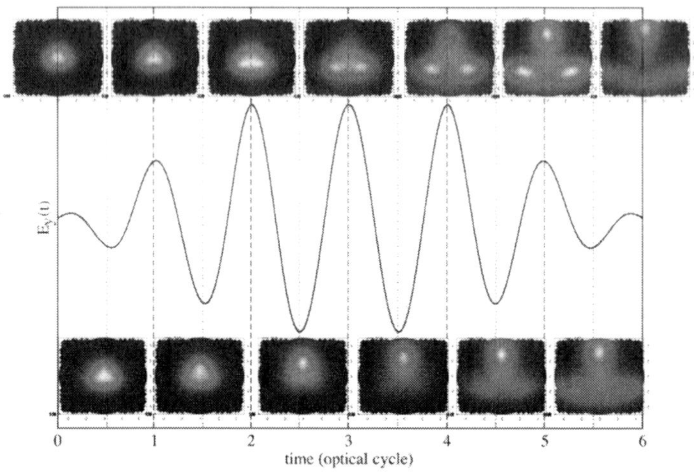

Figure 2. Electric laser pulse, E_y, as a function of time. Snapshots of the electron density in the xy plane. From left to right: upper panels $t = 0, T, 2T, 3T, 4T, 5T, 6T$; lower panels $t = 0.5T, 1.5T, 2.5T, 3.5T, 4.5T, 5.5T$, where $T = 2\pi/\omega$.

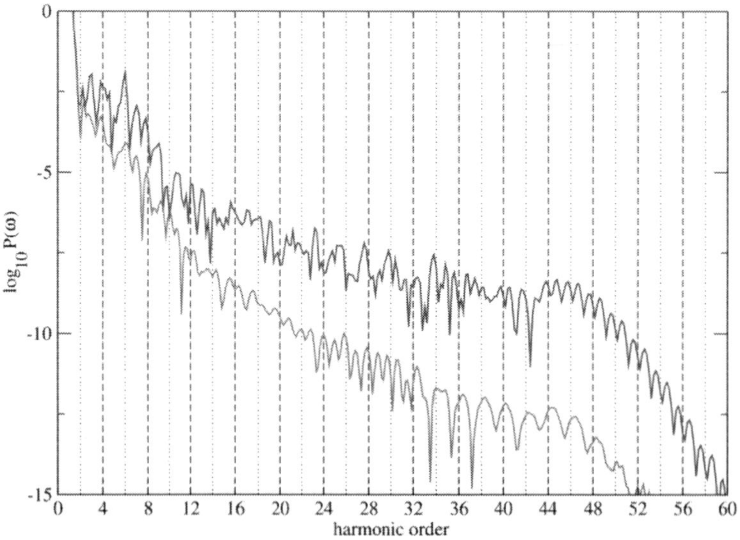

Figure 3. Molecular high-order harmonic (MHOHG) spectra of H_3^{++} with moving nuclei (red curve) and fixed nuclei ($r = R = 5\ a.u.$) (purple curve) at $I = 2 \times 10^{14}\ W/cm^2, \omega = 0.057\ a.u, (\lambda = 800\ nm), T = 2.7\ fs$.

Journal of Physics: Conference Series **341** (2012) 012005 doi:10.1088/1742-6596/341/1/012005

On the Reconstruction of the Convection Pattern Below an Active Region of Solar Corona.

Dorian Pirot, Jonathan Gaudet and Alain Vincent

Université de Montréal, Département de Physique, CP. 6128, Succursale Centre-ville, Montréal (Québec) H3C 3J7, Canada

E-mail: pirot@ASTRO.UMontreal.CA, vincent@ASTRO.UMontreal.CA

Abstract. In order to better understand magneto-convective patterns and flux emergence, we use the Nudging Back and Forth, a data assimilation method with an anelastic convection model to reconstruct the convection zone below a solar active region from observed solar surface magnetograms. To mimic photosphere, vector magnetograms are computed using force free hypothesis. We find that the observed arcade system of AR9077-20000714 ("the slinky") of magnetic lines is actually formed by Ω and U loops generated in the convection zone. We generate temperature maps at top of the convective zone and find that high magnetic fields on either sides of the neutral line produce a local cooling by impeding the overturning motions.

1. Introduction

Convection in the Sun is at the origin of magnetic dynamo, and thus, ultimately of coronal activity. The theoretical problem of turbulent compressible convection has been studied by many authors (e.g. Brummell *et al.*, 2006; Bushby *et al.*, 2008) including radiation effects (Nordlund, 2008; 2009; Nordlund and Stein, 2007). In fact, since the convective envelope is strongly stratified and of relatively low Mach number, the flow is anelastic. The advantage here is that the equations do not take acoustic waves into account. There are indeed acoustic waves in the Sun. Helioseismology is using surface signature to attempt the reconstruction of the interior of the Sun and in particular the convective envelope (Hanasoge *et al.*, 2006; Christensen-Dalsgaard, 2002). But there is also a numerical advantage: time steps (CFL) do not need to be small enough to follow acoustic waves. Anelastic approximation for convection has been studied in details for a long time (e.g. Gough, 1969) and improved recently for astrophysical convective envelopes with variable physical properties (Lantz and Fan, 1999). If one wants to predict the evolution of coronal activity, and in particular a solar flare, it would be necessary to simultaneously compute the MHD flows inside the Corona and also the convective layer below. This has been attempted (Abbett and Fisher, 2002; Hurlburt and De Rosa, 2004) for realistic but for purely theoretical flows. To simulate *realistic* convection including granulation is possible, but to reproduce observed magnetograms and temperatures in a particular area of about $175,000km \times 175,000km$ using simulation of convection remains a challenge.

In this study, we try to reconstruct the convective layer below a particular active region AR9077-20000714. For this purpose, we need to know the actual temperature, velocity and magnetic field at top of convection layer (photosphere) continuously in time. Temperature and magnetic fields are observed by satellite and ground observatories and publicly accessible in the form of heliocentered 2D pixel maps (Fig.1). However, many problems arise from the datasets.

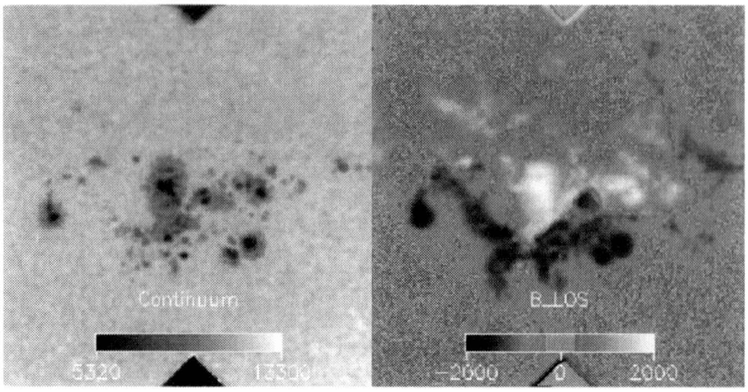

Figure 1. (Left) Temperature map (Kelvin). (Right) Magnetogram observed along the line of sight (Gauss). Observed from Mees Solar Observatory, Haleakala, of the solar active region AR9077-20000714. Sunspots are formed by strong magnetic fields at subphotospheric levels. Courtesy University of Hawaii institute for Astronomy http://www.solar.ifa.hawaii.edu/IVM/Archive/2000/AR9077_20000714.1633_ivm.gif

Firstly, magnetograms are usually not given at the same times and cadence as temperature maps: typical SOHO-MDI magnetograms are given at a 96 minute cadence. Also velocities are not easily observed and must be computed by correlation techniques (e.g. Rieutord *et al.*, 2000; Longcope, 2004). To surmount this, we use Nudging Back and Forth (NBF) (Auroux *et al.*, 2008), a data assimilation technique (Daley, 1991; Lewis *et al.*, 2006; Bouttier and Courtier, 2003), akin to Quasi-Linear Inverse 3D (Kalnay *et al.*, 1999). This is needed to solve the state of the convection zone that matches available observations taken as upper boundary conditions.

Our manuscript is organized as follows. In next section, we set the anelastic equations for solar convection and we explain the data assimilation technique for an active region. Then, in section 3, we assimilate observed vector magnetograms to reconstruct the convection pattern below the active region.

2. Anelastic convection model and NBF data assimilation

We are using the Back and Forth Nudging algorithm (NBF) (Auroux *et al.*, 2008) to reconstruct the convective zone below the photosphere given a pair of observed magnetograms.

2.1. Anelastic convection below an active region.

The convective layer of the Sun is strongly stratified in density and mostly we can regard the Mach number as being small. In this regime, magnetohydrodynamic anelastic approximation allowing for depth dependent parameters is valid (Lantz and Fan, 1999). The advantage is that we do not need to deal with acoustic waves that would impose a very small time-step. A realistic set of anelastic equations for convection in the sun is given in Fan *et al.* (1999). Computed quantities are fluctuations around given mean vertical profiles (Fig. 2) denoted by a z subscript, i.e. it can only vary with depth (Sect. 2.2). The system of conservation equations is in non dimensional units (Lantz and Fan, 1999):

$$\nabla \cdot (\rho_z \mathbf{V}) = 0 \tag{1}$$

$$\rho_z[\frac{\partial \mathbf{V}}{\partial t} + (\mathbf{V} \cdot \nabla)\mathbf{V}] = -\nabla p - \rho g \mathbf{e_z} - \frac{1}{R_o}\rho_z(\Omega \times \mathbf{V}) + (\nabla \times \mathbf{B}) \times \mathbf{B} + \frac{1}{R_e}\nabla \cdot \vec{\vec{\tau}} \tag{2}$$

$$\rho_z T_z[\frac{\partial s}{\partial t} + (\mathbf{V} \cdot \nabla)(s_z + s)] = \frac{1}{P_e}\nabla \cdot (\kappa \rho_z T_z \nabla s) + \frac{1}{R_m}\eta(\nabla \times \mathbf{B})^2 + \frac{1}{R_e}(\vec{\vec{\tau}} \cdot \nabla) \cdot \mathbf{V} \tag{3}$$

$$\nabla \cdot \mathbf{B} = 0 \tag{4}$$

$$\frac{\partial \mathbf{B}}{\partial t} = \nabla \times (\mathbf{V} \times \mathbf{B}) - \frac{1}{R_m}\nabla \times (\eta \nabla \times \mathbf{B}) \tag{5}$$

$$\frac{\rho}{\rho_z} = \frac{p}{P_z} - \frac{T}{T_z} \tag{6}$$

$$\frac{s}{c_p} = \frac{T}{T_z} - \frac{\gamma - 1}{\gamma}\frac{p}{P_z} \tag{7}$$

Here the gas is a polytrope with adiabatic parameter γ, s specific entropy, κ is the diffusivity of small scales entropy fluctuations, η is magnetic diffusivity, c_p is specific heat at constant pressure and $\vec{\vec{\tau}} = \mu \left[\frac{\partial V_i}{\partial x_j} + \frac{\partial V_j}{\partial x_i} - \frac{2}{3}\delta_{ij}\nabla .V\right]$ is the viscous rate of strain tensor.
Boundary conditions are non-penetrating stress-free top and bottom and periodic in the horizontal directions (Fan *et al.*, 1999). Non dimensional numbers are set in our simulations as listed in table.1. Corresponding dimensional physical parameters $g^*, \eta^*, \kappa^*, \mu^*, c_p^*, \Omega^*$ have been computed and listed in table.2 .

R_e	R_m	P_e	R_o
750	1000	1	1

Table 1. Non dimensional numbers used in our simulations.

$g^*(m.s^{-2})$	$\eta^*(m^2.s^{-1})$	$\kappa^*(m^2.s^{-1})$	$\mu^*(kg.m^2.s^{-1})$	$c_p^*(J.kg^{-1}.K^{-1})$	$\Omega^*(s^{-1}))$
260	10^9	1.3×10^9	1.3×10^{10}	3×10^4	1.6×10^{-3}

Table 2. Reference values for physical parameters in our simulations.

Since Rossby number is 1, corresponding Ω is three orders of magnitude higher than it should. We do not have radiation model and we do not simulate the photosphere.
As a time scale, we use the Free Fall Time (FFT) defined by: $t = \frac{H_0}{v_r} = \sqrt{\frac{H_0}{\delta_r g}} \approx 7.10^2 s$ where $\delta_r = 0.4$ is the superadiabaticity (Stix, 2002) and H_0 is the pressure height scale.

2.2. Mean profiles (convection):

Mean vertical polytropic profiles for pressure, entropy, density and temperature P_z, s_z, ρ_z, T_z (Fig.2) are given as follows:

$$P_z(z) = P_o \left[1 - \frac{z}{(m+1)H_o}\right]^{m+1} \tag{8}$$

$$\rho_z(z) = \rho_o \left[1 - \frac{z}{(m+1)H_o}\right]^{m} \tag{9}$$

$$s_z(z) = (m+1)\log \left[1 - \frac{z}{(m+1)H_o}\right] \tag{10}$$

$$T_z(z) = T_o \left[1 - \frac{z}{(m+1)H_o}\right] \tag{11}$$

The temperature gradient β in hydrostatic equilibrium, polytropic index m and pressure height scale H_o are related by:

$$\beta = -\frac{1}{(m+1)H_o} \tag{12}$$

In our problem, a reasonable value is: $m = \frac{1}{\gamma-1} \approx \frac{1}{5/3-1} \approx 1.5$. where $\gamma = \frac{5}{3}$ for an atomic ideal gas and isentropic process (Fan *et al.*, 1999). It can differ for an ionized gas (e.g., Lantz, 1992). Here gravity g and pressure scale height H_o are taken at the middle of the convective zone. Mean profiles are shown Fig.2. Due to our large numerical aspect ratio, we could not take exact profiles in our direct numerical simulation.

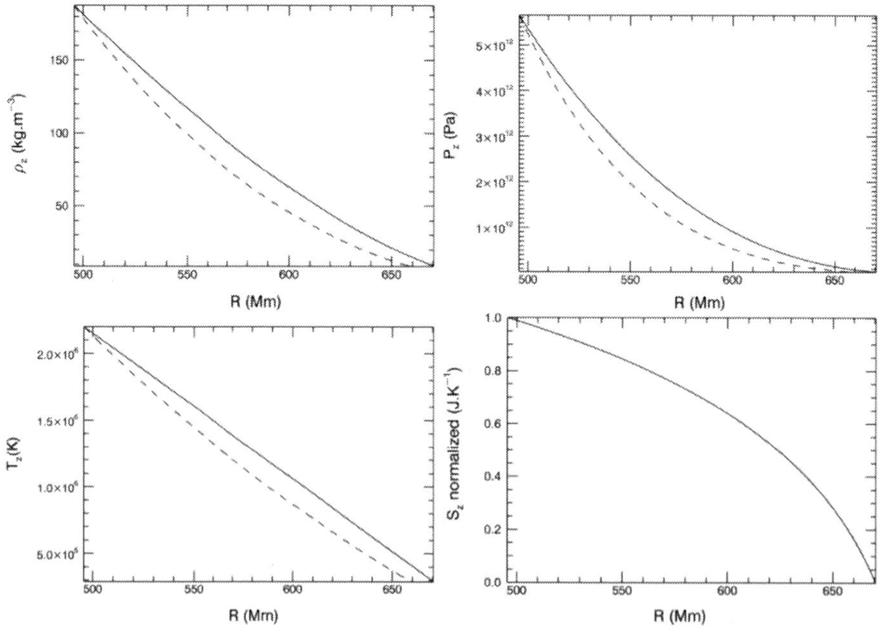

Figure 2. Density, pressure, temperature and entropy mean profiles from our simulation (——) and from a standard solar model (Bahcall *et al.*, 2004) (- - - -). Pressure scale is about 55 Mm at top of the convective zone. Theoretical profiles are slightly deeper than computed.

The code has been parallelized using Open-MP and, with 8 processors, the speed-up is increased by almost a factor 3 (Fig.3). The software we used is "AN-MHD" written by Solar Muri team (Fan *et al.*, 1999; Abbett *et al.*, 2001; *http://solarmuri.ssl.berkeley.edu/public/software/Anmhd/*). The machine we used is a cluster of 1024 cores (128 nodes 2-processors Intel Xeon E5462 quad-cores 3:0 GHz, 16 GB memory-node) at RQCHP (Réseau Québecois de Calcul Haute-Performance). We did also used extensively a 160 nodes cluster at GRPS (Groupe de Recherche en Physique Solaire de l'Université de Montréal).

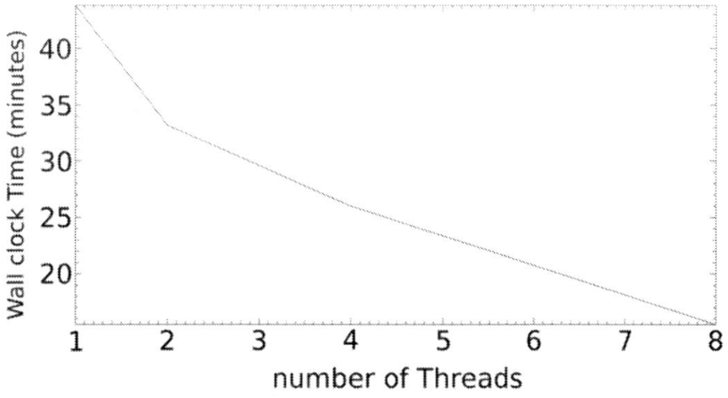

Figure 3. Computational performance of the code. Wall clock time is almost decreased by a factor 3 when the number of cores is increased from 1 to 8. It takes about a few hours to simulate a free fall time.

Numerical resolution is $256 \times 256 \times 64$ grid points corresponding to $175 \times 175 \times 175$ Mm with an 4:1 numerical aspect ratio. We use a periodic boundary condition. With this resolution, we can reach $Re = 750$, $Pr = 1$, $Ro = 1$ and $R_m = 1000$: In solar interior Reynolds number is about 10^6 and magnetic Reynolds number is about 10^9, Rossby number is about 2×10^2. No current simulations from literature approach real solar Reynolds number (Stein, 2011).

2.3. Recreating the convective zone below an active region.
We assimilate the first magnetograms with the NBF algorithm at times when they are available (01:39 and 03:15 on 14/07/2000). This enables to reconstruct a 3-D initial condition in T , P, ρ, $\mathbf{V}$and $\mathbf{B}$. To do this, we add a term in the equation for magnetic induction conservation (eq. 5):

$$\begin{aligned}\frac{\partial \mathbf{B}}{\partial t} &= \nabla \times (\mathbf{V} \times \mathbf{B}) - \frac{1}{R_m} \nabla \times (\eta \nabla \times \mathbf{B}) \\ &+ K \times \delta(z - z(L_z - 2\delta z)) \left\{ (B_x^{obs} - B_x^{cal})\mathbf{e_x} + (B_y^{obs} - B_y^{cal})\mathbf{e_y} + (B_z^{obs} - B_z^{cal})\mathbf{e_z} \right\}\end{aligned} \quad (13)$$

where $\delta(z - z(L_z - 2\delta z))$ is a Dirac function and δz is the vertical grid-step. K is the Nudging parameter, possibly a matrix, equivalent to a Lagrange multiplier (Auroux *et al.*, 2008). We take $K = 0.6$ (free fall time units)$^{-1}$ and $Lz = 64$. Thus, the NBF is a Newtonian relaxation technique. $B^{obs}_{x,y,z}$ and $B^{cal}_{x,y,z}$ are observed and computed vectorial magnetograms. We did also assimilate the two other components B_x and B_y of the vector magnetograms reconstructed from B_z using Linear Force Free approximation.

2.4. Simulation of a single buoyant flux tube, a test for validation.
To validate, the whole procedure can be done in three steps with a purely numerical test. First, by mean of a direct simulation of the anelastic code, we run a simple but realistic case of U-shaped turbulent flux tube. The resulting data are regarded as "the true state" or observations. We then extract top layers of this true state at various times and these are our observed "magnetograms". Third, we run the same code and same parameters but this time in NBF mode and with a random thus different initial state. To compare computed and "observed data", we use a linear Pearson correlation coefficient

$$C(B_z^{cal}, B_z^{obs}) = \frac{\sigma^{(cal.obs)}}{\sigma^{(cal)}\sigma^{(obs)}} = \frac{\sum\limits_{x,y}(B_z^{cal} - \bar{B}_z^{cal}).(B_z^{obs} - \bar{B}_z^{obs})}{\sqrt{\sum\limits_{x,y}(B_z^{cal} - \bar{B}_z^{cal})^2}\sqrt{\sum\limits_{x,y}(B_z^{obs} - \bar{B}_z^{obs})^2}} \tag{14}$$

where B_z^{cal} and B_z^{obs} are computed and observed magnetograms (top layer). A result close to one indicates that the two magnetograms are strongly correlated whereas a zero value indicates uncorrelated magnetograms. In this experiment, we found a good agreement between predictions and "observations" over a few Free Fall Times ($FFT \approx 11min$). This is shown Fig.4 where $B_z^{cal}(x_0, y_0, t)$ and $B_z^{obs}(x_0, y_0, t)$ are plotted. Here we take $x_0 = \frac{L_x}{5}$ and $y_0 = \frac{L_y}{5}$.

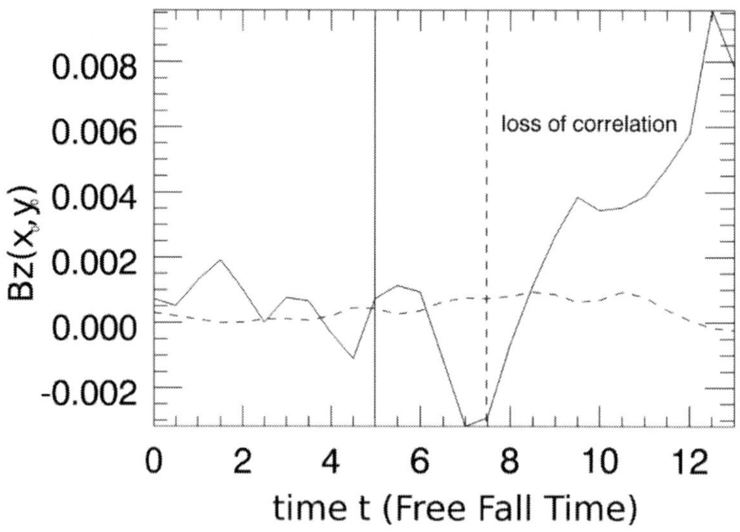

Figure 4. Local predictability using the NBF algorithm. Time evolution of the vertical component of magnetic field B_z at one place ($x = \frac{L_x}{5}, y = \frac{L_y}{5}$) of observed (dashed line) and computed (full line) magnetograms. Observed and computed $B_z(x_0, y_0)$ remain close each other up to a few Free Fall Times.

How far in time can we predict magnetograms could be estimated using the RMS error function $< \sqrt{(B_z^{cal} - B_z^{obs})^2} > (z)$ at top ($z = L_z - 4\delta z$) that is directly related to the Nudging forcing term in eq.13 but here we used correlation function defined above (eq. 14) and averaged over the entire magnetogram (Fig.5). Simulation is restarted at t=0 with initial condition obtained after data assimilation over t=0,5 FFT (the assimilation window). From t=0,12 FFT, computed and observed magnetograms remain correlated within 8 Free Fall Times. This includes the data assimilation window and a part of the prediction window.

Journal of Physics: Conference Series **341** (2012) 012005 doi:10.1088/1742-6596/341/1/012005

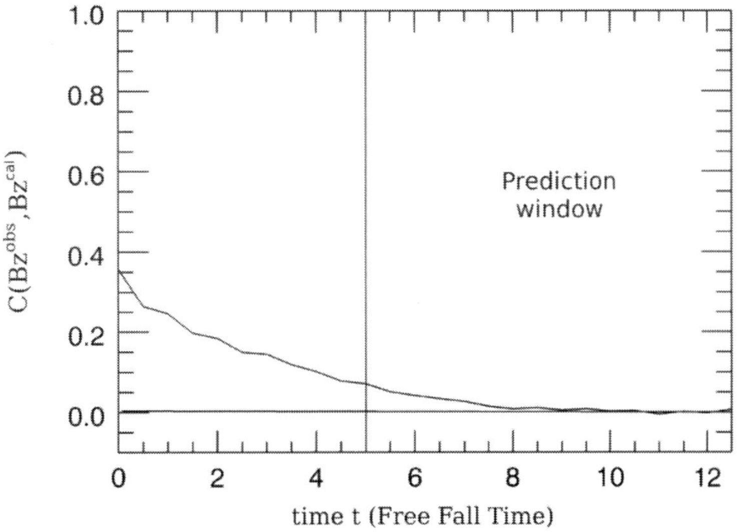

Figure 5. Test of validity. Decreasing of Pearson linear correlation coefficient between computed and observed vertical magnetograms with time. $C(B_z^{obs}, B_z^{cal})$ at end of NBF was optimal at 0.4 in this case. We restarted from t=0.

This shows that an anelastic model for solar convective zone combined with the NBF can be used for a prediction purpose over at least a few Free Fall Time (here about 30 minutes).

2.5. Unicity of solution using regularization

Since this problem is under-determined, we would not get a unique solution. This is the case for stiff problems like inversion problems and we have to use here an additional constraint called a regularization function (see Appendix)

$$\phi(x, y, z) = \frac{\int_{3D} \|\nabla \mathbf{B}(x, y, z)\|^2 dV}{\int_{3D} dV}. \tag{15}$$

Practically, we add to eq.13:

$$\begin{aligned} B_z^{cal}(x, y, z)^{\{t+\delta t\}} = B_z^{cal}(x, y, z)^{\{t\}} + \lambda \times \Big[& B_z^{cal}(x + \delta x, y, z) + B_z^{cal}(x - \delta x, y, z) \\ & + B_z^{cal}(x, y + \delta y, z) + B_z^{cal}(x, y - \delta y, z) + B_z^{cal}(x, y, z + \delta z) \\ & + B_z^{cal}(x, y, z - \delta z) - 6 \times B_z^{cal}(x, y, z) \Big]^{\{t\}} \end{aligned} \tag{16}$$

The solution is then unique but the regularization functional that we used (Tikhonov and Goncharsky, 1987) arbitrarily maximizes correlation between neighbouring points. It could be seen as a minimum entropy method. It is highly reasonable to do this because direct simulations of convection in sun-like conditions show a high degree of coherent structures. Not only we get a solution but also it is smoother than without regularization. By trial and error, we found that an optimal value for the Lagrange multiplier is $\lambda = 10^{-6}$. A largest value would distort magnetic patterns.

2.6. *Continuous top boundary condition using time interpolation*

During each run, we assimilate two vector magnetograms. The first corresponding to initial time and the second to final time. At each time step t, we interpolate linearly to get a continuous variation of magnetograms (here B_z).

$$B_z^{interp}(t) = (1 - \frac{t - t_0}{t_1 - t_0}) \times B_z^{obs}(t_0) + (\frac{t - t_0}{t_1 - t_0}) \times B_z^{obs}(t_1) \tag{17}$$

where t_0 and t_1 are initial and final times. There is an approximation here. Linear order is efficient to interpolate in time the largest scales but not the smallest scales that de-correlate much faster ($< 10min$) than about $1H30$, the time between two SOHO magnetograms.

However, in the tests with ascending flux tubes, we noted that there is no large distortion between the simulated continuous time varying sequence of "observed magnetograms" and the sequence of interpolated magnetograms. The anelastic set of equations (eqs. 1-7) is not equivalent to a linear interpolator but we found that if the time interval between two magnetograms is small enough ($\approx 15min$ or a free fall time) the differences are not large.

The final test of validity would be to compare real observations of magnetograms and temperatures with predicted B and T computed with our code but we do not do this here.

3. Results.

3.1. *Magnetic induction.*

We have reconstructed magnetic patterns below active region AR9077-20000714. Application of a local Nudging term has modified the whole convective zone and not only a thin layer at surface. Fig.6 shows a comparison between a snapshot of B_z component of magnetic field at top of our simulation (left) and the corresponding assimilated magnetogram (right). To extract only the largest scales on the graphics of Figs. 9,10,11 we have used a second order low pass filter (Roberts and Roberts, 1978).

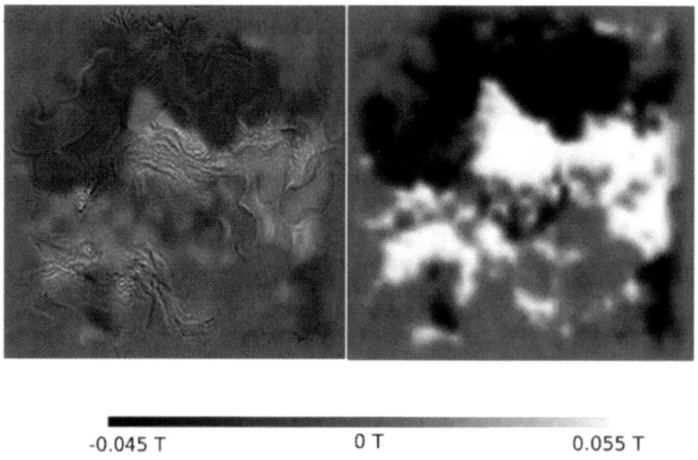

Figure 6. (Left) Computed B_z component of magnetic field at $z = L_z - 2\delta_z$. (Right) Assimilated data from *SOHO/MDI*. Time is a fraction of a free fall time. We see that largest scales are well reproduced.

Our results (Fig.7, bottom) show plume like structures commonly observed in simulations of magneto-convection of the quiet sun (Stein and Nordlund, 2006; Vögler *et al.*, 2005; Abbett *et al.*, 2004) but also an intense concentration of magnetic flux in the upflows (Fig.7, top). This is often observed in magneto-convective simulations of active regions (Stein, 2011). The maximal magnetic flux intensity is located near the neutral line, where gradients of magnetic field are higher.

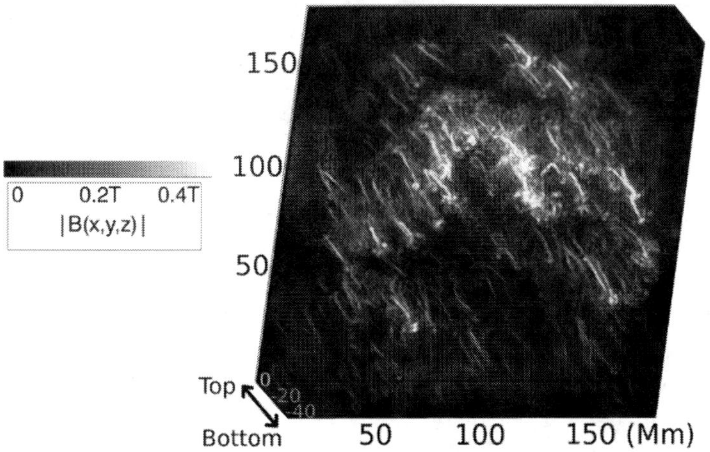

Figure 7. Data assimilated magnetic intensity: $|\mathbf{B}(x, y, z)|$. View from top. Time is a fraction of a free fall time. Magnetic induction at top corresponds to Figure 6 (left). Application of a local Nudging term has modified the whole convective zone and not only a thin layer at surface. At bottom of the figure, magnetic field is concentrated in downflows, showing plumes-like structure observed in magneto-convective simulations of the quiet sun. On the opposite, at top of the figure, magnetic field is highly concentrated in the upflows, forming an active region.

3.2. Temperature fluctuations.

Temperature fluctuations at top of the convective zone shows cool and hot places (Fig.8 left). Coldest areas correspond to very high magnetic field amplitude (Fig.8 right). High amplitude magnetic field should decrease convective heat transfer from inside layers and produce a local cooling (Stein, 2011). We observe that plasma located at the neutral line is hotter and indeed there are high electric currents there (Fig.9) taking the form of a sheet at top of the convective zone. This is because the horizontal magnetograms B_x, B_y we use have been reconstructed from the vertical magnetogram B_z using a linear force free algorithm. As a result, electric currents are parallel to the magnetic field. We did this on purpose because we have no photosphere in our anelastic equations. Additional heating of the neutral line may also be due to viscous friction as significant velocity shears develop there (Fig. 10, right).

Current lines are displayed Fig. 9. High currents are indicated by a lighter color. We see a direct correspondance between this high current line and the magnetic arcade system around it (Fig.10, right:top).

3.3. Magnetic field lines throughout the entire simulated region.

Although we do not simulate the corona or the photosphere, magnetic field lines from our simulations show often typically Ω/U loop structures (Fig. 10, right). Magnetic lines do reconnect but outside of the computational domain, in the corona, to form Ω loops. Magnetic field lines have been computed (VAPOR; Clyne *et al.*, 2005, 2007) assuming horizontal periodicity. Simulations of flux tubes emergence in the quiet sun and in active regions is known

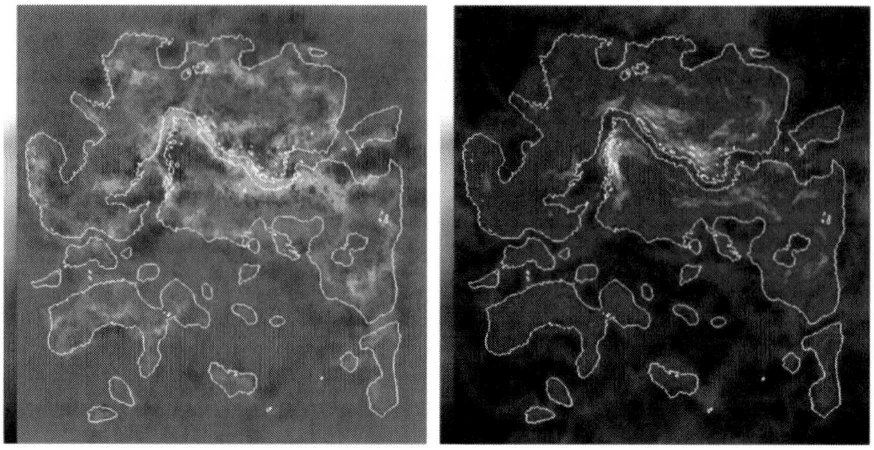

Figure 8. Left: Simultaneous display of temperature fluctuations and contour of the vertical magnetic field showing the polarity inversion line (neutral line). Temperature is lower where magnetic field is high. Right: Simultaneous display of magnetic field intensity and contour of the vertical magnetic field. Strongest magnetic fields are located on either sides of neutral line where magnetic gradients are maximal.

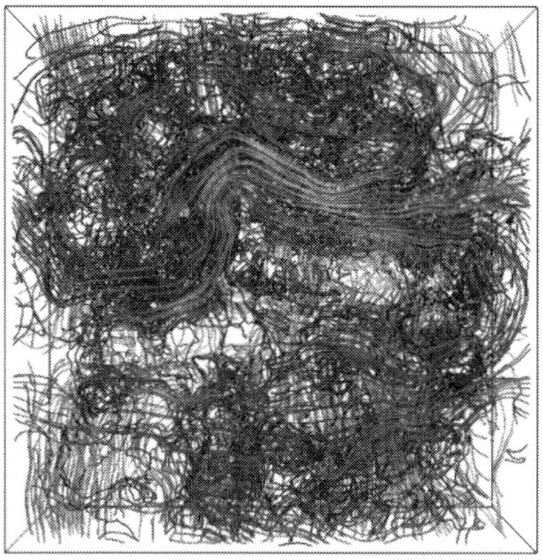

Figure 9. Comparison between magnetic field lines (red/yellow) and plasma current lines (blue). Intensity of the current increases from blue to white. Current intensity is high inside the Ω/U loop and maximal around the "polarity inversion line" where magnetic gradients are higher (Fan and Gibson, 2004).

to produce this kind of structures (Cheung *et al.*, 2010; Abbett *et al.*, 2004). Slow diverging upflows raise the flux and make magnetic lines emerging from the convective zone to form Ω loop, while fast downflows pin it down to produce U loops (Fig.10 left). Simultaneous upward and downward flows result in a vertical velocity shear.

Observations of the active region AR9077-20000714 have shown an arcade system in the corona "the slinky" (SOHO/EIT). Our results show that velocity shears due to convection are at the origin of this pattern (Figs.10,11).

Figure 10. Left: Comparison between magnetic field lines and vertical component's velocity. Magnetic field lines go from red to yellow. Blue areas indicate that the plasma is moving outward (upflows) whereas red indicates an inward movement (downflows). Magnetic field lines emerge from the convective zone in upflows. A low pass filter has been used to generate this graphics. Right: Vertical slice from left panel showing Ω/U loop structure for magnetic field above an active region. Magnetic field lines show an Ω loop structure above the computational domain and a U loop structure in it due to vertical velocity shearing.

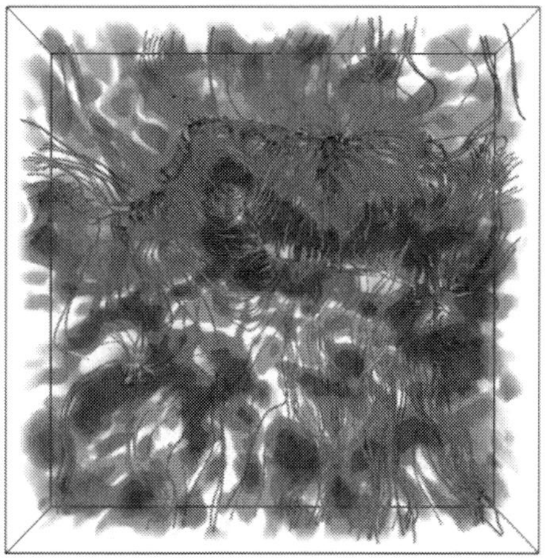

Figure 11. Simultaneous display of magnetic field lines and vertical components of magnetic field. Blue indicates that magnetic field is pointing outward whereas red indicates an inward vertical component. Arcade system of AR9077-20000714 of magnetic lines is actually formed by the Ω and U loops generated in the convection zone below.

4. Conclusion and future work

We have reconstructed the interior of solar convection zone below active region AR9077-20000714 by using the Nudging Back and Forth algorithm (Auroux *et al.*, 2008). To get a unique solution we added a regularization function that is also acting as a smoothing function. This simple data assimilation technique allows us to get the velocity and magnetic 3D fields as well as pressure, density, entropy and temperature inside the computed volume by assimilating vector magnetograms at top of convective zone. One could add more observations like temperature maps for instance. This is realistic because photospheric temperature is also observed by satellite. It would be included in a future work.

In our simulations, we see coherent magnetic patterns as well as places where magnetic fields are very weak. High magnetic intensity at top of the convective zone corresponds to low temperature. We see local Ohmic heating, heat diffusivity from underlying layers as well as viscous friction. We found that velocity shears between slow diverging upflows and fast turbulent downflows produce Ω and U-shaped magnetic field loops. This proves that that the coronal arcade system of AR9077-20000714 has its origin inside the convective zone below.
Our long-term goal is to find a possible 3D convective pattern that can be used as an initial condition to predict in real time, the evolution of temperature, magnetic field and velocity at photospheric levels based on magnetohydrodynamical modeling. Such reconstruction would be an important step toward flare prediction.

4.1. Appendix: Regularization

This problem is made stiff because of the discontinuous nature of the top boundary (Fig.6). One way to get smoother 3-D fields is to add a regularization. This step is under way and is not shown here. The idea is to minimize a functional Φ called the deviation function which quantifies the deviation of the calculated unknown quantities from the observed ones:

$$\begin{aligned}\Phi((\mathbf{B},T)_{3D}) = \;& \int_{2D} \; [R(B,T)_{obs} - R(B,T)_{cal}]^2 dS \\ + \;& \lambda\mathcal{R}(\mathbf{B},T)\end{aligned} \tag{18}$$

where $(\mathbf{B},T)$ is the unknown 3D field, R is a projection $(B,T)_{3D}$ at top of the convective layer. λ is a numerical Lagrange multiplier (small enough) and the second part of Φ (the deviation function) is a regularization functional $\mathcal{R}(\mathbf{B},T)$ because we have here a stiff problem. We use Tikhonov regularization functional that maximizes correlation between neighboring points (Tikhonov and Goncharsky, 1987):

$$\mathcal{R}(\mathbf{B},T) = \frac{\int_{3D} ||\nabla(B,T)||^2 dV}{\int_{3D} dV} \tag{19}$$

Acknowledgments

Graphics have been computed with IDL and VAPOR softwares and with the technical help of Nicolas Lawson from GRPS. We would like to thank professor Paul Charbonneau for his invaluable advices. Computations have been carried out at Réseau Québécois de Calcul Haute Performance, Montréal, Québec. GRPS's computing cluster funded by a CFI grant. We thank Jacques Richer from RQCHP for his help with code optimization. Alain Vincent has an NSERC Individual Research Grant. Finally, we thank the reviewer for constructive remarks.

References

[1] ANMHD *http://solarmuri.ssl.berkeley.edu/public/software/Anmhd/*
[2] Abbett, W.P., Fisher, G.H., Fan, Y., Bercik, D.J.: 2004, *The Astrophysical Journal*, **612**, 557-575.
[3] Abbett, W.P., Fisher, G.H.: 2002, *The Astrophysical Journal*, **582**, 475-485.
[4] Abbett, W.P., Fisher, G. H., Fan, Y.: 2001, *The Astrophysical Journal*, **546**, 1194.
[5] Auroux, D., Bansart, P., Blum, J.: 2008, *Journal de Physique: Conference Series* **135**, 012-011.
[6] Bahcall, J.N., Pinsonneault, M.H.: 2004: *arXiv:astro-ph*, **0402114v2**.
[7] Bouttier, F., Courtier, P.: 2003, "Data assimilation concepts and methods", *ECMWF training course lecture notes*, **59 pp**.

[8] Brummell, N.H.: 2006, *Proceedings IAU Symposium*, **239**, 417-424 . eds. F. Kupka, I.W. Roxburgh and K.L. Chan.
[9] Bushby, P.J.: 2008, *Philosophical Transactions of the Royal Society A*, **366**, 4465-4476.
[10] Christensen-Dalsgaard, J.: 2002, *arXiv:astro-ph/0207403*, **74**, 1073-1129.
[11] Clyne, J., Mininni, P., Norton, A., and Rast, M.: 2007, *New Journal of Physics*, **9**, 301.
[12] Clyne, J. and Rast, M.: 2005, in **proceedings of Visualization and Data Analysis 2005**, pp. 284-294.
[13] Cheung, C.M.,Rempel, M., Title, A.M., Schüssler, M.: 2010, *The Astrophysical Journal*, **720**, 233-244.
[14] Daley, R.:1991, "Atmospheric Data Analysis", *Cambridge University Press.*
[15] Fan, Y., Gibson, S.E.: 2004, *The Astrophysical Journal*, **609**, 1123-1133.
[16] Fan, Y., Zweibel, E.G., Linton, M.G., Fisher, G.H.: 1999, *The Astrophysical Journal*, **521**, 460-477.
[17] Fang, F., Manchester, W., Abbett, W. P., and Van der Holst, B.: 2010, *The Astrophysical Journal*, **714**, 1649-1657.
[18] Gough, D. O.: 1969, *Journal of Atmospheric Sciences*, **26**, 448-456.
[19] Hanasoge, S. M. and Larsen, R. M. and Duvall, Jr., T. L. and De Rosa, M. L. and Hurlburt, N. E. and Schou, J. and Roth, M. and Christensen-Dalsgaard, J. and Lele, S. K.: 2006, *The Astrophysical Journal*, **648**, 1268-1275.
[20] Hurlburt, N.E., De Rosa, M.L.: 2004, *Proceedings IAU Symposium: Multi-Wavelength Investigations of Solar Activity*, **223**, 253-254.
[21] Kalnay, E., Park, S.K., Pu, Zh.-X., Gao, ,J.: 1999, *Monthly Weather Review*, **128**, 864-875.
[22] Lantz, S.R., Fan, Y.: 1999, *The Astrophysical Journal Supplements Series*, **121**, 247-264.
[23] Lantz, S. R.:, 1992, *Dynamical Behavior of Magnetic Fields in a Stratified, Convecting Fluid Layer*, Phd Thesis, Cornell University.
[24] Lewis, J.M., Lakshmivarahan, S., Dhall, S.: 2006, "Dynamic Data Assimilation: A Least Squares Approach", Encyclopedia of Mathematics and its Applications 104, Cambridge University Press.
[25] Longcope, D. W.: 2004, *The Astrophysical Journal*, **612**, 1181-1192.
[26] Nordlund, Å,: 2008, '*Physica Scripta*, **vol T133**, pp. 014002.
[27] Nordlund, Å., Stein, R.F.: 2007, *Bulletin of the American Astronomical Society*, **38**, 125-+.
[28] Nordlund, Å., Stein, R. F., 2006, *The Astrophysical Journal*, **642**, 1246-1255.
[29] Nordlund, Å, Stein, R.F., Asplund, M.: 2009, *Living Review Solar Physics*, **6**, 2-+.
[30] Roberts, J., Roberts, T. D.: 1978, *Journal of Geophysical Research*, **83**, 5510-5514.
[31] Rieutord, M. and Roudier, T. and Malherbe, J. M. and Rincon, F.: 2000, *Astronomy and Astrophysics* **357** 1063-1072.
[32] Stein, R.F.: 2011, preprint, *Article submitted to Royal Society.*
[33] Stix, M., 2002, *The sun : an introduction* – 2nd ed. /Michael Stix. Berlin : Springer, 2002. QB 521 .S75.
[34] Tikhonov, A.N., Goncharsky, A.V.: 1987, "Ill-posed Problems in the Natural Sciences", *Ed. Mir, Moscow.*
[35] Vögler, A., Shelyag, S., Schüssler, M., Cattaneo, F., Emonet, T.,Linde, T.: 2005, *Astronomy and Astrophysics*, **429**, 335-351.
[36] Vögler, A., Schüssler, M.: 2007, *Astronomy and Astrophysics*, **465**, L43-L46.

Relativistic ground state of diatomic molecules from the numerical solution of the Dirac equation on parallel computers

F. Fillion-Gourdeau[1], Emmanuel Lorin de la Grandmaison[2,1], André D. Bandrauk[3,1]

[1] Centre de Recherches Mathématiques, Université de Montréal, Montréal, Canada, H3T 1J4
[2] School of Mathematics and Statistics, Carleton University, Ottawa, Canada, K1S 5B6
[3] Laboratoire de chimie théorique, Faculté des Sciences, Université de Sherbrooke, Sherbrooke, Canada, J1K 2R1

E-mail: filliong@CRM.UMontreal.ca

Abstract.
The parallel implementation of a variational method for the numerical solution of the time-independent Dirac equation is presented. This is utilized in the calculation of diatomic molecule relativistic wave functions and ground state energy, in the Born-Oppenheimer approximation. Following [1], a functional equation predicting the same spectrum as the Dirac equation is discretized over a set of basis functions. We find that the combination of B-spline basis functions and prolate spheroidal coordinate systems facilitate the vectorization of the technique using domain decomposition. This procedure results in a non-linear eigenvalue problem which is solved by iteration using Brent's method. We validate our methodology by comparing with existing results and show that it yields very accurate ones. We also analyze the parallelization and demonstrate that our implementation has a good scaling performance. Finally, the utilization of this method for laser-matter interaction and other physical applications will be discussed.

1. Introduction

The advent of new experimental facilities with lasers reaching intensities above 10^{19} W·cm^{-2} brings new challenges and opportunities for scientists. In this intensity regime, the motion and energy of an electron subjected to such electromagnetic fields approaches the relativistic domain. Thus, the Schroedinger equation loses its validity and should be replaced by its relativistic counterpart: the Dirac wave equation. The numerical solution of this equation is known to be a challenging task due to the presence of negative energy levels, which forbids the utilization of the Rayleigh-Ritz variational method [2], and the appearance of spurious fermionic states, which occurs in the discretization process [3]. To solve these issues, a new numerical technique has recently been developed for the time-independent case by Esteban *et al.* [1, 4–6]. Their method is based on a variational min-max principle and allows the computation of relativistic eigenstates of any bound scalar potentials.

In this work, we present our implementation of this method on a parallel computer architecture and use it to compute the ground state energy of diatomic molecules. Note here that similar calculations were performed in [1, 7–9] but our implementation uses different basis functions and/or coordinates and is parallelized. This work is separated as in the following. In Section 2, the numerical method is presented which includes the variational formulation of the Dirac equation and the choice of basis functions. Then, in Section 3, the details of the implementation of this method on a parallel computer architecture are shown. The results can be found in Section 4 where the scalability of the algorithm and the bound state of diatomic molecules are investigated.

2. The numerical method

The Dirac equation describes the relativistic dynamics of spin-$\frac{1}{2}$ particles (fermions) like the electron. Because the fermions are ubiquitous in nature, this equation is very important in many areas of physics and chemistry. In this analysis, we consider specifically the static Dirac equation without vector potential given by[1]

$$\hat{H}\psi(x) = E\psi(x) \ \text{ with } \ \hat{H} \equiv -ic(\boldsymbol{\alpha}\cdot\boldsymbol{\nabla}) + mc^2\boldsymbol{\beta} + V(x)\mathbb{I}, \tag{1}$$

where $\hat{H}$ is the Hamiltonian operator, m is the electron mass, E is the electron energy and $\psi(x) \in L^2(\mathbb{R}^3)\otimes\mathbb{C}^4$ is a four-spinor. The Dirac matrices are given by

$$\boldsymbol{\alpha} = \begin{bmatrix} 0 & \boldsymbol{\sigma} \\ \boldsymbol{\sigma} & 0 \end{bmatrix} \ \text{ and } \ \beta = \begin{bmatrix} \mathbb{I}_2 & 0 \\ 0 & -\mathbb{I}_2 \end{bmatrix} \tag{2}$$

written in terms of the usual Pauli matrices defined as

$$\sigma_x = \begin{bmatrix} 0 & 1 \\ 1 & 0 \end{bmatrix} \ , \ \sigma_y = \begin{bmatrix} 0 & -i \\ i & 0 \end{bmatrix} \ \text{ and } \ \sigma_z = \begin{bmatrix} 1 & 0 \\ 0 & -1 \end{bmatrix}. \tag{3}$$

This equation gives a consistent description of electrons bound in molecules in the Born-Oppenheimer approximation, i.e. when the atoms are fixed in space. This is a valid approximation when the mass of the atom is much larger than the mass of the atom, which will always be the case for the systems considered in this study.

The main goal of this work is to calculate approximate solutions of Eq. (1). To achieve this, it is convenient to write the four-spinor as $\psi(x) \equiv [\phi(x), \chi(x)]^{\mathrm{T}}$ where $\phi(x), \chi(x) \in L^2(\mathbb{R}^3)\otimes\mathbb{C}^2$ are two bi-spinors [1, 4, 5]. The Dirac static equation can then be decomposed into two coupled equations as

$$\hat{R}\chi = (E - mc^2 - V)\phi \ \text{ and } \ \hat{R}\phi = (E + mc^2 - V)\chi \tag{4}$$

where the operator $\hat{R} \equiv -ic(\boldsymbol{\sigma}\cdot\boldsymbol{\nabla})$. These two equations can be decoupled by substituting the solution of the second one into the first one, so that we get

$$\chi = (E + mc^2 - V)^{-1}\hat{R}\phi \ \text{ and } \ \hat{R}\left(\frac{\hat{R}\phi}{E + mc^2 - V}\right) = (E - mc^2 - V)\phi. \tag{5}$$

Multiplying the latter by $\phi^\dagger$ on the left and integrating on space (using integration by parts) yields a functional equation given by [1, 4, 5]

$$A[E,\phi] \equiv \int d^3x \left[\left(\frac{|\hat{R}\phi|^2}{E + mc^2 - V}\right) - (E - mc^2 - V)|\phi|^2\right] = 0 \tag{6}$$

[1] All the calculations will be performed in atomic units (a.u.) where $m = 1$, $\hbar = 1$ and $c = 1/\alpha$ where $\alpha \approx 1/137$. is the fine structure constant. In all the equations however, we are keeping the mass explicitly, allowing to switch easily from atomic to natural units.

for any $\phi(x) \in L^2(\mathbb{R}^3) \otimes \mathbb{C}^2$. This is obtained by using the divergence theorem and by assuming that the wave function vanishes faster than $\sim \frac{1}{r^2}$ at infinity. This is the case when $V(x)$ is a Coulomb-like potential because the corresponding wave function vanishes as $\phi \sim e^{-r}$ when $r \to \infty$ [10].

It can be shown that the last equation predicts the same spectrum in the mass gap $[-mc^2, mc^2]$ as that of the Dirac equation [1, 4, 5, 11, 12]. Moreover, this formulation allows discretization schemes that use basis functions without the "variational collapse". The latter is related to the fact that the Dirac operator is not bounded from below because of the existence of the negative energy continuum. Therefore, the usual variational formulations which approximate the energy by minimizing a functional will fall into this lower part of the spectrum and will not be able to find the bound states, unless complicated projection operator techniques are used [2]. The method presented in this work avoid these complications and allows the calculation of bound states in the mass gap without the negative energy continuum contribution. Thus, as shown in the following, it is possible to develop a numerical method that solves Eq. (6) by discretizing over a set of basis functions.

2.1. Formulation in cylindrically symmetric system

For a diatomic molecule, the case considered in this study, the static potential is

$$V(x) = -\frac{Z_1}{|\mathbf{x} + R\hat{\mathbf{z}}|} - \frac{Z_2}{|\mathbf{x} - R\hat{\mathbf{z}}|}. \tag{7}$$

where $Z_{1,2}$ are the atomic electric charges, $2R$ is the inter-atomic distance and $\hat{z}$ is a unit vector in the z-coordinate direction. This potential is axially symmetric so it is possible to reduce by one the number of dimensions; the azimuthal coordinate dependence can be treated analytically by factorization. Thus, the four-spinor in cylindrical symmetry reads [7, 13]

$$\psi(x) = \begin{bmatrix} \phi(\xi, \eta, \theta) \\ \chi(\xi, \eta, \theta) \end{bmatrix} = \begin{bmatrix} \phi_1(\xi, \eta) e^{i(j_z - 1/2)\theta} \\ \phi_2(\xi, \eta) e^{i(j_z + 1/2)\theta} \\ i\chi_1(\xi, \eta) e^{i(j_z - 1/2)\theta} \\ i\chi_2(\xi, \eta) e^{i(j_z + 1/2)\theta} \end{bmatrix}. \tag{8}$$

where j_z is the angular momentum projection on the z-axis and η, ξ are prolate spheroidal coordinates. As will be discussed later, this choice of coordinate system is very convenient for the numerical implementation. They are defined as

$$x = R\left[(\xi^2 - 1)(1 - \eta^2)\right]^{\frac{1}{2}} \cos\theta, \tag{9}$$

$$y = R\left[(\xi^2 - 1)(1 - \eta^2)\right]^{\frac{1}{2}} \sin\theta, \tag{10}$$

$$z = R\xi\eta \tag{11}$$

where $\xi \in [1, \infty]$, $\eta \in [-1, 1]$ and $\theta = [0, 2\pi]$ (azimuthal angle).

The discretization of Eq. (6) with the potential Eq. (7) proceeds by expanding the wave function over a set of basis functions, in the same spirit as in Galerkin's Methods. Thus, the bi-spinor can be written as

$$\phi_{1,2}(\xi, \eta) = \sum_{n=1}^{N} a_n^{(1,2)} B_n^{(1,2)}(\xi, \eta) \tag{12}$$

where $a_n^{(1,2)}$ are the coefficient of the basis expansion and $B_n^{(1,2)}(\xi, \eta)$ are the basis functions (to be defined later), for components 1 and 2 respectively. The explicit expression of Eq. (6)

in discretized form depends on the potential considered, on the coordinate choice and is a complicated functional of basis functions (some examples can be found in [1, 5]). For cylindrically symmetric systems expressed in prolate spheroidal coordinates, the case considered in this study, the expression can be found in Appendix A.

Generally, the equation $A[E, \phi] = 0$, once discretized, have the form

$$\sum_{i,j=1}^{2N} a_i A_{ij}(E) a_j = \mu(E) \tag{13}$$

where $\mu(E)$ is the eigenvalue of the A matrix and where

$$a_i = \begin{cases} a_i^{(1)} & \text{for} \quad i \leq N \\ a_{i-N}^{(2)} & \text{for} \quad i > N \end{cases} \tag{14}$$

Therefore, solving the last non-linear eigenvalue problem with $\mu(E) = 0$ gives an approximation of the energy E and eigenfunctions: the energy is a solution of $\mu(E) = 0$ where μ is an eigenvalue of A while wave function coefficients in the basis expansion are the A matrix eigenvector coefficients.

2.2. Basis functions

Throughout this work, the B-spline basis functions are used (a description of these functions can be found in [14]). This choice is favored over other techniques because it can be easily implemented and because the B-spline basis has compact support, leading to a sparse matrix structure. This allows the use of powerful numerical routines for the eigenvalue determination.

The B-splines are fully determined from their order $k_{\xi,\eta}$ and knot vector using the iterative formula [14, 15]

$$b_i^k(x) = \frac{x - t_i}{t_{i+k-1} - t_i} b_i^{k-1}(x) + \frac{t_{i+k} - x}{t_{i+k} - t_{i+1}} b_{i+1}^{k-1}(x) \tag{15}$$

where t_i's are knots coordinates chosen arbitrarily in the domain under consideration and $x = \eta, \xi$ is the coordinate. The number of knots at a given coordinates essentially determines the continuity conditions at that point. Therefore, the number of knots points should be maximal at singular points (at the Coulomb singularity position for instance) to allow for a stiff behavior. Throughout this work, the knot vectors are given by the sequences

$$1 = \xi_1 = \ldots = \xi_{k_\xi} < \xi_{k_\xi+1} < \ldots < \xi_{n_\xi+1} = \ldots = \xi_{n_\xi+k_\xi} = \xi_{\max} \tag{16}$$

$$-1 = \eta_1 = \ldots = \eta_{k_\eta} < \eta_{k_\eta+1} < \ldots < \eta_{n_\eta+1} = \ldots = \eta_{n_\eta+k_\eta} = 1 \tag{17}$$

Here, $n_{\xi,\eta}$ are the number of spline functions in ξ and η coordinates respectively. The coordinates of the knot vectors can be chosen arbitrarily in the domain under consideration. However, to improve accuracy, an exponential sequence with smaller intervals close to the singularities is used in this study. The knot sequences and domain structure for diatomic molecules are depicted in Fig. 1.

The basis function can then be written as the tensor product of B-spline functions. To handle the singularity of the Coulomb potential, this is multiplied by a global function $G^{(1,2)}(\xi, \eta)$ that accounts for the singular behavior, which results in

$$B_n^{(1,2)}(\xi, \eta) = G^{(1,2)}(\xi, \eta) b_i^{k_\xi}(\xi) b_j^{k_\eta}(\eta) \tag{18}$$

Journal of Physics: Conference Series **341** (2012) 012006 doi:10.1088/1742-6596/341/1/012006

where $n = [i, j] \in \mathbb{Z}^2$, $i \in [1, n_\xi]$ and $j \in [1, n_\eta]$. The overall factor is defined as [7, 16]

$$G^{(1,2)}(\xi, \eta) = G_1^{(1,2)}(\xi, \eta) G_2(\xi, \eta) \tag{19}$$

where

$$G_1^{(1,2)}(\xi, \eta) = [(\xi^2 - 1)(1 - \eta^2)]^{\frac{m_{(1,2)}}{2}} \text{ with } m_{(1)} = j_z - \frac{1}{2} \text{ and } m_{(2)} = j_z + \frac{1}{2} \tag{20}$$

accounts for the angular momentum dependence (j_z is the angular momentum projection on the z-axis). The second factor deals with the singular behavior and is given by

$$G_2(\xi, \eta) = r_1^{-1+\gamma_1} r_2^{-1+\gamma_2} \tag{21}$$

with

$$r_1 = (\xi + \eta)R, \; r_2 = (\xi - \eta)R \text{ and } \gamma_{1,2} = \sqrt{\left(|j_z| + \frac{1}{2}\right)^2 - \alpha^2 Z_{1,2}^2}. \tag{22}$$

This form of the overall factor is based on the analytical solution of the single atom: as one gets closer to one of the atom in the molecule, the potential tends to the single atom potential and the wave function of the molecule and its constituent ion coincide in that region.

The boundary conditions are chosen as $\phi(\xi_{\max}, \eta) = 0$. Using B-splines, this can be implemented easily by setting $b_{n_\xi}(\xi) = 0$ and by considering only $n_\xi - 1$ B-spline functions in ξ coordinates. The other boundaries are free. It was shown in [1] that the choice of boundary conditions at $\xi = \xi_{\max}$ has negligible effect on the final result if $\xi_{\max}$ is large enough. Thus, we give $\xi_{\max}$ the values already determined in [1, 7].

3. Parallel Implementation

The code implementing the previously described numerical method for the solution of the time-independent Dirac equation is written in the C++ programming language. The parallelization is performed by using a domain decomposition strategy and the inter-processor communications are done with the Message Passing Interface (MPI). The details of the parallelization and algorithms are presented in the following.

The algorithm can be separated into four main steps:

(i) Guess an initial value for the bound state energy E.

(ii) Build the A matrix.

(iii) Diagonalize A to evaluate $\mu(E)$.

(iv) Solve the equation $\mu(E) = 0$.

The initial guess for the energy is given by $E_{\text{init}} = E_1 + E_2$ where $E_{1,2}$ are the energies of the single atoms, which have well-known analytical expressions [10].

The next step is the construction of the A matrix and is the more demanding in terms of computation time. It consists in the numerical evaluation of Eq. (A.1) which is an integral equation. This part of the resolution is an embarrassingly parallel problem: each region can be integrated independently. Each element is sent to a different processor and the PPPack library (written in FORTRAN 90) [15] is used to evaluate all the non-zero B-splines in that region. The integration is computed by the rectangle rule; the elements are subdivided into smaller domains as shown in Fig. 1 and the functions are evaluated at the center of these subregions. The result of the integration on each processor is added on the master processor to get the sparse A matrix. The data flow between processors during this step is shown in Fig. 2.

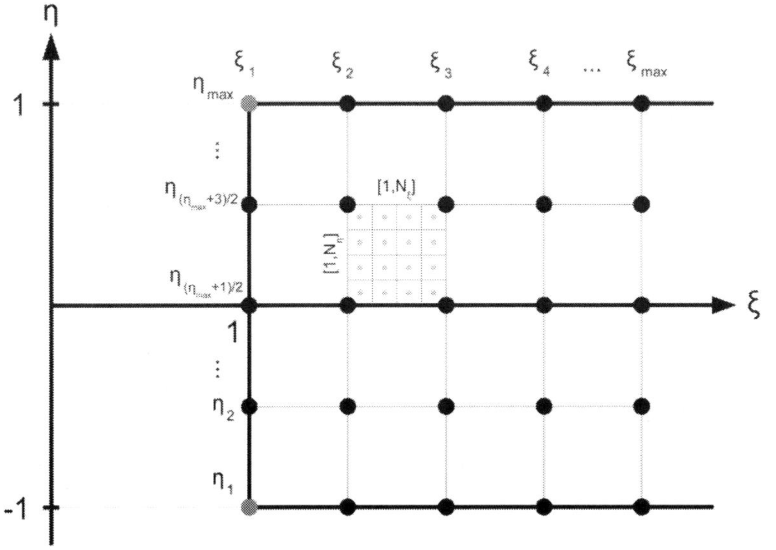

Figure 1. Discretization of the domain. The black dots represent knot points, the red dot are knot points where there is a Coulomb singularity and the small blue dots are integration points. The domain under consideration is discretized into a certain number of elements while each elements is subdivided into smaller region for the space integration. In this picture, a constant element size is shown whereas in the actual computation, we are using an exponential size distribution with smaller elements close to the Coulomb singularities.

The latter is diagonalized using the PRIMME software [17] which is an efficient preconditioned eigensolver for symmetric (and hermitian) matrices. This allows to compute the eigenvalues and eigenvectors of A and consequently, to evaluate $\mu(E)$. This step is performed sequentially because the matrix is not too large (approximately 400×400) for the needed accuracy and the diagonalization is very efficient.

The root-finding algorithm of the GNU Scientific Library (GSL) based on Brent's method [18] is used to solve the equation $\mu(E) = 0$. This method is a combination of the bisection and interpolation methods. Its main advantages is that it is guaranteed to converge once the root is bracketed, it does not necessitate the evaluation of function gradients and it has a better order of convergence than the bisection method [18].

4. Results

In this section, the results of our implementation are presented. We consider first the parallel performance to test the efficiency of our implementation. Then, we apply our method to the computation of the ground state of diatomic molecules.

4.1. Scaling properties

The parallel performance of the implementation is evaluated by looking at the efficiency defined as $e \equiv \frac{T_1}{N_p T_{N_p}}$ where T_{N_p} is the computation time on N_p processors and T_1 is the computation time on one processor. As a benchmark problem, we compute the ground state energy of diuranium U_2^{183+} ($Z_{1,2} = 92$) with a semi inter-atomic distance of $R = \frac{1}{92} \approx 0.010869565$ a.u.. We split the domain into 12×12 elements and consider B-splines of order 7. We then consider two calculation parameters: in the first one, each element is subdivided into 25×25 regions

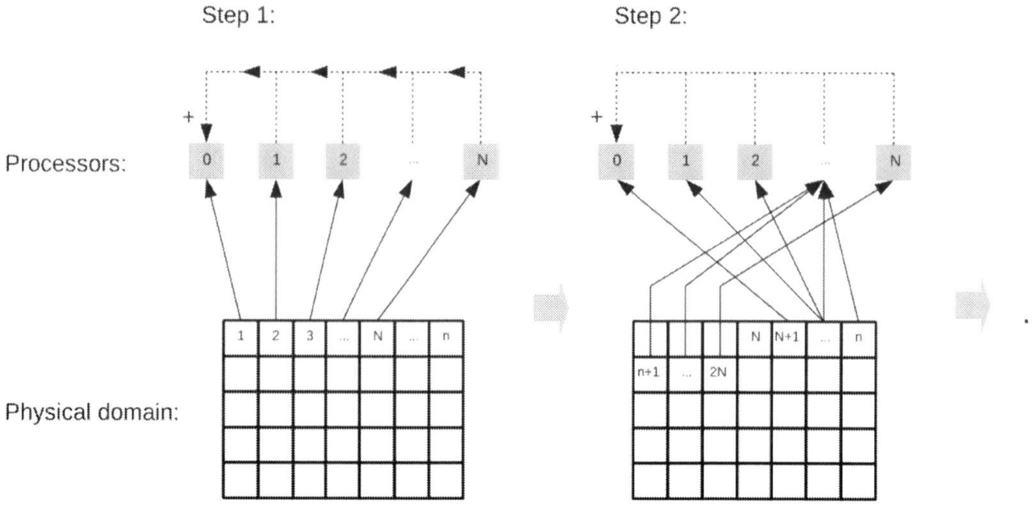

Figure 2. Data flow between processors during the calculation of the A matrix. Here, N is the number of processors while n is the number of regions. In this example, we have $n > N$. During step 1 (on the left), regions 1 to N of the physical domain are sent to different processors where their A matrix contributions are computed (continuous arrows). Then, the results are sent to the master processor where they are added together (dashed arrows). The same procedure is done in step 2 for regions $N + 1$ to $2N$. This continues until all region contributions are calculated.

while in the second one, we have 50×50 regions. The bound state energy obtained in this way is $E_{\mathrm{bound}} \approx -9965.1878$ a.u. for the first one and $E_{\mathrm{bound}} \approx -9964.8066$ a.u. for the second one. All the computation were performed on the MAMMOUTH cluster (which has Intel Xeon 64bit 3.6 GHz CPUs and an Infiniband non-blocking (800 MB/sec) network).

The results for the parallel efficiency in these computations are shown in Fig. 3. For low number of processors ($\lesssim 32$ for the first one and $\lesssim 64$ for the second one), the efficiency is slightly above 1. The most probable explanation for this result is the cache effect: as the number of processors is increased, the percentage of the problem fitted in cache memory is also enhanced. Because the cache memory is faster than normal memory, the computation time is reduced. Moreover, the more demanding part of the computation is embarrassingly parallel, which results in a very good parallelization performance. Above these number of processors however, the efficiency diminishes due to inter-communication between different processors. This diminution is more important in the smaller problem considered, suggesting that larger problems can be parallelized more efficiently on a large number of processors. This happens because the code spends more time to perform the embarrassingly parallel part (the construction of the A matrix) in larger problems.

4.2. Diatomic molecules

In this section, we are investigating the convergence of our numerical method. More specifically, we study and calculate the ground state of di-thorium (Th_2^{179+} which have $Z_{1,2} = 90$) and dihydrogen (H_2^+ which have $Z_{1,2} = 1$)). Note that both have only one electron. The semi inter atomic distance is set to $R = \frac{1}{90} \approx 0.011111$ a.u. for dithorium and to $R \approx 1.000$ a.u. for dihydrogen. The results for the calculation of the ground state binding energy for different simulation parameters are shown in Table 1.

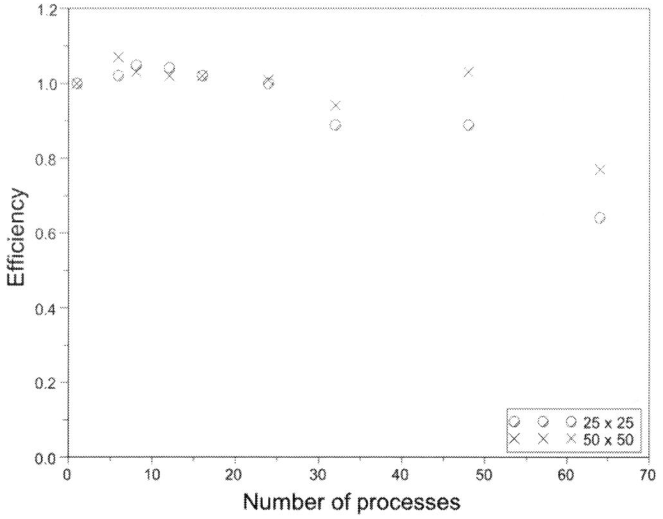

Figure 3. Efficiency as a function of the number of processor for the diuranium molecule computation.

The results presented in this table show the convergence of the method as the number of elements and B-spline order is increased. In Table 2, a comparison with results obtained in the literature is presented. This shows that the results obtained are very accurate, although there is a small difference ($\approx 0.03\%$) between our results and the results presented in [7] and [8]. This difference can be explained by a different choice of boundary conditions and element formulation.

Table 1. Results of the numerical computation for the ground state of H_2^+ and Th_2^{179+} for different calculation parameters.

Order	n_ξ	n_η	# elements/cell	# of cells	$E_{H_2^+}$ (a.u.)	$E_{Th_2^{179+}}$ (a.u.)
5	8	8	625	16	-1.1071293	-9493.2409
3	10	10	625	64	-1.1100043	-9496.9237
5	12	12	625	64	-1.1024476	-9503.4105
7	14	14	625	64	-1.1025155	-9503.6125
5	16	16	625	144	**-1.1026076**	**-9504.2743**

5. Conclusion

In this work, we presented a numerical method to solve the time-independent Dirac equation based on a variational principle and using a B-spline discretization. The Coulomb singularity was dealt with by adding an overall factor to the basis functions. We applied this method to the computation of diatomic molecule ground state energies. Because of their axial symmetry, it was convenient to use prolate spheroidal coordinates. The method was implemented in parallel and the code showed an efficient scaling with the number of processors. It was used to compute

Table 2. Comparison with the results from [7] and [8] for point-like nuclei.

Reference	$E_{H_2^+}$ (a.u.)	$E_{Th_2^{179+}}$ (a.u.)
This work	-1.1026076	-9504.2743
Kullie *et al.* [7]	-1.10264158	-9504.756746922
Artemyev *et al.* [8]	-1.1026409	-9504.752

the ground state energy of dihydrogen, dithorium and diuranium. A comparison with results in the literature demonstrated that our method yields very accurate results.

This method can then be utilized in many applications. Among others, this can be used to investigate relativistic laser-matter interaction: the solution obtained from this code can be used as an initial condition for the solution of the time-dependent Dirac equation. This will be the subject of future investigations.

Appendix A. Explicit expression of Eq. (6)

The explicit expression of Eq. (6) is obtained by using the Dirac equation in cylindrical coordinates given in [13]. Then, combining the ansatz in Eq. (8) with the basis function expansion, we get

$$\begin{aligned}
A(E) \equiv & \sum_{i=1}^{N}\sum_{j=1}^{N}\int d^3x\Bigg\{ \\
& a_i^{(1)}a_j^{(1)}\Bigg[(\partial_z B_i^{(1)})(\partial_z B_j^{(1)}) + (\partial_r B_i^{(1)})(\partial_r B_j^{(1)}) + \frac{(j_z-\frac{1}{2})^2}{r^2}B_i^{(1)}B_j^{(1)} \\
& -\frac{(j_z-\frac{1}{2})}{r}B_i^{(1)}(\partial_r B_j^{(1)}) - \frac{(j_z-\frac{1}{2})}{r}(\partial_r B_i^{(1)})B_j^{(1)}\Bigg]\frac{c^2}{E+mc^2-V} \\
+ & a_i^{(2)}a_j^{(2)}\Bigg[(\partial_z B_i^{(2)})(\partial_z B_j^{(2)}) + (\partial_r B_i^{(2)})(\partial_r B_j^{(2)}) + \frac{(j_z+\frac{1}{2})^2}{r^2}B_i^{(2)}B_j^{(2)} \\
& +\frac{(j_z+\frac{1}{2})}{r}B_i^{(2)}(\partial_r B_j^{(2)}) + \frac{(j_z+\frac{1}{2})}{r}(\partial_r B_i^{(2)})B_j^{(2)}\Bigg]\frac{c^2}{E+mc^2-V} \\
+ & a_i^{(1)}a_j^{(2)}\Bigg[(\partial_z B_i^{(1)})(\partial_r B_j^{(2)}) + \frac{(j_z-\frac{1}{2})}{r}B_i^{(1)}(\partial_z B_j^{(2)}) \\
& -(\partial_r B_i^{(1)})(\partial_z B_j^{(2)}) + \frac{(j_z+\frac{1}{2})}{r}(\partial_z B_i^{(1)})B_j^{(2)}\Bigg]\frac{c^2}{E+mc^2-V} \\
+ & a_i^{(2)}a_j^{(1)}\Bigg[(\partial_z B_j^{(1)})(\partial_r B_i^{(2)}) + \frac{(j_z-\frac{1}{2})}{r}B_j^{(1)}(\partial_z B_i^{(2)}) \\
& -(\partial_r B_j^{(1)})(\partial_z B_i^{(2)}) + \frac{(j_z+\frac{1}{2})}{r}(\partial_z B_j^{(1)})B_i^{(2)}\Bigg]\frac{c^2}{E+mc^2-V} \\
+ & a_i^{(1)}a_j^{(1)}(V+mc^2-E) + a_i^{(2)}a_j^{(2)}(V+mc^2-E) \qquad \text{(A.1)} \\
\equiv & \sum_{i=1}^{2N}\sum_{j=1}^{2N} a_i A_{i,j}(E) a_j \qquad \text{(A.2)}
\end{aligned}$$

The last expression can then be expressed in prolate spheroidal coordinates by using

$$\partial_r = \frac{\sqrt{(\xi^2 - 1)(1 - \eta^2)}}{R(\xi^2 - \eta^2)} [\xi\partial_\xi - \eta\partial_\eta] \quad (A.3)$$

$$\partial_z = \frac{(\xi^2 - 1)}{R(\xi^2 - \eta^2)}\eta\partial_\xi + \frac{(1 - \eta^2)}{R(\xi^2 - \eta^2)}\xi\partial_\eta \quad (A.4)$$

and the integration measure is given by

$$d^3x = R^3(\xi^2 - \eta^2)d\xi d\eta d\theta. \quad (A.5)$$

References

[1] Dolbeault J, Esteban M J and Séré E 2003 *International Journal of Quantum Chemistry* **93** 149 – 155
[2] Datta S N 1980 *Chemical Physics Letters* **74** 568 – 572
[3] Stacey R 1982 *Phys. Rev. D* **26** 468–472
[4] Dolbeault J, Esteban M J and Séré E 2000 *Journal of Functional Analysis* **174** 208 – 226
[5] Dolbeault J, Esteban M J, Séré E and Vanbreugel M 2000 *Phys. Rev. Lett.* **85** 4020–4023
[6] Bandrauk A, Delfour M and Bris C (eds) 2007 *High-dimensional partial differential equations in science and engineering* (*CRM proceedings & lecture notes* vol 41) (American Mathematical Society)
[7] Kullie O, Kolb D and Rutkowski A 2004 *Chemical Physics Letters* **383** 215 – 221
[8] Artemyev A N, Surzhykov A, Indelicato P, Plunien G and Stohlker T 2010 *Journal of Physics B: Atomic, Molecular and Optical Physics* **43** 235207
[9] Sundholm D, Pyykko P and Laaksonen L 1987 *Physica Scripta* **36** 400
[10] Itzykson C and Zuber J B 1980 *Quantum Field Theory* (Mcgraw-hill)
[11] Dolbeault J, Esteban M J and Séré E 2000 *Calculus of Variations and Partial Differential Equations* **10**(4) 321–347
[12] Dolbeault J, Esteban M J, Loss M and Vega L 2004 *Journal of Functional Analysis* **216** 1 – 21
[13] Schluter P, Wietschorke K H and Greiner W 1983 *Journal of Physics A: Mathematical and General* **16** 1999
[14] Bachau H, Cormier E, Decleva P, Hansen J E and Martin F 2001 *Reports on Progress in Physics* **64** 1815
[15] De Boor C 1978 *A practical guide to splines* (Springer-Verlag, New York :)
[16] Kullie O, Dusterhoft C and Kolb D 1999 *Chemical Physics Letters* **314** 307 – 310
[17] Stathopoulos A and McCombs J R 2010 *ACM Trans. Math. Softw.* **37**(2) 21:1–21:30
[18] Brent R 1973 *Algorithms for Minimization without Derivatives* (Prentice-Hall)

Journal of Physics: Conference Series **341** (2012) 012007 doi:10.1088/1742-6596/341/1/012007

Optimization of the Kinetic Activation-Relaxation Technique, an off-lattice and self-learning kinetic Monte-Carlo method

Jean-François Joly[1], Laurent Karim Béland[1], Peter Brommer[1], Fedwa El-Mellouhi[2] and Normand Mousseau[1]

[1] Département de Physique and Regroupement Québécois sur les Matériaux de Pointe (RQMP), Université de Montréal, C.P. 6128, Succursale Centre-Ville, Montréal, Québec, Canada H3C 3J7

[2] Science Program, Texas A&M at Qatar, Texas A&M Engineering Building, Education City, Doha, Qatar

E-mail: jean-francois.joly.1@umontreal.ca; normand.mousseau@umontreal.ca

Abstract. We present two major optimizations for the kinetic Activation-Relaxation Technique (k-ART), an off-lattice self-learning kinetic Monte Carlo (KMC) algorithm with on-the-fly event search THAT has been successfully applied to study a number of semiconducting and metallic systems. K-ART is parallelized in a non-trivial way: A master process uses several worker processes to perform independent event searches for possible events, while all bookkeeping and the actual simulation is performed by the master process. Depending on the complexity of the system studied, the parallelization scales well for tens to more than one hundred processes. For dealing with large systems, we present a near order 1 implementation. Techniques such as Verlet lists, cell decomposition and partial force calculations are implemented, and the CPU time per time step scales sublinearly with the number of particles, providing an efficient use of computational resources.

1. Introduction

Diffusion in the solid phase is dominated by rare events, with high activation energy barriers compared to system temperature. The simulation of such activated events is demanding: simulation schemes linear in time, such as molecular dynamics, often cannot attain the time scales on which those events take place. Low observed rates, however, make it possible to view the events as independent steps in a Markov chain, allowing methods such as the kinetic Monte Carlo (KMC) algorithm [1] to be used.

In the standard implementation of KMC, a predefined catalog of possible events is required, which usually dictates confining the atomic positions to a lattice. Atomic motions and the associated energy barriers can then be determined *a priori* based on local environments. However, in many systems of interest, off-lattice positions and long-range interactions play an important role in the dynamics, and standard KMC cannot adequately describe these contributions.

We recently introduced the kinetic Activation Relaxation Technique (k-ART), an on-the-fly, self-learning, off-lattice KMC method [2, 6]. K-ART combines a topological classification of local

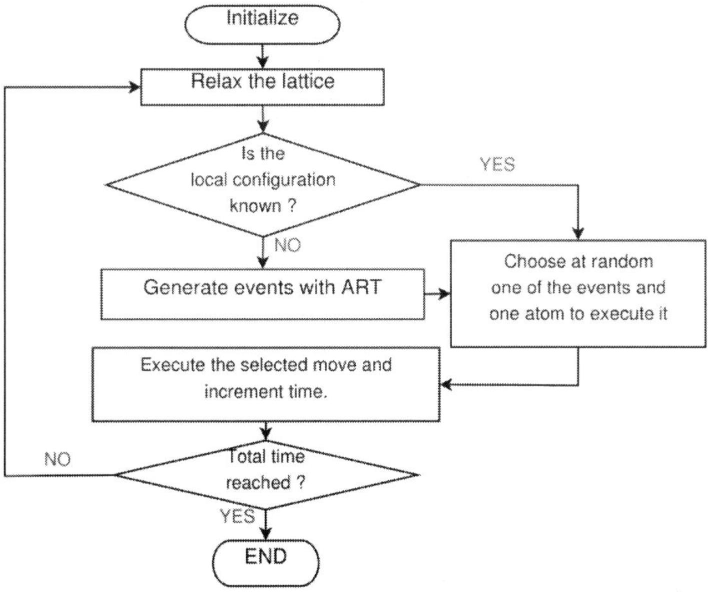

Figure 1. Flowchart showing the main structure of k-ART.

neighborhoods and events using NAUTY [3] with the *ART nouveau* method [4, 5] to find diffusion pathways as new environments are visited. Not constrained to simple geometries, k-ART has been successfully applied to a number of complex systems such as interstitials in Fe, self-defect annihilation in Si and relaxation of amorphous Si [6].

The computational complexity of the method dictates an implementation that is both effective and efficient. In this context, effectiveness means that results can be determined quickly (many results per time unit), while efficiency means that no computational resources are wasted (low computational cost per result). The nature of events and the structure of the algorithm make it possible to significantly improve the simulation through two avenues. First, the scaling of event generation with the number of particles N, which was initially order N, can be reduced to almost order one. The computational cost of k-ART becomes therefore linked only to the complexity of the structure and not its size. Second, the catalog-building part of the algorithm can be parallelized efficiently using the Message Passing Interface (MPI) [7]. These improvements are the core of this paper.

In the next sections, we describe briefly the k-ART method and its application to an ion-implanted silicon box. More details can be found in Ref. [6]. We then focus on the sublinear scaling in the number of particles and the parallelization implementation, evaluating the scaling properties both with respect to system size and number of processors used.

2. Algorithmic overview

K-ART, like conventional KMC methods, uses an event catalog to compute the rate of escape from a local minimum and to move the simulation forward in time. What is different is its self-learning and off-lattice capabilities. The algorithm consists of four steps (see also Fig. 1):

(i) At the beginning of a step, k-ART looks at each atom in the system and analyses its local topology (see Sec. 2.1). Using the program NAUTY [3], a numerical key is obtained to identify uniquely that topology. This method enables k-ART to quickly differentiate between local atomic configurations without being constrained to a fixed lattice.

(ii) For each new topology encountered, a series of event searches, using the latest version of the ART nouveau method [5, 8, 9] is launched. Generated events are analyzed and new events are added to the database. The rate associated with the event i is given by $r_i = \tau_0 \exp(-\Delta E_i/k_B T)$, where τ_0, the attempt frequency, is fixed at the onset and, for simplification, assumed to be the same for all events (10^{13} s^{-1}). ΔE_i is the barrier height, i.e., the energy difference between saddle point and initial minimum. k_B and T are, as usual, the Boltzmann constant and the temperature respectively. These *generic events* are linked to a given topology, meaning that all atoms sharing the same local topology are considered to have access to the same set of events.

(iii) Once the event database is up to date, all active events are ordered by their activation barrier in 0.1 eV bins. Going through the bins from low to high barriers, generic events in the bin are refined on each individual atom sharing the initial topology of that event, until events representing more than 99.9% of the total rate of the system are refined. This means that low barrier events are fully relaxed in response to the actual geometry. This enables k-ART to include effects of long-ranged elastic forces. A refined event is called a *specific event* since it is associated with an individual atom. With these adjustments, the total rate catalogue is constructed.

(iv) Finally, the standard KMC algorithm [1] is applied to select an event and advance the clock. The elapsed time to the next event is computed as $\Delta t = -\ln \mu / \sum_i r_i$ where μ is a random number in the $[0, 1[$ interval and r_i is the rate associated with event i. While the rate of each event depends on a constant attempt frequency, its value was chosen to correspond to an average phonon frequency. This gives the correct order of magnitude in the calculated rates and timesteps. The clock is pushed forward, an event is selected with the proper weight and the atoms are moved accordingly, after a geometrical reconstruction. The simulation then goes back to step one and looks if any new topologies are in the system, etc.

In KMC, the residence time and the next state picked are dominated by the lowest energy barrier separating the current state from its neighbours. These low barriers can cause flickers, i.e., rapid back-and-forth movements across one or more barriers. This is generally undesirable, as it costs CPU time without leading to structural evolution, and traps the simulation in a small part of configuration space. States linked by such low barriers are called a *basin*. We developed the basin auto-constructing Mean Rate Method (bac-MRM) [6], based on the MRM by Puchala *et al.* [10], to handle these flickers and overcome the resulting dramatic slowing down of the algorithm.

The bac-MRM builds basins on-the-fly, i.e., a basin is expanded as long as the trajectory visits new states separated by low barriers. It ensures that no basin state is visited twice. While kinetics in basin are averaged out, the absorbing states are picked with the correct probability. In this way, bac-MRM permits the simulation of systems with low-energy barriers and basins, which would otherwise be inaccessible to KMC methods.

2.1. Event and topology tables

For most physical systems, the number of topologies and events can become large quite rapidly as the simulation progresses. For this reason, both the topology and the event information are kept in large and mostly sparse tables. In both cases, the table element is given by a hash key constructed from the topological information given by NAUTY.

Identifying the local topologies is the most important step in k-ART. To do so, a local cluster surrounding the atom of interest is selected (generally, all atoms situated within a sphere of given radius centered on the atom is interest) and by using a well defined distance cutoff value to determine which atoms are linked together, a connected graph is constructed. The NAUTY software takes the connected graph as input and returns an ordered set of three integers identifying the graph. The topology hash key is then calculated from the three integers characterizing a topology given by NAUTY.

For the event table, we use the hash keys of the initial, saddle and final topologies associated with an event to construct the event hash key. This way, information from both tables can be accessed quickly. In the case of hash collisions (meaning that two different topologies/events have the same final key), we just increment the key by one until a free value is found. This does not happen so often as to have a noticeable effect on the speed of the method.

This mapping supposes that there is a one to one correspondence between geometry and topology. It is normally the case, due to the constraints imposed by embedding the local graph into the global system. As discussed in Ref. [6], by construction the algorithm automatically finds out when this is not the case, allowing us to modify the graph until the correspondence holds again.

2.2. Event analysis

Since k-ART is off-lattice, a robust method for identifying diffusion pathways is needed. Every new generic (specific) event is compared to the list of already known events for that initial topology (atom). If an event with the same activation energy is already in memory, the scalar product of displacement vectors from the reference state to the final states is used to distinguish between two similar diffusion events taking place in two different directions. A second scalar product of the displacement vectors between the initial and the saddle configurations is determined, so that two events having the same end configurations, but crossing over different saddle points, can be distinguished and also treated as separate events.

3. Example: Relaxation of an ion-bombarded semiconductor

To demonstrate the versatility of k-ART, we apply the algorithm to the relaxation of a large Si box after ion implantation. In a previous work by Pothier *et al.* [11], a Si atom was implanted at 3 keV in a box containing 100 000 *c*-Si atoms at 300 K. This system was then simulated for 900 ps wit molecular dynamics. From the final configuration of this simulation, we extracted a smaller box of 26.998 atoms which contains the vast majority of induced defects and imposed periodic boundary conditions. We show the initial configuration in Fig. 2.

K-ART is applied at 300 K on this complex system. Figure 3 reports the evolution of the simulation over 282 events, which corresponds to a simulated time of 0.63 µs, at a cost of 3500 CPU-hour. During this time, the systems relaxes by 30 eV as numerous defects crystallize.

Most of the CPU time used for this run is spent on building the catalog for this new system. Even though the initial number of topologies is relatively small (839 topologies), as the system first evolves, most events generate new topologies that must also be sampled and catalogued. After 282 steps, almost 60 000 topologies (including the saddle and final minima topologies associated with potential events that are never visited but are added to the catalog) have been explored. With time, however, the catalog becomes more exhaustive and the algorithm moves faster through the phase space. Since catalogs can be merged, the information gained in preliminary simulations or runs on related systems, such as amorphous silicon, for example, can be re-used, resulting in a significant speed-up for new simulations and an optimal use of computational resources.

While standard KMC simulations easily run for millions of steps, regularly reaching time scale of minutes or more, fundamental restrictions have limited them to simple lattice-based problems such as metal on metal growth. The possibility, with k-ART, to now apply these accelerated approaches to complex materials such as ion-bombarded Si, over simulated timescale many orders of magnitude longer than can be reached by MD, opens the door to the study of numerous systems that were long considered out of reach as is demonstrated by this example.

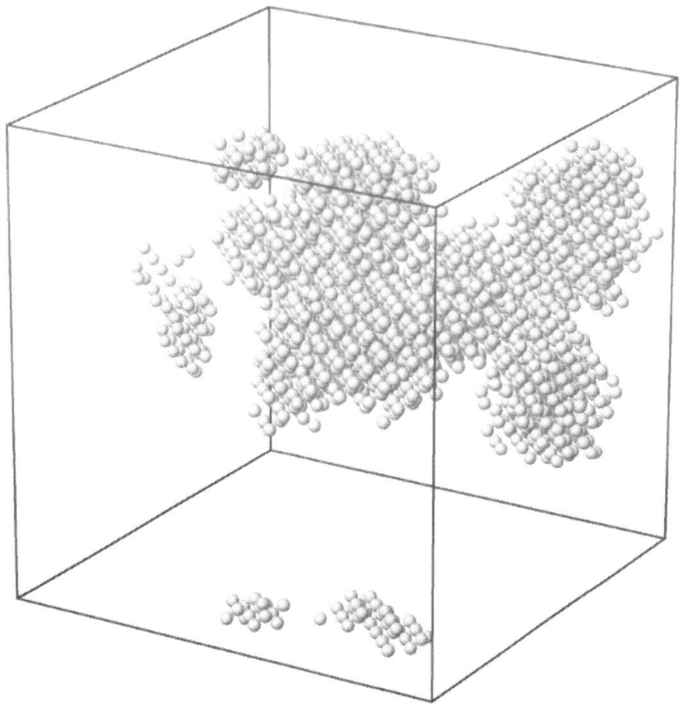

Figure 2. The initial state of our ion bombardment damage cascade simulation. We show atoms with a topology different than the perfectly crystalline topology in silicon.

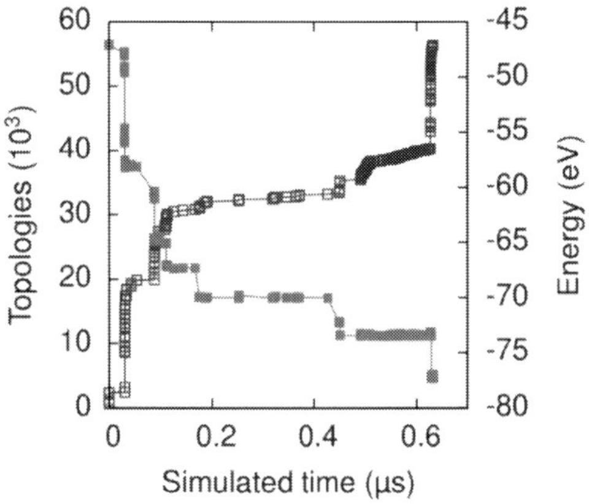

Figure 3. The number of topologies which were encountered during our simulation (□) and the energy of the system in eV (■) as a function of the simulated time

4. Optimization

4.1. Dealing with large system sizes

As shown in the previous example, systems with many tens to many hundreds of atoms to adequately capture the physics of interest. We can take advantage of the local nature of the activated dynamics to handle very efficiently these large configurations. Indeed, away from any phase transition, diffusive motion typically involves regions composed of a few tens to a few hundreds of atoms, and the forces induced by this displacement typically propagate up to a few nanometers. Because k-ART focuses on these events while ignoring the thermal vibration, we can restrict most force calculations to atoms directly involved in each event, bringing the algorithm from order N a sub-linear order..

To scale-down the computational cost associated to such large structures, we combine standard cell lists [13] and the Verlet algorithm [14] for the construction of neighbor lists with local force calculations. Starting from a table containing energy and forces for all atoms, every force call first identifies atoms upon which forces exceed a given threshold. Forces and energy are then updated only on those *active* atoms and their first neighbours, as deformations propagate locally. With this adaptive local-force calculation, the cost of generating an event becomes almost system-size independent.

We see in Fig. 4 that the system behaves phenomenologically with an order $N^{0.4}$ scaling. While this is not quite order 1, it means that an event in a 25 000 atom box costs only 3.5 times more CPU time than one in a 1000 atom cell.

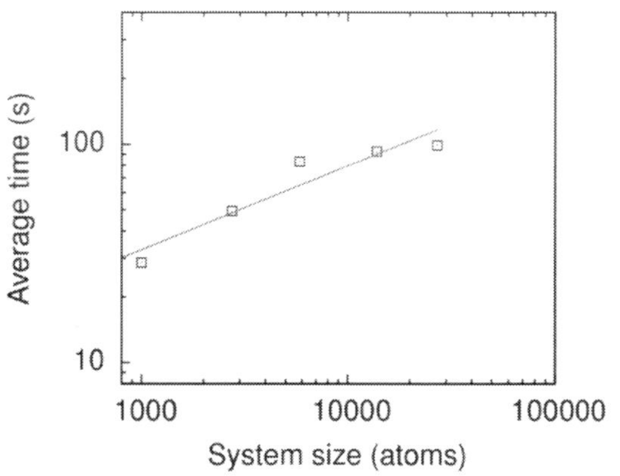

Figure 4. The mean time to attempt a generic event search for systems of various sizes. The initial configuration for each simulation was a c-Si box containing 1000 to 27000 atoms with four vacancies in each. The simulations were performed at 500 K and run on a 2.66 GHz single core on a quad-core Xeon processor. The blue squares correspond to the data, while the green line is a least-square fit of $f(N) = 2.3\,N^{0.4}$

We also ensure that computational efforts spent in a region of the system which stays unchanged from one KMC step to another are not lost. A modest amount of bookkeeping allows the program to *recycle* specific events when the local environment associated with these is kept untouched. To take into account elastic deformation, all specific saddle points are reconverged, but this step is considerably cheaper than refining from a generic conformation since the starting configuration is much closer to the real transition state.

4.2. MPI implementation

In the serial version of our algorithm, the program spends most of its time generating and refining events that are independent from each other. Our current implementation of the parallel version of k-ART focuses on optimizing this process by using a master node to perform the topological analysis and apply the KMC algorithm (steps 1 and 4) while the search for generic events and the refining of specific events (steps 2 and 3) are dispatched to workers.

For the event generation phase, the master process loops through the list of topologies to be searched and sends a search task to each available worker. The workers then apply the ART nouveau algorithm for a given atom. If the search for a saddle point and associated final state is successful, the worker checks whether the event is acceptable. Finally, the result (success or failure) is sent back to the master process. The main thread receives the information and checks whether the event is new or not. New events are then added to the event database. Every time a worker sends its results to the master, another task is sent back. This is done until the search list has been exhausted.

For the event refinement phase, the master process goes through the rate histogram as in the serial version. If events in the current energy bin need to be refined, a refinement task for each atom with the event's topology is sent to the workers. The workers refine the saddle point and send their result back (success or failure) to the master. The more events share the same energy bin (are closer in energy), the more workers processes can be used to refine them concurrently.

In our implementation, the master thread alone is responsible for all bookkeeping (tables of topologies, events, basins); the workers only need to know the starting configuration of the search and the interaction potential, limiting communication and the memory footprint: the hash tables for a 1000-atom system require roughly one gigabyte of memory. Additionally, there is no need to keep databases in sync between master and workers. The downside is that only the master can ultimately decide if an event is accepted or rejected. This bottleneck limits the number of workers a master can control. In the following, we will explore where this limitation sets in. All calculations in this section were performed on a cluster with dual Intel Xeon E5462 quad-core processors at 3 GHz.

4.2.1. Example: bcc iron, 1019 atoms, starting from scratch In Figs. 5 and 6, the efficiency of the parallelization is displayed for the simulation of a 5-vacancy cluster in bcc iron (1019 atoms) with EAM potentials [15], starting from scratch (i.e., without event or topology catalog). In Fig. 5, the parallel efficiency $E_p = T_1/(pT_p)$ for generic event searches and specific event refinements is plotted for two different computing environments. Here, T_i is the time for one search with i processors, and p is the total number of processors. For few CPUs, the master-worker parallelization in k-ART limits the the speedup, as the master is sitting idle while waiting for results from the workers. For generic event searches, the efficiency stays competitive for a larger p, which shows that the serial part of the generic event search (evaluating newly found saddle points, update of event and topology databases) is not limiting for this system. Additionally, an average of 600 ± 40 generic event searches were performed per KMC step, so all CPUs were kept busy. For specific event refinements, the speedup T_p/T_1 saturates at around 4 for 4–8 CPUs. This is due to the fact that only 17 events needed to be refined on average at each KMC step. As the parallelization is over histogram slices, typically less than eight events can be refined in parallel, and so there is not enough work for all CPUs.

The limited efficiency of specific event refinement does not affect significantly the overall parallelization speedup. In Fig. 6, the computational cost (i.e., the total CPU time required) to perform a 40 KMC step simulation under the aforementioned conditions is displayed, normalized to the time for a serial simulation. As it is impossible to repeat identical multiprocessor simulations (the activation process is highly chaotic so even using the same seeds, the fact that we do not control on which node runs which event is sufficient to create diverging simulations), the number of generic and specific event searches fluctuates wildly: the system may remain in known areas of configuration space or it might by chance move into unexplored territory, requiring many generic event searches. To compare different runs, a typical simulation time was calculated from the average time per search and the average number of searches per run, to which the time not spent on searches was added. These numbers fluctuate much less, although there is still some randomness involved. From Fig. 6, it can be seen that, for this simulation, 8–16 CPUs will result in less than 20 % overhead compared to the serial simulation. With fewer CPUs the master process is idle for too long, while using more CPUs results in a bottleneck in the analysis on the main process. Specific events become important only for large CPU numbers (at 64 CPUs they account for about 10 % of the time spent). The program only spends a negligible amount in serial code of time outside of the event search.

4.2.2. Example: amorphous silicon, 1000 atoms, from scratch and with catalogue Amorphous silicon (a-Si) is a disordered solid that provides an interesting test model for testing parallelization. With no two atoms of our 1000-atom model sharing the same topology, a-Si represents the most challenging configuration possible for k-ART, since generic event searches need to be launched for every atom before even taking the first KMC step and convergence of the catalog is likely to require many thousands of steps.

For our test, we choose 20 searches per topology, resulting in 20 000 generic event searches for building the initial catalog. To emphasize the event search phase, we only show the results for a single KMC step. As expected, the computation cost (Fig. 8) is completely dominated

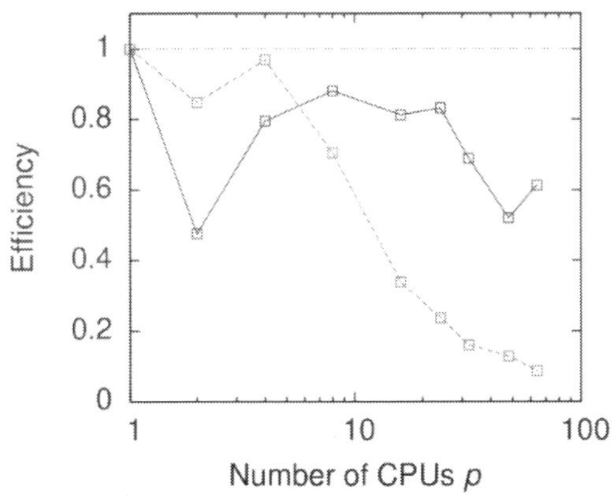

Figure 5. Parallel efficiency of event search. —— generic event search; - - - - specific event search; $\cdots\cdots$ ideal efficiency.

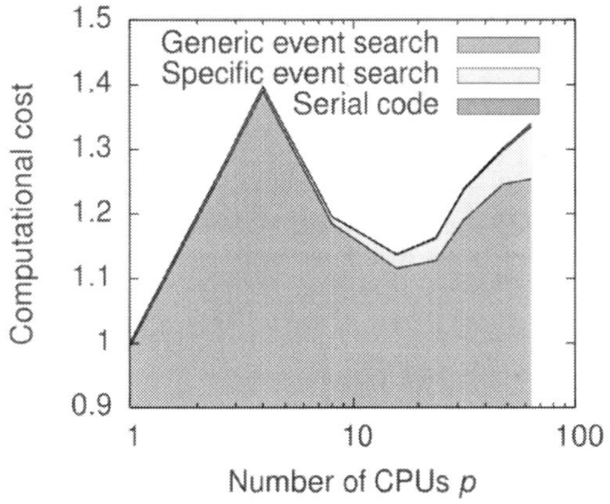

Figure 6. Computational cost (number of CPUs × per CPU time) to simulate 40 KMC events, relative to single CPU performance. Lower field: generic event search. Central field: specific event search. Top field: serial code. The relative computational cost displayed here is the inverse of the parallel efficiency.

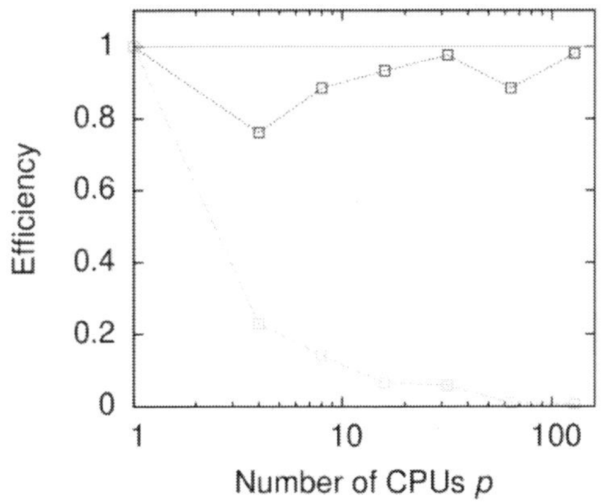

Figure 7. Parallel efficiency of event search for the first KMC step. —— generic event search; - - - - specific event search; $\cdots\cdots$ ideal efficiency.

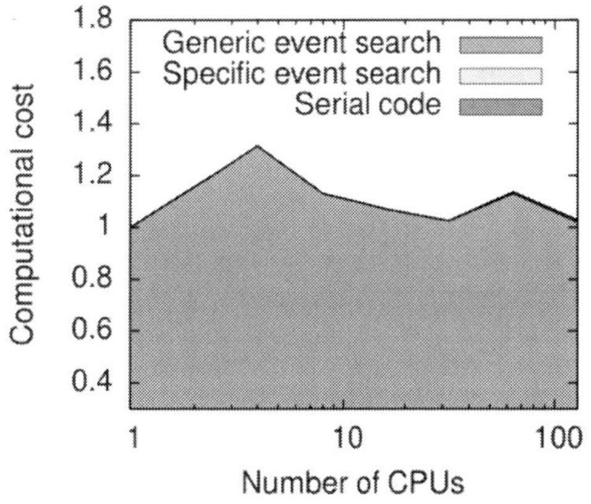

Figure 8. Computational cost to simulate the first KMC step in *a*-Si without catalog, relative to serial performance.

by the generic event searches and the cost of both the specific event searches and serial code is completely negligible (the small number of low energy events explain the decreasing efficiency of the specific event portion of the code after 4 CPUs). The total computational cost remains relatively constant even for a large number of CPUs. This can be seen in Fig. 7, where we see that parallelization efficiency of generic event searches is almost perfect for up to 128 CPUs. Since the efficiency is highly dependent on the ratio of the number of searches to the number of CPUs available, the result is due to the very large number of searches. This confirms that a more complex system, with more topologies to explore, will parallelize better than a simple one.

We note, moreover, that even with 128 CPUs, the event analysis bottleneck on the master thread for this 1000-atom system is not reached; the master thread was able to cope with the increased number of searches performed per second without problem and the speed-up for this example is 125 for 128 CPUs. More CPUs could probably be added without having a large detrimental effect on the efficiency for generating the initial catalog.

5. Conclusion

We have shown that k-ART is an efficient tool to perform KMC simulations for complex system where atoms are not constrained to lattice position or where long-range elastic effects are important. By using local force calculations, our implementation scales in a near order 1 with the system size. This is necessary if the method is to be applied to larger systems with tens or hundreds of thousands of atoms. We also show that the task of performing hundreds of independent searches for events can be efficiently distributed over several CPUs.

The efficiency of the parallelization depends on the simulation phase. As the search for generic events scales better in parallelization than refining specific events, the parallel efficiency is better while there are many topologies to be explored. This will be the case either when a new simulation is started without a pre-existing catalog, or if the system evolves over time. In contrast, simulations for which an adequate catalog is available will not be able to profit similarly from parallelization. This means that parallelization is most efficient for the hardest systems, making k-ART particularly useful for characterizing the kinetic of complex problems that cannot be addressed by standard tools.

Acknowledgments

This work has been supported by the Canada Research Chairs program and by grants from the Natural Sciences and Engineering Research Council of Canada (NSERC) and the *Fonds Québécois de la Recherche sur la Nature et les Technologies* (FQRNT). The authors thank Jean-Christophe Pothier for providing the ion bombardment simulation snapshot. We are grateful to *Calcul Québec* (CQ) for generous allocations of computer resources.

References

[1] Bortz A B, Kalos M H and Lebowitz J L 1975 *J. Comp. Phys.* **17** 10–18
[2] El-Mellouhi F, Mousseau N and Lewis L J 2008 *Phys. Rev. B* **78** 153202
[3] McKay B D 1981 *Congressus Numerantium* **30** 45–87
[4] Barkema G T and Mousseau N 1996 *Phys. Rev. Lett.* **77** 4358–4361
[5] Malek R and Mousseau N 2000 *Phys. Rev. E* **62** 7723–7728
[6] Béland L K, Brommer P, El-Mellouhi F, Joly J F and Mousseau N 2011 Kinetic activation relaxation technique submitted to Phys. Rev. E
[7] Gropp W, Lusk E and Skjellum A 1999 *Using MPI - 2nd Edition* (Cambridge, MA, USA: The MIT Press) ISBN 0–262–57132–3
[8] Marinica M C, Willaime F and Mousseau N 2011 *Phys. Rev. B* **83** 094119
[9] Machado-Charry E, Caliste D, Genovese L, Mousseau N and Pochet P 2011 *Phys. Rev. E* Accepted for publication
[10] Puchala B, Falk M L and Garikipati K 2010 *J. Chem. Phys.* **132** 134104
[11] Pothier J C, Schiettekatte F and Lewis L J 2011 *Phys. Rev. B* **83** 235206
[12] Plimpton S 1995 *Journal of Computational Physics* **117** 1–19
[13] Allen M P and Tildesley D J 1987 *Computer simulation of liquids* (New York, NY, USA: Clarendon Press) ISBN 0-19-855375-7
[14] Verlet L 1967 *Phys. Rev.* **159** 98
[15] Ackland G J, Mendelev M I, Srolovitz D J, Han S and Barashev A V 2004 *J. Phys.-Condens. Mat.* **16** S2629–S2642

New Population Synthesis Techniques in the Analysis of Interacting Binaries

Lorne Nelson

Department of Physics, Bishop's University
2600 College, Sherbrooke, QC, Canada J1M 1Z7

E-mail: lnelson@ubishops.ca

Abstract. Novel approaches to understanding the observed properties of interacting binaries containing compact accretors such as neutron stars and white dwarfs are examined. Explaining the evolution of these systems is a computationally challenging problem because the vector space of initial conditions that describes the progenitor binaries is wide-ranging. There are large variations in the chemical abundance (e.g., metallicity), binary mass correlations, and assumed input physics. In this paper we compare two very different strategies to synthesize a specific subset of the currently observed population of compact binaries. Both involve the pre-computing a large grid of representative models. In the first case, the grid of initial conditions is densely packed thereby allowing us to identify the spectrum of initial conditions and the most probable evolutionary channels leading to the formation of the observed binaries. In the second, the grid is accurately interpolated to provide us with the ensemble properties of the currently observed population of interacting binaries (e.g., Cataclysmic Variables). As an example of the utility of the first approach, we have taken advantage of the multicore processing power of the fast, new stellar evolution code known as MESA to compute an extensive grid of binary evolution tracks for low- and intermediate-mass X-ray binaries. The grid is about two orders of magnitude larger than any previous computation of X-ray binary evolution and includes more than 40,000 models. It comprises 60 initial donor masses over the range of 1 to 4 $M_{\odot}$ and, for each of these, 700 initial orbital periods over the range of 10 to 250 hours were chosen. Using a 'traceback' analysis, we show how the extremely massive neutron star (1.97 $M_{\odot}$) in the binary pulsar PSR J1614-2230 is likely to have evolved. We find that the initial donor stars which produce the closest relatives to PSR J1614-2230 are likely to have had a mass of between approximately 3.4 to 3.8 $M_{\odot}$. Nonetheless, we conclude that it is difficult to form high-mass neutron stars unless they are born with masses larger than the 1.4 $M_{\odot}$ canonical value.

1. Introduction

Interacting binaries are stellar systems composed of two stars in mutual orbit that are undergoing (or have undergone) mass transfer or systemic mass loss. They are generally very rich in physics, especially with respect to the study of radiative hydrodynamics, nuclear physics, and high-energy physics. An important sub-class of these systems is comprised of interacting binaries containing compact accretors (i.e., white dwarfs [WDs], neutron stars [NSs], and black holes [BHs]). If the separation between the stars is sufficiently small, the companion star (donor) can become very tidally and rotationally distorted. If the separation is decreased, it will eventually overflow its Roche lobe (corresponding the critical gravitational equipotential surface) and matter will generally flow towards the compact accretor (see figure 1). The matter that is stripped from the donor has too much angular momentum to be directly captured by the accretor and thus an accretion disk forms around it (see Hilditch [1] for more details).

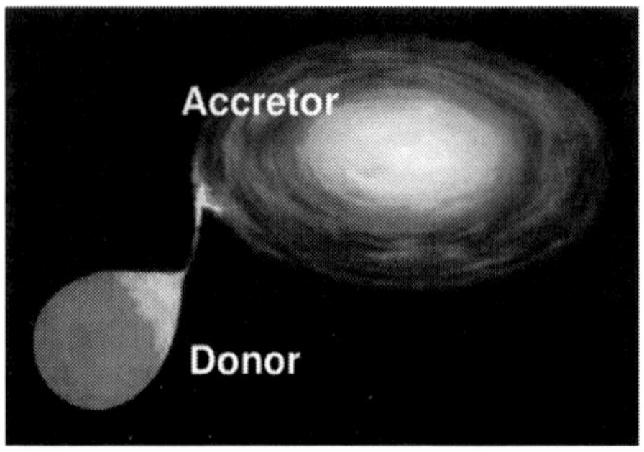

Figure 1. Artist's rendition of an interacting binary.

There are several sub-populations of interacting binaries containing compact objects that are delineated based on the nature of the accretor (see, e.g., Eggleton [2] and Warner [3]). If the accretor is a white dwarf, then the binary is known as a Cataclysmic Variable (CV). If it is a neutron star, then the binary is referred to as an X-ray binary and this terminology is further refined depending on whether the donor is a low-mass or intermediate-mass star. While it is true that we generally have good models to describe the properties of these systems, it is important to understand their ensemble properties. In particular we would like to understand how these systems formed and subsequently evolved thereby producing the relative numbers of the various classes of interacting binaries that we actually observe at this epoch in the Galaxy's history. Only when we can synthesize all of the classes of interacting binaries in the correct numbers and with the correct properties can we say that we truly have a robust, unified picture for their formation and evolution.

In this paper I discuss two distinctly different approaches to addressing this issue. The first approach takes advantage of the tremendous increases in computational speed; it is now possible to calculate extensive grids of evolutionary models for a multidimensional parameter space of initial conditions. Although this approach can be viewed as somewhat of a brute force method, it does have its advantages. For example, if the grid is sufficiently-well resolved, the models can produce new and interesting types of evolution that require further investigation (e.g., ultracompact accreting millisecond pulsars). Moreover, it may be possible to trace the progenitor evolution of astrophysically important systems such as the binary millisecond pulsar PSR J1640-2230. For the second approach, population synthesis techniques are used to generate a Galactic-disk population of CVs (see, e.g., Howell, Nelson and Rappaport [4], and references therein). In this case a grid of evolutionary models is pre-computed and then interpolated based on the actual initial properties of individual primordial binaries that comprise the ensemble. The strengths and weaknesses of both of these approaches are also discussed.

2. Methodology

In order to analyze the evolution of interacting binaries containing compact objects, the detailed structure and response of the donor star to mass loss must be calculated. The analysis of stellar structure and evolution was originally carried out using analytic and semi-analytic methods (see, e.g., Chandrasekhar [5] and Schwarzschild [6]). However, the advent of computers greatly facilitated the solution of the highly coupled and non-linear partial differential equations describing their evolution. For the calculations presented in this paper, the evolution and response of the donor to mass-loss is calculated in tandem with the orbital dynamics and the quasi-hydrostatic rules that govern the mass-loss rate. They are briefly described in the two following sections.

2.1. Stellar Structure and Evolution of the Donor

Aside from changes in the chemical composition, the structure and evolution of a spherically symmetric (one-dimensional) donor star is governed by four coupled partial differential equations. Adopting a Lagrangian viewpoint, we take the independent variables to be t, the elapsed time, and m_r, the mass contained within a sphere of radius r. These equations ensure mass continuity, energy and momentum conservation, and account for energy transport by either radiation, conduction and/or convection. They can be expressed as:

$$\begin{aligned}
\frac{\partial P}{\partial m_r} &= -\left(\frac{G}{4\pi}\right)\frac{m_r}{r^4} - \frac{1}{4\pi r^2}\left(\frac{\partial^2 r}{\partial t^2}\right)_{m_r} \\
\frac{\partial r}{\partial m_r} &= \frac{1}{4\pi r^2 \rho} \\
\frac{\partial L}{\partial m_r} &= \varepsilon_{nuc} - \varepsilon_{\nu} - T\left(\frac{\partial S}{\partial t}\right)_{m_r} \\
\frac{\partial T}{\partial m_r} &= -\left(\frac{3}{64\pi^2 ac}\right)\frac{L\kappa}{r^4 T^3} \quad \text{[Radiative/Conductive]} \\
\frac{\partial T}{\partial m_r} &= \left(\frac{T}{P}\right)\left(\frac{\partial P}{\partial m_r}\right)\nabla_{\text{ad}} \quad \text{[Convective]} \ ,
\end{aligned} \tag{1}$$

where P is the pressure, T is the temperature, L is the luminosity, S is the specific entropy, ρ is the density, ε is the specific energy generation/degradation (nuclear energy generation and neutrino losses, respectively), κ is the radiative/conductive opacity, and ∇_{ad} is the adiabatic temperature gradient.

In order to solve these equations, the appropriate boundary conditions and physical constraints must be applied. Specifically, we enforce L=0 and r=0 at the centre of the star, and match an atmosphere at the surface that indirectly gives us T and P there. The input physics serves to provide us with the equations of constraint. The set of constraints can be written as:

$$\begin{aligned}
P &= P(\rho, T, X_i) \\
\varepsilon_{\text{nuc}} &= \varepsilon_{\text{nuc}}(\rho, T, X_i) \\
\varepsilon_{\nu} &= \varepsilon_{\nu}(\rho, T, X_i) \\
\kappa &= \kappa(\rho, T, X_i) \ .
\end{aligned} \tag{2}$$

The equations represent the equation of state, the nuclear energy generation, neutrino losses, and the radiative/conductive opacities, respectively. Note that X_i is the abundance fraction (by mass) of the ith chemical constituent. The temporal evolution of X_i is computed using an auxiliary set of differential equations.

The method of solution that is most widely used, yielding both high accuracy and good computational speed, is the Henyey Method (described by Kippenhahn, Weigert and Hofmeister [7]). The algorithm used to solve the differential equations falls into the category of 'relaxation methods'. The method requires an initial guess for the values of the physical variables at the shell interfaces (i.e., a 'block start') at a particular instant in time. Once a set of approximate solutions has been applied to the difference equations, a more exact solution is computed. This iterative process continues until the desired accuracy is achieved. Since a temporal sequence of evolutionary models is being computed, the properties of the model computed at t_0 are then used as an approximation for the interior properties of the next model in the sequence (t_0+δt). The physical structure of the new model is allowed to relax to the prescribed accuracy.

2.2. Binary Dynamics

For the interacting binaries that we are considering, the separation is so small that the binary system circularizes and the donor synchronizes on very short timescales. Thus, to a first approximation, we can treat the orbit as Keplerian and assume that the donor (of mass M_2) is completely constrained by Roche geometry. The latter constraint implies that there is an approximate analytic relationship between the orbital separation A and the spherical-volume-equivalent Roche lobe radius R_L. Furthermore, if the donor star is filling its Roche lobe, then we can assume that its radius R_2 is equal to R_L. This yields a relation between the intrinsic binary properties (A, M_1, and M_2) and the macro-properties of the donor star being modeled (R_2 and M_2):

$$\frac{R_2}{A} = \frac{R_L}{A} \simeq \frac{0.49q^{2/3}}{0.6q^{2/3} + \ln\left(1+q^{1/3}\right)} , \qquad q \equiv \frac{M_2}{M_1} , \qquad M_T \equiv M_1 + M_2 \quad . \tag{3}$$

Mass loss from the donor is typically driven by the expansion of the donor on its nuclear timescale or the shrinking of the orbit as a result of orbital angular momentum dissipation. Assuming a circular orbit, we can write the orbital angular momentum ($J_{\rm orb}$) and its time derivative as:

$$J_{\rm orb} = M_1 M_2 \sqrt{\frac{GA}{M_1+M_2}} \quad \Rightarrow \quad 2\left(\frac{\dot{J}_{\rm orb}}{J}\right) = 2\left(\frac{\dot{M}_2}{M_2}\right) + 2\left(\frac{\dot{M}_1}{M_1}\right) - \left(\frac{\dot{M}_T}{M_T}\right) + \left(\frac{\dot{A}}{A}\right) \tag{4}$$

Since the contributions to the orbital angular momentum dissipation can be written as

$$\left(\frac{\dot{J}_{\rm orb}}{J}\right) = \left(\frac{\dot{J}_{GR}}{J}\right) + \left(\frac{\dot{J}_{\rm MB}}{J}\right) + \left(\frac{\dot{J}_{\delta M_T}}{J}\right), \qquad \left(\frac{\dot{J}_{\delta M_T}}{J}\right) = \alpha(1-\beta)\left(\frac{M_T}{M_1}\right)\left(\frac{\dot{M}_2}{M_2}\right),$$

$$\left(\frac{\dot{J}_{\rm GR}}{J}\right) = -1.3\times10^{-8}\left(\frac{M_T}{M_\odot}\right)^{-1/3}\left(\frac{M_1}{M_\odot}\right)\left(\frac{M_2}{M_\odot}\right)\left(\frac{P_{\rm orb}}{1\ \rm hr}\right)^{-8/3} \rm yr^{-1}, \text{ and} \tag{5}$$

$$\left(\frac{\dot{J}_{MB}}{J}\right) = -7.1\times10^{-6}\left(\frac{M_T}{M_\odot}\right)^{1/3}\left(\frac{M_1}{M_\odot}\right)^{-1}\left(\frac{R}{R_\odot}\right)^{\gamma}\left(\frac{P_{\rm orb}}{1\ \rm hr}\right)^{-10/3} \rm yr^{-1} ,$$

where $P_{\rm orb}$ is the orbital period, we have a mathematically well-posed algorithm to calculate the evolution of the binary system once the response of the donor to mass-loss has been computed. Note that the subscripts GR and MB correspond to orbital angular momentum losses due to gravitational radiation (as predicted by General Relativity) and magnetic braking, respectively. The magnetic braking formula is a parameterization of the Verbunt-Zwaan law as described by Rappaport, Verbunt and Joss [8]. The parameter γ is normally assigned a value between 2 and 4. The third contribution to angular momentum losses can be ascribed to systemic mass-loss. For example, if the accretor is a neutron star and if the mass transfer rate exceeds the Eddington limit, the excess mass may be lost from the binary system in a wind that carries away orbital angular momentum ($\delta J_{\delta M_T}$). The parameter α is a measure of the specific angular momentum that is carried away by the mass (e.g., normally a 'fast Jeans' mode' corresponding to the specific angular momentum of the accretor), and β is the fraction of the mass that is transferred from the donor that is permanently captured by the accretor.

In a typical evolutionary sequence, the structure of the donor is calculated first and then its radius relative to the Roche lobe radius is compared thereby allowing the mass-loss rate to be calculated for a given δt. As long as the timesteps are sufficiently small compared to the relevant physical timescales of the donor (e.g., its Kelvin time), the evolution of the orbital separation can be calculated explicitly.

3. First Approach: Dense Grid of Initial Conditions

One strategy that can be adopted to understand the overall behavior of a particular class of interacting binaries is to create a set of initial conditions that is sufficiently well-resolved that most of the salient details can be determined from the evolutionary calculations. This approach was adopted by Lin et al. [9] who chose to explore the evolution of low-mass x-ray binaries (LMXBs) and intermediate-mass x-ray binaries (IMXBs) that contain NS accretor. We computed a grid of 42,000 evolutionary tracks each of which contained thousands of models corresponding to specific ages of the donor star. We chose 60 different initial donor masses in the range of 1 to 4 $M_{\odot}$, and for each mass, 700 evenly (logarithmically) spaced initial orbital periods in the range of 10 – 250 hours were selected.

The initial conditions were used as an input into the newly developed stellar evolution program known as `MESA star` [10] which simultaneously solves equations (1) and (2). A brief description of the `MESA` integrator and other numerical details are given in the next section. An application of this approach to the astrophysically important binary millisecond pulsar PSR J1614-2230 will be described in section 3.3.

3.1. Numerical Method (`MESA`)

The core integrator which was used to compute the evolution and response of the donor to mass-loss was developed by Paxton et al. [10] as part of an open-source suite of numerical simulation modules known as `MESA` (Modules for Experiments in Stellar Astrophysics). `MESA star` was used to solve the equations of stellar structure and evolution using adaptive mesh techniques and advanced timestep controls. The code finds its origins in the work of Eggleton [11] who developed the EZ code and made it publicly available. It has been extensively modified through the incorporation of superior input physics (such as the equation of state) and it also employs sophisticated software engineering tools that can take full advantage of cutting-edge HPC architectures. In particular the `MESA` modules are 'thread-safe' thus allowing applications to use multicore processors (i.e., it supports the shared-memory model based on OpenMP). The `MESA` codebase is always being updated as the physics and numerical algorithms are being improved. The suite of packages can be downloaded from the project webserver at http://mesa.sourceforge.net/. A complete description of the code (and relevant links) are also provided at the site.

The code that drives the binary evolution is very similar to that described in the paper by Madhusudhan et al. [12]. One of the big differences concerns the treatment of mass transfer near the Eddington limit. For sub-Eddington rates the capture fracture was taken to be 90% (i.e., β=0.9). For super-Eddington rates, the mass accretion by the neutron star is simply set equal to the rate corresponding to the Eddington limit. The evolution proceeds until either an age of 10 Gyr is reached or until mass transfer becomes dynamically unstable. For all the cases shown in the next section, the initial mass of the neutron star was taken to be the 'canonical' mass of 1.4 $M_{\odot}$.

3.2. Results

The evolution of L/IMXBs can best be understood by examining how their observables (orbital period and mass of the donor) change with time. The various subclasses of evolution are clearly evident in figure 2. A set of representative tracks are labelled as CV, UC, and G. In each figure the red box denotes the region occupied by systems at the onset of their evolution (i.e., the set of initial conditions). The CV tracks are very analogous to the evolution that is expected for Cataclysmic Variables (interacting binaries with WD accretors). The systems start out having largely unevolved (H-rich) donors and evolve towards shorter P_{orb}s until magnetic braking ceases resulting in an orbital period gap between approximately 2 and 3 hours (mass transfer ceases temporarily). As the donor's mass decreases, the system reaches a minimum orbital period (about 70 minutes) at which point the donor is becoming degenerate and expanding (leading to a larger P_{orb}). The donor continues to lose mass as it becomes a brown dwarf. This type of evolution has been discussed extensively by Rappaport, Verbunt, and Joss [8], and Howell, Nelson, and Rappaport [4].

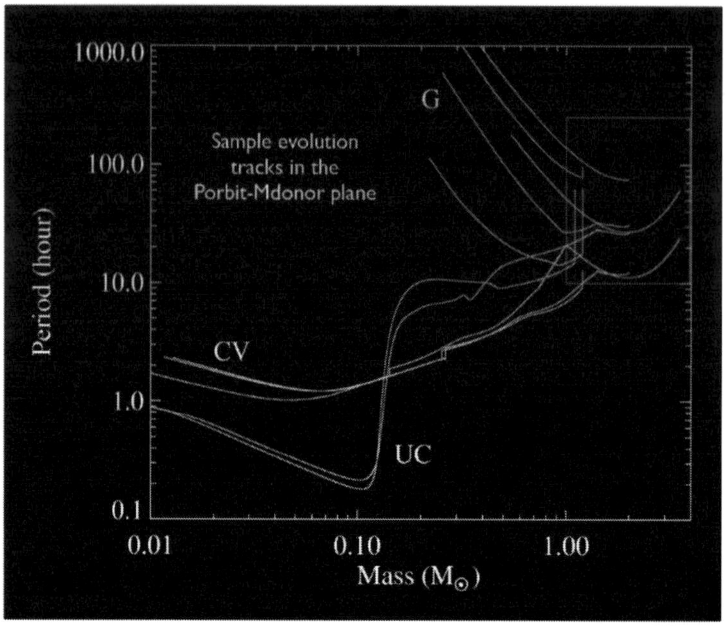

Figure 2. Representative L/IMXB evolutionary tracks in the M_2-P_{orb} plane. The various cases are denoted as CV, UC, and G.

If the donors are able to evolve on a short enough nuclear timescale, it is possible for them to become sub-giants and/or red giants before too much mass is stripped away. In this case (tracks denoted by G in figure 2) P_{orb} increases as mass is stripped away from the giant. This occurs because the radius of red giants is strongly dependent on the masses of their He-rich degenerate cores for $M_2 <$ 2.2 $M_\odot$ (as opposed to their own mass) and because the orbital period is constrained by the radius of the red giant (and its mass). Eventually so little mass is contained in the envelope of the giant (i.e., outside its He-rich core) that it collapses leaving behind a cooling helium WD in a detached binary system. Since the accretor is a neutron star that has been spun up by accretion torques (thus making it a pulsar), the system is properly referred to as a binary millisecond pulsar (see Podsiadlowski, Rappaport, and Pfahl [13], and Nelson, Dubeau, and MacCannell [14], and references therein).

The dividing line between these two extremely diverse sets of behaviors is known as the bifurcation (Pylyser and Savonije [15]). The bifurcation is extremely sensitive to the assumed input physics (for example, magnetic braking) because it depends sensitively on the relative magnitudes of the mass-loss timescale and the nuclear timescale. For systems whose initial conditions place them extremely close to the bifurcation, yet below it, it is possible for them to evolve to extremely short orbital periods (P_{orb} as small 6 minutes) before their orbital period starts increasing again. These evolutionary tracks are shown as the ultracompact (UC) cases in figure 2. The reason why these systems can reach such short orbital periods can be explained by the fact that the donor stars had almost formed a helium-degenerate core before mass loss precluded any further growth. Thus the donors in the systems are very helium rich and consequently much smaller than their hydrogen-rich counterparts. Given the constraint of Roche geometry, this guarantees that their orbital periods will be extremely short (see Fedorova and Ergma [16], and Nelson and Rappaport [17] for further details).

A pictorial presentation of all 42,000 systems is shown in the upper-left panel of figure 3. Each panel is a pixel map that is divided into 1300 X 1200 cells. As each of the tracks passes through a pixel, the cumulative time that is spent by all of the systems evolving through that particular pixel is recorded. The cumulative time that the ensemble of systems spends in each pixel varies from 10^{10} yr (red pixels) to 10^3 years (violet pixels). Thus the M_2-P_{orb} plane in figure 3 can be viewed as a (logarithmic) relative probability density plot. The other three panels correspond to the types of evolution discussed above. One of the benefits of this type of analysis is that we can clearly see the subset of initial conditions that leads to the formation of each type of system.

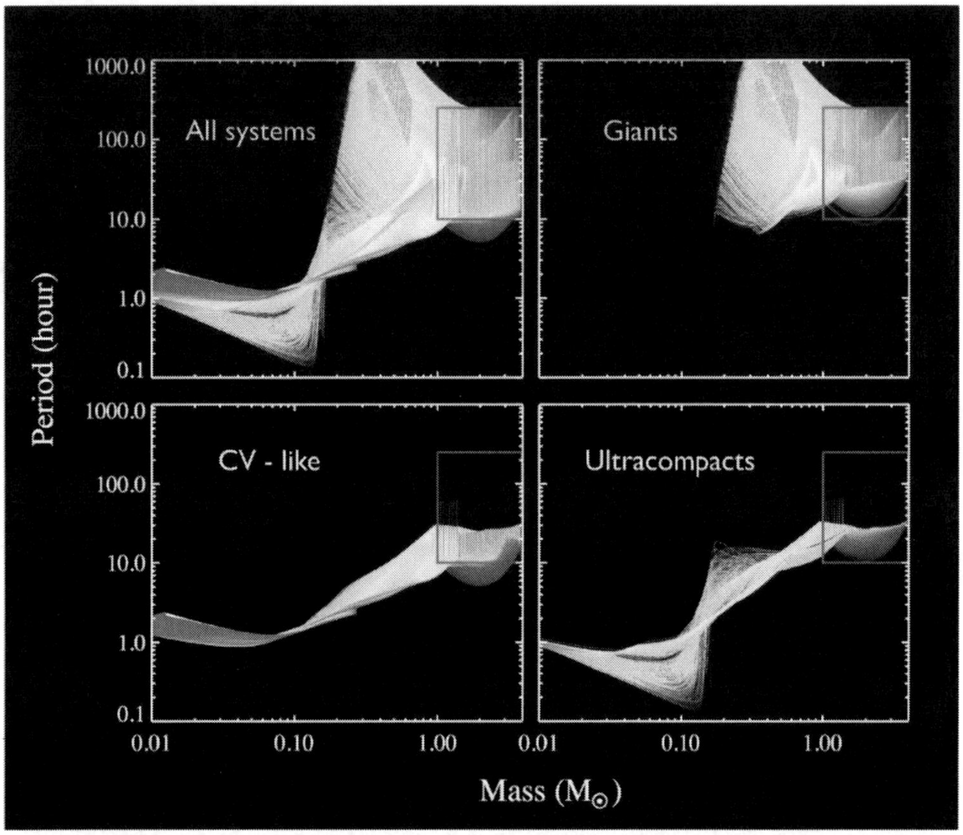

Figure 3. Relative color probability plots for the ensemble of models. The initial grid of models (i.e., initial conditions) is visible in the upper right portion of each panel, outlined by a red box. All 42,000 systems are superposed in the upper-left panel of the figure. Specific subsets are plotted in the other three panels. See Lin et al. [9] for more details.

Other potentially observable properties of the binary system (such as the mass-transfer rate) can also be plotted in order to gain insight into the evolution of these systems. Figure 4 shows the interrelationship of the orbital period, mass of the donor and the mass-transfer rate. Note that the mass-transfer rates of the ultracompact become very high just as they reach their minimum periods.

3.3. Application to PSR J1644-2230

Demorest et al. [18] recently made a very precise measurement of the mass of the neutron star ($1.97 \pm 0.04 M_\odot$) in the binary millisecond pulsar J1644-2230. This is the highest mass ever measured for a neutron star and has extremely important consequences for our understanding of high-energy physics. There has been considerable debate as to the equation of state that prevails inside neutron stars. Some models invoke exotic forms of matter such as strange quarks and kaon condensates (soft equations of state) to construct neutron star models. These models do not allow for the existence of high-mass neutron stars ($\cong 2\ M_\odot$ or larger). Clearly this new determination of the mass rules out the softer equations of state (see figure 5).

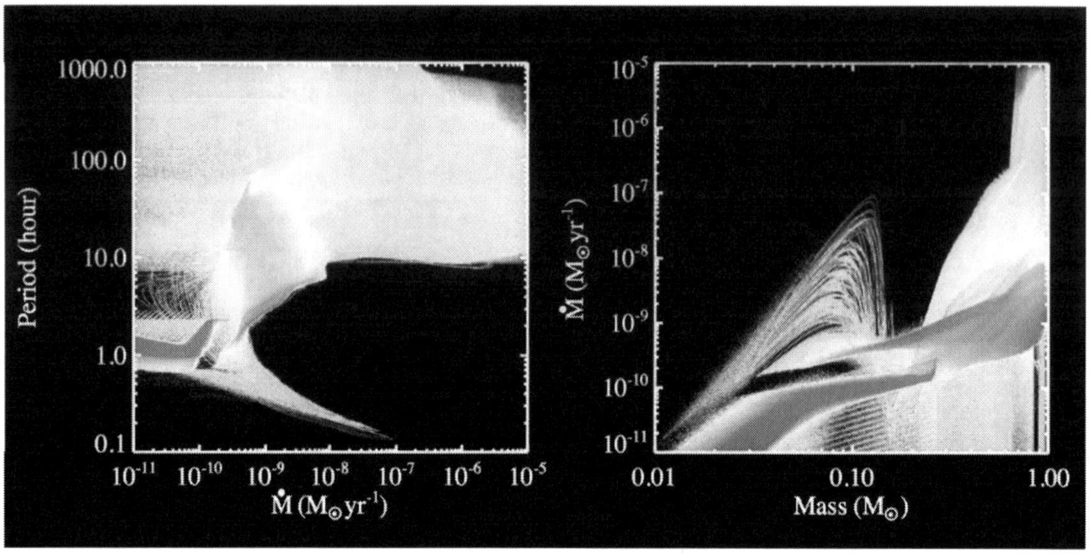

Figure 4. Color probability plots showing the dependence on the mass-transfer rate ($\dot{M}$).

One of the questions that can be addressed by computing a dense grid of models concerns the initial conditions that lead to the formation of this binary. Could this system have formed from an L/IMXB? According to our results (Lin et al. [9]), the closest relatives of J1614-2230 had an initial donor mass of 3.4 - 3.8 $M_\odot$ (assuming the natal mass of the neutron star to be 1.4 $M_\odot$). But we found that it is difficult to form neutron stars as massive as 1.97 $M_\odot$ unless their initial masses are somewhat higher (1.6 ± 0.1$M_\odot$). In this case, the initial donor mass and P_{orb} would likely have been ≅ 4.25 $M_\odot$ and ≅ 49 hours, respectively.

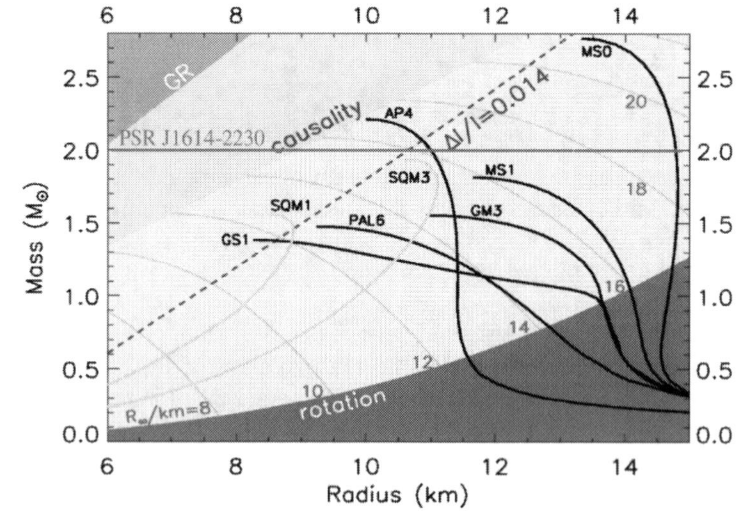

Figure 5. Mass-radius diagram for several equations of state used to compute the structure of neutron stars. The black curves correspond to "normal-matter" equations of state while the green curves (labelled SQM) correspond to strange-quark equations of state. Regions excluded by general relativity (GR), causality, and rotational constraints are also indicated. The figure is reproduced from the article by Lattimer and Prakash [19].

Journal of Physics: Conference Series **341** (2012) 012008 doi:10.1088/1742-6596/341/1/012008

4. Second Approach: Population Synthesis using Grid Interpolation

Population synthesis techniques are used in an attempt to reproduce all of the various types of stars and multiple systems that would be detected at any epoch in the galaxy's history. The details associated with this method as it pertains to interacting binaries can be found in de Kool [20], Politano [21], and Howell, Nelson, and Rappaport [4] (and references). Based on empirical probability rules, a set of primordial binaries is created. The separation and masses of the two components are completely specified. For the case of Cataclysmic Variables, the individual evolutions of each system then have to be computed. In the past this difficulty limited the number of primordial binaries that could be evolved. Moreover, many important (but low probability) phases could be accidentally excluded.

The large number of dimensions of parameter space often forced workers in the field to make simplifications such as ignoring the effects of nuclear evolution on the donor star (e.g., Nelson, MacCannell, and Dubeau [22]). However another strategy was recently employed by Goliasch and Nelson [23] wherein a grid of carefully constructed evolutionary tracks is pre-computed (starting from Zero-Age Cataclysmic Variables [ZACVs]) and then interpolated in order to obtain the synthesized population at the current epoch (i.e., Present-Day Cataclysmic Variables [PDCVs]). The advantage of this method is that if the grid is well constructed, the only computational cost comes from the Monte Carlo calculation to obtain the set of primordial binaries and then from the multidimensional interpolation of the evolutionary grid.

4.1. Population Synthesis

A Monte Carlo technique was used to seed the Galaxy with the observed numbers and types of stars (see Goliasch and Nelson [23] for detailed information). The parameters that were adopted are listed in table 1. We started by creating 10^8 primordial binaries. The primary masses (corresponding to the component that evolves to form the WD accretor) were chosen according to the Miller and Scalo [24] initial mass function (IMF). The donor mass was then chosen according to a specific correlation function (generally had a weak dependence on M_1). The separation was chosen based on either a uniform distribution in log P_{orb} or log A. The birthrate function was assumed to be either constant (0.6 stars/yr with $M > 0.8\ M_{\odot}$) or exponentially decaying for an assumed galactic (disk) age of 10 Gyrs.

Table 1. Parameters used in the Population Synthesis.

Parameters	Adopted Values
Primary mass M_1 (0.8 - 10 $M_{\odot}$)	Miller & Scalo IMF
Initial mass ratio correlation (M_2)	Uncorrelated OR $q^{1/4}$ OR q^1
Primordial separation (A)	Uniform in Log P_{orb} OR Log A
Common Envelope efficiency (α_{CE})	1/3 OR 1
Birthrate Function (BRF)	Constant OR $\propto e^{-t/\tau}$ (τ=3.5Gyr)
Age of the Galaxy	10^{10} years

The most problematic parameter to ascertain is the so-called common envelope efficiency factor which is a measure of the energetic efficiency for the removal of the envelope of the primary during the common envelope phase ($\alpha_{CE}=\Delta E_{binding}/\Delta E_{orbital}$). The 'standard model' assumed a mass correlation probability proportional to $q^{1/4}$, $\alpha_{CE} = 1$, and a constant birthrate function. A 1-D Henyey code was then used to calculate a full evolutionary grid of about 300 models (the rudiments of the code are described in section 2.1). The grid was then interpolated for each set of initial binary parameters (M_1, M_2, A) and a probability density function was computed for all present-day observable parameters of the (ensemble) population.

Journal of Physics: Conference Series **341** (2012) 012008 doi:10.1088/1742-6596/341/1/012008

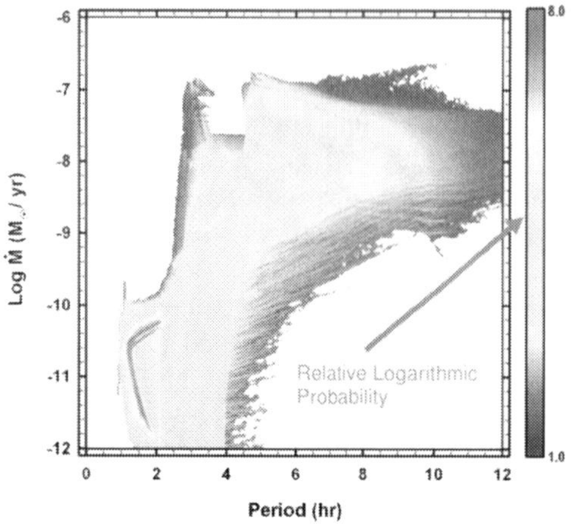

Figure 6. Computed population of Present-Day CVs for our standard model in the log $\dot{M}$ -$P_{\rm orb}$ plane. The colors represent the relative logarithmic probability of observing a CV with these properties (the corresponding scale is given on the right side of the plot).

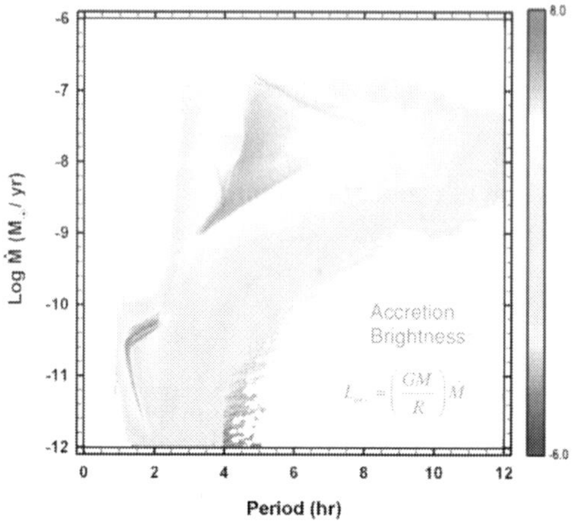

Figure 7. In order to approximately take into account observational selection effects, the probabilities derived for figure 6 have been weighted by the system's mass-transfer rate, $\dot{M}$ (the accretion luminosity is roughly proportional to the mass-transfer rate).

4.2. The Present-Day Population of Cataclysmic Variables

Figure 6 shows the computed probability density for the intrinsic population of PDCVs in the orbital period-mass transfer rate plane. There are 1000 pixels along the abscissa and 1000 pixels along the ordinate and the colors represent the relative logarithmic probability of finding a CV with those properties. It should be noted that the highest probability density has arbitrarily been normalized to a value of 8.0 on the logarithmic probability scale. Figure 7 attempts to crudely take into account observational selection effects by assuming that they are dominated by the accretion brightness (which is proportional to the mass-transfer rate). In this context, the importance of the systems with very high mass-transfer rates can be easily seen.

Also apparent in both figures is a 'period gap' that occurs for orbital periods of between approximately 2 to 3 hours. According to the interrupted magnetic braking paradigm that has been implemented in this model, when the donor's interior becomes fully convective, magnetic braking is greatly attenuated. Since this angular momentum dissipation mechanism is largely driving mass transfer, mass transfer ceases (the donor shrinks beneath its Roche lobe because it is thermally bloated) until gravitational radiation losses can drive the donor back into Roche-lobe contact at a $P_{\rm orb}$ of about 2 hours. An orbital-period minimum can also be seen for a $P_{\rm orb}$ of about 70 minutes. The value of the minimum period is smeared out because of the presence of both He and CO white dwarf accretors, and because of the nuclear evolution of the donor before it became a ZACV. Although this effect was never previously investigated, Goliasch and Nelson [23] show that it is not highly significant. In fact, the population synthesis shows that the volume of the phase space of initial conditions is so small that it is very unlikely that nature can produce the ultracompact AM CVn systems via this channel. They are more likely produced either as a result of the formation of a double degenerate or a helium-star progenitor system.

Figure 8 shows a histogram of the presently observed CVs in the galaxy. It should be noted, however, that the observations are subject to many serious selection effects. Nonetheless, the salient features of the distribution are very evident: there is a period gap between roughly 2 and 3 hours, and there is a minimum orbital period at about 80 minutes. Note that the observed period gap is approximately reproduced in the folded data based on the population synthesis results for the standard model (see figure 9). Moreover, the relative number of long-period CVs (< 12 hours) is a reasonable facsimile of what is observed based on figure 8. Note that very few AM CVn systems are synthesized. Also note that there is an enhancement in the number of systems near the minimum period as is observed (the "period spike") but that it tends to be smeared out by the phenomena noted above.

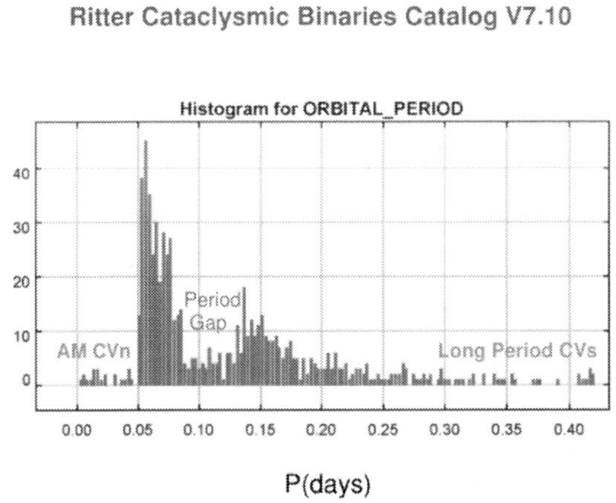

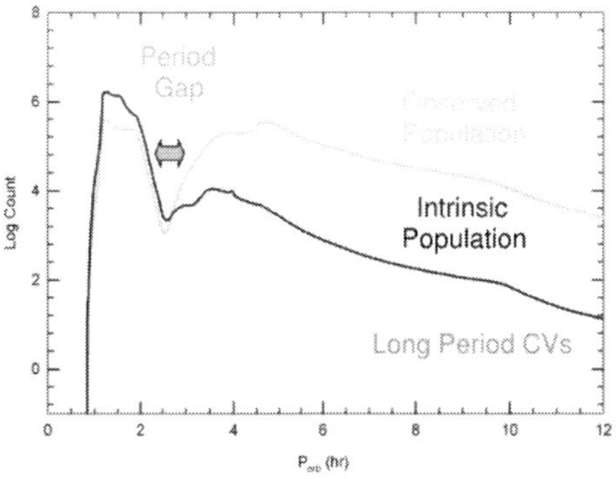

Figure 8. Orbital-period histogram showing the currently observed population of CVs that was generated using the Ritter Cataclysmic Binaries Catalog (version 7.10). The label AM CVn denotes the existence of several short-period AM Canum Venaticorum systems that typically have helium-rich donors.

Figure 9. Folded orbital-period histogram of PDCVs for the standard model. The **black curve** corresponds to the actual (intrinsic) number of systems that are found in a particular P_{orb} bin. The green curve corresponds to the "observed" population that was weighted by $\dot{M}$.

5. Conclusion

Two different strategies for analyzing a subset of the present-day population of compact binaries have been compared. Both require that grids of evolutionary tracks be computed. The computations for both approaches were massively scalar and required about 400000 and 250000 core-hr, respectively, on MS2 (RQCHP). The main advantage of the second approach is that the primordial population of binaries can be easily recomputed and thus the sensitivity of the synthesized results to various physical assumptions affecting the formation process (such as α_{CE}) can be reasonably quantified. The disadvantage is that the grid needs to be carefully constructed so that difficult regions in initial condition space (e.g., dynamical instabilities) are properly covered. The first approach provides greater physical insight into how the more complicated evolutionary channels can lead to the formation of specific binary systems. As was shown for the case of PSR J1644-2230, the grid can also be used to identify the spectrum of initial conditions corresponding to the most probable evolutionary channels leading to the formation of the observed system (i.e., traceback analysis).

Some of the issues that are currently being addressed include: (1) the inclusion of the effects of X-ray irradiation on the donors in L/IMXBs; (2) an examination of the effects of disk instabilities in those same systems; and, (3) an expansion of the previous population synthesis which takes into account the chemical evolution of stars in the disk and halo of our galaxy (i.e., the inclusion of low-metallicity Population II stars).

Acknowledgments
I would like to acknowledge the work of my collaborators, namely J. Goliasch (Bishop's), S. Rappaport (MIT), J. Lin (MIT), B. Paxton (UCSB), Ph. Podsiadlowski (Oxford), and P. Todorov (Observatoire de Paris) who carried out many of the computations. I am particularly indebted to J. Goliasch and S. Rappaport for allowing me to use some of the figures presented in this paper. I also acknowledge the Réseau québécois de calcul de haute performance (RQCHP) for providing the computational facilities to carry out the work presented here and I thank the Natural Sciences and Engineering Research Council (NSERC) of Canada for financial support. Finally I wish to thank an anonymous referee for useful comments.

References

[1] Hilditch R W 2001 *An Introduction to Close Binary Stars* (Cambridge: Cambridge University Press)
[2] Eggleton P 2006 *Evolutionary Processes in Binary and Multiple Stars* (Cambridge: Cambridge University Press)
[3] Warner B 1995 *Cataclysmic Variable Stars* (Cambridge: Cambridge University Press)
[4] Howell S, Nelson L and Rappaport S 2001 *Ap. J.* **550** 897
[5] Chandrasekhar S 1938 *An Introduction to the Study of Stellar Structure* (New York: Dover)
[6] Schwarzschild M 1958 *Structure and Evolution of the Stars* (New York: Dover)
[7] Kippenhahn R, Weigert A and Hofmeister E 1967 *Methods in Computational Physics* **7** 129
[8] Rappaport S, Verbunt F and Joss P 1983 *Ap. J.* **275** 713
[9] Lin J, Rappaport S, Podsiadlowski Ph, Nelson L, Paxton B and Todorov P 2011 *Ap. J.* **732** 70
[10] Paxton B, Bildsten L, Dotter A, Herwig F, Lesaffre P and Timmes F 2011 *Ap. J. S.* **192** 3
[11] Eggleton P 1971 *MNRAS* **151** 351
[12] Madhusudhan N, Rappaport S, Podsiadlowski Ph, and Nelson L 2008 *Ap. J.* **688** 1235
[13] Podsiadlowski Ph, Rappaport S, and Pfahl E 2002 *Ap. J.* **565** 1107
[14] Nelson L A, Dubeau E, and MacCannell K 2004 *Ap. J.* **616** 1124
[15] Pylyser E and Savonije G-J 1988 A&A **191** 57
[16] Fedorova, A and Ergma, E 1989 Ap&SS 151 125
[17] Nelson L A and Rappaport S 2003, *Ap. J.* **598** 431
[18] Demorest P, Pennucci T, Ransom S, Roberts M, and Hessels J W T 2010 Nature **467**, 1081
[19] Lattimer J and Prakash M 2004 Science **304** 536
[20] de Kool M 1992 *A. & A.* **261** 188
[21] Politano M 1996, *Ap. J.* **456** 338
[22] Nelson L A, MacCannell K, and Dubeau E 2004 *Ap. J.* **602** 938
[23] Goliasch J and Nelson L A 2011 *Ap. J.* (to be submitted)
[24] Miller G and Scalo J 1979 *Ap. J. S.* **41** 513

Computer simulation study of folding thermodynamics and kinetics of proteins in osmolytes and denaturants

Apichart Linhananta, Gianluca Amadei, and Timothy Miao

Department of Physics, Lakehead University, Thunder Bay, Ontario, Canada

E-mail: apichart.linhananta@lakeheadu.ca

Abstract. In living cells, the presence of macromolecular crowders, such as osmolytes or denaturants, strongly affects the stability of folded proteins. The overall effects depend on the size, concentration, and chemical properties of the crowding agents. This work uses an all-atom Gō model of the Trp-cage in spherical solvents to probe the physical origin of protein stabilization/destabilization by osmolytes/denturants. The solvent quality is controlled by the solvent-protein contact ε_{PS} that can represent repulsive osmolytes ($\varepsilon_{PS} > 0$) or attractive denaturants ($\varepsilon_{PS} < 0$). The model is used to show that protein stabilization by osmolytes proceeds by an excluded volume, entropy-driven mechanism. Protein destabilization by denaturant is shown to be driven by changes in enthalpy. It is found that small osmolytes are the most effective stabilizer of proteins. Folding simulations of the Trp-cage in osmolytes observe a two-fold increase in folding rates, for small osmolytes. This is due to an osmolyte-induced shift to more compact unfolded protein conformations.

1. Introduction

Osmolytes are intracellular organic molecules that stabilize proteins against unfolding under environmental stresses such as high temperatures, desiccation, or the presence of chemical denaturants such as urea.[1] The stabilizing property of osmolytes has been correlated with the exclusion of osmolytes from protein domains, resulting in the accumulation of water near protein domains.[2, 3] The exclusion of osmolytes from proteins arise from repulsive interactions between osmolytes and the backbone of proteins.[4, 5] A possible stabilization mechanism suggests an osmolyte-induced loss of protein conformational entropy, with the greater entropic loss by the unfolded (U) state, leading to an overall shift in equilibrium towards the native (N) state. The entropy loss mechanism is consistent with experimental works that observed increased compactness in unfolded states of cutinase[6], protein S6,[7] and Rnase S[8] due to osmolytes.

In contrast to protecting osmolytes, denaturants (such as urea) are nonprotecting osmolytes that denature proteins from folded to unfolded state. The attractive protein-denaturant interactions lead to the accumulation of denaturants near proteins.[9] This lowers the free energy of both the N and U states, but due to its larger solvent exposed surface area the free energy of the U state is lowered by a greater amount.[5] Consequently, the addition of denturants to protein solutions results in an enthalpy-driven mechanism that shifts the equilibrium to the unfolded state.

There has been relatively few theoretical or computer simulation investigations of proteins in aqueous solution of osmolytes and denturants. Daggett *et al.*[10, 11] performed all-atom molecular dynamics (MD) simulations of proteins in aqueous solution of osmolytes and urea to show that the stabilization property of the osmolyte trimethylamine N-oxide (TMAO) arises from enhance water structures due to their interactions with TMAO. All-atom MD simulations have been performed to show that urea destabilizes proteins by direct and water-assisted urea-protein interactions.[9, 12, 13] There have been simulations of polymer and hydrocarbons,[14, 15] and of RNA hairpin,[16] in solutions of osmolytes or urea. However, the high computational cost of all-atom MD simulations of proteins in explicit solvents has thus far prevented a complete thermodynamics study required to fully understand the mechanisms of protein stabilization/destabilization by osmolyte/denaturant. The exception is recent landmark study by Canchi *et al.*[17] MD simulation of Trp-cage in solution of urea, which found an enthalpy-driven protein denaturation mechanism - a result that is supported by experiments.[18]

A universal description of protecting and denaturing (urea) osmolytes has been proposed by Bolen.[5] Based on a solvent quality paradigm, it classifies solvents as good or poor. In poor solvents (solvophobic) protein intramolecular interactions dominate, which favors compact folded native states that minimizes solvent exposed protein surface area. In good solvents (solvophilic), protein-solvent interactions dominate, which favors unfolded states that maximize protein-solvent contacts. Water is a poor solvent since the effective water-protein interactions lead to the hydrophobic effects, one of the major forces that fold proteins. Aqueous osmolytes and aqueous denaturant solution are poor and good solvent, respectively. The solvent quality paradigm led to several molecular free energy transfer models, which calculate free energy change of proteins transferred from pure water to aqueous osmolyte/urea solutions.[19, 20] A recent study combines protein conformations from simulation data of $G\bar{o}$ model simulations with transfer free energy models to infer the thermodynamics property of proteins in solutions of osmolytes and urea.[21] The study predicted that solution of osmolyte and urea raises and lowers, respectively, the folding temperatures of the $G\bar{o}$ model of proteins. However, free energy transfer models do not take into account the size of solvents, and may lead to an incomplete description of the change in protein conformational entropy due to solvents. Excluded volume due to solvents can reduce the amount of accessible protein conformations, changing the unfolded protein state to a more compact native-like unfolded state, as observed in experiments.[6-8] In addition, the binding of solvent to protein depends on the size of water, osmolytes and urea, and solvent size must be included in theoretical or MD models to fully assess the enthalpy change due to solvents.[22]

In this work we construct an all-atom $G\bar{o}$ model protein immersed in solvent molecules to investigate how osmolyte/urea stabilizes/destabilizes proteins. The previously studied Trp-cage Go model is selected for its small size (i.e. low computational cost) and cooperative behavior.[23] The Trp-cage model is immersed in spherical solvents, where of the solvent-protein interaction is controlled by the solvent-protein contact energy ε_{PS}. ε_{PS} is the control parameter of the solvent quality paradigm discussed earlier. Positive/negative value of ε_{PS} corresponds to bad/good solvents, which mimics the effects of aqueous osmolyte/denaturant solutions. In the next section, the detail of the discontinuous molecular dynamics (DMD) model and the methods used to calculate the free energies and entropy are discussed. The model is used to demonstrate thermal stabilization/destabilization by osmolyte/denaturant. It will be shown that the protein stabilization by osmolytes ($\varepsilon_{PS} > 0$) is driven by entropy change, and that protein destabilization (denaturation) by denaturant is driven by enthalpy change. Solvent size effects and the effects of osmolyte/denaturant on folding kinetics are also presented.

2. Methods

2.1. Models of Protein in Solvents

Discontinuous Molecular Dynamics (DMD) is an efficient method that has been used to study protein folding, protein aggregation,[24-26] and *ab initio* protein structure prediction[27]. This work is based on a previous DMD all-atom model of the Trp-cage protein.[23] A brief description of DMD is presented here. The initial heavy-atom positions are obtained from the NMR structure (structure 1 of PDB ID 1L2Y) and the missing polar hydrogen molecules are constructed as before. The resulting structure is comprised of 189 heavy atoms and polar hydrogen atoms, which in this "discontinuous" model are represented as hard spheres. Two bonded atoms i and j, as well any 1,3 angle-constrained pair and 1,4 aromatic pair, are constrained by an infinite square-well potential

$$u_{ij}^{bond} = \begin{cases} \infty, \; r < 0.9\sigma_{ij} \\ 0, \; 0.9\sigma_{ij} < r < 1.1\sigma_{ij} \\ \infty, \; r > 1.1\sigma_{ij}, \end{cases} \quad (2.1)$$

where σ_{ij} is the separation distance of the bonded i,j pair in the NMR structure. The model also includes discontinuous improper dihedral potential, to maintain chirality about tetrahedral heavy atoms (such as α carbon without explicit hydrogen), and certain planar atoms. The potential has the form

$$u_{ij}^{improp} = \begin{cases} \infty, \; \omega < \omega_0 - 20^\circ \\ 0, \; \omega_0 - 20^\circ < \omega < \omega_\circ + 20^\circ \\ \infty, \; \omega > \omega_0 + 20^\circ, \end{cases} \quad (2.2)$$

where ω represents the dihedral angles of the constrained atoms, which are restricted to values near $\omega_0 = 35.26439^\circ$ for tetrahedral heavy atoms, and $\omega_0 = 0^\circ$ for planar atoms. Two non-bonded atoms i,j pair interacts by a hard-core and square-well potential

$$u_{ij}^{non-bond} = \begin{cases} \infty, \; r < 0.8\sigma_{ij}^{vdW} \\ B_{ij}\varepsilon_{Go}, \; 0.8\sigma_{ij}^{vdW} < r < 1.2\sigma_{ij}^{vdW} \\ 0, \; r > 1.2\sigma_{ij}^{vdW}, \end{cases} \quad (2.3)$$

where σ_{ij}^{vdW} are the van der Waals (vdW) diameters of the CHARMM potential set 19 and B_{ij} and ε_{Go} are the G$\bar{o}$ interaction strength parameters. Using these parameters we performed a short discontinuous molecular dynamics (DMD) to remove bad contact from the initial experimental structure to produce the ground-state structure of the model Trp-cage, as shown in **Figure 1**A. The G$\bar{o}$ model potential is implemented by setting the non-bonded square-well depth to $-\varepsilon_{Go}$ ($B_{ij} = -1$) for ij pairs with van der Waals overlap in the ground-state structure, otherwise it is set to 0 ($B_{ij} = 0$).

The G$\bar{o}$ model in solvents system is implemented by placing the Trp-cage protein in a 40Å X 40Å X 40Å box, with randomly generated (without overlaps) spherical solvent molecules. Standard periodic boundary conditions are implemented. Solvent molecules interact with each other by a hard-core and square-well potential $U_{ij}^{solvent-solvent}$

$$\frac{u_{ij}^{solvent-solvent}}{\varepsilon_{Go}} = \begin{cases} \infty, \; r < 0.8\sigma_{ij}^{SS} \\ \frac{\varepsilon_{SS}}{\varepsilon_{Go}} = \varepsilon_{SS}^*, \; 0.8\sigma_{ij}^{SS} < r < 1.2\sigma_{ij}^{SS} \\ 0, \; r > 1.2\sigma_{ij}^{SS}, \end{cases} \quad (2.4)$$

where σ_{ij}^{SS} is the vdW diameter of the solvent. ε_{SS}^* is the solvent-solvent square-well depth in units of the G$\bar{o}$ contact energy ε_{Go}. Solvent-protein molecules interact with each other by a hard-core and square-well potential $U_{ij}^{solvent-protein}$

$$\frac{u_{ij}^{protein-solvent}}{\varepsilon_{Go}} = \begin{cases} \infty, \; r < 0.8\sigma_{ij}^{PS} \\ \frac{\varepsilon_{PS}}{\varepsilon_{Go}} = \varepsilon_{PS}^{*}, \; 0.8\sigma_{ij}^{PS} < r < 1.2\sigma_{ij}^{PS} \\ 0, \; r > 1.2\sigma_{ij}^{PS}, \end{cases} \quad (2.5)$$

where $\sigma_{ij}^{PS} = \left(\sigma_i^{vdW} + \sigma_j^{S}\right)/2$ is the average vdW diameter of the protein-solvent (i,j) pair, where σ_i^{vdW} is the CHARMM potential set 19 vdW diameter of the ith atom of the protein and σ_j^{S} is the vdW diameter of the jth solvent molecule. ε_{PS}^{*} is the protein-solvent square-well depth in unit of ε_{Go}.

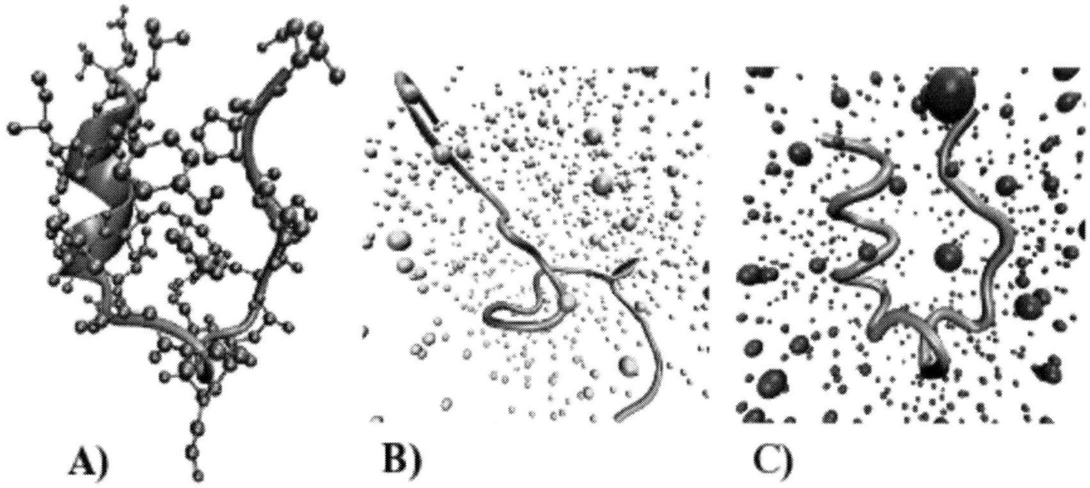

Figure 1. A) Ground-state native state structure of the all-atom Trp-cage protein (note the short helix and the hydrophobic core.); B) Unfolded Trp-cage in denaturants; C) Folded Trp-cage in osmolytes.

ε_{PS}^{*} is the solvent quality scale and is varied from a minimum value of $\varepsilon_{PS}^{*} = -0.8$, to represent a strongly denaturing denaturant solution (**Figure 1**B), up to a maximum value of $\varepsilon_{PS}^{*} = +1.0$, to represent a strongly protein-stabilizing osmolyte solution (**Figure 1**C). The solvent-solvent square-well depth is usually set $\varepsilon_{SS}^{*} = -1.0$ to reflect the solvent particles overall preference to interact with itself rather than with the protein ($\varepsilon_{SS}^{*} < \varepsilon_{PS}^{*}$). In some simulations it is set to zero.

As in previous DMD studies the energy scale is set by the Gō contact energy, with native Gō contact energy $-\varepsilon_{Go}^{*} = -1$.[23, 28-31] Free energies (such as $E^{*} = E/\varepsilon_{Go}$) and temperature ($T^{*} = k_B T/\varepsilon_{Go}$) are scaled in units of ε_{Go}. A reduced time unit $t^{*} = t\sqrt{\varepsilon_{Go}/M\sigma_L^2}$, where M is the average atomic mass of the protein and σ_L = 1 Å, is used.

2.2. Implementation of Discontinuous Molecular Dynamics

Standard continuous molecular dynamics (MD) are usually performed by numerical integration of Newton's equation in femtosecond time step. In most algorithms, the potential energy and force acting on the system of particles are calculated at every time step. DMD avoid these time consuming steps, since the discontinuous particles interact only at the instant of collision. The algorithm involves calculating all collision times of the model protein.[32, 33] The collision time, $t_{collision}$, of bonded (**Equation** 2.1) and non-bonded (**Equation** 2.3 and 2.4) pairs are found by solving a quadratic equation in time, with solution,

$$t_{collision} = \frac{2\vec{v}_{ij} \bullet \vec{r}_{ij} \pm \sqrt{4\left(\vec{v}_{ij} \bullet \vec{r}_{ij}\right)^2 - 4v_{ij}^2 r_{ij}^2}}{2v_{ij}^2} \quad (2.6)$$

where $\vec{v}_{ij} = \vec{v}_i - \vec{v}_j$ is the relative velocity, and $\vec{r}_{ij} = \vec{r}_i - \vec{r}_j$ is the current relative position of the ij pair. The calculation of the time of a dihedral collision (**Equation** 2.2) involves solving a cubic equation by a Newton-Raphson scheme, to find the smallest positive root, as documented in a previous publication.[34] The collision times are compare to find the next collision time, t_{next} and to identify the colliding pair, or the colliding quartet in the case where the next collision is a dihedral collision. All atoms are propagated for the duration of t_{next}, at which point the colliding pair or quartet collide, and the new velocities are determined by conservation of momentum. The velocities of non-colliding particles are unchanged. New collision times for all pairs or quartets that include the colliding particles are recalculated as above. The collision times of pairs or dihedral quartets that do not include the colliding particles do not require recalculations, but are adjusted by $t_{collision}^{new} = t_{collision}^{old} - t_{next}$. A neighbour link list is used to further reduce the computational cost.[33]

The DMD simulation is conducted in the canonical ensemble with the temperature fixed by the Andersen ghost particle method, where the atoms of the proteins and solvent particles experience a random collision with heat-bath ghost particles. The rate of collision follows a distribution $\exp(-\nu t)$ with the collision frequency given by standard kinetic theory $\nu = n\sigma^2\sqrt{16\pi k_B T / m}$, where σ and m are the average size and mass, respectively, of the system, and n is the ghost particle density. For the current protein-solvent model a ghost density of 0.1 per cubic angstrom gives satisfactory temperature equilibration, without altering folding behaviour. The ghost collision time of all particles are calculated by the algorithm $t_{ghost} = -(1/\nu)\ln R$, where R is a random number. The ghost collision times are compared with other collision times to determine the next collision time. When the next collision is a ghost collision the velocity of the colliding atom is recalculated according to the Maxwell-Boltzman distribution $\propto \exp\left(-\frac{1}{2}\frac{m\text{v}^2}{k_B T}\right) = \frac{1}{2}\left(\sqrt{\frac{m}{\varepsilon}}\vec{\text{v}}\right)^2\left(\frac{1}{k_B T/\varepsilon}\right)$, using the algorithm presented in reference[32].

The efficiency of the DMD model enables the folding of an all-atom Trp-cage model in solution in about 12 hr to 72 CPU hours by a serial job submission on a 2.2 GHz Opteron processor. In all, roughly 10000 Trp-cage all-atom models were folded to the native state under a wide variety of thermodynamic conditions and solutions. Taking into account the $4\mu s$ folding time of the Trp-cage[35], this corresponds to about 40 ms in real time. In contrast, the recent landmark study by Garcia et al.[17] on an all-atom model of the Trp-cage in solution of urea, involves a total sampling of about $35\mu s$. They used the TIP3P model for water, and the Kirkwood-Buff model for urea. The model presented here takes an alternative approach by using a coarse-grain solvent quality model that reflects the solvent quality paradigm of Bolen.[5] In the supplementary section of a previous work,[36] it is shown how the effect of water on proteins in an osmolyte/urea solution can be included implicitly into an effective interaction between a united spherical solvent particle and an atom of the protein, controlled by the effective solvent-protein contact energy ε_{PS}^*. Setting $\varepsilon_{PS}^* > 0$ captures some of the effective repulsion between a protein and osmolytes due to the interaction between osmolytes and the water molecules that hydrates the surface of the protein. Setting $\varepsilon_{PS}^* < 0$ captures the favourable protein-denaturant interactions that some researchers believe drives protein denaturation. Though such united-solvent models cannot examine specific aspects of protein denaturation/stabilization by urea/osmolyte, its computational efficiency and simple structure facilitate the study of universal properties, that, as discussed earlier, are still computationally unfeasible by more sophisticated models. Of particular interest are the effects of solvent size, concentration, and interaction strength, on protein stability and cooperativity, and protein folding kinetic.

2.3. Free Energy and Entropy Calculations

The free energy calculations are based on standard multiple weight histogram method (WHAM), [37-39] which calculates thermodynamics quantities by approximating the density of states $g(\varepsilon)$ (i.e. the number of states with energy ε) from simulation data. This defines the partition function $Z=\sum g(\varepsilon)\exp(-\beta\varepsilon)$, the free energy $F=-k_BT\ln Z$, the internal energy $U=\langle E\rangle=\sum \varepsilon g(\varepsilon)\exp(-\beta\varepsilon)/Z$, and the entropy $S=(U-F)/T$. This is generalized by defining the free energy functional as a function of an order parameter x:

$$f(x,\Delta x)=-k_BT\ln\sum_{\varepsilon}{}^{*} g(\varepsilon,x,\Delta x)\exp(-\beta\varepsilon), \tag{2.7}$$

where * indicates a summation over all energy states within the order parameter range from x to $x+\Delta x$, and $g(\varepsilon,x,\Delta x)$ is the density of state of energy ε in the order parameter range x to $x+\Delta x$. The probability that the protein systems occupy states with order parameter range from x to $x+\Delta x$, $P(x,\Delta x)$ can be determined directly from the protein configurations recorded during simulations, and it obeys the relation

$$P(x,\Delta x)=\frac{\sum_{\varepsilon}{}^{*} g(\varepsilon,x,\Delta x)\exp(-\beta\varepsilon)}{\sum_{\varepsilon} g(\varepsilon)\exp(-\beta\varepsilon)}=\frac{\sum_{\varepsilon}{}^{*} g(\varepsilon,x,\Delta x)\exp(-\beta\varepsilon)}{Z}, \tag{2.8}$$

where Z is the partition function. Some manipulations lead to the relation for the free energy density

$$f(x,\Delta x)=-k_BT\ln P(x,\Delta x)-k_BT\ln Z=-k_BT\ln P(x,\Delta x)+F, \tag{2.9}$$

where F is the free energy calculated by standard WHAM. We also defined the internal energy functional $u(x,\Delta x)$, the average energy for order parameter range x to $x+\Delta x$, which can be obtained directly from the simulation data. This leads to the entropy functional:

$$s(x,\Delta x)=\frac{u(x,\Delta x)-f(x,\Delta x)}{T}. \tag{2.10}$$

In this work the non-local native fraction Q, defined previously in reference [23], is employed as the order parameter, and varies from 0 (completely unfolded) to 1 (completely folded). For the free energy functional calculation, a bin size $\Delta Q=0.02$ is used. As mentioned earlier, the computed functional of energy $u^*(Q,\Delta Q)$, free energy $f^*(Q,\Delta Q)$, and entropy $s^*(Q,\Delta Q)$ are in units ε_{GO}.

3. Results and Discussions

To study the thermodynamic, we performed equilibrium simulations of Trp-cage in1000 spherical solvents, of vdW diameter $\sigma_j^S=3.0$ Å (roughly the size of a water molecule) in a 40Å X 40Å X 40Å box. The solvent-solvent interaction is set to $\varepsilon_{SS}^*=-1.0$, and the protein-solvent contact energy varies from $\varepsilon_{PS}^*=-0.8$ (strong denaturants) to $\varepsilon_{PS}^*=1.0$ (repulsive osmolytes). All simulations start from the ground-state native structure of **Figure 1**a, and is equilibrated for a scaled time of $t^*=20000$, follow by $t^*=60000$ of simulations. A serial run takes up to two days on a SHARCNET cluster. To ensure good statistic for WHAM, five independent runs at a fixed temperature are performed, for T* = 3.0 to 6.0, in steps of $\Delta T^*=0.2$. **Figure 2**a plots the heat capacity vs. temperature of the simulations. As expected the folding temperature, T_f^*, defined as the peak of the heat capacity, increases with ε_{PS}. **Figure 2**b is the Temperature vs. solvent quality, ε_{PS}, phase diagram constructed by assuming that T_f^* defines the phase boundary. As ε_{PS} increases, and the solvent quality transforms from denaturant-like ($\varepsilon_{PS} < 0$) to osmolyte-like ($\varepsilon_{PS} > 0$), the protein remains in the native (N) state up to higher temperatures. Hence, the model reproduces thermal stabilization by osmolytes. In the following subsections, the physical origins of protein stabilization by osmolytes are discussed.

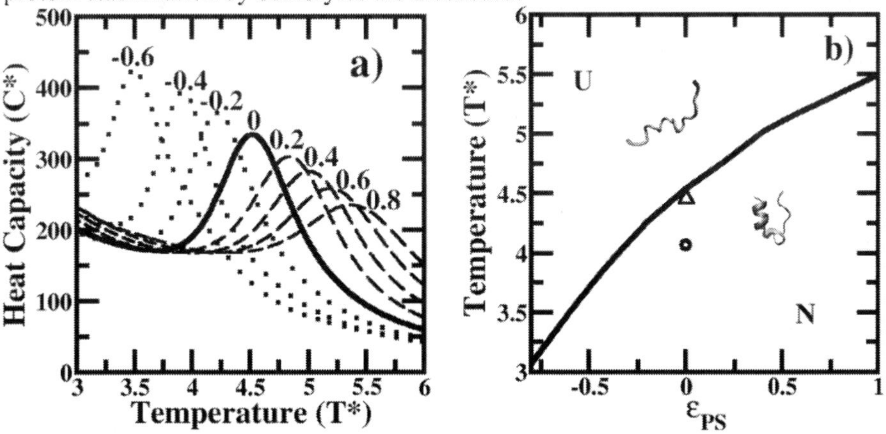

Figure 2. a) Heat Capacity vs. Temperature for Trp-cage and 1000 spherical solvents in a 40Å X 40Å X 40Å box with $\varepsilon_{PS} = -0.6$ to 0.8; b) Temperature vs. solvent quality (ε_{PS}) phase diagram.

3.1. Stabilization Mechanism

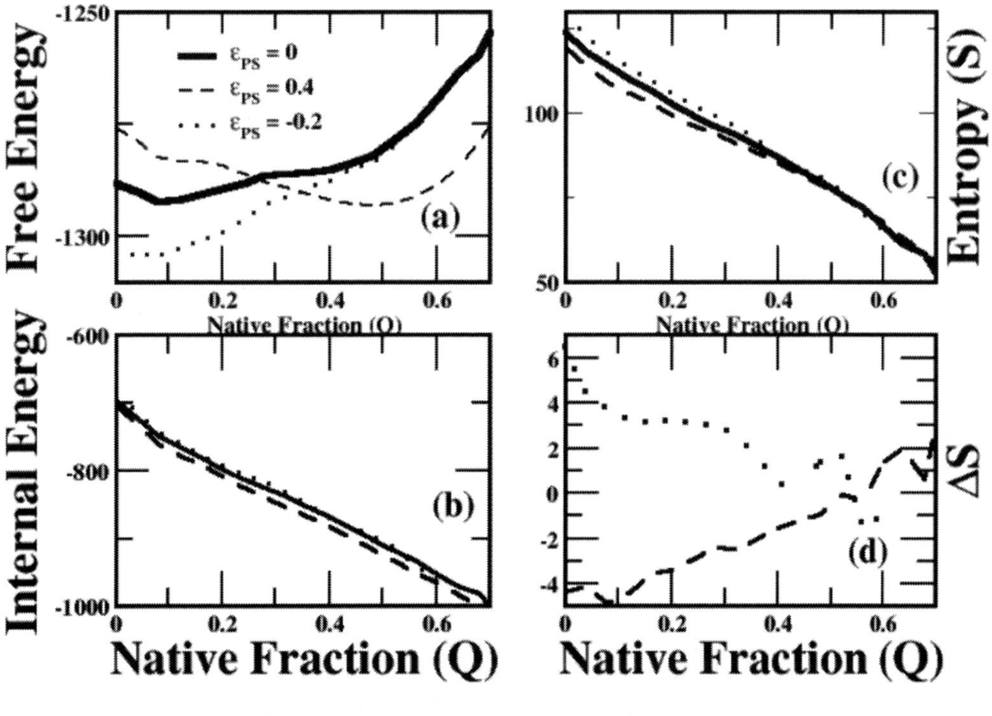

Figure 3. a) Free energy functional (f^*) vs. native fraction (Q) at temperature $T^* = 4.8$, with solvent quality $\varepsilon_{PS} =$ 0 (solid), 0.4 (dashed), -0.2 (dotted); b) Internal energy functions (u^*) vs. Q; c) Entropy

functional (s^*) vs. Q; d) Relative entropy ΔS vs. Q, $\Delta S = s^*(\varepsilon_{PS}) - s^*(0)$, where $s^*(0)$ is the entropy of the model with $\varepsilon_{PS} = 0$, and $s^*(\varepsilon_{PS})$ is for $\varepsilon_{PS} = 0.4$ (dashed) and $\varepsilon_{PS} = -0.2$ (dotted).

To determine the stabilization mechanism, analysis that compare solvent models near their folding temperatures are done. Here comparison of three models at $T^* = 4.8$ are presented: $\varepsilon_{PS} = 0$ (neutral, $T_f^* = 4.5$); $\varepsilon_{PS} = -0.2$ (denaturant-like, $T_f^* = 4.3$); $\varepsilon_{PS} = 0.4$ (osmolyte-like, $T_f^* = 5.0$). **Figure 3**a plots the free energy functional obtained by combining WHAM with Eqs.(2.8) and (2.9). The neutral solvent and denaturant-like models have free-energy minima at low Q (< 0.2) indicating an unfolded (U) state, while the osmolyte-like model has a minimum at high Q (> 0.5) consistent with a stable folded native (N) state. The free energy is divided into the internal energy and entropic (Eq.(2.10)) components. **Figure 3**b shows that the difference between the internal energy of the three models do not vary significantly with Q, and hence the stabilization mechanism is not driven by enthalpy (internal energy). **Figure 3**c,d shows that the entropy differences between the models are substantial only for unfolded states (low Q), and that, compare to the neutral solvent ($\varepsilon_{PS} = 0$), the Trp-cage in osmolyte-like solvent ($\varepsilon_{PS} = 0.4$) shows a substantial loss of entropy for U state (Q < 0.3, $\Delta S < 0$), but not for the N state (Q > 0.5, $\Delta S \approx 0$). This agrees with our hypothesis that the repulsive osmolytes ($\varepsilon_{PS} > 0$) reduce the entropy of unfolded states more than folded states, resulting in a free-energy bias that produces a more stable native state. This entropy-driven stabilization mechanism is consistent with several experiments.[6-8] For the weak denaturant-like system ($\varepsilon_{PS} = -0.2$), **Figure 3**c,d shows significant entropy gain ($\Delta S > 0$) only for the unfolded state, also consistent with an entropy-driven denaturing mechanism. Other analysis (not shown) also found entropy-driven mechanism for osmolyte-like or weak denaturant solvents ($\varepsilon_{PS} \geq -0.2$).

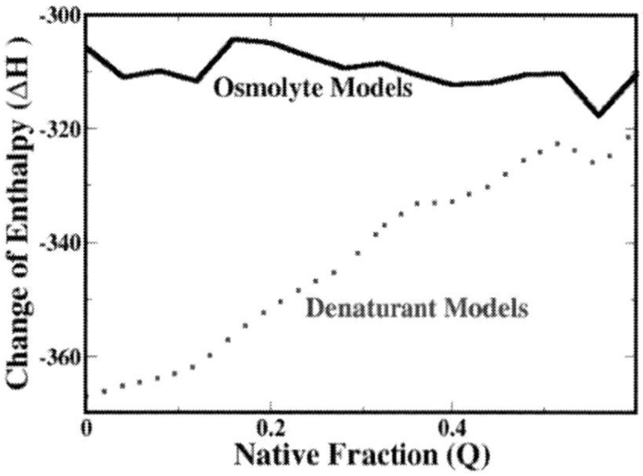

Figure 4. ΔH vs. Q, $\Delta H = u^*(\varepsilon_2) - u^*(\varepsilon_1)$, where $u^*(\varepsilon_i)$ is the internal energy functional of the solvent model ε_i. The osmolyte models (solid line) compares the internal energy of two osmolyte models $\varepsilon_1 = 0$ and $\varepsilon_2 = 0.4$. The denaturant models (dotted) compares the internal energy of two strongly denaturing solvent models $\varepsilon_1 = -0.4$ and $\varepsilon_2 = -0.6$. As defined above, ΔH is the internal energy difference. The use of ΔH facilitates comparisons with other works.

For strong denaturant ($\varepsilon_{PS} < -0.4$) the attractive solvent-protein interactions changes the stabilization/destabilization mechanism to enthalpy driven. In **Figure 4**, the osmolyte models' curve, $\varepsilon_1 = 0$ and $\varepsilon_2 = 0.4$ (see **Figure 3**c), shows no variation of $\Delta H = u^*(\varepsilon_2) - u^*(\varepsilon_1)$ with Q. In contrast, the denaturant models' curve, $\varepsilon_1 = -0.4$ and $\varepsilon_2 = -0.6$, shows ΔH decreasing with Q, which indicates an energetic preference for unfolded protein conformations in the presence of strong denaturants. The entropy difference ΔS (not shown) between the strong denaturant models does not show the clear variation with Q found in the osmolyte models. This finding is consistent with the work of the Garcia group that performed all-atom AMBER simulation of Trp-cage in urea [17], and also observed an enthalpy-driven protein denaturation mechanism. In that work, the thermodynamics are obtained by fitting simulation data to a phenomenological Hawley-type free energy surface. In contrast, in the work presented here, the free energy, enthalpy (in our model $\Delta U = \Delta H$), are directly calculated by the method outlined in subsection 2.3.

3.2. Size effects and cooperativity

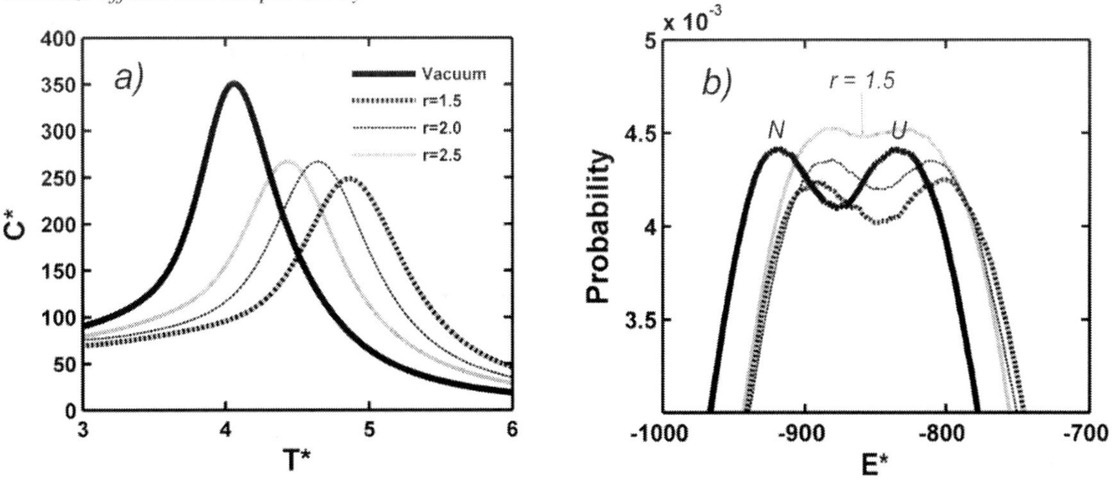

Figure 5. a) Heat capacity (C*) vs. Temperature (T*) for Trp-cage, with no solvents (vacuum), and in hard-sphere solvent of radius 1.5, 2.0, 2.5 Å, in a periodic box of 40Å X 40Å X 40Å; b) Probability distribution vs. Energy (E*), where the Trp-cage in vacuum is the thick solid line, and the peaks corresponding to the N and U states of the Trp-cage in vacuum are indicated. The distribution for solvent radius 1.5Å is also indicated. Both graphs were obtained by WHAM.

In view of the depletion-type (entropy driven) stabilization mechanism of osmolyte-like solvent, $\varepsilon_{PS} > 0$, it is expected that smaller solvents are better protein stabilizer. To examine this, simulations of Trp-cage in solvents are performed at solvent volume fraction of 30% for solvent vdW radii $r_{vdW} = 1.5, 2.0, 2.5$, and 3.0 Å, which, for a 40Å X 40Å X 40Å box, correspond to 1358, 573, 293, and 170 solvent molecules, respectively. The solvent-solvent and solvent-protein contact energy are set to zero ($\varepsilon_{SS} = 0$, $\varepsilon_{PS} = 0$). Hence the solvents are pure hard spheres. **Figure 5**a shows that the folding temperature increases from $T_f^* = 4.0$, for the Trp-cage in vacuum, to $T_f^* = 4.4$, 4.7 and 5.0 for solvent radius $r_{vdW} = 2.5$, 2.0, and 1.5 Å, respectively. This shows that smaller osmolytes are more effective at stabilizing the native state of proteins. Simulations of Trp-cage in denaturants, $\varepsilon_{PS} < 0$, also found that smaller denaturants are more effective at destabilizing proteins (data not shown).

It is well established that proteins are cooperative, in that they exist only in two states: native or unfolded.[40] In a previous study[23] the Trp-cage Gō in vacuum was shown to be cooperative. This is evident in the **Figure 5**b, where at its folding temperature $T_f^* = 4.0$, the energy probability distribution is bimodal, with the low and high energy peaks corresponding to the N and U state, respectively. In contrast, the Trp-cage in solvent systems show a progressive loss of two-state behavior, with a near complete loss of cooperativity for solvent size $r_{vdW} = 1.5$ Å, at its folding temperature $T_f^* = 5.0$, where the molten globular state (a state intermediate to the N and U state) is stable. The cooperative behavior of proteins in osmolytes and denaturants is a complex topic, and has been partially dealt with in a previous publication.[36]

3.3. Folding Kinetics

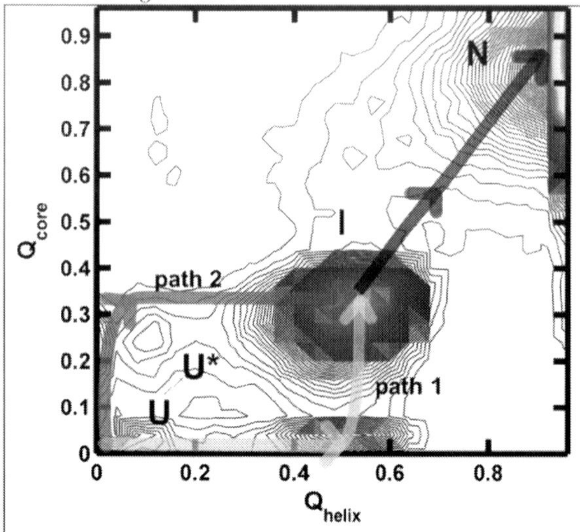

Figure 6. Contour map of the folding kinetic of Trp-cage cage in vacuum. All trajectories start from the U state ($Q_{core} \approx 0 \approx Q_{helix}$). The diffusion-collision path 1, involves rapid formation of helix ($Q_{helix} > 0$), follow by partial core collapse ($Q_{core} > 0$) to the intermediate state, I, before folding to the N state. The hydrophobic core collapse path 2 begins with formation of core collapse, follows by slow formation of helical structure to reach the I state follow by the N state. Q_{core} is the fraction of hydrophobic core contact and varies from 0 (no core structure) to 1 (all core contacts are formed). Similarly Q_{helix} is the fraction of helical structure. U* illustrates the shift that occurs in the unfolded state in the presence of osmolytes.

Folding simulations is performed for Trp-cage in pure HS solvent systems, by performing 100 unfolding simulations at $T^* = 6.0$ for a period of $t^* = 10000$ to produce 100 distinct unfolded conformations. The conformations are the initial structures of 100 folding simulations at $T^* = 3.0$, each lasting $t^* = 100000$. The trajectories are analyzed for the first passage time, defined as the time it takes for the protein to fold to its native state. A protein is defined to be folded if it has a main chain root-mean-square deviation (rmsd) of less than 1.5Å and an all-heavy atom rmsd of less than 2.0 Å. For the Trp-cage in vacuum the 100 folding simulation gives an average folding time, defined as the average first passage time, of $\tau^*_{fold} = 17350$. **Figure 6** plots the Q_{core} vs. Q_{helix} contour plot, which shows that the folding follows two basic paths: path 1, the fast diffusion-collision path; path 2, is the hydrophobic collapse path. In a previous publication[23], it was shown that path 2 is much slower because the core contacts hinders helical formation. That over 90% of trajectories fold by path 1 is the reason the Trp-cage is one of the fastest-folding proteins.[35] Folding simulations were performed on Trp-cage in a 40Å X 40Å X 40Å box with 20% solvent volume fraction of solvent radius $r_{vdW} = 1.5$, 2.0, 3.0, 4.0 and 6.0 Å, which correspond to 906, 382, 113, 48 and 14 spherical solvent molecules, respectively. For larger solvents $r_{vdW} \geq 3.0$ Å, the folding times are $\tau^*_{fold} \approx 16000$, similar to that of the Trp-cage in vacuum. For smaller solvents $r_{vdW} = 1.5$ and 2.0 Å, the folding times are 8890 and 13150, respectively. Similarly for system of 30% solvent fraction of size $r_{vdW} = 2.0$ and 2.5 Å, which correspond to 573 and 293 solvents, the folding times are 7760 and 9920. It is clear that smaller repulsive osmolytes, as well as higher osmolyte concentration, leads to higher protein folding rates (decrease of τ^*_{fold}) compare to the vacuum system. Analysis found (not shown here) that the Trp-cage in osmolytes still fold mainly by path 1 (about 90%), just like for Trp-cage in vacuum. The increase in folding rates is due mainly to osmolyte-induced entropy reduction of unfolded state, to a more compact native-like unfolded state, represented by U* in **Figure 6**. This conclusion is supported by experiments.[41]

4. Conclusions

This work present a DMD model of Trp-cage in spherical solvents, constructed by the spherical solvent paradigm[5], where the solvent quality is controlled by the protein-solvent contact energy ε_{PS} to produce good (denaturant, $\varepsilon_{PS} < 0$) and bad solvents (osmolyte, $\varepsilon_{PS} > 0$) . Equilibrium simulations observe thermal stabilization, where proteins remains in the native state up to a higher temperature, as ε_{PS} increases, and the solvents become more osmolyte-like. Using a modified WHAM method the protein stabilization mechanism is determined to be entropy driven, where the repulsive osmolytes reduce the entropy of unfolded conformations more than that of folded conformations. This mechanism is supported by experiments.[6-8] For strong denaturants ($\varepsilon_{PS} < -0.4$), it is demonstrated that the destabilization of the native state proceeds by enthalpy-driven mechanism. The result is largely consistent with the much more sophisticated model of the Garcia group[17], which suggests that generic aspects are important in protein stabilization/destabilization by osmolyte/denaturant. It is found that smaller osmolytes stabilize proteins more effectively, which is expected, in view of the depletion (entropy) driven mechanism. The model produces a loss in cooperative behavior due to osmolytes, which is surprising since osmolytes stabilizes the native state proteins. The cooperative behavior of proteins in solvent is a complex topic, which has been partially dealt with in a different publication,[36] and is a topic of future study. DMD folding simulations found a substantial increase in folding rates, in the presence of small osmolytes, at osmolyte volume fraction of 20% and 30%. In the experiment of Silow *et al*.[41], folding rates do increase at low or moderate osmolyte concentration, but decreases at very high concentration. Reproducing this result with DMD simulation at high osmolyte concentration is a challenging future project.

Acknowledgement

AL acknowledges financial support of the National Science and Engineering Research Council of Canada and the Shared Hierarchical Academic Research Computing Network (SHARCNET), as well as the computing resources provided by SHARCNET and Compute/Calcul Canada.

References

[1] P. H. Yancey *et al.*, Science **217**, 1214 (1982).
[2] G. F. Xie, and S. N. Timasheff, Biophys. Chem. **64**, 25 (1997).
[3] G. F. Xie, and S. N. Timasheff, Protein Sci. **6**, 211 (1997).
[4] D. W. Bolen, and L. V. Baskakov, J. Mol. Biol. **310**, 955 (2001).
[5] D. W. Bolen, and G. D. Rose, Annu. Rev. Biochem. **77**, 339 (2008).
[6] R. P. Baptista *et al.*, Biopolymers **89**, 538 (2008).
[7] L. Chen *et al.*, J. Mol. Biol. **351**, 402 (2005).
[8] G. S. Ratnaparkhi, and R. Varadarajan, J. Biol. Chem. **276**, 28789 (2001).
[9] B. J. Bennion, and V. Daggett, Proc. Natl. Acad. Sci. USA **100**, 5142 (2003).
[10] Q. Zou *et al.*, J. Am. Chem. Soc. **124**, 1192 (2009).
[11] B. J. Bennion, and V. Daggett, Proc. Natl. Acad. Sci. USA **101**, 6433 (2004).
[12] D. Klimov, J. E. Straub, and D. Thirulamai, Proc. Natl. Acad. Sci. USA **101**, 14760 (2004).
[13] L. Hua *et al.*, Proc. Natl. Acad. Sci. USA **105**, 16928 (2008).
[14] M. V. Athawale, J. S. Dordick, and S. Garde, Biophys. J. **89**, 858 (2005).
[15] R. Zangi, R. Zhou, and B. J. Berne, J. Am. Chem. Soc. **131**, 1535 (2009).
[16] D. L. Pincus, C. Hyeon, and D. Thirulamai, J. Am. Chem. Soc. **130**, 7364 (2009).
[17] D. R. Canchi, D. Paschek, and A. E. Garcia, J. Am. Chem. Soc. **132**, 2338 (2010).
[18] L. N. R. Wafer, W. W. Streicher, and G. Makhatadze, Proteins-Structure Function and Bioinformatics **78**, 1376 (2010).
[19] T. O. Street, D. W. Bolen, and G. D. Rose, Proc. Natl. Acad. Sci. USA **103**, 13997 (2006).
[20] J. Rosgen, B. M. Pettitt, and D. W. Bolen, Biophys. J. **89**, 2988 (2005).
[21] E. O'Brien *et al.*, Proc. Natl. Acad. Sci. USA **105**, 13403 (2008).

[22] R. Mountain, and D. Thirulamai, J. Phys. Chem. B **108**, 6826 (2004).
[23] A. Linhananta, J. Boer, and I. MacKay, Journal of Chemical Physics **122** (2005).
[24] H. Jang, C. K. Hall, and Y. Q. Zhou, Biophys. J. **86**, 31 (2004).
[25] F. Ding, and J. J. LaRocque, J. Biol. Chem. **280**, 40235 (2005).
[26] F. Ding *et al.*, Structure **14**, 5 (2006).
[27] F. Ding *et al.*, Structure **16**, 1010 (2008).
[28] Y. Q. Zhou, and A. Linhananta, Proteins-Structure Function and Genetics **47**, 154 (2002).
[29] Y. Q. Zhou, and A. Linhananta, Journal of Physical Chemistry B **106**, 1481 (2002).
[30] A. Linhananta, and Y. Q. Zhou, Journal of Chemical Physics **117**, 8983 (2002).
[31] A. Linhananta, H. Y. Zhou, and Y. Q. Zhou, Protein Science **11**, 1695 (2002).
[32] M. P. Allen, and D. J. Tildesley, *Computer Simulation of Liquids* (Oxford University Press, Oxford, 1987).
[33] D. Frenkel, and B. Smit, *Understanding Molecular Simulations* (Academic Press, 2002).
[34] Y. Q. Zhou, and M. Karplus, Journal of Molecular Biology **293**, 917 (1999).
[35] J. Kubelka, J. Hofrichter, and W. A. Eaton, Curr. Opin. Struct. Biol. **14**, 76 (2004).
[36] A. Linhananta, S. Hadizadeh, and S. S. Plotkin, Biophys. J. **100**, 459 (2011).
[37] A. M. Ferrenberg, and R. M. Swendsen, Phys. Rev. Lett. **61**, 2635 (1988).
[38] J. Lee, and J. M. Kosterlitz, Phys. Rev. Lett. **65**, 137 (1990).
[39] S. Kumar *et al.*, J. Comput. Chem. **13**, 1011 (1992).
[40] H. S. Chan, Proteins-Structure Function and Genetics **40**, 543 (2000).
[41] M. Silow, and M. Oliveberg, J. Mol. Biol. **326**, 263 (2003).

Journal of Physics: Conference Series **341** (2012) 012010 doi:10.1088/1742-6596/341/1/012010

Large-scale first-principles molecular dynamics simulations: application to the microsolvation of biologically-relevant ions in aqueous clusters

Chun C Mak, Denise M Koch and Gilles H Peslherbe*

Centre for Research in Molecular Modeling and Department of Chemistry and Biochemistry, Concordia University, 7141 Sherbrooke St. W., Montréal, Québec, Canada, H4B 1R6

*gilles.peslherbe@concordia.ca

Abstract. First-principles molecular dynamics simulations are employed to investigate the solvation structure of the biologically-relevant paradigm guanidinium-ion in water clusters, $Gdn^+(H_2O)_n$. In these simulations, the intermolecular interactions are evaluated along the trajectories with the self-consistent-charge density-functional tight-binding (SCC-DFTB) model with the on-site third-order correction and modified effective Coulomb interaction, an approximate model of density-functional theory. For $Gdn^+(H_2O)_{21}$, the simulated probability distribution of the ion-to-water cluster centre-of-mass distances and water angular coordinates suggests that Gdn^+ is primarily localised at the surface of the water cluster, with the surface of the ion relatively devoid of water molecules. Estimates of the computational resources required for first-principles molecular dynamics simulations of larger $Gdn^+(H_2O)_n$ ($n \leq 100$) indicate that, with a modern supercomputer, such simulations can readily be performed in a matter of days.

1. Introduction

Proteins play numerous important roles in biology, serving for instance as enzymes to catalyze vital chemical reactions, transporters of chemical substances between different parts of an organism, and as regulators of biological processes. The function of a given protein is closely related to its conformation, and alterations to its native conformation may result in loss of function and in disease. It has been recognised for some time that molecular interactions involving the guanidinium ends of the side chains of arginine residues may play a crucial role in the determination of the structure of proteins and protein complexes [1]. Therefore, a molecular-level understanding of such interactions may be necessary to develop a clearer picture of the determinants of structure in biomolecules.

The guanidinium ion, $[C(NH_2)_3]^+$ or Gdn^+, is an excellent model of the positively-charged guanidinium group of arginine and, since Gdn^+ is widely believed to be responsible for the activity of several powerful protein denaturant salts, much experimental and theoretical work has been carried out to understand its hydration structure and its interaction with other dissolved solutes in aqueous solutions [2-4]. The general picture arising from previous work on aqueous solutions of Gdn^+ is that Gdn^+ is very weakly hydrated and it exhibits a high propensity to aggregate with itself, other ions, as well as small aromatic molecules; the latter may have implications for the role of the Gdn^+ group in protein folding, protein-protein interactions and protein denaturation [2]. The existence of these ionic aggregates in aqueous Gdn^+ solutions has not been undisputed [3]; however, calculations of the potential of mean force of association of Gdn^+ with itself or with small aromatic molecules in aqueous solutions generally suggest that such aggregates are thermodynamically stable [4].

The investigation of gas-phase guanidinium-water clusters of finite size, $Gdn^+(H_2O)_n$, may offer a complementary perspective on the hydration of Gdn^+, which could potentially result in a more in-depth understanding of the seemingly unusual behaviour of Gdn^+ in aqueous media. On the experimental side, Hof and co-workers have used energy-dependent electro spray ionisation mass spectrometry to investigate the dehydration process of $Gdn^+(H_2O)_n$ and various methylated derivatives $Me_mGdn^+(H_2O)_n$ (m=0-4 and 6; n=1-50) [5]. Their work demonstrated that methylation of Gdn^+ results in increased hydrophobicity of the ion, highlighting the possible role of hydration/dehydration in the modulation of biologically-relevant intermolecular interactions involving the Gdn^+ group. In this contribution, the use of reliable, first-principles molecular dynamics simulations to investigate the hydration of Gdn^+ in aqueous clusters is first described, and the approach is applied to characterise the solvation structure of small $Gdn^+(H_2O)_n$. It will then be shown that, with a modern supercomputer, an investigation of large $Gdn^+(H_2O)_n$ (n<100) using the same first-principles molecular dynamics approach is readily attainable.

2. Computational Procedure

Molecular dynamics simulations were used to investigate the structural properties of $Gdn^+(H_2O)_n$ and determine the preferential position of Gdn^+ within the cluster. Two coordinates, defined in figure 1, the ion-to-water cluster centre-of-mass separation r_{cm} and the angles θ between a given water molecule, the centre of mass of Gdn^+ and the centre of mass of the water cluster were used to discriminate between the predominant solvation structures, interior or surface, of $Gdn^+(H_2O)_n$, as proposed in earlier work [6].

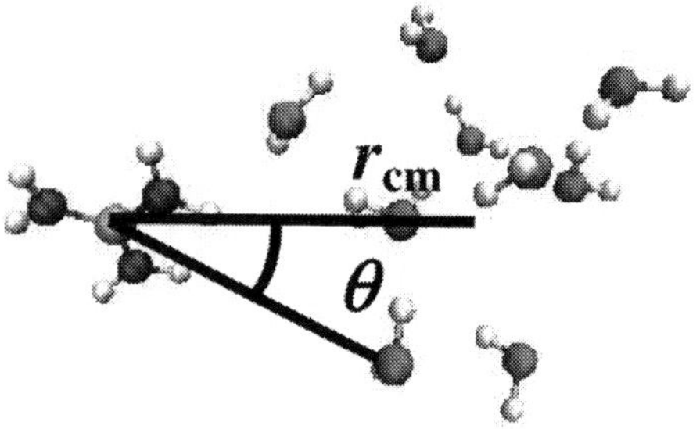

Figure 1. Definition of the ion-to-water cluster centre-of-mass distance, r_{cm}, and θ, the angle between a given water molecule, the ion and the water cluster centres of mass.

In order to ensure adequate sampling of $Gdn^+(H_2O)_n$ configurations, including interior and surface-solvated cluster structures, the umbrella sampling method was employed [7]. In this method, a series of molecular dynamics simulations are performed, each in the presence of an added umbrella potential U_{umb} centred on a different value of r_{cm} (denoted r^0_{cm}) to restrain the ion-to-water cluster centre-of-mass separation to values near r^0_{cm}.

$$U_{umb} = \frac{1}{2}k(r_{cm} - r^0_{cm})^2$$

For $Gdn^+(H_2O)_{21}$, the value of the force constant k of the umbrella potential was chosen to be 7.0 kcal·mol^{-1}Å^{-2}, while the r^0_{cm} values ranged from 0.0 Å to 8.0 Å, with each value spaced by 1.0 Å; this prescription ensured efficient generation of a smooth potential of mean force for this cluster size. Each molecular dynamics simulation was run for 2 nanoseconds using the velocity Verlet algorithm [8], with 1 nanosecond of equilibration followed by an additional nanosecond of simulation for data collection. The energies and energy gradients of $Gdn^+(H_2O)_n$ used in the molecular dynamics simulations were computed using approximate density-functional theory, namely the self-consistent-charge density-functional tight-binding (SCC-DFTB) model with the on-site third-order correction and a modified effective Coulomb interaction [9]. This model, which is a modification of the traditional SCC-DFTB model to provide an improved description of hydrogen-bonding interactions and proton affinities, appears to adequately describe the relative strength of Gdn^+-H_2O and H_2O-H_2O interactions, a crucial requirement for reliable simulation results [10]. The approximate density-functional theory calculations were performed with the DFTB+ computer code [11] A Nosé-Hoover thermostat was applied to maintain the temperature of the system at 298 K [12]. Furthermore, to prevent the escape of water molecules from $Gdn^+(H_2O)_n$ and thus ensure that a proper canonical ensemble of cluster configurations is sampled, a restraining potential U_{rstr} was added whenever the distance d between any given water molecule and the centre of mass of the water cluster exceeded d_{max}.

$$U_{rstr} = \frac{1}{2} K (d - d_{max})^2$$

The optimal values of K and d_{max} are dependent on the cluster size n, and in general, are chosen such that the cluster geometries are only slightly affected by the presence of U_{rstr}. For $Gdn^+(H_2O)_{21}$, optimal values of K and d_{max} were found to be 200 kcal·mol^{-1} Å^{-2} and 15 Å, respectively.

The calculation of the potential of mean force for $Gdn^+(H_2O)_n$ requires energy and gradient calculations for a total of 1.8×10^7 cluster configurations, which represents a significant investment of computational resources. Preliminary test simulations were performed to determine the optimal distribution of the required calculations on a cluster of Intel® Xeon® X5560 processors consisting of nodes that each have 8 cores. Results of these preliminary simulations, shown in figure 2, indicate that the elapsed time for a given simulation decreases dramatically as the number of cores used is increased from one to four, with further significant decreases up to 8 cores.

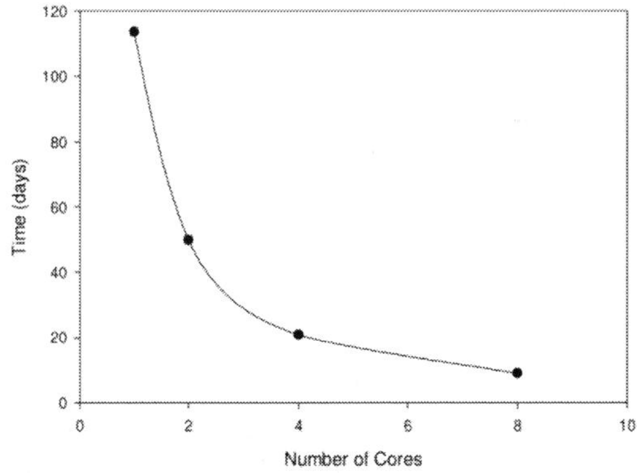

Figure 2. Real time elapsed for 2-nanosecond molecular dynamics simulations of $Gdn^+(H_2O)_{21}$ using different numbers of cores

The simulations for each value of r^0_{cm} were therefore distributed on distinct nodes of the computer cluster, with 8 cores used for every simulation. All 9 independent simulations were carried out simultaneously, minimising the elapsed time needed for the entire set of simulations, which were completed in around 10 days, requiring a total of 17280 cpu×hours (the equivalent of 2 years on one core processor). Once the simulations were completed, the weighted histogram analysis method (WHAM) was used to remove the effects of the biasing potential on the probability distribution of sampled configurations and to reconstruct the complete probability distribution of r_{cm} values from the 9 individual simulations [13].

3. Results and Discussion

The probability distribution $P(r_{cm},\theta)$ for $Gdn^+(H_2O)_{21}$ obtained from the molecular dynamics simulations is shown in figure 3; it exhibits one single peak at r_{cm} = 4 Å, $\theta \approx 30°$. For a spherical cluster of uniform density in which the Gdn^+ is internally solvated, $P(r_{cm}, \theta)$ should peak at $r_{cm} \approx 0$ Å and exhibit a dependence on θ proportional to $\sin(\theta)$ *i.e.* an isotropic angular distribution. By contrast, $P(r_{cm}, \theta)$ for similar spherical clusters in which the ion is located at the surface would exhibit a peak at values of r_{cm} that approach the radius of the cluster, with very low $P(r_{cm}, \theta)$ at values of θ larger than 90°. Since $P(r_{cm}, \theta)$ for $Gdn^+(H_2O)_{21}$ peaks at approximately r_{cm} = 4 Å, which is about half the average radius (around 8 Å) of the water cluster, it is difficult to draw firm conclusions on the preferential solvation structure of $Gdn^+(H_2O)_{21}$ from the most probable value of r_{cm} alone. The plot of $P(r_{cm}, \theta)$ indicates that there is a notable absence of water molecules located at angles θ larger than 90° for all values of r_{cm} for which the value of $P(r_{cm}, \theta)$ is non-negligible, indicating that the water molecules are not evenly distributed about Gdn^+. Typical $Gdn^+(H_2O)_{21}$ configurations at r_{cm} = 4 Å are shown in

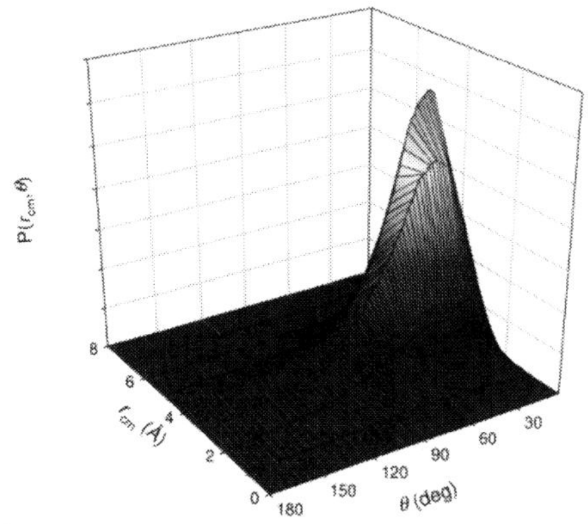

Figure 3. Probability distribution $P(r_{cm}, \theta)$ for $Gdn^+(H_2O)_{21}$

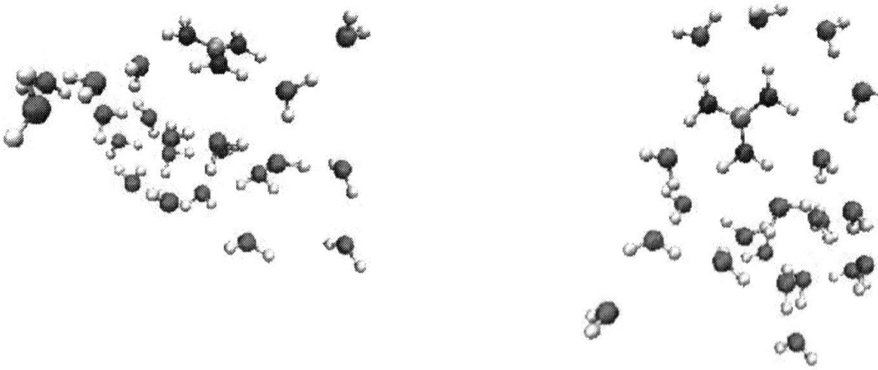

Figure 4. Typical $Gdn^+(H_2O)_{21}$ structures at r_{cm} = 4 Å

figure 4. Despite the fact that r_{cm} is much smaller than the radius of the cluster, Gdn^+ is obviously located at the surface due to the non-spherical geometry of these $Gdn^+(H_2O)_{21}$ configurations. The surfaces of Gdn^+ appear to be relatively devoid of water molecules in these configurations, with most Gdn^+-H_2O interactions involving the NH_2 groups of the ion. Figure 5 shows two different views of the average spatial distribution of the water molecules around Gdn^+ in $Gdn^+(H_2O)_{21}$, providing a more complete picture of Gdn^+ hydration in small aqueous clusters. These plots were generated by superposing the $Gdn^+(H_2O)_{21}$ configurations sampled from the molecular dynamics simulation centred

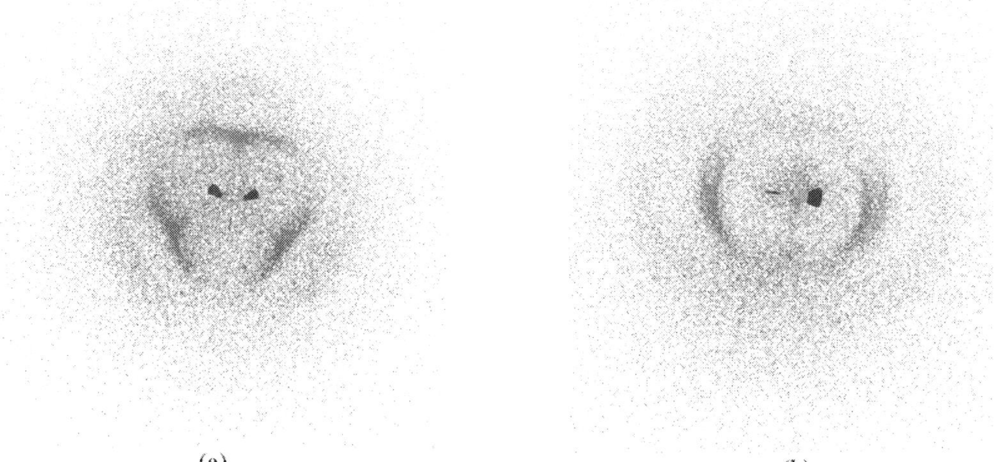

(a) (b)

Figure 5. Average spatial distribution of H_2O around Gdn^+ in $Gdn^+(H_2O)_{21}$ at r_{cm} = 4 Å. (a) view from above the Gdn^+ plane (b) view along the Gdn^+ plane

at r_{cm} = 4 Å, with Gdn^+ fixed in space and identically oriented in all configurations. Hydrogen atoms are omitted for clarity. Regions around Gdn^+ with a heavier density of points correspond to more hydrated regions *i.e.* regions of the cluster where water molecules are more likely to be located. These plots show that there is a higher probability for water molecules to be located in three "islands" around Gdn^+ outwards from the "pockets" defined by adjacent NH_2 groups, suggesting that there may be an increased propensity for water molecules to interact with these groups. There is, on the other hand, a lower probability for water molecules to be found on the surfaces of Gdn^+, a possible reflection of the seemingly hydrophobic nature of Gdn^+ in certain environments. The results from these reliable first-principles molecular dynamics simulations of $Gdn^+(H_2O)_{21}$ corroborate the findings of previous simulations of Gdn^+ and its salts in aqueous solution using various model potentials, which indicate that Gdn^+ is extremely weakly hydrated with surfaces relatively devoid of water molecules [2]. These findings may have important implications for understanding interactions involving Gdn^+ groups in biological systems, where the dehydrated face of the planar ion may be available for binding with non-polar molecules.

While the present findings already shed new light onto the hydration of Gdn^+, a thorough theoretical investigation of $Gdn^+(H_2O)_n$ for various cluster sizes may be necessary to develop a sound understanding of the role of hydration in the modulation of biologically-relevant interactions involving Gdn^+ groups. With powerful, modern supercomputers, the first-principles molecular dynamics approach employed in the present work to investigate the solvation structure of $Gdn^+(H_2O)_{21}$ is readily applicable to larger $Gdn^+(H_2O)_n$ ($n \leq 100$) . Preliminary molecular dynamics simulations of various $Gdn^+(H_2O)_n$ ($20 \leq n \leq 100$) using the SCC-DFTB model with the on-site third-order correction and modified effective Coulomb interaction have been performed to assess the feasibility of using large-scale first-principles molecular dynamics simulations to investigate the solvation structure of larger clusters. For each cluster size, a single simulation was run on one node of a cluster of Intel® Xeon® X5560 processors using 8 cores with the wall-time limit set to 8 hours, and the number of simulation steps completed within this time was determined. The data were then used to estimate the time required to complete a 2-nanosecond simulation for each $Gdn^+(H_2O)_n$, and the results are summarised in figure 6. The time required to complete a single 2-nanosecond simulation increases almost linearly with cluster size, with one simulation requiring 10 days for $Gdn^+(H_2O)_{20}$ and the same run requiring 50

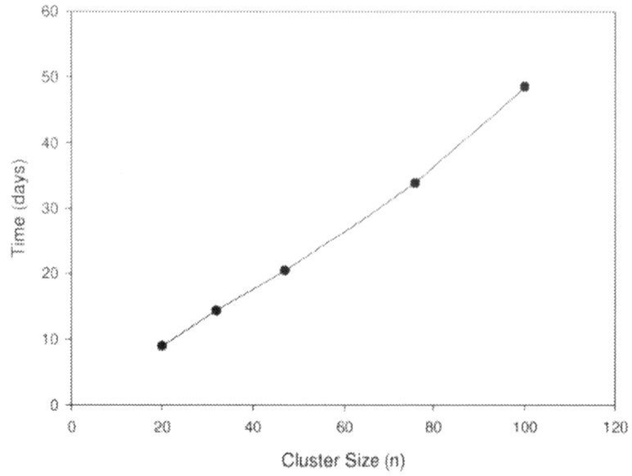

Figure 6. Real elapsed time for 2-nanosecond first-principles molecular dynamics simulations of $Gdn^+(H_2O)_n$ ($20 \leq n \leq 100$)

days for $Gdn^+(H_2O)_{100}$. In all fairness, the computational cost of the potential of mean force calculation will not scale linearly with cluster size. Firstly, a larger number of umbrella sampling windows might be needed as the cluster grows in size and its radius becomes larger. Secondly, a larger number of configurations might need to be sampled as the density of cluster states increases along with cluster size, a point that is, however, not grounded on a solid theoretical foundation. Assuming that a complete $P(r_{cm}, \theta)$ could be obtained using the WHAM method with data from 12 independent simulations restrained around different values of r_{cm}, the solvation structure of $Gdn^+(H_2O)_{100}$ could be characterised in 50 days under ideal conditions in which uninterrupted calculations are possible on 12 8-core compute nodes. Such a simulation would entail energy and gradient calculations for a total of 2.4×10^7 cluster configurations and require the use of ~115200 cpu×hours (~13 years of a single core processor) of computing time, which while demanding is still feasible on most modern supercomputers. Finally, as the number of cores per node tends to increase dramatically on modern clusters, we anticipate the real-time cost of the calculations to decrease even further on new cluster infrastructures.

4. Summary and Conclusions

In this contribution, first-principles molecular dynamics simulations using the SCC-DFTB model with the on-site third-order correction and modified effective Coulomb interaction, which reliably describes the relative strength of the Gdn^+-H_2O and H_2O-H_2O interactions when compared with high-level *ab initio* electronic structure theory calculations, were used to investigate the solvation structure of $Gdn^+(H_2O)_{21}$. The simulation results tentatively indicate that, in small $Gdn^+(H_2O)_n$, Gdn^+ is preferentially located at the surface of the water cluster, with the ion surface essentially dehydrated. These preliminary findings, which are consistent with the conclusions of previous work on the hydration structure of Gdn^+ in aqueous solutions, provide complementary, molecular-level insight into the apparent "hydrophobicity" of Gdn^+. The feasibility of applying first-principles molecular dynamics simulations to the investigation of larger $Gdn^+(H_2O)_n$ was explored, laying the groundwork for a comparative investigation of Gdn^+ hydration in aqueous clusters of varying sizes that may lead to a more in-depth understanding of the influence of hydration on biologically important intermolecular interactions involving Gdn^+ groups.

Acknowledgements
This work was funded by research grants from the Natural Sciences and Engineering Research Council (NSERC) of Canada. C.C.M. is the recipient of an NSERC award. Computational resources

were provided by the Réseau Québecois de Calcul Haute Performance (RQCHP) and the Centre for Research in Molecular Modeling (CERMM), which was established with the financial support of the Concordia University Faculty of Arts & Science, the Ministère de l'Education du Québec (MEQ) and the Canada Foundation for Innovation (CFI). G.H.P. holds a Concordia University Research Chair.

References

[1] Pednekar D, Tendulkar A and Durani S 2009 *Proteins* **74** 155
Crowley P and Golovin A 2005 *Proteins* **59** 231
Bogan A and Thorn K 1998 *J. Mol. Biol.* **280** 1
Magalhaes A, Maigret B, Hoflack J, Gomes J and Scheraga H 1994 *J. Protein Chem.* **13** 195

[2] Mason P, Neilson G, Dempsey C, Barnes A and Cruickshank J 2003 *Proc. Natl. Aca. Sci. USA* **100** 4557
Mason P, Dempsey C, Neilson G, Kline S and Brady J 2009 *J. Am. Chem. Soc.* **131** 16689
Mason P, Neilson G, Kline S, Dempsey C and Brady J 2006 *J. Phys. Chem. B* **110** 13477
Mason P, Dempsey C, Neilson G and Brady J 2005 *J. Phys. Chem. B* **109** 24185
Mountain R and Thirumalai D 2004 *J. Phys. Chem. B* **108** 19711
Mason P, Neilson G, Enderby J, Saboungi M, Dempsey C, Mackerell A and Brady J 2004 *J. Am. Chem. Soc.* **126** 11462

[3] Hunger J, Niedermayer S, Buchner R and Hefter G 2010 *J. Phys. Chem. B* **114** 13617

[4] Soetens J, Millot C, Chipot C, Jansen G, Ángyán J and Maigret B 1997 *J. Phys. Chem. B* **101** 10910
Saigal S and Pranata J 1997 *Bioorg. Chem.* **25** 11
Duffy E, Kowalczyk P and Jorgensen W 1993 *J. Am. Chem. Soc.* **115** 9271
Boudon S, Wipff G and Maigret 1990 *J. Phys. Chem.* **94** 6056

[5] McQuinn K, McIndoe J and Hof F 2008 *Chem. Eur. J.* **14** 6483

[6] Peslherbe G, Ladanyi B and Hynes J 2000 *Chem. Phys.* **258** 201

[7] Torrie G and Valleau J 1974 *Chem. Phys. Lett.* **28** 578

[8] Swope W, Andersen H, Berens P and Wilson K 1982 *J. Chem. Phys.* **76** 637

[9] Yang Y, Yu H, York D, Cui Q and Elstner M 2007 *J. Phys. Chem. A* **111** 10861
Elstner M, Porezag D, Jungnickel G, Elsner J, Haugk M, Frauenheim T, Suhai S and Seifert G 1998 *Phys. Rev. B* **58** 7260

[10] Mak C, Koch D and Peslherbe G Article to be submitted

[11] Aradi B, Hourahine B and Frauenheim T 2007 *J. Phys. Chem. A* **111** 5678

[12] Nose S 1984 *J. Chem. Phys.* **81** 511
Hoover G 1985 *Phys. Rev. A* **31** 1695

[13] Kumar S, Bouzida D, Swendsen R, Kollman P and Rosenberg J 1992 *J. Comput. Chem.* **13** 101

Grid Computing Application for Brain Magnetic Resonance Image Processing

F Valdivia [1], B Crépeault [1], and S Duchesne [1,2]

[1] Centre de Recherche de l'Université Laval Robert-Giffard, 2601 de la Canardière, Québec (QC) G1J 2G3, Canada
[2] Département de radiologie, Université Laval, Québec (QC) G1K 7P4, Canada

E-mail: fernando.valdivia@crulrg.ulaval.ca, burt.crepeault@gmail.com, simon.duchesne@crulrg.ulaval.ca

Production Editor, *Journal of Physics: Conference Series*,
IOP Publishing, Dirac House, Temple Back, Bristol BS1 6BE, UK

Abstract: This work emphasizes the use of grid computing and web technology for automatic post-processing of brain magnetic resonance images (MRI) in the context of neuropsychiatric (Alzheimer's disease) research. Post-acquisition image processing is achieved through the interconnection of several individual processes into pipelines. Each process has input and output data ports, options and execution parameters, and performs single tasks such as: a) extracting individual image attributes (e.g. dimensions, orientation, center of mass), b) performing image transformations (e.g. scaling, rotation, skewing, intensity standardization, linear and non-linear registration), c) performing image statistical analyses, and d) producing the necessary quality control images and/or files for user review. The pipelines are built to perform specific sequences of tasks on the alphanumeric data and MRIs contained in our database.

The web application is coded in PHP and allows the creation of scripts to create, store and execute pipelines and their instances either on our local cluster or on high-performance computing platforms. To run an instance on an external cluster, the web application opens a communication tunnel through which it copies the necessary files, submits the execution commands and collects the results.

We present result on system tests for the processing of a set of 821 brain MRIs from the Alzheimer's Disease Neuroimaging Initiative study via a nonlinear registration pipeline composed of 10 processes. Our results show successful execution on both local and external clusters, and a 4-fold increase in performance if using the external cluster. However, the latter's performance does not scale linearly as queue waiting times and execution overhead increase with the number of tasks to be executed.

1. Introduction

The primary objective of the Medical Data, Information and Knowledge (**MEDICS**) laboratory is to find early signs of neuropsychiatric diseases through novel techniques for medical image analysis. Early applications of the research lie in the establishment of biomarkers for Alzheimer's disease (**AD**) based on population analyses of voxel-based features [1], high-dimensional morphometry characteristics [2, 3] and structural volumetry [4, 5], all extracted from magnetic resonance images (**MRI**) of human subjects.

To achieve the necessary statistical power in the conduct of this research, we collaborate with various research organizations to collect wide and comprehensive patient and image databases. Examples include research consortia (e.g. the Alzheimer's Disease Neuroimaging Initiative (**ADNI**)[6]; European ADNI [7], International Consortium for Brain Mapping [8] or individual laboratories, either through open (e.g. Open Access Series of Imaging Studies [9]) or close-source access (e.g. Laboratory of Epidemiology, Neuroimaging, & Telemedicine [10]).

However, irrespective of the many problems related to processing multicentric images *per se*, this pooling of data from multiple sources presents a number of difficulties, hindering the timely conduct of research. Those relate to:

(1) formatting: patient and image data formats vary widely from one organization to another, making it difficult to perform similar analyses on different sets of data;
(2) searching: it is difficult, sometimes impossible, to comprehensively sift through large, incohesive data stacks based on specific patient criteria or image characteristics;
(3) cross-referencing: moreover, most dataset are decoupled, e.g. patient data tables are typically disconnected from the image files and must be reassembled; and
(4) processing: data (e.g. images) are typically not processed, and require dedicated application suites to extract relevant biomarker information.

2. Objective

In this article we report the design, development and use statistics of the integrated MEDICS database and processing platform, built to address these issues by complying with the following requirements:

(1) ability to accept different input formats, and to transpose to a common format for processing;
(2) ability to house massive amounts of data, in image form or otherwise in relational format;
(3) capacity for remote, web-access, search and processing;
(4) capacity for flexible data processing, with complete traceability; and
(5) ability to use grid computing to accelerate the processing of large-scale datasets.

3. Methods

3.1. Software Architecture

The MEDICS system is composed of three major software modules (Figure 1):

Database - Module B. The alphanumeric and image database is a relational database system, able to accept all prevalent medical image formats, starting with DICOM [11] and other reconstructed formats which are eventually all converted and stored as MINC[1] images. Any number of scalar or text files can be accepted for alphanumerical data (e.g. neuropsychological testing). These files are accessed directly

[1] Medical Image NetCDF format; http://www.bic.mni.mcgill.ca/ServicesSoftware/MINC.

through the *User Interface* or the *Pipeline Processing* modules (see below). The data has been organized in a semi-formal domain-specific ontology [12] assembled from various ontological sources (e.g. anatomical [13-15]); medical image data (DICOM reference guide 2001); data processing [16, 17]; pathological [18, 19] and controlled terminologies (e.g. quality control [20]; validation procedure and inter-relationships [21, 22]).

Pipeline Post-Processing - Module EX. The Post-Processing module allows the definition of processing pipelines (i.e. individual processes assembly into serial/parallel execution streams) and accepts user requests through the web-compatible *User Interface* module for their execution (**EX**). Module EX scripts manage concurrent pipeline executions, event logging and job submission to local or external computer clusters, using a queue management system as distributed resource management.

User Interface – Module WX. The Web-eXchange (WX) User Interface represents the interface between the user of the MEDICS system and the system itself. Its main functions are: user creation and authentication, data search, dataset creation, pipeline creation and execution, image quality control and rating [1].

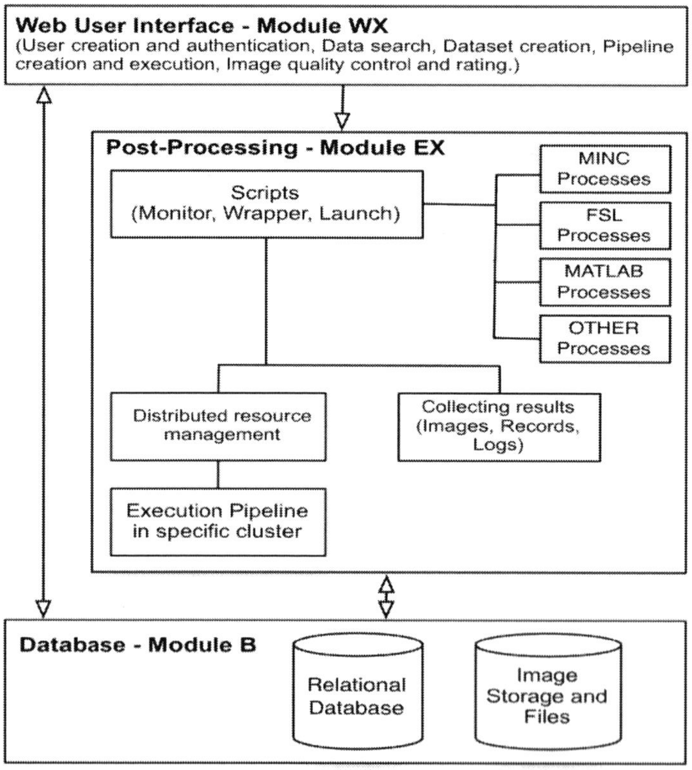

Figure 1. MEDICS System Software Architecture [2].

3.2. Hardware Architecture

The MEDICS system rests on the following major hardware components (Figure 2):

Journal of Physics: Conference Series **341** (2012) 012011 doi:10.1088/1742-6596/341/1/012011

User Workstation: Users can access the MEDICS system via any computer through an Internet connection and standard web browsers.

WX server: The server is a computer running web server software. The server will stand inside a demilitarized zone in order to be accessible through HTTPS.

EX and B server: The EX server is a computer running the MEDICS Pipeline Post-Processing EX module as well as grid control software when present; the B server houses the relational database management system. The B and EX servers stands behind a firewall in the secure zone. At present, we use the same physical server as both EX and B; they can be decoupled in the future to meet growing demands without loss of functionality.

Local and External Grid: both local MEDICS Laboratory computer grid or external high-performance clusters can be accessed to distribute jobs and resource management for the execution of processing pipelines.

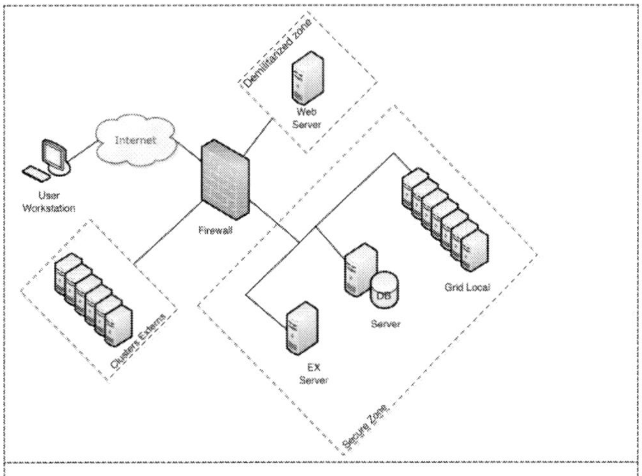

Figure 2. MEDICS System Hardware Architecture [1].

3.2.1. Architectural Decisions

Table 1. Detail of components and products.

Component	Products
Relational database	PostgreSQL
Image storage and files	Network File System (NFS), format MINC image and text files

Image processing librairies [2]	MINC, FSL [3], SPM [4], MATLAB
Development environment	PHP 5.3.4, JavaScript 1.8, Bash 4.0.33, HTML 4.01, Apache 2.2.12, Eclipse 2009
Distributed resource management	Sun Grid Engine
Operational systems	Linux Ubuntu Mac OS Leopard 10.5.8
Servers	DELL R710 APPLE iMAC 24"

3.3. Web User Interface

All instructions to the system are made through one of the WX functions (applications) that include [1]:

- User authentication, where a user enters credentials to access the system;
- Data search, where users can sift through data by entering ad-hoc queries to filter information from database;
- Dataset creation, where users can assemble search results and create datasets based on their search criteria;
- Pipeline creation, where users can assemble series of processes together that will perform a certain image-processing task;
- Pipeline execution, where users can instantiate pipelines with datasets, give values to certain process options (when applicable) and launch the execution; and
- Visual image quality control and rating, where users assign scores based on a scale to medical images produced by a processing pipeline.

3.4. Data processing

3.4.1. Processes

Processes are scriptable elements that perform concrete actions on data:

- Manipulate data arrays within the application;
- Perform data transformations on scalars and images;
- Perform statistical comparisons on scalars and images;
- Perform quality control of input data (e.g. scalars, files, images), of other processes, of whole pipelines; and
- Perform quality assurance of output data (e.g. scalars, files, images) from other processes, or whole pipelines.

Each process has distinct characteristics and options (mandatory and optional), recorded in the alphanumeric database. Processes can be modelled in a fashion represented in Figure 3, whereby each process has any number of inports and outports of any type, as well as a number of options. Options can

[2] The MEDICS system is compatible with any scriptable image processing library
[3] FMRIB Software Library ; http://www.fmrib.ox.ac.uk/fsl/
[4] Statistical Parametric Mapping ; http://www.fil.ion.ucl.ac.uk/spm/

be viewed as the arguments of a process, e.g. the various commands that can make a process execute in a certain way. The execution of a process will produce a return value that indicates if it completed successfully or not.

Ports can be defined as the entry or exit points for data entering or leaving processes (and by extension, pipelines). In our model so far we have defined entities that correspond to the port types depicted in Figure 3, along with their graphical notation.

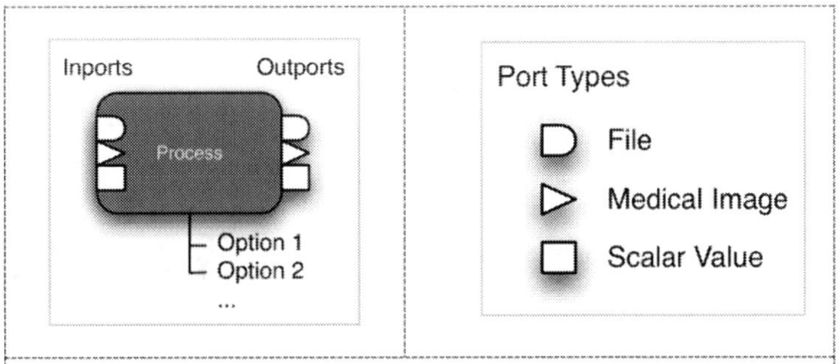

Figure 3. Process Model and Port Types [1].

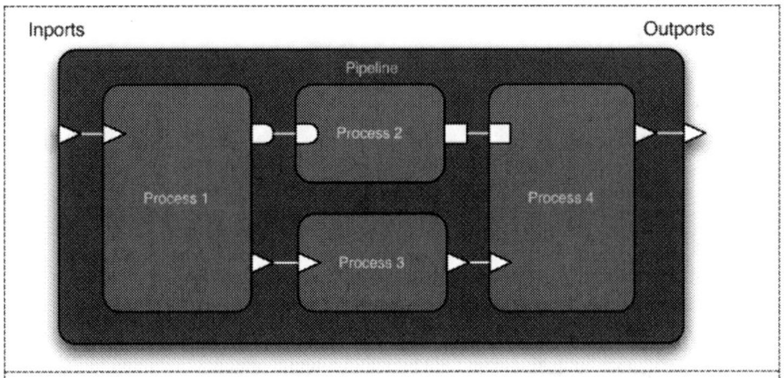

Figure 4. Pipeline Model [1].

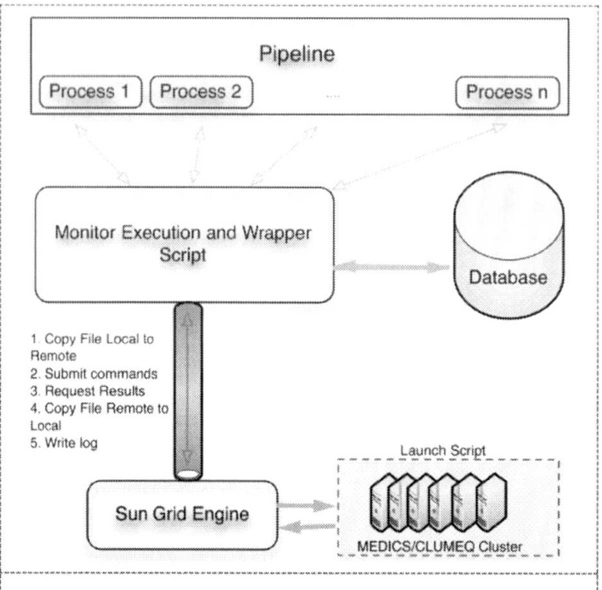

Figure 5. Data and pipeline processing control.

3.4.2. Pipelines

Data processing is performed as a set of processes running within a pipeline [1] (Figure 4). Pipelines are collections of processes, each one performing a single task, that are assembled to perform a series of scalar and/or image processing steps that lead to the desired result(s).

In our model, we can observe that the pipeline has ports, representing the means by which data is fed into the pipeline for processing (inports) or produced as a result of the pipeline execution (outports). Inside the pipeline are chained processes that also have inports and outports that are linked to one another according to their type, graphically represented with different symbols. Outports of a process can only be linked to the inports of the following process if their types are the same [1].

The creation and execution of pipelines requires a set of parameters, stored in the database, which include: (i) mandatory and optional individual process options; (ii) mandatory and optional input and output individual process ports; (iii) individual process source/versions to execute; (iv) individual estimated execution time per process; (v) number of processes to execute in parallel per node; (vi) general execution cluster data (type local/remote, home path, user, max job in queue, commands, etc.).

3.4.3. Control scripts

A set of PHP (Hypertext Pre-processor) and BASh (Bourne Again Shell) scripts are responsible for monitoring execution of pipeline instances, build process execution commands, submit jobs to the queue, retrieve results, and copy files local to remote and remote to local, if necessary; all the while running from behind the web application. The three main scripts are: (i) **Wrapper**: Prepares the data and builds the basic commands with process execution options; (ii) **Ex-monitor**: Manages pipeline instances execution, builds final commands, opens communication tunnel to external/local cluster, copies files, submits jobs, retrieves results and creates log files; and (iii) **Launch command**: Interacts with Ex-monitor to invoke the execution of the processes in clusters.

4. Experiments and results

4.1. Databases

The alphanumeric database is relational and implemented in PostgreSQL. The database stores a number of scalar-type data on patients, such as demographics, family and past medical history; neurological, neuropsychiatric and neuropsychological test results; genotyping information; biomarkers; and results from 137 tables. Further, a number of original and processed MRI are stored and linked to database records.

Table 2. Summary records of the database.

Origin	**Quantity**	
	Patients	**Original Images**
ADNI	822	15 799
E-ADNI	3	79
ICBM	150	300
LENITEM	1 038	1 698
OASIS	416	1 608
MEDICS	19	68
TOTAL	**2 448**	**19 552**

4.2. High-performance clusters

At present the MEDICS system can interface with two clusters: a local cluster ("***MEDICS***"), composed of all individual workstations and servers within the laboratory; and the external high-performance computer cluster *Colossus*, from the Consortium Laval UQAM McGill and Eastern Quebec [5] ("***CLUMEQ***"). Basic features for each cluster are shown in Table 3.

Table 3. *MEDICS* and *CLUMEQ* basic features

Description	***MEDICS***	***CLUMEQ***
Processors	32 (10 workstations (WS))	7664
Core types	Dual core 2.8 Ghz – 3.06 Ghz	Core Intel Nehalem 2.8 Ghz.
Memory RAM	2WS with 2GB, 6WS with 4GB and 2WS with 8 GB	1/3 node 12GB and 2/3 nodes 24GB
Jobs in execution or in queue	Limited by hardware	200
Operational systems	Ubuntu release 9.10	CentOS release 5.4
Distributed resource management	Sun Grid Engine 6.2	Sun Grid Engine

[5] http://www.clumeq.mcgill.ca/

4.3. Execution Time Analysis

For testing purposes we generated a single pipeline focused on spatial registration of image. The purpose of this pipeline was the computation of a nonlinear deformation grid exactly mapping one volume to another. At the core of this pipeline is the ***MINC Tracc*** process, which estimates the spatial transformation. Special attention is given to this process and this pipeline because it is one of the most expensive to run in terms of time and required resources.

Table 4 compares the execution times for different processes in the MEDICS and CLUMEQ clusters for a small dataset of 12 images. Table 5 compares the execution times for the whole pipeline on a set of 821 images taken from the ADNI database.

We tested every execution parameter on both clusters and observed that all specifications were maintained regardless of its nature [23, 24]. Regarding execution times, results show that *CLUMEQ* outperforms *MEDICS* in both experiments. This is unsurprising, given obvious hardware differences between the two clusters. First, while *MEDICS* normally executes one job for every workstation, we were able to assign eight jobs per node in *CLUMEQ*. Secondly, there existed an execution bottleneck in MEDICS within a single workstation with limited RAM (2GB), which severely limited its ability to handle large deformation volumes.

Interestingly however, we notice that the performance increase between the two clusters falls from a factor of 8.1 (small dataset) to 4.1 (large dataset). This diminution is due to the increase in queue waiting time at *CLUMEQ*, and the delays associated with overhead communications between our in-house EX server and the grid comptroller at *CLUMEQ*.

Table 4. Execution times for the **Nonlinear Registration** pipeline by cluster for a dataset of 12 images.

Process	**Execution time by Cluster**		**Speed-Up**
	MEDICS (1 job per node)	**CLUMEQ** (8 job per node)	
MINC Blur 8 mm[d]	11 sec.	6 sec.	14.66
MINC Blur 4 mm	15 sec	6 sec.	20
MINC Blur 2 mm	16 sec.	7 sec.	18.24
MINC Tracc 8 mm	4 min.	3 min.	10.64
MINC Tracc 4 mm	14 min. (8GB RAM)[a] 16 min. (4GB) 17 min. (2GB)	13 min.	4.89 10.38 11.08
MINC Tracc 2 mm	16.5 min. (8GB)[a] 18 min. (4GB) 6 hours 6 min. (2GB)	14 min. [b]	2.35 5.16 104.57
MINC Resample	4 min.	4 min.	8
MINC Blob	38 sec.	21 sec.	14.47
MINC Resample	1 min.	19 sec.	28.21

Total execution time	**7h: 15m: 50s**	**0h: 34m: 59s**
Global total processing time[c]	**8h: 55m: 59s**	**1h: 6m: 55s**

[a] Two jobs par node.
[b] With this option (2 mm) only 4 job per node.
[c] Time execution + Waiting time in queue + Transfer time images files.
[d] mm = Millimetres.

Table 5. Execution times for the **Nonlinear Registration** pipeline by cluster for a dataset of 821 images.

Cluster	Total execution time [a]
MEDICS	3d: 5h: 23m: 16s
CLUMEQ	0d: 19h: 0m: 58s

[a] d=days, h=hours, m=minutes and s=seconds.

5. Discussion

We report our work towards the conceptualization, design and implementation of a web application that allows: a) construction of pipelines dynamically from a set of processes; b) interaction with an alphanumeric database and file system; and c) execution of high performance computing on different clusters. The parallel programming and parameterization of options in the execution of processes greatly improves application performance, but increases the complexity of the application.

Grid computing is shown to reduce significantly the execution time of post-processing of brain magnetic resonance images, even though the effect is mitigated by queue waiting times and the necessity for individual process control. Quite possibly, for applications such as ours, there exists a niche for a dedicated micro-cluster (e.g. 200 cores total capacity) that would result in optimal throughput, midway between low-cost *ad hoc* assemblies of workstations and high-cost high-performance supercomputers.

Our application was built to answer the specific needs of our laboratory, but compares in principle to the cluster-based CBRAIN system (with LORIS database)(see the AddNeuroMed study for an applicative example [25]) or the LONI pipelining and execution system [26]. Both of the latter systems however were designed to be either exported or support a large community of users.

6. Acknowledgments

We thank the Ministère du Développement Économique, de l'Innovation et de l'Exportation du Québec and the National Science and Engineering Research Council for financial support towards the creation of the processing system and the MEDICS hardware architecture ; the Fonds de Recherche en Santé du Québec (Réseau de Bio-Imagerie du Québec) for funding towards the MEDICS database ; Dr. G. B. Frisoni from the IRCCS Fatebenefratelli (Brescia, Italy) for access to LENITEM and E-ADNI data ; the ADNI, OASIS, and ICBM consortia for similar reasons ; and finally CLUMEQ for access to its Université Laval-based infrastructure.

7. References

1. Canu, E., et al., *HOXA1 A218G polymorphism is associated with smaller cerebellar volume in healthy humans*. J Neuroimaging, 2009. **19**(4): p. 353-8.
2. Duchesne, S., et al., *Amnestic MCI future clinical status prediction using baseline MRI features*. Neurobiology of Aging, 2010. **31**(9): p. 1606-17.
3. Duchesne, S., et al., *MRI-Based Automated Computer Classification of Probable AD Versus Normal Controls*. IEEE Trans Med Imaging, 2008. **27**(4): p. 509-20.
4. Mouiha, A. and S. Duchesne, *Multi-decade hippocampal and amygdala volume analysis: Equal variability and limited age effect*. Neuroscience Letters, 2011. **499**(2): p. 93-8.
5. Mouiha, A. and S. Duchesne, *Hippocampal atrophy rates in Alzheimer's disease: automated segmentation variability analysis*. Neuroscience Letters, 2011. **495**(1): p. 6-10.
6. Mueller, S.G., et al., *Ways toward an early diagnosis in Alzheimer's disease: The Alzheimer's Disease Neuroimaging Initiative (ADNI)*. Alzheimers Dement, 2005. **1**(1): p. 55-66.
7. Frisoni, G.B., et al., *The pilot European Alzheimer's Disease Neuroimaging Initiative of the European Alzheimer's Disease Consortium*. Alzheimers Dement, 2008. **4**(4): p. 255-64.
8. Mazziotta, J.C., et al., *A probabilistic atlas of the human brain: theory and rationale for its development. The International Consortium for Brain Mapping (ICBM)*. Neuroimage, 1995. **2**(2): p. 89-101.
9. Marcus, D.S., et al., *Open Access Series of Imaging Studies (OASIS): Cross-sectional MRI Data in Young, Middle Aged, Nondemented, and Demented Older Adults*. J Cogn Neurosci, 2007. **19**(9): p. 1498-507.
10. Galluzzi, S., et al., *The Italian Brain Normative Archive of structural MR scans: norms for medial temporal atrophy and white matter lesions*. Aging Clin Exp Res, 2009. **21**(4-5): p. 266-76.
11. *DICOM reference guide*. Health Devices, 2001. **30**(1-2): p. 5-30.
12. H. Houmani, B. Crépeault, and S. Duchesne. *Ontological Support for Alzheimer's Disease Aid to Diagnostic*. in *Society for Imaging Informatics In Medicine*. 2009. Minneapolis, Minnesota.
13. Bowden, D.M., M. Dubach, and J. Park, *Creating neuroscience ontologies*. Methods Mol Biol, 2007. **401**: p. 67-87.
14. Bowden, D.M. and M.F. Dubach, *NeuroNames 2002*. Neuroinformatics, 2003. **1**(1): p. 43-59.
15. Hole, W.T. and S. Srinivasan, *Adding NeuroNames to the UMLS Metathesaurus*. Neuroinformatics, 2003. **1**(1): p. 61-3.
16. Temal, L., et al., *Towards an ontology for sharing medical images and regions of interest in neuroimaging*. J Biomed Inform, 2008.
17. Aubry, F. and A. Todd-Pokropek, *Mimos: a description framework for exchanging medical image processing results*. Stud Health Technol Inform, 2001. **84**(Pt 2): p. 891-5.
18. Ciccarese, P., et al., *The SWAN biomedical discourse ontology*. J Biomed Inform, 2008.
19. Clark, T. and J. Kinoshita, *Alzforum and SWAN: the present and future of scientific web communities*. Brief Bioinform, 2007. **8**(3): p. 163-71.
20. Ihalainen, T., O. Sipila, and S. Savolainen, *MRI quality control: six imagers studied using eleven unified image quality parameters*. Eur Radiol, 2004. **14**(10): p. 1859-65.
21. Jannin, P., et al., *Validation of medical image processing in image-guided therapy*. IEEE Trans Med Imaging, 2002. **21**(12): p. 1445-9.
22. Jannin, P., C. Grova, and C.R. Maurer, *Model for designing and reporting reference based validation procedures in medical image processing*. IJCARS, 2006. **1**(2): p. 1001-1115.
23. Crepeault, B., *ADNI+ICBM – ABC Pipeline Verification Procedure*, 2009, Centre de recherche Université Laval - Robert Giffard: Quebec City, Canada. p. 1-105.

24. Crepeault, B., *MEDICS Image processing system - Verification procedure*, 2009, Centre de recherche Université Laval - Robert Giffard: Quebec City, Canada. p. 1-40.
25. Simmons, A., et al., *The AddNeuroMed framework for multi-centre MRI assessment of Alzheimer's disease: experience from the first 24 months*. Int J Geriatr Psychiatry. **26**(1): p. 75-82.
26. Dinov, I.D., et al., *Efficient, Distributed and Interactive Neuroimaging Data Analysis Using the LONI Pipeline*. Frontiers in neuroinformatics, 2009. **3**: p. 22.

Journal of Physics: Conference Series **341** (2012) 012012 doi:10.1088/1742-6596/341/1/012012

Monte Carlo simulation on low-energy electrons from gold nanoparticle in radiotherapy

James C L Chow[1,3], Michael K K Leung[2], Sean Fahey[3], Devika B Chithrani[3] and David A Jaffray[1,2]

[1] Department of Radiation Oncology, University of Toronto and Radiation Medicine Program, Princess Margaret Hospital, University Health Network, Toronto, Ontario M5G 2M9, Canada; [2] Department of Medical Biophysics, University of Toronto, Toronto, Ontario, Canada; [3] Department of Physics, Ryerson University, Toronto, Ontario, Canada.

james.chow@rmp.uhn.on.ca

Abstract. This study investigated the low-energy electrons (LEEs) produced when a gold nanoparticle (GNP) is irradiated by photon beams. The secondary electrons emitted from a GNP (diameter = 100 nm), interacting with photon beams with energies equal to 35, 73.3 and 600 keV, were simulated using the Geant4 Monte Carlo code. The phase spaces of the secondary electrons were then used to simulate the LEEs in water using the NOREC Monte Carlo code. All secondary electrons emitted by the GNP, and all LEEs produced by each secondary electron were tracked in Monte Carlo simulations. It is found that the energy distributions of the LEEs from the GNP do not vary significantly between different photon beam energies. Moreover, the 660 keV photon beam produced more LEEs travelling to a longer range than photon beams of lower energies (35 and 73.3 keV). This higher energy deposition and longer range LEEs produced by the 660 keV photon beam can enhance the cell kill. Based on our Monte Carlo results, it is concluded that the unexpected close of the radiosensitization enhancement factors of the 35 (1.66) and 660 keV (1.18) photon beams from our previous measurements is because of the cell kill enhancement with the increased LEE yield and range in the 660 keV photon beam.

1. Introduction

It is generally believed that cancerous cell can be killed by ionizing photon beams that are energetic enough to detach electrons from atoms in the strand of the deoxyribonucleic acid (DNA). This may induce a damage of DNA strand called double-strand break (DSB), which separates the DNA into two pieces.[1,2] Since DNA is essential for the cell reproduction process, damaging the DNA means stopping the growth of cancerous cell or cell kill. On the other hand, electromagnetic radiations are ionizing when the photon energy is larger than 124 eV. Therefore, for the electron to be ionizing, it must have sufficient energy (i.e. > 124 eV), and must interact with atoms of the target (i.e. DNA strand).

There are studies suggesting that most energy deposited by the ionizing photon beams is transferred to a large number of low-energy electrons (LEEs) in the cell medium. These LEEs are in the energy range of 3 – 20 eV and produced by a cascade effect of photon beams in the medium.[3-5] These LEEs, which may have energy lower than the ionization thresholds, are found to yield substantial DNA damages such as single-strand break (SSB) and DSB.[3,4] This finding prompts us to think about the cancerous cell killing algorithm caused by the ionizing photon beams, and challenges the general understanding that cell kill only happens when the DNA interacts with secondary electrons with energy higher than the onset of ionization.

Gold nanoparticle-enhanced radiation therapy employs gold nanoparticle (GNP) uptaken by a tumour under a photon beam irradiation to enhance the biological effective dose or treatment outcome.[6-11] Hainfeld et al[6] used the 1.9 nm GNPs with concentration of gold up to 7 mg Au/g in tumours of mice for radiation therapies using the 250 kVp photon beam. They found that the one-year survival rate was 86% versus 20% with and without using GNP in the treatment. This great improvement of treatment outcome is because of the achievement of the high metal content in tumours necessary for the significant high atomic number (*Z*) radioenhancment. As more photoelectrons interact with DNA strands in the cancerous cell to produce DNA damage, the yield of cell kill is increased and hence a better treatment outcome can be achieved. The addition of GNP to the tumour increases the biological effective dose, which is the required dose reaching the cells where adverse effect occurs.[8,11]

When the photon beam energy is in the range of 100 keV – 10 MeV, Compton interaction is dominant to produce secondary electrons. Since the cross-section of Compton scattering does not depend on the *Z* as in the case of the photoelectric effect, adding GNP to the tumour should not benefit the biological effective dose for photon beams in the Compton energy range compared to photoelectric (energy < 50 keV). This can be seen from our previous works of Monte Carlo simulations,[12,13] where we found that for the 35 keV photon beam, adding GNP to water or tumour can increase the number of photoelectric interaction over 2900 times. While for the higher energy of 660 keV photon beam in which Compton interaction is dominant, adding GNP to water or tumour can only increase the number of Compton interaction by 6 times. Therefore, a combination of GNP addition and photon beam with low energy such as 35 keV should produce more secondary electrons compared to that of the higher photon beam energy of 660 keV. However, according to our measurements, the radiation sensitization enhancement factor (REF), defined as the ratio of dose (without GNPs)/dose (with GNPs) at 10% survival for HeLa cells,[11] for the 35 keV photon beam is 1.66, while that for the 660 keV photon beam is 1.18. The increase of the REF from decreasing the photon beam energy from 660 to 35 keV is just about 1.4 instead of 483 times higher based on the increased number of interactions in Monte Carlo simulations.[12,13] The disagreement between the Monte Carlo results and measurements shows that more work should be done in order to understand the relationship between the dosimetry/energy deposition of the GNP and cell kill due to the DNA damage.

One potential explanation of the above disagreement of REF is that the cell kill is heavily dependent on the LEE mentioned above. As the photon beam with higher energy (i.e. 660 keV) should be able to produce much larger number of LEE than the low energy, it results in the REF value of the 660 keV photon beam within the same order of magnitude compared to the 35 keV beam. In order to justify this point of view, Monte Carlo simulations were used to study the interactions of LEE produced by an irradiated GNP using photon beams with different energies. The aim of this study is to predict the yield and energy distribution of the LEE from the irradiated GNP to understand the relationship between the cell kill in terms of the REF and photon beam energy.

2. Methods

The LEEs generated by the secondary electrons from the GNP were simulated through two phases. Phase 1 involved the Monte Carlo simulation on the secondary electrons emitted by the irradiated GNP using photon beams with energies equal to 35, 73.3 and 660 keV. The Geant4 Monte Carlo code was used to determine the phase spaces of the secondary electrons.[14] Phase 2 was the Monte Carlo simulation on the LEEs based on the phase spaces of the secondary electrons from Phase 1. The NOREC Monte Carlo code was used to track all electrons produced by every secondary electron with energy down to the LEE range.[15,16] Simulations of the Geant4 and NOREC were performed on the General Purpose Cluster (GPC) supercomputer at the SciNet from the University of Toronto. A single core of the GPC was assigned to a simulation of specific photon beam energy. Therefore, three simulations (35, 73.3 and 660 keV) were run in parallel. A low statistic error can be achieved by using a large number of histories. In this work, two hundred and fifty million of histories were run using the Geant4 code followed by the NOREC.

2.1 Simulation geometry and setup

The simulation geometry is shown in Fig. 1. A spherical GNP with a diameter of 100 nm was centered inside a cubic tracking volume consisting of water. The NP was irradiated by photon beams with energies of 35, 73.3, and 660 keV. The cross-sectional diameter of the photon beam was taken to be 100 nm, and the axis of the beam was aligned with the center of the NP. With increasing beam energy, the linear dimension of the cubic tracking volume was increased from 0.3 to 1.5 mm to compensate for the longer electron path. Two hundred and fifty million photon histories were used at each beam energy. In Geant4, production thresholds are defined in terms of distance, which is then internally converted into an energy threshold below which no secondary particles are generated. A universal threshold of 100 nm is used for all particle types. For each electron emitted from the irradiated GNP, the total energy deposited, and the range were tracked in water. The simulation time ranged from 10 to 50 hours depending on the energy of the photon beam.

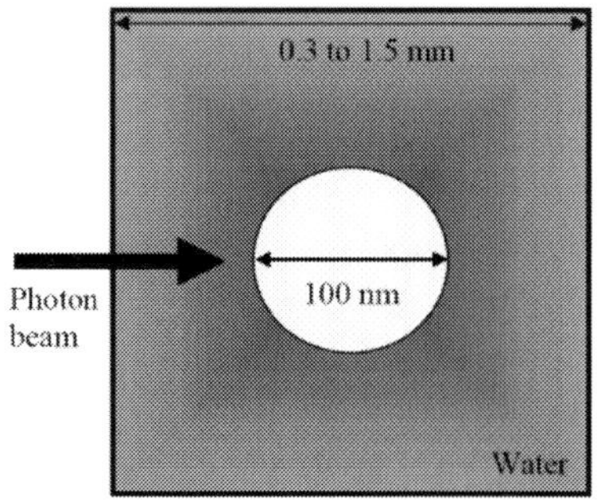

Figure 1. Schematic diagrams (not to scale) showing the simulation geometry of the GNP in a water cube irradiated by a photon beam.

2.2 Monte Carlo simulation using Geant4

Geant4 (GEometry ANd Tracking 4) is a Monte Carlo toolkit for the simulation of the transports and interactions of photons and electrons through matter. The algorithm of particle transport in Geant4 contains four levels: "Run", "Event", "Track" and "Step". The level "Run" defines the initialization and termination of a particle and the number of history. "Event" is the simulation of interaction in one history. "Track" is an intermediate level between "Event" and "Track", while "Step" defines the particle advancing one step in the medium.

The support for low-energy interactions is necessary for small target volumes (of the order of nanometers in this study), to ensure that the particle trajectories are accurately modelled. In our simulation we used the Penelope-based Geant4 low-energy library,[17,18] which has been verified to be in excellent agreement with the original PENELOPE code and is more suitable for microdosimetry applications.[19,20] Geant4 includes routines that allow for the precise definition of complex particle transport volumes and tools to visualize and analyze the simulation results. Release 9.0 was used to simulate the energy distributions of the secondary electrons emitted from a GNP irradiated by the photon beams.

2.3 Monte Carlo simulation using NOREC

Since Geant4 can only track electron down to energy of 250 eV, another specific Monte Carlo code, should be used to track energy spectral results of secondary electrons, generated by the Geant4 with energy down to about 10 eV. NOREC is a package containing a single-scattering Monte Carlo electron track-structure code for water.[16] NOREC can simulate electron histories in an energy range from 7.4 eV to 1 MeV and is based on the earlier Oak Ridge Electron Code (OREC).[21] The algorithm of NOREC employs the ELAST database, developed at the National Institute of Standards and Technology, for the differential cross sections of elastic scattering of electrons by atoms. Phase shifts are calculated by the Dirac equation using the WKB approximation. For tracking the LEEs with energy between 10 eV and 7.4 eV, a subroutine is called in the code to terminate the event-by-event transport and select a single additional step to bring the electron below the subexcitation energy threshold. The remaining energy is then recorded at the location of the last inelastic collision.

Although this code only works in a homogeneous medium of water, it does not affect calculations of the energy spectrum and yield of the LEEs because only the inhomogeneity of GNP has been considered in the Geant4. The idea is to obtain the secondary electron phase spaces using the Geant4 when the incident photon beam interacts with the GNP, and input the energy spectral result to NOREC to obtain the energy spectrum and yield of the LEEs. Modifications of the NOREC were done to allow the code to run more than one history. With both the Geant4 and NOREC, secondary electrons generated by the photon beams can be tracked with energy down to the level of the LEE.

3. Results and discussion

Figures 2(a), 2(b) and 2(c) show the energy spectra of the secondary electrons emitted from the GNP when irradiated by photon beams with energies equal to 35, 73.3 and 660 keV, respectively. The intensities of the energy distributions in Figs. 2 are normalized to the maximum intensities in the figures. Fig. 2(a) shows that the energies of the secondary electrons emitted from a GNP irradiated by a 35 keV photon beam are mainly in the range from 21 to 33 keV. The energy spread increases with increasing photon beam energy as shown in Figs. 2(b) and 2(c). In Fig. 2(c), the continuous energy distribution between 0 to 500 keV represents the Compton spectrum with the full energy peak at 580 keV. This shows that Compton interaction is dominant in the 660 keV photon beams, while photoelectric interaction is dominant in the 35 and 73.3 keV photon beams.

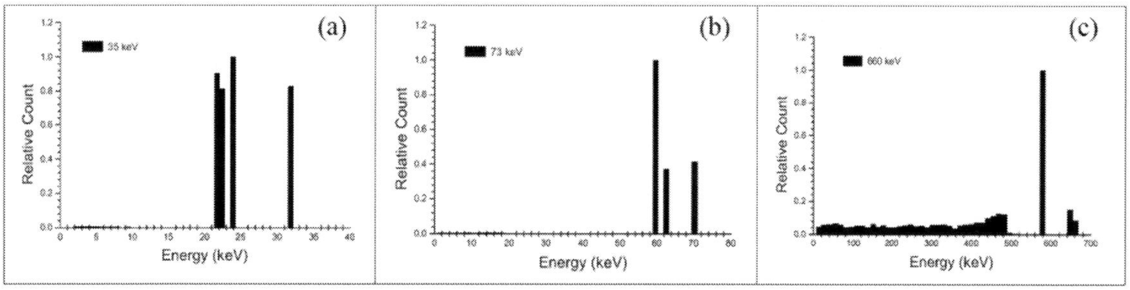

Figure 2. Energy spectra of the secondary electrons emitted by the GNP when irradiated by the (a) 35 keV, (b) 73.3 keV and (c) 660 keV photon beams. All intensities in the spectra are normalized to the maximum values in the figures.

Figures 3(a), 3(b) and 3(c) show the energy distributions of the LEEs produced by the secondary electrons with energy distributions in Figs. 2(a), 2(b) and 2(c), respectively. Figures 3(a), 3(b) and 3(c) represent the total energy distributions of the LEEs in water and all intensities of the energy distributions are normalized to their maximum values in the figures. It can be seen in Figs. 3(a), 3(b) and 3(c) that the energy distributions of the LEEs for the total interactions are very similar between different energies. This is further supported by the distributions of interaction type of the LEEs with different energies as shown in Fig. 4 that the distribution of the LEE interaction does not change significantly as the photon beam energy is changed. This shows that the dependence of the secondary electron spectrum from the GNP on the energy distribution of the LEEs is not significant.

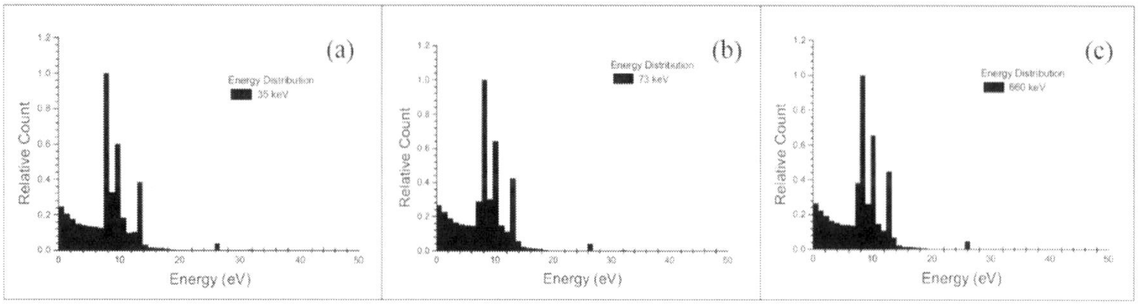

Figure 3. Total energy distribution of the LEEs from the irradiated GNP using the (a) 35, (b) 73.3 and (c) 660 keV photon beams.

Table 1 shows the summary of simulation data using the NOREC. Since the incident electrons deposit all energy within the simulation tracking volume, the total energy deposited is equal to the total energy of the incident electrons. The total number of interaction or event is highly dependent on the incident energies per number of history, and higher energy electrons will go through many more events to deposit all of their energy. With the same number of history, it can be seen that the 660 keV photon beam can deposit about 15 times more energy than the 35 keV beam from the generated LEEs. Table 2 shows the volume in which the interactions occur for each spectrum, and the maximum range of electrons in each spectrum. The volume is assumed to be rectangular, and is obtained by multiplying the difference of the maximum and minimum x, y and z values. At higher energies the electrons are much more penetrating and the interactions occur in a much larger volume. The energy deposited in all three photon beams in Table 1 is very close, hence suggesting that there is a greater dose enhancement when using lower energy photon beams, which agrees with previous studies.[6,8] However, since it is found in Table 2 that the LEEs produced by the 660 keV photon beams can spread more than 100 times further than the 35 keV, the increased energy deposition (Table 1) and spread from the LEEs should result in a larger cell kill. This may explain why the REF value of the 35 keV photon beam is not about 483 times larger than that of the 660 keV,[11] by considering only the increased number of photoelectron due to the photoelectric enhancement.

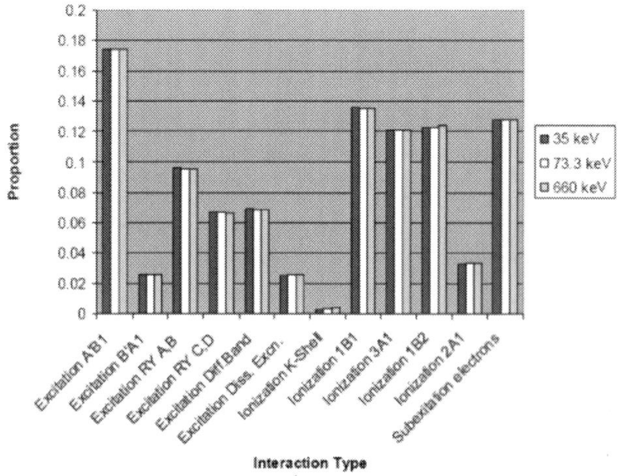

Figure 4. Distributions of the LEE interactions from the GNP irradiated by photon beams with energies equal to 35, 73.3 and 600 keV.

Table 1. Number of histories, events and total energy deposited for the LEEs from the GNP irradiated by the 35, 73.3 and 660 keV photon beams. It can be seen that the number of event and total deposited energy are nearly equal for 3000 absorbed or scatted 35 keV photons and for 200 absorbed or scatted 660 keV photons.

Energy	Number of history	Number of Event	Total energy deposited
35 keV	3000	9256589	7.334×10^7 eV
73.3 keV	1200	9217073	7.307×10^7 eV
660 keV	200	9271881	7.356×10^7 eV

Table 2. Sampling volumes and maximum ranges of the LEEs studied in Table 1.

Energy	Volume (μm^3)	Maximum range (μm)
35 keV	44303	22.064
73.3 keV	1.648×10^6	82.273
660 keV	1.823×10^{10}	2501.9

4. Conclusions

Monte Carlo simulations on the secondary electrons generated by the interaction between the GNP and photon beams, and LEEs produced by the secondary electrons from the GNP were carried out using the Geant4 and NOREC code. It is found that the energy distributions of the LEEs with different photon beam energies do not depend significantly on the energy spectra of the secondary electrons from the GNP. Moreover, increased deposition and spread of energy due to the LEEs were found for the 660 keV photon beam, when compared to the lower photon beam energies of 35 and 73.3 keV. It is concluded that higher photon beam energies of 660 keV can produce more LEEs with longer interaction range, and result in an enhancement of cell kill.

Acknowledgement

The authors would like to acknowledge the contribution from Professor Martin Lee, University of Toronto, who passed away on 21st April 2009. His enthusiastic support of this study is very much missed. The authors would also like to acknowledge the SciNet HPC Consortium in the University of Toronto for providing computing support (SciNet RAC Allocations for 2010).

References

[1] Obe G, Johannes C, and Schulte-Frohlinde D 1992 DNA double-strand breaks induced by sparsely ionizing radiation and endonucleases as critical lesions for cell death, chromosormal aberrations, mutations and oncogenic transformation. *Mutagenesis* **7** 3

[2] Sokolov M V, Smilenov L B, Hall E J, Panyutin I G, Bonner W M, Sedelnikova O A 2005 Ionizing radiation induces DNA double-strand breaks in bystander primary human fibroblasts *Oncogene* **24** 7257

[3] Bodia B, Cloutier P, Hunting D, Huels M A, and Sanche L 2000 Resonant formation of DNA strand breaks by low-energy (3 to 20 eV) *Sci.* **287** 1658

[4] Sanche L 2005 Low energy electron-driven damage in biomolecules *Eur. Phys. J. D* **35** 367

[5] Lu Q B, Kalantari S, and Wang C R 2007 Electron transfer reaction mechanism of cisplatin with DNA at the molecular level *Molecular Pharmaceutics* **4** 624

[6] Hainfeld J F, Slatkin D N, Smilowitz H M 2004 The uses of gold nanoparticles to enhance radiotherapy in mice *Phys. Med. Biol.* **49** N309

[7] Hainfeld J F, Dilmanian F A, Slatkin D N, and Smilowitz H M 2008 Radiotherapy enhancement with gold nanoparticles *J Pharm Pharmacol.* **60** 977

[8] Cho S H 2005 Estimation of tumour dose enhancement due to gold nanoparticles during typical radiation treatments: a preliminary Monte Carlo study *Phys. Med. Biol.* **50** N163

[9] Roa W, Zhang X, Guo L, Shaw A, Hu X, Xiong Y, Gulavita S, Patel S, Sun X, Chen J, Moore R, and Xing J Z 2009 Gold nanoparticle sensitize radiotherapy of prostate cancer cells by regulation of the cell cycle *Nanotechnology* **20** 375101

[10] Rahman W N, Bishara N, Ackerly T, He C F, Jackson P, Wong C, Davidson R, Geso M 2009 Enhancement of radiation effects by gold nanoparticles of superficial radiation therapy *Nanomedicine* **5** 136

[11] Chithrani D B, Jelveh S, Jalali F, van Prooijen M, Allen C, Bristow R G, Hill R P and Jaffray D A 2010 Gold nanoparticles as radiation sensitizers in cancer therapy *Rad. Res.* **173** 719

[12] Chow J C L, Leung M, Chithrani D, Lee M, Oms B, and Jaffray D 2009 Dosimetry on gold nanoparticle: a microscopic and macroscopic study using Monte Carlo simulations *Med. Phys.* **36** 2512

[13] Leung M, Chow J C L, Chithrani D, Lee M, Oms B and Jaffray D 2009 Characterization of the spaial and energy distribution of electrons emitted from a gold nanoparticle irradiated by x-rays using Monte Carlo simulations *Med. Phys.* **36** 2819

[14] Agostinelli S et al 2003 GEANT4 – A simulation toolkit *Phys. Res. A* **506** 250

[15] Semenenko V A, and Steward R D 2004 A fast Monte Carlo algorithm to simulate the spectrum of DNA damages formed by ionizing radiation *Rad. Res.* **161** 451

[16] Semenenko V A, Turner J E, and Borak T B 2003 NOREC, a Monte Carlo code for simulating electrons tracks in liquid water *Radiat. Environ. BioPhys.* **42** 213

[17] Spiga J, Siegbahn E A, Brauer-Krisch E, Randaccio P, and Bravin A 2007 The GEANT4 toolkit for microdosimetry calculations: Application to microbeam radiation therapy (MRT) *Med. Phys.* **34** 4322

[18] Stewart R D, Wilson W E, McDonald J C, and Strom D J 2002 Microdosimetric properties of ionizing electrons in water: A test of the PENELOPE code system *Phys. Med. Biol.* **47** 79

[19] Jones B L, Krishnan S, Cho S H 2010 Estimation of microscopic dose enhancement factor around gold nanoparticles by Monte Carlo calculations *Med. Phys.* **37** 3809

[20] Mainardi E, Donahue R J, Wilson W E, and Blakely E A 2004 Comparison of microdosimetric simulations using PENELOPE and PITS for a 25 keV electron microbeam in water *Rad. Res.* **162** 326

[21] Turner J E, Hamm R N, Souleyrette M L, Martz D E, Rhea T A, and Schmidt D W 1988 Calculations for beta dosimetry using Monte Carlo code (OREC) for electron transport in water *Health Phys.* **55** 741

Journal of Physics: Conference Series **341** (2012) 012013 doi:10.1088/1742-6596/341/1/012013

Comparison of the computational NMR chemical shifts of choline with the experimental data

C. Alcorn[1], M. Cuperlovic-Culf[1,2] and K. Ghandi[1]*

1. Department of Chemistry and Biochemistry, Mount Allison University
2. Institute for Information Technology, National Research Council of Canada

kghandi@mta.ca; Miroslava.Cuperlovic-Culf@nrc-cnrc.gc.ca

Abstract. One of the main biological markers of the presence of cancer in living patients is an over-expression of total choline (tCho), which is the sum of free choline and its derivatives. ^{1}H Magnetic Resonance Spectroscopy, or H-MRS, enables the quantification of tCho via its proton spectra, and thus has the potential to be a diagnostic tool for the presence of cancer and an accurate early indicator of the response of cancer to treatment. However, it remains difficult to quantify individual choline derivatives, since they share a large structural similarity ($(CH_3)_3$-N^+-CH_2-CH_2-O-), of which the strongest signal detectable by MRS is that of the choline "head group": the three methyl groups bonded to the nitrogen. This work used ACENet, a high performance computing system, to attempt to model the NMR parameters of choline derivatives, with the focus of this report being free choline. Optimized structures were determined using Density Functional Theory and the B3LYP electron correlation functional. The Polarizable Continuum Model was used to evaluate solvent effects. The Gauge-Invariant Atomic Orbital method was found to be the superior method for calculating the NMR parameters of cholines.

1. Introduction

A wide variety of cancers radically alter the metabolism of free choline (Cho) and its derivatives in patients [1-5]. The increase in total choline (tCho) concentration observed in tumours is universal for many types of cancers, and in fact the quantity of tCho present *in vivo* is used as a diagnostic biomarker of the presence and aggressiveness of cancer [1-2]. Quantification of tCho is also useful in differentiating between malignant and benign tumours [4], the effectiveness of cancer treatments, and the targeting of radiation therapy [2].

While tCho is a useful biomarker of cancer, knowledge of the specific choline derivative concentrations present in tumours would give more information about the altered metabolic pathways, and thus clues on tumour sub-types, prognosis and treatment planning. For example, in the case of breast and ovarian cancer cells, it has been shown that phosphocholine (PCho) concentration is 10-fold higher than that of glycerophosphocholine (GPCho) [1-2,4]. In fact, the ratio of PCho to GPCho is altered in most types of cancers [1]. Change in the PCho/GPCho ratio can be correlated with increased activity of choline kinase. Introduction of choline kinase inhibitors have been shown to exhibit anti-cancer activity [4].

Although *in vitro* high field ^{1}H and ^{31}P NMR spectroscopy can be used for measurement of concentrations of individual choline derivatives, it remains difficult to determine their concentrations *in vivo*. Cholines are detected in a clinical setting by Magnetic Resonance Spectroscopy (MRS) techniques [1-3,5-7]. The main proton peak of interest with respect to total choline detection are the nine chemically equivalent protons found on the three methyl groups bonded to the quaternary nitrogen (see Figures 1–3).

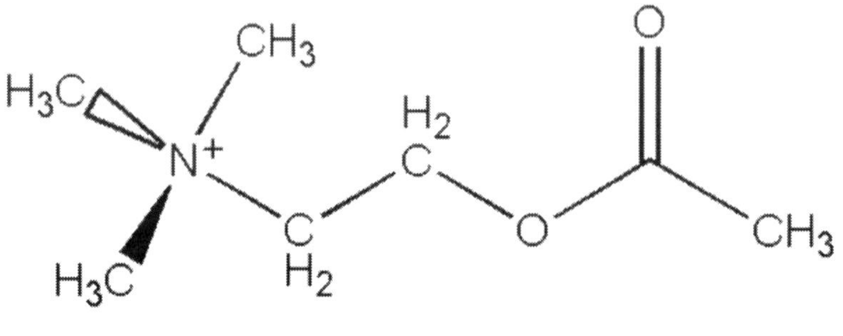

Figure 1. Chemical structure of acetylcholine

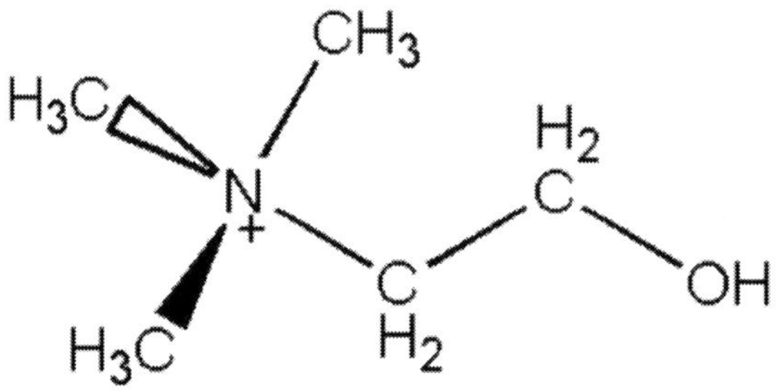

Figure 2. Chemical structure of choline

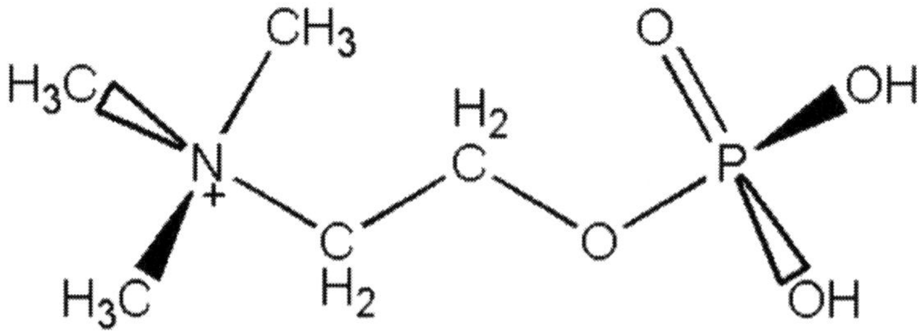

Figure 3. Chemical structure of phosphocholine

All choline derivatives share this structural similarity, making the quantification of specific choline derivatives difficult due to the resolution of current clinical MRS instruments, which stems from technical difficulties for *in vivo* detection such as low field regulations, magnetic field inhomogeneities, sample size, and a sample embedded *in vivo*, as well as inability to run experiments for extended periods of time [6]. Thus, knowledge of subtle differences in the NMR spectra of choline derivatives may help us in developing a model spectrum for a mixed sample of choline derivatives, for direct comparison to an MRS spectrum. This model can then be utilized to decipher the individual choline derivative concentrations present in a mixed sample.

The only structural differences in the three choline derivatives of interest to us (acetylcholine (ACho), Cho, and PCho) are the groups bonded to the oxygen in the CH_2O group, as shown in Figures 1–3 respectively. These differences will affect the NMR parameters in two ways. First, these groups will donate or withdraw unequal amounts of electron density from the remainder of the molecule by simple electronegativity, and secondly, these groups provide several sites for possible hydrogen bonding, which in turn would cause polarization of the electron density by the solvent.

Computational chemistry is a very effective tool in calculating the NMR parameters of molecules [8-10], such as complicated organic molecules [11-12] and molecules of biological importance [13-15]. Accurate geometries are paramount when using computational methods to calculate NMR parameters. In this work we calculated the geometry of Cho in both the gas and liquid phase, and compared our results to experimental crystal structures. We then calculated the NMR parameters and compared our results with the experimental data to gauge the accuracy of NMR calculations. From calculation, it is possible to obtain chemical shifts, coupling constants as well as relaxation parameters; however this work deals only with the chemical shifts.

Computational chemistry is a very effective tool in calculating NMR parameters of ionic molecules [8-10]. In order to use computational methods to calculate NMR parameters, we must have accurate geometries. In this work we calculated the geometry of Cho in the gas and liquid phase, and compared our method to experimental crystal structures. We then calculated the NMR parameters and compared our results with the experimental data to gauge the accuracy of NMR calculations. From calculations it is possible to obtain chemical shifts, coupling constants as well as relaxation parameters; however this work deals only with the chemical shifts.

2. Computational methods

As mentioned above, the first type of calculation performed in this work was gas phase quantum calculations. Moving from the gas to the liquid phase, the simplest method for accounting for solvent effects is through what is known as continuum solvent models. The PCM model functions by creating a cavity in the solvent for the solute, and surrounds the cavity with a continuous medium with dielectric constant ε_r. PCM defines the shape of the cavity as the overlap of several atom centered spheres, whose radii are 1.2 times the van-der-Wall radius of the atom it is centered on [16]. The electrostatic potential and surface point charge, $\phi(\vec{r})$ and Q_k, are initially unknown; they are solved by iteration [17]. Iterations are performed until the electron density is stabilized to within a certain threshold value.

In this work, the Integral Equation Formalism variant of the Polarizable Continuum Model (IEF-PCM) [18] was used exclusively, with water as the solvent medium, having a dielectric constant of ε_r = 78.3553.

It must be mentioned that solvent effects in this study were limited to liquid water to simulate solvation in the cytoplasm of blood-free cells, though the types of environments found *in vivo* are much more complicated than simple water. Other dissolved constituents of the cytoplasm are not considered at this time, nor is the binding of cholines to lipids or cell membranes.

All calculations in this work were performed using the Gaussian 03 package [19]. Structures were optimized using Density Functional Theory (DFT) [20] with the B3LYP correlation function [21-22], at a large Pople's basis set [23] of 6-311++G(2d,2p). DFT [20] defines all of the electrons as a density function with the contributions of the total energy as a sum of kinetic energy (E_T), electron-nuclear attraction (E_{en}), and the electron-electron repulsion which is a sum of the Coulomb (E_C) integral, and an exchange/correlation energy (E_{XC}), [24].

Numerous exchange/correlation energies exist. One of the successful ones, used in this work, is the hybrid B3LYP [21-22] functional derived by Lee, Yang and Parr, which is a linear combination of exchange energies found by other methods.

Another exchange/correlation relation, the PBE0 functional [25], was also used in this work. PBE0 is a variant of B3LYP, which only differs from B3LYP by weighting the exchange and correlation energies by 25% and 75% respectively.

The geometry of Cho was optimized in both the gas phase and by using the IEF-PCM model, and was compared with experimental crystal structures [26] for structural confirmation. From the optimized structures, calculations of the nuclear shielding were done using both the Gauge-Invariant Atomic Orbital (GIAO) [27-28] and Continuous Set of Gauge Transformations (CSGT) methods [29], at the same method and basis set.

A full description of the GIAO and CSGT methods for the calculation of NMR parameters is beyond the scope of this paper. In short, the GIAO method is derived by considering the application of a magnetic field as an energetic perturbation to the original Self-Consistent Field (SCF) equations of the Hartree-Fock (or DFT) method [18]; the first term of the Taylor series expansion of these equations with respect to the magnetic field describes the magnetic shielding tensor $\boldsymbol{\sigma}_B$ [27]. However, the GIAO method introduces a gauge-dependence by multiplying the wavefunctions ϕ_s by a gauge-dependent term; the CSGT method avoids this problem by calculating the current density induced by the perturbing magnetic field for all points, and taking those points as gauge origin [29].

3. Results and discussion

The computational work was done stepwise in the following order. 1) First, structures were optimized to an energy minimum (with and without IEF-PCM for solvent effects). 2) A frequency calculation was performed on the optimized structure for confirmation of a local energy minimum configuration. 3) Calculations of the nuclear shielding were done using both the GIAO and CSGT methods.

The structures of the gas phase and IEF-PCM optimized choline cation is compared to that of the choline chloride crystal structure found by Senko and Templeton [26], with bond lengths, angles and dihedral angles reported in Table 1. Atom labels are referenced from the Cho geometry shown in Figure 4.

Only minor differences are seen in bond lengths for both gas and liquid phase optimized geometries. However, the gas phase geometry reports significant differences in bond angles, and a *very* large difference in the dihedral angle of the gas phase. For the IEF-PCM optimized geometry, the only significant variation from the crystal structure is observed in one of the dihedral angles. However, since we are comparing the solid crystal structure, which has paired cations/anions and very closely packed, to a solvated cation alone, an 18° difference in the dihedral angle is reasonable. The dihedral angle itself is only a "twist" in the backbone of the molecule, which could be expected to be required to form a rigid solid. However, the 84° difference reported in our gas phase calculations is much too large and is a reflection of the fact that the positive charge on the nitrogen will attract the electronegative oxygen atom, which is easy to achieve through torsion. This attraction seems to be adequately dampened by the solvent field.

Table 1. Comparison of the experimental crystal structure of Cho [26] to our calculated geometries

Parameter	Crystal Structure of Cho [26]	Gas Phase	IEF-PCM
$r(N_1\text{-}C_5)$	1.52	1.51	1.50
$r(N_1\text{-}C_9)$	1.51	1.51	1.50
$r(N_1\text{-}C_{13})$	1.54	1.51	1.51
$r(N_1\text{-}C_2)$	1.60	1.52	1.53
$r(C_2\text{-}C_{17})$	1.56	1.53	1.51
$r(C_{17}\text{-}O_{20})$	1.45	1.41	1.42
$\llcorner(O_{20}\text{-}C_{17}\text{-}C_2)$	112	103	111
$\llcorner(C_{17}\text{-}C_2\text{-}N_1)$	111	117	117
$\llcorner(C_5\text{-}N_1\text{-}C_2\text{-}C_{17})$	172	180	168
$\llcorner(O_{20}\text{-}C_{17}\text{-}C_2\text{-}N_1)$	84	0	66

Bond lengths shown in angstroms, angles in degrees.

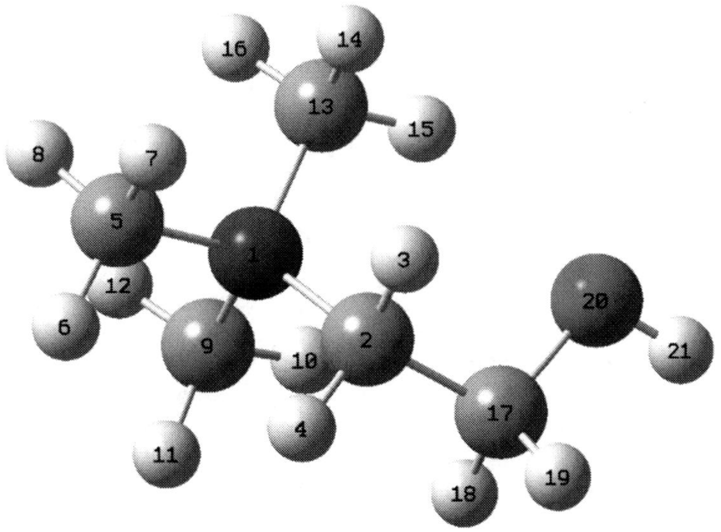

Figure 4. The optimized structure of Cho (with IEF-PCM, using B3LYP/6-311++G(2d,2p)) with atom labels for reference to the geometry in Table 1.

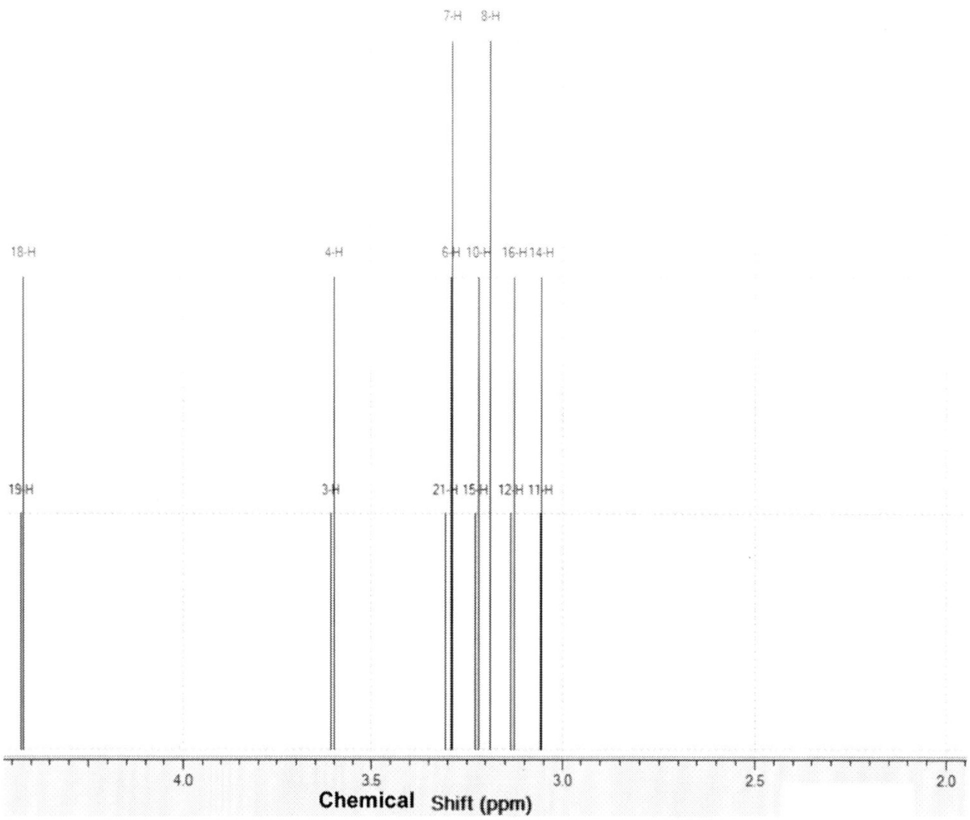

Figure 5. Gaussian NMR spectrum of the choline cation. Label assignments are in Figure 4.

Considering that our IEF-PCM method gives an acceptable geometry, we feel confident to use it to calculate the NMR parameters for free choline. However, the NMR parameters were also calculated from our gas phase geometry for a comparison of the difference between calculated NMR parameters method by method.

Calculated chemical shifts are referenced to a separate calculation of TMS, at the same method and basis set. The rotational averaging of chemically equivalent protons is not represented in these types of calculations, since a geometry optimization leads to a rigid structure. Thus, a GIAO NMR output can have a different chemical shift for chemically equivalent protons, as can be seen in Figure 5. For this reason, the values of chemically equivalent protons reported in this work have been averaged.

Table 2 lists the experimental ^{1}H NMR signals of Cho, obtained from the spectrum of choline chloride from the Human Metabolome Data Base (HMDB) [30]. The spectrum is shown in Figure 6. The calculated NMR chemical shifts of gas phase optimized Cho are listed in Table 3, which are compared to experimental results. As can be seen in Table 3, the chemical shifts calculated by the GIAO method have a lower mean absolute deviation than those calculated by the CSGT method for both the PBE0 and B3LYP correlation functions. For this reason, only the GIAO method is considered further in this work. Lastly, when incorporating the IEF-PCM model, we can see from Table 3 that the B3LYP functional gives the most accurate chemical shifts, though the mean absolute deviation is in very close in agreement to that of the PBE0 functional.

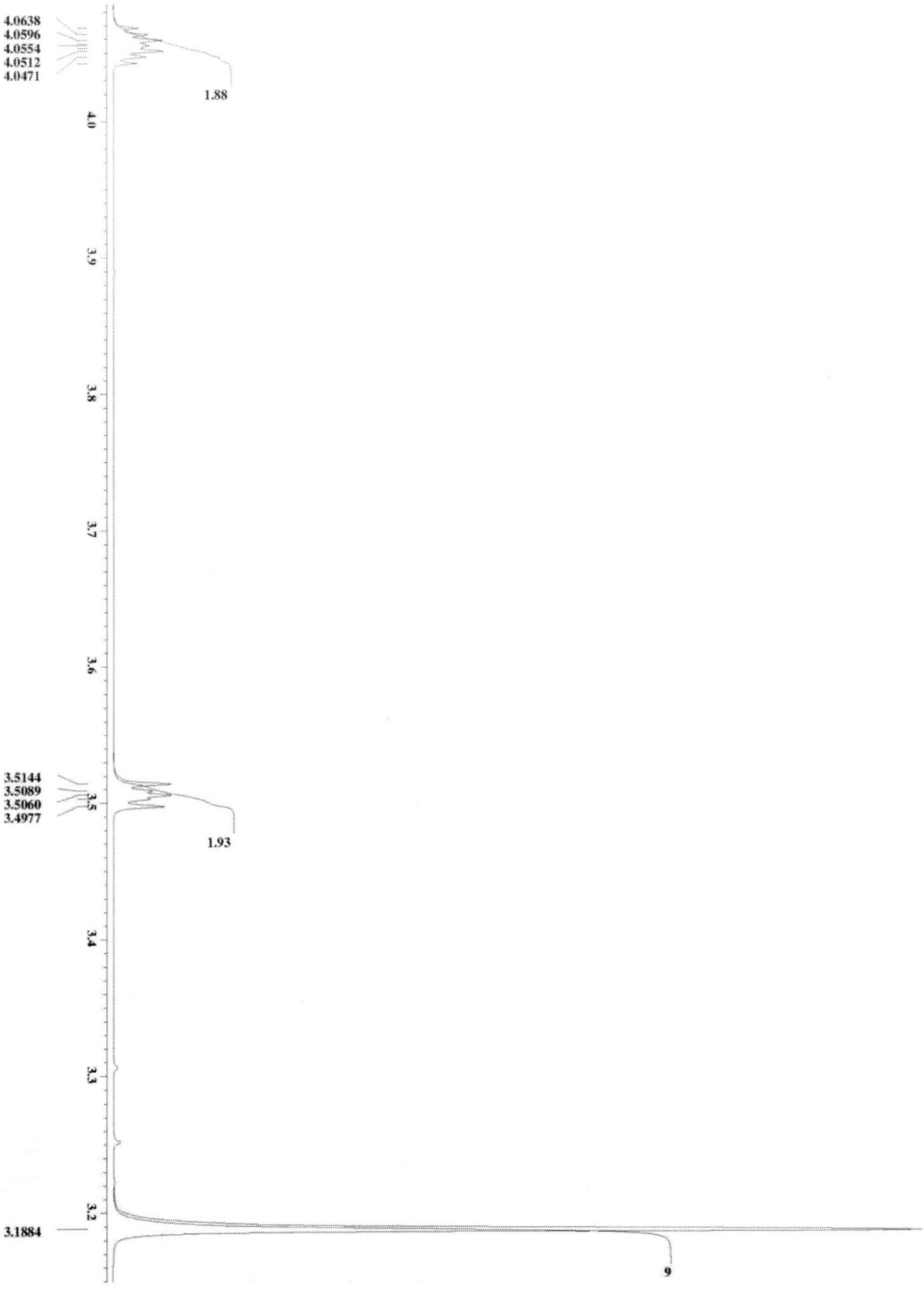

Figure 6. ^{1}H NMR spectrum of 115 mM Cho [11]. See Table 2 for assignment of peaks.

Table 2. Experimental chemical shifts of Cho in aqueous solution

^1H Signal	Cho (ppm)
$(CH_3)_3N$	3.19 (<u>9</u>)
NCH_2	3.51 (1.93)
CH_2O	4.06 (1.88)

Relative areas of integration are shown in parenthesis, with normalized values underlined.

Table 3. Calculated chemical shifts of gas phase optimized Cho by method

Proton Group	Gas Phase				IEF-PCM		Exp. [30]
	CSGT, PBO	CSGT, B3LYP	GIAO, PB0	GIAO, B3LYP	GIAO, PB0	GIAO, B3LYP	
NCH_2	3.57	3.59	3.37	3.40	3.45	3.51	3.51
CH_2O	4.44	4.48	4.21	4.26	4.26	4.33	4.06
$N(CH_3)_3$	3.22	3.22	3.04	3.05	3.06	3.09	3.19
O-H	1.12	1.28	1.10	1.26	2.34	3.21	N/A
Mean Absolute Deviation	0.163	0.184	0.145	0.152	0.130	0.126	-

The chemical shifts listed in Table 3 are calculated from our gas phase geometry, which could be less accurate than our geometry optimized with IEF-PCM. However, the values listed in Table 3 are helpful in gauging the mean absolute deviation in chemical shifts from method to method. The largest difference in mean absolute deviation reported in Table 3 is 0.06 ppm, which is on par with the absolute errors between calculated and experimental chemical shifts. Thus, each method is reasonably accurate. However, continuing our analysis using all the methods reported in Table 3 would be computationally expensive, and thus we rely on the GIAO method, using both IEF-PCM and the B3LYP functional.

Table 4 lists the IEF-PCM/GIAO/B3LYP calculated chemical shifts of the IEF-PCM optimized structure of Cho listed in Table 2. As can be seen in Table 4, the mean absolute deviation is higher than the lowest value in Table 3. However, the mean absolute deviation is increased mainly by the larger error on the CH_2O peak, which is also overestimated by all methods reported in Table 2. Furthermore, the remarkable accuracy of the calculated chemical shift of the $(CH_3)_3N$ peak, which is the main peak of interest in H-MRS, is quite promising.

As mentioned in our introduction, explicit hydrogen bonding can also alter the chemical shifts of molecules in solution. This type of solvation has been neglected in this work. Thus, the future work in this area remains to see if the values reported in Table 4 can be further refined by the addition of explicit water molecules to the optimized structures of cholines.

Table 4. Calculated chemical shifts of IEF-PCM optimized Cho by IEF-PCM/GIAO/B3LYP

Signal	Cho (ppm)
NCH_2	3.56 (3.51)
CH_2O	4.45 (4.06)
$N(CH_3)_3$	3.19 (3.19)
OH	3.18 (N/A)
Mean Absolute Deviation	0.153

Experimental chemical shifts shown in parenthesis.

4. Conclusion

Optimized structures using DFT, with both the B3LYP functional and IEF-PCM for evaluation of solvent effects gives a reasonable geometry compared to the crystal structure of choline. The small differences between our computational results and the crystal structure (mostly the dihedral angles) are most probably due to the very different nature of the crystalline and aqueous environments. The GIAO method was shown to lead to reasonable agreement with the experimental chemical shifts when combined with IEF-PCM and the B3LYP functional. Slight differences in chemical shifts could be due to explicit solvation effects (hydrogen bonding). Our method is a first step towards future calculation of NMR parameters and line shapes for choline and choline derivatives under different conditions present in vivo.

5. Acknowledgements

Computational work was performed on the Atlantic Computational Excellence Network (ACENet) within the consortium of Compute Canada, who has our thanks for their support. This research was financially supported by the Natural Sciences and Engineering Research Council of Canada and the National Research Council Institute for Information Technology (NRC-IIT). Their support is gratefully acknowledged. Special thanks go to Joey Bernard from ACENet for extensive technical support in the area of parallel computing.

References

[1] Glunde K and Bhujwalla Z 2011 *Semin. Oncol.* **38** 26-41
[2] Glunde K, Jacobs M and Bhujwalla Z 2006 *Expert Rev. Mol. Diagn.* **6**, 821-29
[3] Morita N, Harada M, Otsuka H, Melhem E and Nishitani, H 2010 *Magn. Reson. Med. Sci.* **9**, 167-75
[4] Katz-Brull R, Seger D, Rivenson-Segal D, Rushkin E and Degani H 2002 *Cancer Research* **62**, 1966-70
[5] Haddadin I, McIntosh A, Meisamy S, Corum C, Styczynski-Snyder A, Powell N, Nelson M, Yee D, Garwood M and Bolan, P 2009 *NMR Biomed.* **22**, 65-76
[6] Soares D and Law M 2009 *Clin. Radiol.* **64**, 12-21
[7] Ramírez de Molina A, Rodríguez-González A, Gutiérrez R, Martínez-Piñeiro L, Sánchez J, Bonilla F, Rosell R and Lacal J 2002 *Biochem. Biophys. Res. Commun.* **296**, 580-83
[8] Kaupp M, Buhl M and Malkin V 2004 *Calculation of NMR and EPR Parameters: Theory and Applications*, Wiley-VCH, Verlag GmbH & Co. KGaA, Weinheim, Germany
[9] Helgaker T, Jaszunski M and Ruud K 1999 *Chem. Rev.* 99 293-352
[10] Bagno A and Saielli G 2007 *Theor. Chem. Acc.* 117, 603-19
[11] Dwan J, Durant D and Ghandi K 2008 *Cent. Eur. J. Chem.* 6(3), 3447-58
[12] Rychnovsky S 2006 *Org. Lett.* 8, 2895-8.

[13] Clore M and Schwieters C 2002 *Curr. Opin. Struc. Biol.* 12, 146-53.
[14] Gervais, C, Dupree, R, Pike, K, Bonhomme, C, Profeta, M, Pickard, C and Mauri, F 2005 *J. Phys. Chem. A*, 109, 6960-9
[15] Allouche A, Aubert-Frecon Mand Raveron-Demilly D 2007 *Phys. Chem. Chem. Phys.* **9**, 3098-103
[15] Allouche A, Aubert-Frecon Mand Raveron-Demilly D 2007 *Phys. Chem. Chem. Phys.* **9**, 3098-103
[16] Tomasi J and Persico M 1994 *Chem. Rev.* **94**, 2027-94
[17] Levine I 2008 *Quantum Chemistry 6th Ed.*, Prentice-Hall, Englewood Cliffs, NJ
[18] Tomasi J, Mennucci B and Cammi R 2005 *Chem. Rev.* **105**, 2999-3093
[19] Gaussian 03, Revision C.02, Frisch M, Trucks G, Schlegel H, Scuseria G, Robb M, Cheeseman J, Scalmani G, Barone V, Mennucci B, Petersson G, Nakatsuji H, Caricato M, Li X, Hratchian H, Izmaylov A, Bloino J, Zheng G, Sonnenberg J, Hada M, Ehara M, Toyota K, Fukuda R, Hasegawa J, Ishida M, Nakajima T, Honda Y, Kitao O, Nakai H, Vreven T, Montgomery, Jr J, Peralta J, Ogliarom F, Bearpark M, Heyd J, Brothers E, Kudin, K, Staroverov V, Kobayashi R, Normand J, Raghavachari K, Rendell A, Burant J, Iyengar S, Tomasi J, Cossi M, Rega N, Millam N, Klene M, Knox J, Cross J, Bakken V, Adamo C, Jaramillo J, Gomperts R, Stratmann R, Yazyev O, Austin A, Cammi R, Pomelli C, Ochterski J, Martin R, Morokuma K, Zakrzewski V, Voth G, Salvador P, Dannenberg J, Dapprich S, Daniels A, Farkas Ö, Foresman J, Ortiz J, Cioslowski J and Fox D 2003 *Gaussian Inc.*, Wallingford, Connecticut
[20] Parr R and Yang W 1989 *Density Functional Theory of Atoms and Molecules*, Oxford University Press, Oxford, United Kingdom
[21] Becke A 1993 *J. Chem. Phys.* **98**, 5648-52
[22] Lee C, Yang W and Parr R 1988 *Phys. Rev. B* **37**, 785-89
[23] Foresman J and Frisch A 1996 *Exploring Chemistry with Electronic Structure Methods, 2nd Ed.* Gaussian Inc, Pittsburgh, PA
[24] Engel T and Reid P 2006 *Physical Chemistry 2nd Ed.*, Pearson Education, San Francisco, CA
[25] Adamo C and Barone V 1999 *J. Chem. Phys.* **110**, 6158-69
[26] Senko M and Templeton D 1960 *Acta Cryst.* **13**, 281-85
[27] Ditchfield R 1974 *Mol. Phys.* **27**, 789-807
[28] Schreckenbach G and Ziegler T 1995 *J. Phys. Chem.* **99**, 606-11
[29] Keith T and Bader R 1993 *Chem. Rev. Lett.* **210**, 223-31
[30] Wishart D, Tzur D, Knox C, Eisner R, Chi Gio A, Young N, Cheng D, Jewell K, Arndt D, Sawhney S, Fung C, Nikolai L, Lewis M, Coutouly M.-A, Forsythe I, Tang P, Shrivastava S, Jeroncic K, Stothard P, Amegbey G, Block D, Hau D, Wagner J, Miniaci J, Clements M, Gebremedhin M, Guo N, Zhang Y, Duggan G, MacInnis G, Meljie A, Dawlatabadi R, Bamforth F, Clive D, Greiner R, Li L, Marrie T, Sykes B, Vogel H and Querengesser L 2007 *Nucleic Acids Res.* **35**, D521-6

Journal of Physics: Conference Series **341** (2012) 012014 doi:10.1088/1742-6596/341/1/012014

Computational Poromechanics of Human Knee Joint

Mojtaba Kazemi and **LePing Li**
Department of Mechanical and Manufacturing Engineering, University of Calgary, Calgary, Alberta, Canada
E-mail: LePing.Li@ucalgary.ca

Abstract. Extensive computer modeling has been performed in the recent decade to investigate the mechanical response of the healthy and repaired knee joints. Articular cartilages and menisci have been commonly modeled as single-phase elastic materials in the previous 3D simulations. A comprehensive study considering the interplay of the collagen fibers and fluid pressurization in the tissues in situ remains challenging. We have developed a 3D model of the human knee accounting for the mechanical function of collagen fibers and fluid flow in the cartilages and menisci. An anatomically accurate structure of the human knee was used for this purpose including bones, articular cartilages, menisci and ligaments. The fluid pressurization in the femoral cartilage and menisci under combined creep loading was investigated. Numerical results showed that fluid flow and pressure in the tissues played an important role in the mechanical response of the knee joint. The load transfer in the joint was clearly seen when the fluid pressure was considered.

1. Introduction

An essential mechanism in the load support of the knee joint is the multiple mechanical contacts between the hydrated cartilaginous tissues, which are articular cartilages and menisci. These tissues can be highly pressurized with the interstitial fluid under knee compression and therefore the load response of the knee is influenced by the fluid flow consequently produced by the gradients of the fluid pressure. In the normal joint, articular cartilages provide smooth surfaces for the relative movements of the bones and minimize stress concentrations in the knee. The menisci are located between the tibial and femoral cartilages and minimize the direct contact between the cartilages [8]. Depending on the loading conditions, some joint load is supported by the four ligaments, i.e. anterior cruciate, posterior cruciate, medial collateral and lateral collateral ligaments. These ligaments also provide the joint stability by restricting its motion in the corresponding directions.

Articular cartilages and menisci are often considered as biphasic materials. The solid phase is mainly composed of proteoglycan matrix and the collagen network. The fluid phase is the most prevalent component of the cartilaginous tissues [9]. These knee tissues exhibit viscoelastic behavior, e.g. creep and stress relaxation, under external loadings. The viscoelasticity is mainly due to the effect of interstitial fluid flow and intrinsic viscoelasticity of the collagen fibers.

Computational models of the knee joint and its individual tissues (articular cartilages, menisci and ligaments) have been extensively used in the recent decades to investigate the load response of the knee in different loading or healthy conditions. The possible medical applications of these models include understanding the mechanics of cartilage damage, osteoarthritis, meniscectomy, ligament reconstruction, joint abnormality and alignment [1-3]. Invaluable information on the mechanical functions of the knee joint can be obtained from such studies. The past computational studies used either simplified geometrical models with fairly complex constitutive relationships [4,5] or real three-dimensional (3D) knee geometries with simple elastic solid constitutive behavior only [6,7]. Biphasic models have been used extensively to study the viscoelasticity of the cartilages and menisci using two-dimensional (2D) and axisymmetric geometries, or experimental explants geometries [4]. However, cartilages and menisci have been modeled as single-phase materials when anatomically accurate 3D geometries of the knee joint were used in the finite element modeling [3,7].

A 3D model of the knee joint with fluid pressurization in the soft tissues remains challenging because of the multiple mechanical contacts involving viscoelastic response. Recently, we have proposed a 3D model that includes fluid pressurization in the cartilages and menisci, as well as proper fiber orientations in these tissues [10,11]. In the present study, our recent 3D finite element model was used to investigate knee mechanics under combined creep loading. The model includes femur, tibia and fibula as bony structures, cartilages and menisci as fluid-saturated tissues and the four major ligaments as elastic tensile elements.

2. Methods

The geometry of the knee joint was previously constructed using MRI data of a male's right knee. The subject had no previous records of knee injury or surgical operations. The finite element model was generated using ABAQUS software v6.8-2 (Simulia Inc., Providence, RI, USA). The bony structures of the knee (femur, tibia, and fibula) were considered as rigid. This assumption was based on the fact that bones are much stiffer as compared to the soft tissues. In order to account for the effect of collagen fibers in cartilages, menisci and ligaments, these tissues were modeled as fibril-reinforced materials. Cartilages and menisci were modeled using porous elements to account for fluid pressurization in these tissues. Solid elastic elements were used for the ligaments. The same finite element meshes that were tested previously were used in the present study [10].

The material model for the cartilages and menisci was previously developed [14]. It consists of a fluid-saturated matrix that is reinforced by a fibrillar matrix (fibril-reinforced modelling). The fluid pressure and flow are described by a linear Darcy law, while the permeability of the tissue can be dependent on the void ratio which changes with tissue compression. The fibrillar matrix is considered nonlinear and orthotropic.

The fiber orientation in different tissues was chosen based on available data in the literature. The site specific orientation based on split line patterns [12] was used for the femoral cartilage and the depth dependency of the collagen orientations was ignored. Circumferential direction was taken to be the primary fiber direction in the menisci [13]. Fibers in ligaments were aligned in the longitudinal direction. Due to lack of data about the tibial cartilages, random orientation was used for these tissues.

The required mechanical properties for each tissue are: the Young's modulus and Poisson's ratio of the proteoglycan matrix (non-fibrillar part of the solid phase), void ratio and permeability of the solid phase and the Young's modulus of the collagen network. It was assumed that the collagen fibers only support tension and have zero stiffness in compression. Material properties of different tissues were obtained from the literature and can be found in our recent study [11]. A user defined FORTRAN subroutine was used to implement the fiber properties into the ABAQUS software.

The loading protocol was a combined creep loading of 300-N compressive force and 1500-N.mm valgus torque. Both the force and torque were simultaneously applied in one second and remained constant thereafter. Loads were applied to the femur, while tibia and fibula were constrained in all directions. Femur was constrained in lateral-medial and anterior-posterior translations as well as flexion-extension and internal-external rotations. The implicit finite element using transient *Soil Consolidation* method was used for the simulations. The Newton method was used to solve the nonlinear equations. The maximum time increments were controlled manually and in the meantime by given convergence criteria. They were chosen so that the numerical results we obtained did not undergo significant changes when the values of the maximum time increments were reduced further.

The mechanical contact between cartilaginous tissues was modeled using small sliding frictional contact between femoral cartilage-menisci, menisci-tibial cartilages, and femoral cartilages-tibial cartilages. For each contact pair, one surface is selected as the master surface and the other one as the slave. The contact constraint can be enforced using different methods. In the present study, the linear penalty method was selected for the contact enforcement. Using penalty method, the contact constraint is approximately enforced and very small penetrations of the master surface into the slave surface may occur in some areas. This method results in less convergence difficulties as compared to strict enforcement of the contact constraint.

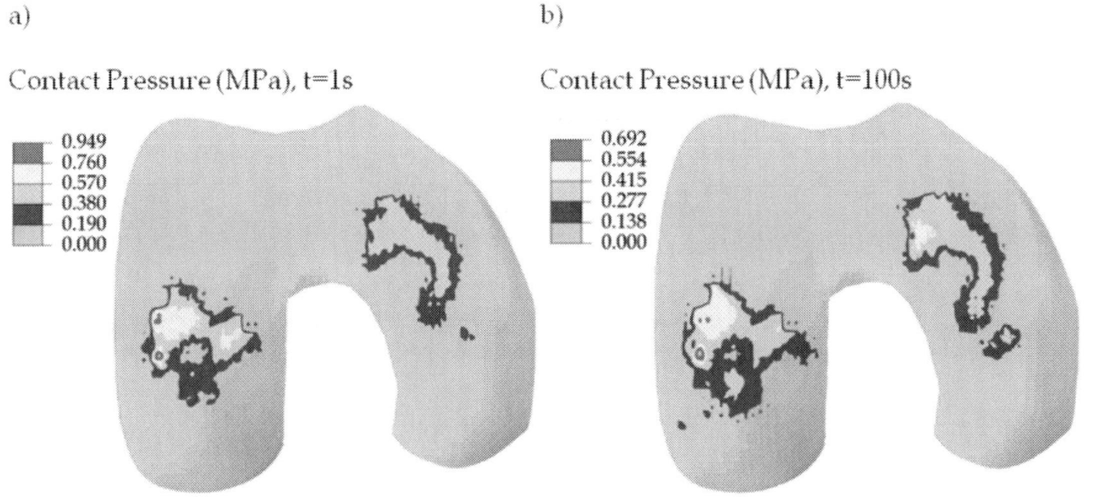

Figure 1. Contact pressure on the articulating surface of the femoral cartilage for (a) t = 1s and (b) t = 100s, respectively. The lateral condyle is on the left and the anterior site on the top of the figure (the inferior view of the right knee).

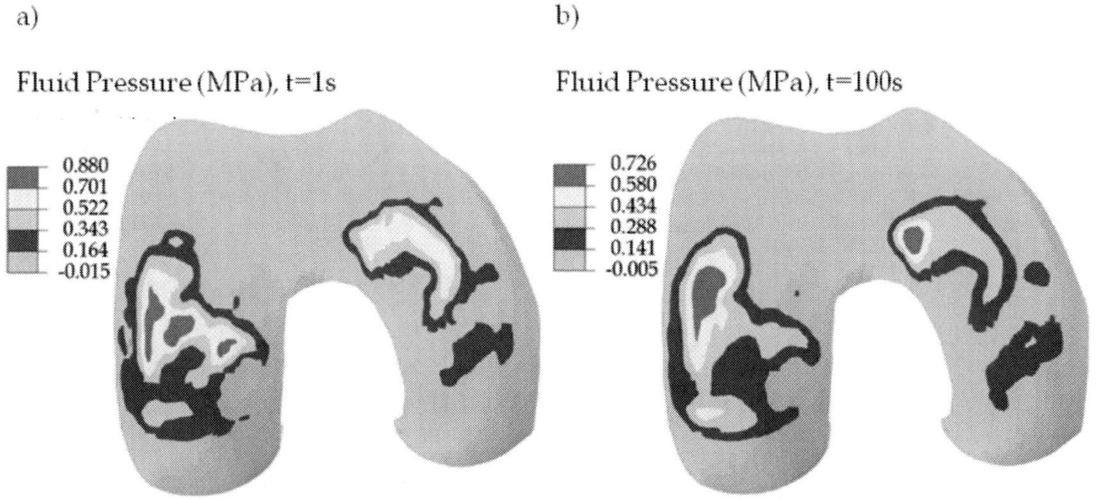

Figure 2. Fluid pressure in the deep layer of the elements of the femoral cartilage for (a) t = 1s and (b) t = 100s, respectively (the inferior view of the right knee).

3. Results

For the combined loading considered, the maximum contact pressure on the articulating surface of the femoral cartilage was observed in the lateral condyle (Figure 1a, t = 1s). As time increased or creep developed, the contact pressure in the femoral cartilage decreased (Figure 1b, t = 100s) and more loading was transferred to the menisci (figures follow). The maximum fluid pressure in the femoral cartilage was also observed in the lateral condyle (Figure 2a). As creep developed, the fluid pressure decreased and the high pressure regions moved towards the anterior sites of the femoral condyles (Figure 2b). During the entire creep, the lateral meniscus was much more pressurized as compared to the medial meniscus (Figure 3). At early times, the highly pressurized regions were observed in the central regions of the inner third of the lateral meniscus (Figure 3a). As time increased, the high pressure regions concentrated at the anterior part of the inner third (Figure 3b).

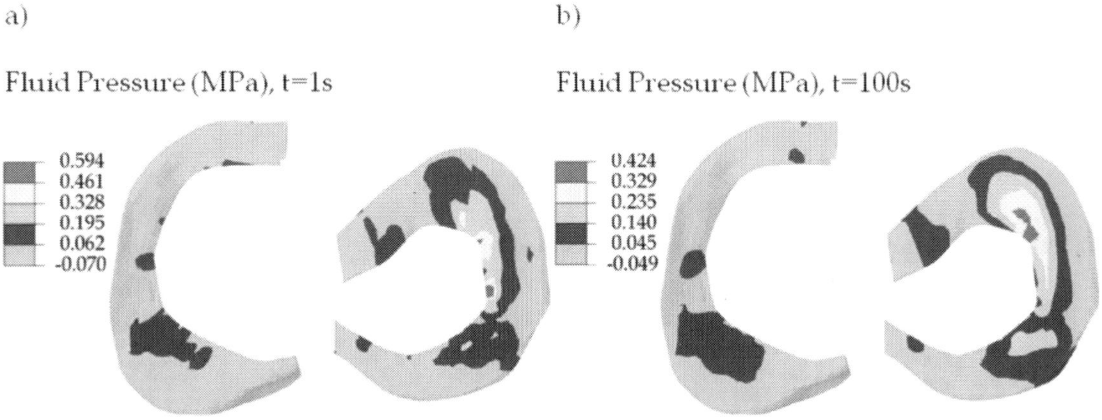

Figure 3. Fluid pressure in the menisci for (a) t = 1s and (b) t = 100s, respectively. The lateral meniscus is on the right and the anterior site on the top of the figure (the top view of the right knee).

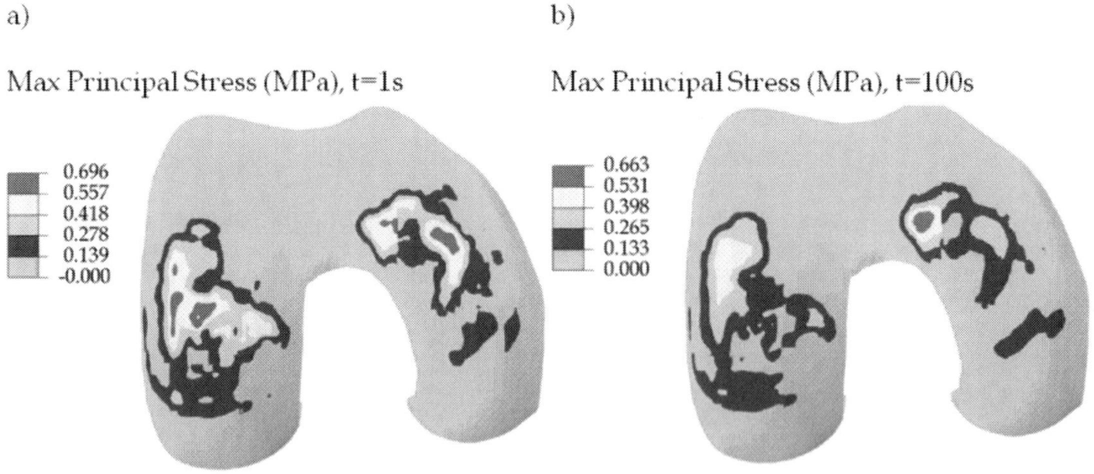

Figure 4. The first principal stress in the deep layer of the elements of the femoral cartilage for (a) t = 1s and (b) t = 100s, respectively (the inferior view of the right knee).

The first principal stress (i.e. the greatest principal stress that was tensile in this case) in the femoral cartilage seemed to have two maxima in the same range in both condyles at early times (Figure 4a). However, the lateral compartment encountered significantly more stresses as creep developed (Figure

4b). The high stress regions in both condyles moved toward the anterior sites as creep developed (Figure 4b vs. 4a) and the stress values decreased as time increased.

The lateral meniscus was more stressed as compared to the medial meniscus (Figure 5). The latter was almost stress-free under the given small loading (Figure 5). The high stress region was observed at the central part of the inner third of the meniscus and this location remained almost stationary during the creep period considered. Furthermore, the stress level in the menisci increased with creep, which was in contrast to the fluid pressurization in the femoral cartilage.

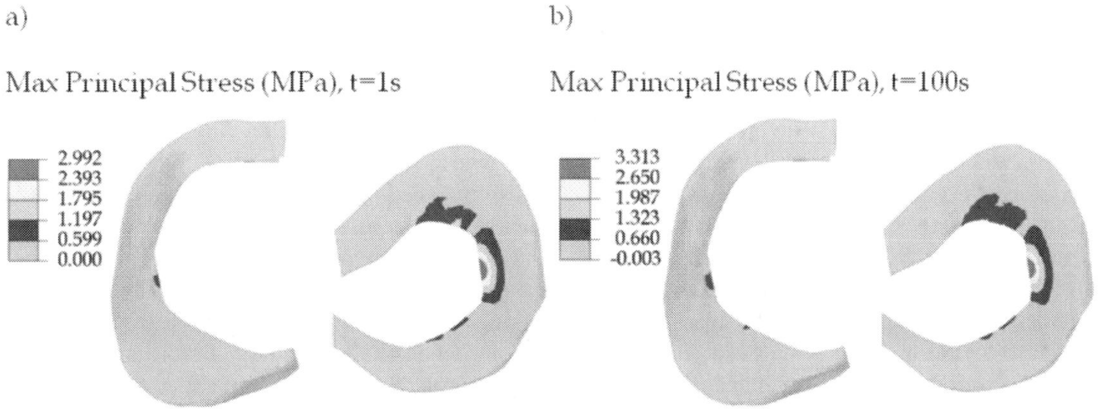

Figure 5. The first principal stress in the menisci for (a) t = 1s and (b) t = 100s, respectively (the top view of the right knee).

4. Discussions

A fibril-reinforced poromechanical model of the cartilages and menisci was used to study the tibio-femoral contact configuration under a combined creep compression and valgus torque. The results of the healthy and meniscectomy joints under simple compression have been obtained earlier [11]. As compared to the joint under simple compression, the valgus torque significantly changed the stress/fluid pressure patterns of the cartilages and menisci. In particular, the lateral compartment supported more stresses/pressures compared to the medial compartment (Figures 1 to 3). While in the case of knee joint under simple compression [11], the medial compartment supported more loading.

Due to effect of fluid flow and pressure, the mechanical responses were time-dependent. For instance, the contact pressure in the femoral cartilage changed as time passed. In particular, the maximum value of the contact pressure decreased as creep developed (Figure 1). This was due to two main reasons: increase in the contact area, and more load transfer from the femoral cartilage to the menisci. Comparison between the time-dependent first principal stress in the cartilage and menisci (Figure 4 vs. 5) indicates the latter reason. The stress in the cartilage decreased as creep developed (Figure 4). However, the stress in the menisci increased with creep, which showed that the menisci supported more loading as the contact area increased (the applied force/torque were held constant after t = 1s, i.e. creep after 1s).

The fluid pressure distribution in the femoral cartilage was also time-dependent. Due to fluid exudation and changes in contact configuration, the fluid pressure decreased as creep developed, which was expected. During the entire creep, the maximum values of fluid pressure were observed at the lateral condyle of the cartilage, which was because of the effect of valgus torque and constrained motion of the femur in lateral-medial and posterior-anterior directions. However, the location of the high pressure region shifted within the condyle as creep developed. As time increased, the high pressure regions moved from the central area of the lateral condyle to the anterior sites (Figure 2b).

The fluid pressure distributions in the medial and lateral menisci were considerably different. During the entire creep, the lateral meniscus was highly pressurized, especially in the central areas of the inner third (Figure 3). The medial meniscus, however, was almost pressure-free during the entire creep. This was due to the effect of valgus torque that imposed more loading on the lateral compartment and less on the medial side. The location of the high pressure region in the lateral meniscus was time-dependent. It moved from the central areas at early times to the anterior site as time increased.

The maximum first principal stress in the lateral condyle of the femoral cartilage decreased more rapidly as compared to that in the medial condyle (Figure 4). Again, this was due to the fact that the lateral condyle was pushed into lateral meniscus by both compressive load and the valgus torque. This established a better contact condition between the lateral contact pairs as compared to the loose contact between the medial pairs (i.e. medial condyle and the medial meniscus). However, such an effect should not been seen under higher physiological loadings in the presence of pre-stresses in ligaments. The distribution pattern of the first principal stress in the menisci remained virtually unchanged during the entire creep (Figure 5). The lateral meniscus was more stressed as compared to the medial one and its stress level increased as creep developed (Figure 5).

This study further showed the feasibility of a fibril-reinforced poromechanical model to describe the time-dependent response of the cartilages and menisci under different loadings. It has also been observed how the loading in the joint transfers from one tissue to another tissue. The major limitation of this modelling was small deformation assumption, which prevented us from analyzing the joint under higher physiological loadings. However, using small deformation assumption the complex contact situation of the tissues can be studied more comprehensively. The large deformation study will be performed after we get better insight about the convergence difficulties associated with the contact of the cartilages and menisci as well as the fluid pressurization. The model must be finally validated with new experimental results. One possibility is to use cadaveric knees to measure the reaction forces and contact pressures in the joints. Another possibility is to measure the tissue deformation in the knee using MRI, when the knee is loaded with a non-metal device. In either case, the validation will remain challenging.

References

[1] Yang N, Nayeb-Hashemi H and Canavan PK, 2009, *Annals of Biomel. Eng.* **37**, 2360
[2] Zielinska B and Donahue TL, 2006, *J. Biomech. Eng.* **128**, 115
[3] Peña E, Calvo B, Martínez MA and Doblaré M, 2008, *Comput. Biol. Med.* **38**, 69
[4] Mow VC, Kuei SC, Lai WM and Armstrong CG, 1980, *J. Biomech. Eng.* **102**, 73
[5] Ateshian GA, Lai WM, Zhu WB and Mow VC, 1994, *J. Biomech.* **27**, 1347
[6] Li G, Gil J, Kanamori A and Woo SL, 1999, *J. Biomech. Eng.* **121**, 657
[7] Shirazi R, Shirazi-Adl A and Hurtig M., 2008, *J. Biomech.* **41**, 3340
[8] Shrive NG, O'Connor JJ and Goodfellow JW, 1978, *Clin. Orthop. Relat. Res.* **131**, 279
[9] Mow VC and Ratcliffe A, 1990, *Biomechanics of Diarthrodial Joints* (Springer, New York)
[10] Gu KB and Li LP, 2011, *Med. Eng. Phys.* **33**, 497
[11] Kazemi M, Li LP, Savard P and Buschmann MD, 2011, *J. Mechanical Behavior of Biomedical Materials* **4**, 1351, doi:10.1016/j.jmbbm.2011.05.004
[12] Below S, Arnoczky SP, Dodds J, Kooima C and Walter N, 2002, *J. Arthros. Relat. Surg.* **18**, 613
[13] Aspden RM, Yarker YE and Hukins DW, 1985, *J. Anat.* **140**, 371
[14] Li LP, Cheung JTM and Herzog W, 2009, *Medical & Biological Engineering & Computing* **47**, 607

Journal of Physics: Conference Series **341** (2012) 012015 doi:10.1088/1742-6596/341/1/012015

Holographic multiscale method used with non-biased atomistic forcefields for simulation of large transformations in protein

L Dupuis[1] and N Mousseau[2]

[1] Département de Biochimie et Centre Robert-Cedergren, Université de Montréal, C.P. 6128, Succursale Centre-Ville,
Montréal, Québec, Canada H3C 3J7

[2] Département de Physique, Regroupement Québécois sur les Matériaux de Pointe (RQMP) et Centre Robert-Cedergren, Université de Montréal, C.P. 6128, Succursale Centre-Ville, Montréal, Québec, Canada H3C 3J7

E-mail: `lilianne.dupuis@UMontreal.CA`

E-mail: `normand.mousseau@UMontreal.CA`

Abstract.
We present a multiscale approach for simulating protein flexibility. The originality of our method is its ability to perform dynamic multiscaling based on continuous revaluation of overlapping areas. The holographic multiscale method overcomes the limitations of motions determined by predefined and fixed high-level descriptions and allows the reproduction of residue-specific impact on large scale motion. The method is tested with two different non-biased all atom implicit solvent forcefields. These show stretched proteins A and G, with maintained secondary structure, folding back near native states in a small number of transition events, demonstrating the advantages of this multiscale approach.

1. Introduction

The success of macromolecular computer simulations rests on describing appropriately two crucial aspects of these complex systems: the interactions between particles, provided by forcefields of various descriptive levels, and the physical geometrical realization of the structure, which can be extracted from experimental data, such as X-ray diffraction or NMR, or constructed by methods for sampling the system's configurational space. In spite of ever growing computational resources, it is still necessary to find the simplest level of description and sampling that will provide the best balance for answering the problem we are interested in.

This right balance is particularly difficult to find when looking at large-scale protein motions, that are often controlled by atomic-scale details but occur on time scale beyond current's general sampling methods. To overcome these opposite requirements, it is necessary to consider multiscaling approaches that can focus on crucial aspects of both of interactions and motions.

Traditional molecular dynamics, for example, describes exactly, within the forcefield limits, motion at all scales, from thermal vibrations to docking and folding. However, because no bias is provided to favor large motion, the large majority of the computing efforts go to describing thermal oscillations, providing little new understanding. Monte Carlo approaches benefit from considerably more freedom in selecting the appropriate level of motion [1, 2, 3]. However, since these do not generally take forces into account, the size of moves is typically limited by collisions and steric constraints. These are avoided by methods such as Normal Mode Analysis (NMA) [4, 5, 6, 7], a method that identifies directly the softest directions of change by first mapping the protein as an elastic network to compute the lowest-frequency modes of the dynamical matrix, a remarkably good approximation. As low frequencies only depends on the general shape of the molecule [8], NMA can be used with simple coarse-grained level representations [9, 10, 11, 12].

There exists many problems and phenomena, however, where large scale motion of a molecule is affected by mutation of single residues [13, 14, 15, 16] and that cannot be described by coarse-grained harmonic state analysis [17]. For such movements, it is essential to adopt a multilevel representation that adapts to the role of the various molecular structures and incorporates non-harmonic motions. The latter can be achieved with activated methods that propose a non-biased approach for the study of complex transformations affecting the general shape of proteins. These approaches perform continuous evaluations of force curvatures in order to discover energetically feasible transition passages, providing a natural way to leave local minima and identify physically-possible pathways [18, 19, 20, 21]. Focusing on the atomic scale, however, these methods rapidly become dominated by small, local rearrangements that contribute very little to overall configurational changes.

Here, we propose a multiscale algorithm that allows the activated method to focus on large scale motion while retaining the capacity to adapt this representation to atomistic constraints and motion. The holographic multiscale method (HMM) introduces a dynamical definition of the various representation scale that increases considerably the versatility of this approach compared with standard multiscale methods [3, 9, 10, 22, 23]. More precisely, the HMM combines the activation-relaxation technique (ART nouveau), an activated method that has been used extensively for protein folding and aggregation, both in cartesian [20, 24, 25, 26] and in internal coordinates [27], with an adaptive structure-dependent multiscale representation that focuses on large cooperative ensemble motions while allowing, at every step, atomistic relaxation that fully incorporates the role of specific residues and interactions on large scale dynamics.

The structure of this paper is as follows. We begin with a brief overview of multilevel characteristic of proteins, on which the holographic view is based. We then present the details of the holographic multiscale method, including the multiscale projection into reduced internal coordinates. We briefly discuss our in-house implicit-solvent reduced-representation extended OPEP forcefield and then present a few test applications that demonstrate the strength of the algorithm.

2. Methodology

2.1. Basic protein motion

A protein is a chain of a few tens to many thousands of amino acids characterized by a common part composed of a short sequence of a nitrogen and a carbon atom positioned on each side of a central carbon (Cα) and by a lateral chain, or residue, of varying length and composition, rooted at the Cα. Amino acids are linked by their common part through a series of covalent bonds called peptidic links, which form the protein's main-chain or backbone. The presence of covalent bonds restricts effectively the possible motion of each amino acid with respect to its neighbors to the torsion angles situated on both sides of the Cα. With these degrees of freedom, the main-chain can adopt a number of conformations that determine the protein secondary structures — α-helices, β-sheets, loops, random coils, etc. — which are further packed into a three-dimensional organization called tertiary structure. This well-defined organization is guided by delicate balance of backbone–backbone, side-chain–side-chain, backbone–side-chain and protein-solvent interactions, controlled by hydrogen bonds, hydrophobic, polar and charge groups.

Large structural changes in the protein necessarily involve large modifications of the main-chain torsion angles, ϕ and ψ, located, respectively, around the N-Cα bond, and the Cα-C bond. For example, significant changes in these angles are required to go from $\phi \approx -55$ and $\psi \approx -45$ degrees for α-helices to $\phi \approx -130$ to -160 and $\psi \approx 130$ to 160 degrees for β-sheets. Even though these angles largely characterize the protein spatial organization, other angles and bond-lengths will adapt slightly to every configuration in order to minimize steric effects and optimize stability. The holographic multiscale method combines these two levels of relaxation. Following other large scale motion techniques [28], the HMM focuses one these angles to generate the activation. Its originality comes from the fact that allows all degrees of freedom to adapt during the relaxation phase.

2.2. Overview of the holographic multiscale method

To focus on large scale cooperative motion, the holographic multiscale method proposes a multilevel approach that starts with the protein in a local energy minimum and goes as follows:

(i) The protein conformation is described on the basis of a list of flexible Cα pivots. It is at this moment that a multiscale description is introduced, by the identification of flexible (or *active*) regions, typically associated with loops and random-coil conformations, and rigid regions, those with stable secondary structures. At each active Cα pivot the protein is split into three blocks: the chain fragment preceding the pivot, that following it, and the lateral chain. Each pivot possesses six degrees of freedom associated with the rotation of the last two sections with respect to the first one and to each other. After each displacement step, the all-atom Cartesian space representation of the molecule is regenerated in order to compute the atomistic forces which are projected back onto each overlapping block definition delimited by the various freed pivots (Fig. 3). This generates a smoothened configurational subspace, eliminating small, but non-diffusive, barriers.

(ii) A first activated event is generated, using ART nouveau, on this reduced configurational subspace, leading to a first-order saddle point associated with large-scale cooperative motion.

(iii) From this saddle-point associated with a transition state, the protein is brought into a local-energy minimum using damped molecular dynamics applied to the all-atom Cartesian representation with all degrees of freedom allowed to move.

These three steps represent an ART event, going from one local energy minimum to another, passing through an activated state and following a physically-possible pathway, that allows an efficient sampling of the configurational energy landscape. The next sections describe in more details each of these steps.

2.3. Construction of elastic blocks and projected flexible regions

As mentioned in the previous section, the crucial step in HMM is the construction of the multiscale representation. For this, we first construct a spherical representation of positions and forces projected onto the peptidic plane angular degrees of freedom that determine the rotation of the ϕ and ψ angles. For each of these planes, we compute an orthogonal basis, using the Cα to Cα axis direction as first axis,

$$\vec{a_1} = \frac{\vec{r}_{C\alpha 2} - \vec{r}_{C\alpha 1}}{\|\vec{r}_{C\alpha 2} - \vec{r}_{C\alpha 1}\|}, \tag{1}$$

and the perpendicular direction toward the main-chain C-bound oxygen as second axis,

$$\vec{a_2} = \frac{\vec{r}_{oxy-C\alpha_1} - (\vec{r}_{oxy-C\alpha_1} \times \vec{a_1}) \times \vec{a_1}}{\|\vec{r}_{oxy-C\alpha_1} - (\vec{r}_{oxy-C\alpha_1} \times \vec{a_1}) \times \vec{a_1}\|}. \tag{2}$$

The third axis is deduced from a right hand rule

$$\overrightarrow{a_3} = \frac{\overrightarrow{a_1} \times \overrightarrow{a_2}}{\|\overrightarrow{a_1} \times \overrightarrow{a_2}\|}. \tag{3}$$

This forms the absolute base A for each peptidic plane δ. We also need to create its relative base B_δ, which is orthogonal:

$$B_\delta = (A_{\delta-1} \centerdot {A_\delta}^T)^T \text{ giving } \overrightarrow{b_1}, \overrightarrow{b_2}, \overrightarrow{b_3} \tag{4}$$

This relative representation allows use to evaluate the angular position (swiveling) of the following area relatively to the preceding, using two spherical angles: the longitude γ and the latitude λ of the Cα-Cα axis (b_1), relatively to the preceding plane, defined as:

$$\lambda = -\arccos \frac{b_{1,2}}{\|\overrightarrow{b_1}\|} \tag{5}$$

and

$$\gamma = \frac{b_{1,3}}{|b_{1,3}|} \arccos \frac{b_{1,1}}{\|\overrightarrow{b_1}\| \times |\sin \lambda|}. \tag{6}$$

Because the blocks are tree-dimensional, a third angle represents the rotation θ of the plane around its own Cα-Cα axis (b_1). As this axis also swivels in space, the evaluation of its rotation requires a comparison with defined referential which we take to be the x axis of the preceding base. We first determine the axis of rotation c between axis b_1 and the x axis:

$$\overrightarrow{c} = \overrightarrow{x} \otimes \overrightarrow{b_1} = (0, b_{1,3}, -b_{1,2}). \tag{7}$$

We then determine the rotation matrix C around this c axis for the angular course needed between axis b_1 and x axis. We obtain this angle from the arc-cosinus of the scalar product between axis b_1 and x axis. This rotation matrix C is then used to apply a theoretical swivelling of the current matrix B toward the x axis of the referential previous block, giving the matrix S:

$$S = (C \centerdot B^T)^T. \tag{8}$$

The first axis of matrix S coincide with the x axis. The second axis of matrix S is used to determine the current rotation angle of the block, using its angular position relative to the y axis:

$$\theta = \frac{s_{2,3} - s_{2,1}}{|s_{2,3} - s_{2,1}|} \arccos \frac{s_{2,2}}{\|\overrightarrow{s_2}\|} \tag{9}$$

The modification of these tree angles corresponds to a holographic swiveling/rotation around the peptidic junction δ-1,δ. With this representation, the entire protein following a Cα pivot is allowed to swivel and rotate relatively to the preceding part. This representation uses 3 angles instead of the 2 torsion ϕ and ψ angles. This means that some variation in the valence angles is allowed here, in addition of the easier dihedral rotations. The valence angles around the Cα may be stressed at the beginning of the activation, but the convergence method will soon relaxed them and conducted the motions to rotate according to phi and psi dihedral angles. The position of the lateral chain is also express in spherical representation relatively to the preceding peptidic plane, including a rotation around its own Cα-Cβ axis.

Atomic positions must be retrieved at each step of the simulation in order to compute the forces. We first get back axis b_1 of the relative base B:

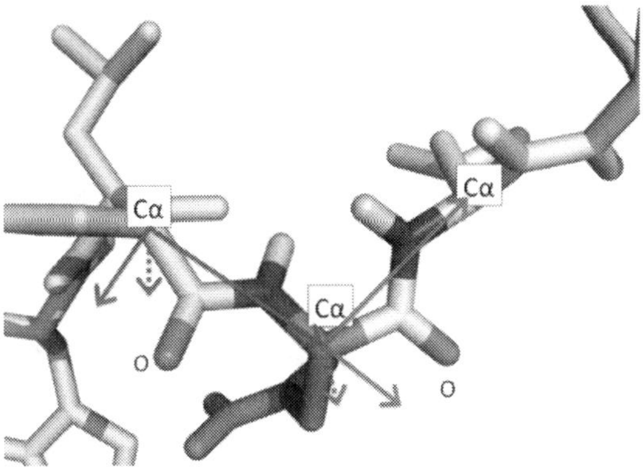

Figure 1. A $C\alpha$ is present at the root of the residue represented in magenta. An orthogonal basis is defined for the two peptidic planes on each side of this pivot, using the two ending $C\alpha$ of the plane and the perpendicular direction within the plane, on the oxygen side. We obtain the absolute bases matrices $A_{\delta-1}$ and A_δ. Relative matrix B_δ is then defined as the representation of matrix A_δ in the orthogonal basis of matrix $A_{\delta-1}$. By projecting the cartesian-defined forces on these bases, we obtain spherical rotating/swivelling forces of the whole cyan area relatively to the whole yellow area. The side-chain represented here in magenta is also evaluated relatively to the two large areas, and also simulated as a rotating/swiveling block.

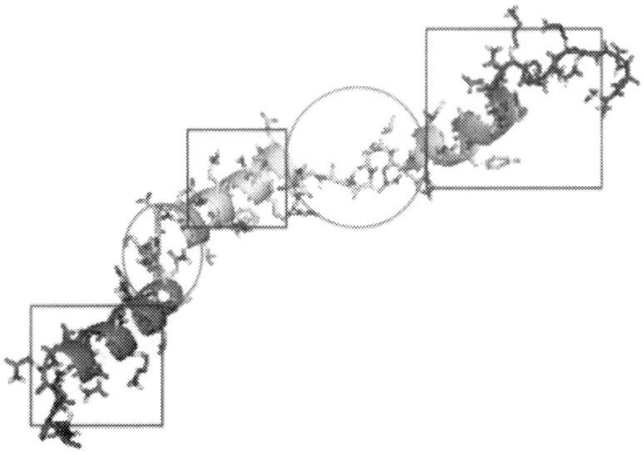

Figure 2. All $C\alpha$'s of a protein may be defined as pivots. For multiscale simulations, we consider rather some part of the molecule (squares) as structurally stable and define pivots only on $C\alpha$'s in the flexible zones (circles). Protein A, for example, is defined in the following way: α-helices are considered stable zones and all $C\alpha$'s in the loops and at the end of α-helices parts bordering them are fully flexible.

$$\overrightarrow{b_1} = (\sin\lambda \times \cos\gamma, \cos\lambda, \sin\lambda \times \sin\gamma). \tag{10}$$

The b_2 axis is obtained using the new θ angle, reversing the steps of equations 7,8,9. From matrix B we get back the absolute base A:

$$A_\delta = ({A^T}_{\delta-1}, {B_\delta}^T)^T. \tag{11}$$

Then the new positions are computed from the initial positions, using the initial absolute base A and the current absolute base A' as transformation matrices.

$$\overrightarrow{r_i}' = (A'^T \cdot A)^T \cdot \overrightarrow{r_i}, \tag{12}$$

where i is the atomic number and $\vec{r_i}$ is a three-dimensional cartesian position vector.

The rotation forces around the different axes are in correspondence with angles around these axes. The force contribution of all the atoms of the molecule is evaluated for each pair of holographic areas. The lateral chain positioned at the concerned $C\alpha$ pivot is also designed as a block that ca rotate relatively to them. The contribution of each atom for the block rotation around a particular axis is inversely proportional to its radius distance to the axis of rotation, and the part of the force that is perpendicular to both the axis and the radius between the axis and the atom.

$$F_{a_l} = \frac{\sum(\overrightarrow{f} \times (\overrightarrow{r} - (\overrightarrow{r} \cdot a_l) \cdot a_l))}{\sum \|\overrightarrow{r}\|} \tag{13}$$

where $l = 1, 2, 3$.

Using the current array of relative angular positions of each overlapping block pair, and the newly computed corresponding arrays of sperical block forces, ART nouveau gives new arrays of relative angular positions. At each step this new array is used to compute back cartesian atomistic positions and then the cartesian atomistic forces from forcefield. Those are used by HMM to update of the spherical holographic block forces used by ART. Fig. 3 summarizes this circular process.

2.4. Sampling method

We use the activation-relaxation technique, ART nouveau [19, 24], to generate energetically favorable transition events from one conformation to the other. Since the method has been reviewed recently [29, 30], we present here only the general scheme as applied into HMM. Starting from a configuration in a local minimum, a random direction of deformation is selected among the available degrees of freedom (pivots). The configuration is slowly deformed. At each step, the lowest eigenvalues of the Hessian matrix are computed using a Lanczos scheme. This random deformation is continued until the lowest eigenvalue becomes negative, indicating the presence of a nearby first-order saddle point. The configuration is then pushed, in the reduced holographic projection, along the eigenvector corresponding to this eigenvalue while the energy is relaxed in the perpendicular hyperplane. This activation phase stops when the total force becomes close to zero and the configuration has converged onto the transition state. During this first phase, motion can take place around all free pivots; the rest of the protein is moved as blocks. For the relaxation phase, the configuration is then nudged over the saddle point and the system is relaxed into a new minimum using, now, the full set of degrees of freedom in cartesian coordinates. The new configuration is then accepted or rejected using a Metropolis criterion.

The advantage of this approach is that, by using the system's forces to direct the activation, collisions are naturally avoided, allowing large collective motions. An example of an ART nouveau step can be seen in Fig. 4 and is discussed in Section 2.6.

2.5. Forcefields

The holographic multiscale method can be used with any non-biased implicit solvation atomistic force field. We present here results with two different force fields: CHARMM19 with the solvation model EEF1 as well as the extended optimized-potentiel for extended proteins (EOPEP) forcefield, an extended version of the OPEP coarse grain forcefield.

For better integration with HMM, we developed an in-house version of CHARMM19, based on Ref. [31]. Following Ponder's approach with Tinker [32], we use our own torsion angle description, based on OPEP's [33], but adapting the prefactors to those of CHARMM19. We combined this potential with the EEF1 solvation model volume exclusion terms [34]. For the latter, we applied ionic neutralization of the side-chains, the N-terminal and the C-terminal, as specified in EEF1.

To distinguish between forcefield and methodological limitations, we also developed our own extended version of the OPEP forcefield. The original OPEP is an implicit solvent reduced-representation forcefield that includes all heavy main-chain atoms as well as a single bead for the lateral chains, with statistically-derived interactions. Among its original features is a four-body interaction term to describe the cooperativity observed in the H-bond formation for stabilizing secondary structure. The quality of this potential has been well-tested in protein folding and protein aggregation simulations for chains of 70 residues or less [33, 27, 35, 20]. While this potential performs very well for secondary and tertiary structure, it fails to describe precisely more flexible regions where long residues often dominate. EOPEP inherits OPEP's ability to stabilize secondary structure using a cooperativity term for the evaluation of hydrogen bridges as well as provide side-chain hydrogen-bridge. In addition to these terms, EOPEP develops fully the lateral chains to ensure better packing for large residues.

More precisely, EOPEP represents all side-chain's carbon, oxygen, nitrogen, sulfur, and polar hydrogen (linked to O or N) atoms. The various carbon configurations are defined with implicit hydrogens. For these atoms, we adopt the van der Waals atomic dimensions of CHARMM19 [31]. Following OPEP, attractive and repulsive interactions between side-chain atoms are described using a screened Lennard Jones 6-12 term, except for the side-chain hydrogen bridges, for which we use OPEP's Lennard Jones 10-12 term used in the N-H–O interactions. This choice ensures that EOPEP can be used in conjunction with OPEP, providing a multilevel description inside a single protein and decreasing significantly the computational costs while adding the appropriate degree of complexity where needed. Because of their small size, however, proteins studied here are described fully by EOPEP.

Table 1. EOPEP parameters for side-chain atoms. These parameters are used in addition to the covalent bonding and main-chain main-chain interactions already defined in OPEP. EOPEP parameters were adjusted according to the procedure described in the text.

Interaction	Implementation	Partial contribution
Side-chain Hydrogen Bridge		
Carboxyl ogygen	3 terms LJ10-12	As OPEP Main-chain O
Other oxygen receptor	3 terms LJ10-12	0.10 vs Main-chain O
Nitrogen and oxygen donor	3 terms LJ10-12	0.20 vs Main-chain N
Side-chain non polar Carbons	2 terms LJ6-12	$0.05(\mathrm{Kcal/mol})^2$
Sulfur	2 terms LJ6-12	As OPEP MET
Non HB interaction from polar atoms		
Oxygen and Nitrogen	2 terms LJ6-12	0.5 of OPEP ASP
Hygrogen	2 terms LJ6-12	0.05 of OPEP ARG

The value for the extended set of parameters is shown in Table 1. To avoid overstabilizing side-chain–side-chain interactions, the hydrogen bridge strength for lateral chain is considerably reduced compared to the main-chain H-bridge. The hydrogen bridge receptor of carboxyl groups (aspartate, glutamate) conserves main-chain interaction strength, while other oxygen receptor see their contribution decreased by a factor ten, and the contribution of side-chain hydrogen bridge donor (nitrogens or some oxygens) by a factor of 20. These choices reproduce the aqueous immersion effect on them, as such contacts are generally only established consecutively to tertiary structure formation, which are directed by hydrophobicity [36]. Hydrophobic side-chain interactions also follow OPEP and are described by a uniform Lennard Jones 6-12 function, with a 0.05 $(\mathrm{kcal/mol})^2$ well-depth contribution for any type of non-polar side chain carbon atom. Sulfur atoms, for their part, are directly copied from the OPEP methionin behavior.

The EOPEP forcefield was calibrated through extensive stability tests of the native state of protein A as well as calmodulin C and N terminals using long holographic ART simulations.

EOPEP was subsequently used unchanged in folding simulations of these three sequences as well as Troponin C NT and Protein G.

2.6. Description of a typical holographic ART transition event

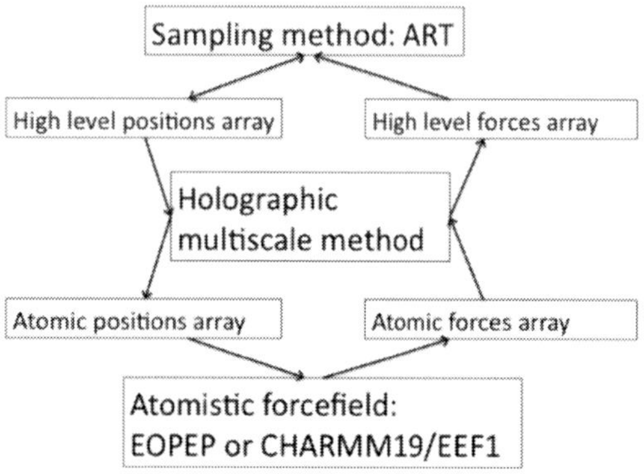

Figure 3. Software organization. The activation and saddle passage steps of ART use the high-level pivot and blocks representation. After each iteration, ART returns new relative angular positions for the various blocks. These are translated back into Cartesian coordinates and the respective force is computed and then projected onto the reduced space corresponding to the high-level description.

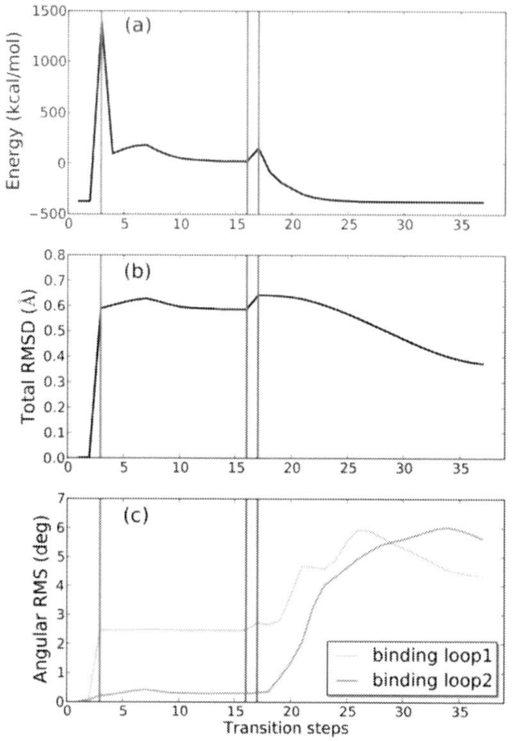

Figure 4. A typical HMM/ART event during an EOPEP folding simulation of Calmodulin NT. (a) Evolution of the total energy as a function of iteration number. (b) Global Cα RMSD measured from iteration zero. (c) Root-mean square angular displacement computed over ϕ and ψ, as measured from iteration zero and averaged over the two binding loops. The open CAM protein model is taken from pdb 1CLL, with residues 1-3 added, totalizing 76 residues. Vertical lines delineates the various phases of an activation/relaxation event. From left to right: push outside of the harmonic basin, convergence to the saddle point, push over the saddle, relaxation into the final minimum.

A typical HMM-ART event takes a few minutes of on a single CPU. Fig. 4 reports details for a typical accepted event during a closing simulation of the open Calmodulin NT model, which undergoes transformation in its binding loops when deprived of its calcium ions [37, 38]. For this simulation, we defined as pivots residues 18 to 31, which includes the first binding loop, and residues 54 to 67, which includes the second binding loop. The event was started from a energy minimum at -372.9 kcal/mol. The ART nouveau steps are identified on the plot and correspond to the exit from the harmonic basin, the convergence to the saddle point, the push over the saddle and the relaxation to a new minimum. Looking at panels (a) and (b), we note that the energy goes up rapidly as the protein is deformed, reaching 1450 kcal/mol, at the exit of the harmonic basin in a random direction (step 3). Since the convergence to the saddle point is accompanied by a relaxation of the N-1 other excited degrees of freedom, the energy falls rapidly during this next phase but the saddle point is still found at 528.6 kcal/mol above the initial minimum (step 17). This unphysically large energy is due to the fact that only angles around pivots are allowed to move, creating considerable strain in the other degrees of freedom, such as bond lengths and various angles. Indeed, most of this strain is released during the first steps of the relaxation to the final minimum, which takes place in Cartesian space, on all coordinates, and the final structure (step 37) is at 0.4 Å away and 4 kcal/mol (with a total energy of -376.7 kcal/mol) below the corresponding initial state values.

Panel (c) of Fig. 4 reports the root mean square average of ϕ and ψ Ramachandran angles within each binding loop,

$$\theta_{\mathrm{RMS}} = \sum_i \sqrt{(\Delta\phi_i)^2 + (\Delta\psi_i)^2} \tag{14}$$

where i runs on the active residues of loop 1 or 2.

The initial activation takes place mostly in the first binding loop, with relatively little deformation in the second loop. Interestingly, most of the angular displacement in this phase takes place during the exit of the harmonic basin and only small angular adjustments are observed during the convergence to the saddle point. Relaxation allows the full protein to react to this transformation and we see the second loop adjusting significantly during relaxation, creating a large cooperative event that changes considerably the protein conformation.

Table 2 reports the averaged RMSD between initial and final states of events for various simulations. Events can be very large, typically between 2.0 and 4.0 Å. The largest moves are unlikely to lead to low energy structures, however, and at low Metropolis temperature, most moves tend to be rejected. Statistics for accepted events at 300 K show that most lead to displacements are around 1 Å, which is still considerable.

Table 2. Mean RMSD between events: statistics for a few simulation examples

Protein model	Forcefield	Sim	Mean RMSD at saddle all events	Mean RMSD at minima all events	at saddle accepted only	at minima accepted only
Protein A	CHARMM19	5	2.29	2.39	1.27	1.28
Protein A	CHARMM19	16	2.93	2.96	0.96	0.77
Protein A	EOPEP	1	4.13	4.38	1.11	1.03
Protein A	EOPEP	20	3.10	3.25	1.16	1.09
Protein G	CHARMM19	2	2.24	2.03	1.26	0.98
Protein G	CHARMM19	11	2.07	1.94	1.12	0.95
Protein G	EOPEP	2	2.74	2.66	1.25	0.83
Protein G	EOPEP	14	1.93	1.52	1.19	0.69

3. Protein models

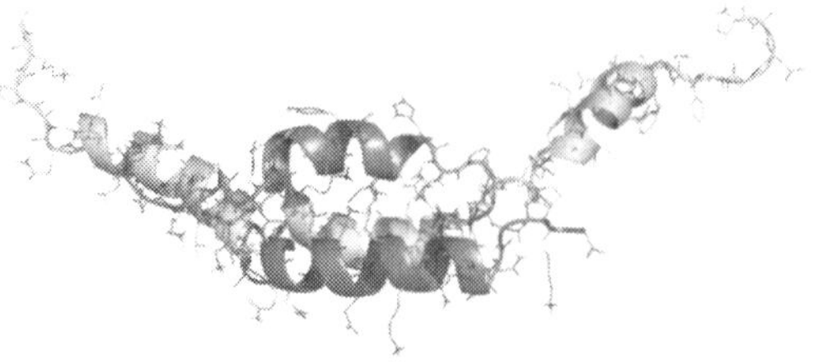

Figure 5. Extended protein A with 3 helices aligned compared to native model. In cyan: the central helix 2. In blue: helices 1 and 3 folded toward helix 2 in the native form. In green: helices 1 and 3 extended away from helix 2, at the start of our simulations.

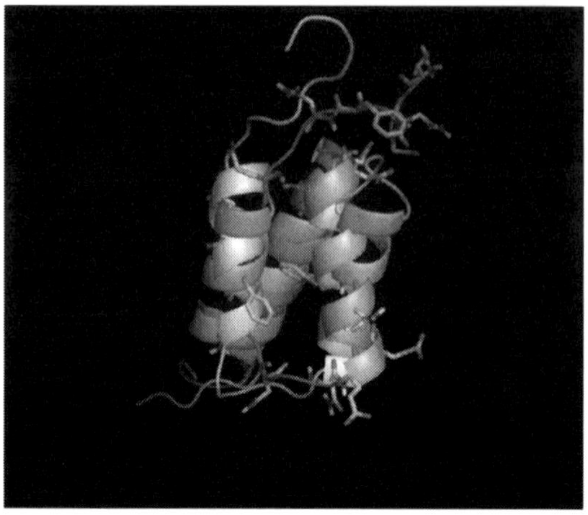

Figure 6. Pink: protein A at end of EOPEP simulation 20. White: the native protein A model.

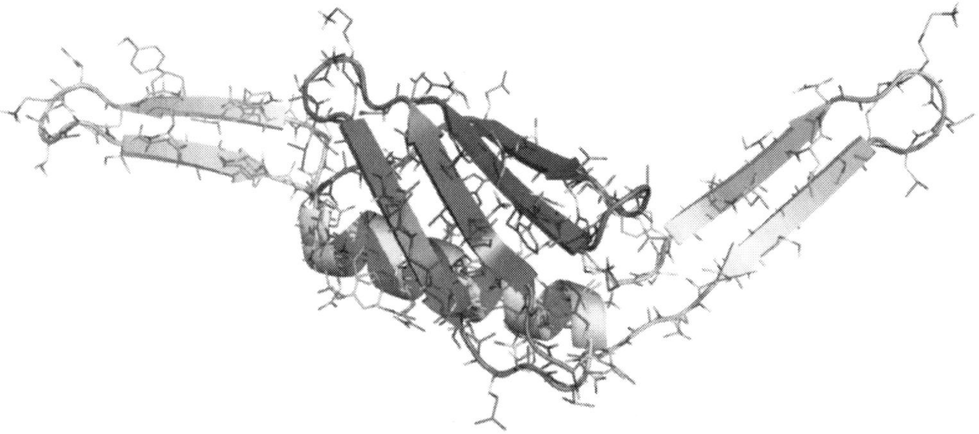

Figure 7. Protein G with beta sheet widely opened, compared to native model. In cyan: the central helix. In blue: the beta sheet formation folded along the central helix in the native form. In green: in our simulations starting model, the beta sheet formation is split away at each end of the central helix

We tested here the ability of the method to perform long distance transformations. To do that, we first stretched two proteins while preserving intact their secondary structure. Folding simulations for protein A domain of protein A (1BDC pdb file), a 60-residue three-α-helix bundle, was started from a configuration with the three helices aligned and a 16.6 Å RMSD with respect to the native configuration (Fig. 5). A 56-residues fragment of protein G model (pdb file 1GB1), composed a α-helix and a four-stranded β-sheet, was also artificially opened to provide a starting conformation at 19.6 Å RMSD away from the native conformation (Fig. 7).

protein A

TADNKFNKEQQNAFYEILHLPNLNEEQRNG

FIQSLKDDPSQSANLLAEAKKLNDAQAPKA

protein G

MTYKLILNGKTLKGETTTEAVDAATAEKV

FKQYANDNGVDGEWTYDDATKTFTVTE

Figure 8. Amino acid sequences of the protein A and protein G domains used in our simulations. α-helices are represented in green, β-strands in orange and the loops in light purple. In the present study, activations and saddle passages were allowed for the Cα pivots underlined in this figure. Relaxations were performed in all-atom mode on the entire sequences.

Figure 2 shows that it is possible to consider some part of the molecule as stable regions and avoid activations in these areas. Figure 8 reports the sequence of the protein A and protein G domains used and the Cα pivots that were allowed for activation. At each activation we performed a random choice of two trios of consecutive C-alpha pivots among the ones underlined in Figure 8. We activate at each of these chosen pivots in random directions for the three peptidic plane angles of the main chain and the for the three rotating block angles of the side chain.

4. Results and Discussion

We performed 24 folding simulations starting from the open version of protein A model, each with 150 to 300 accepted events for CHARMM19 and 300 to 500 for EOPEP at a Metropolis temperature of 300 K (except eop16a, which is discussed below). The evolution of the best runs, both in terms of energy and RMSD are reported in Fig. 9. EOPEP simulations 2 and 20 bring the model under 3.5 Å in respectively 3 hrs and 5.5 hrs CPU time for 50 to 75 events. Simulation 20 reaches 2.8 Å at the lowest-energy state of all simulation after 1.6 day CPU time. Convergence is slightly less for CHARMM19/EEF1, where the best simulation (che21a) comes to 3.8 Å with the native state.

We obtain similar results with protein G. Fig. 10 reports a selection of the 48 folding simulations starting from the stretched protein G model, each running for 200 to 400 accepted events at 300 K. The best RMSD descent is obtained with the CHARMM19/EEF1 simulation number 33, reaching 3.7 Å with the native model. This simulation also displays the lowest energy of all 48 runs using the CHARMM19/EEF1 forcefield. Several EOPEP simulations fall

Journal of Physics: Conference Series **341** (2012) 012015 doi:10.1088/1742-6596/341/1/012015

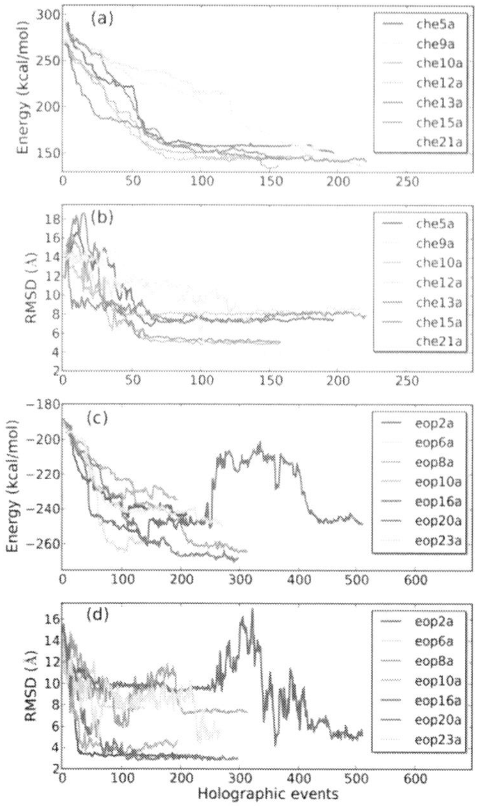

Figure 9. 24 holographic simulations are performed starting from the artificially open model of protein A (three α-helices aligned), using both CHARMM19/EFF1 and EOPEP. From each set we show the four trajectories visiting the lowest-energy structures and the four trajectories visiting structures with the lowest RMSD with respect to the native conformation (only simulation ch5a meets both criteria). (a) Energy and (b) RMSD evolution as a function of the number of accepted ARTn events usign CHARMM1/EEF1. (c) and (d), same but with EOPEP forcefield. The RMSD is measured relative to the Cα of the native form, excluding the unstable ending loops (residues 1-11 and 55-60). All simulations are run at 300 K, except eop16a, which is run at a Metropolis temperature of 4000 K from event 250 to a few events past 400 and then returned at 300 K.

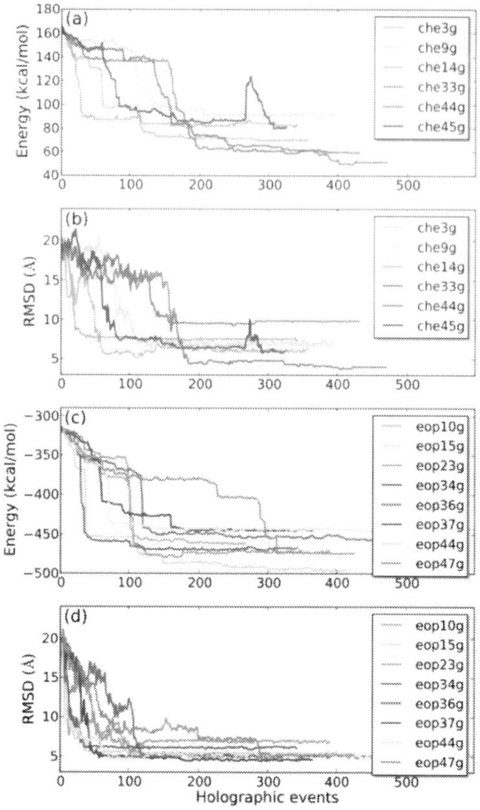

Figure 10. 48 holographic simulations are simulated starting from the artificially open model of protein G. From each set we show the four trajectories visiting the lowest-energy structures and the four trajectories visiting structures with the lowest RMSD with respect to the native conformation (simulations che3g and che33g meet both criteria). (a) Energy and (b) RMSD evolution as a function of the number of accepted ARTn events using CHARMM1/EEF1. (c) and (d), same but with EOPEP forcefield RMSD is relative to all the Cα of the native form. All simulations are run at 300 K, except che45g, which is run at a Metropolis temperature of 4000 K for a few events around event 270 and then returned at 300 K.

below 5 Å RMSD within 7 to 14 hours CPU time.

From HMM point of view, these bests results are excellent as HMM simulations tests of the stability for these two proteins show that the folding runs end up near the effective native state of each potential. Extensive HMM sampling simulations at 300 K starting from the native model for both protein A and protein show a stabilization near 2.5 and 1.7 Å-RMSD for protein A under EOPEP and CHARMM/EEF1 and 2.5 and 3.0 Å-RMSD, respectively, for protein G (data not shown), comparable to the best results of HMM folding.

The fact that only a fraction of all simulations reach the forcefield minima is due to the constant temperature scheme used here. If we simply heat up a structure trapped in a relatively high energy or RMSD basin and allow it to leave this metastable state, it is easy to bring the sequence in more stable state. This is the case for protein A EOPEP simulation number 16 which was trapped in a basin at around 10 Å RMSD after 250 steps. Raising the Metropolis temperature to 4000 K for about 150 events, the run explores configurations up to 17 Å RMSD away from the crystalline state, but rapidly folds to 4.6 Å RMSD after the temperature is brought back to 300 K. A similar but shorter test performed at the end of simulation che45g (protein G with CHARMM19/EEF1) enable the protein to relax from 7 to 6 Å RMSD from the native state.

Comparing the RMSD and energy evolutions in Fig. 9 and 10, we note that the energy drop typically follows the RMSD change, reflecting the existence of periods of local adjustment for minimizing the energy after a large configurational change. While this occurs for the four simulation sets presented in Fig. 9 and 10, the phenomenon is particularly obvious for protein G CHARMM19/EEF1 simulations presented in Fig. 10 (a) and (b): while both energy and RMSD follow a similar drop in value, for each of the 6 presented simulations, the energy relaxation occur systematically tens or more events events after a RMSD abrupt slope.

5. Conclusions and perspectives

This work presents the holographic multiscale method for generating efficiently large range protein motions. Large RMSD and energy steps are generated using the ART nouveau sampling method combined with a multiscale angular representation. In particular, by coupling block motion for the activation to full-scale atomic relaxation, HMM takes full advantages of the secondary structures while allowing it to change in response to large motion. This step is crucial to couple the role of side-chains, for example, with the stabilization of tertiary structure.

Results presented here on two model systems demonstrate the method's efficiency for generating large reorganization along well-controlled pathways. Clearly, however, a description of the energy landscape at the atomistic level increases the competition between local packing and block motion. This makes itdifficult to accept events that get away from local energy basins without destabilizing the whole structure. The use of simulated annealing or other relaxation techniques would greatly help here, and further study will focus on identifying the best algorithm for constructing the proper Metropolis temperature sequence, which could involve an additional set of activated moves on side chains, after each global move, to optimize the local structure before accepting or rejecting the event. The crucial result presented here, however, is that HMM can generate large motions that are not necessarily observed by normal mode analysis but that can nevertheless take place without uncontrolled steric clashes.

In addition to the model systems presented here, HMM has also been applied with success to characterize the folding of two EF-hands proteins, Calmodulin and Troponin C, upon removal of Ca ions. This work, to be published somewhere else, shows that HMM can be used to identify the specific events, atomic interactions and cooperative moves responsible for large scale changes [39].

HMM suffers nevertheless from two limitations. First, the method is not very effective for constrained loops as it requires being able to move a part of a protein with respect to the other, something that cannot easily be done while ensure a closing constraint. An appropriate selection of the initial deformation in ART nouveau step should be able to lift this inefficiency but more work remain to be done to establish a universal approach.

More complex, however, is the limitation on the forcefield. Because HMM works effectively at zero K, i.e., without thermal vibrations, it must be used with implicit solvent schemes, which are less precise than fully explicit models. The inherent flexibility of implicit-solvent model are obvious, especially with the coupling of the coarse-grained OPEP forcefield with the all-atom EOPEP. More work, however, must be done to ensure that these models work well away from the native state but this is beyond the development of a method such as HMM.

Overall, HMM provides an efficient and effective tool for characterizing the multiscale cooperativity associated with folding and large-scale motion. By providing physically possible pathways, HMM generates ensemble motion that are likely to be close to how real proteins work as long as forcefield are sufficiently reliable.

Acknowledgements

This work was funded in part by the BiT bioinformatic scholarship program, the NSERC, FQRNT and the Canada Research Chair Foundation. Calculations were done using resources from the Calcul Québec.

References

[1] Noguti T Go N 1985 *Biopolymers* **3** 527–546
[2] Fichthorn K A and Weinberg W H 1991 *The Journal of Chemical Physics* **95** 1090–1096
[3] Wells S, Menor S, Hespenheide B and Thorpe M F 1983 *Physical Biology* **2** 127–136
[4] Brooks B and Karplus M 1983 *Proceedings of the National Academy of Sciences* **80** 6571–6575
[5] Levitt M, Sander C and Stern P S 1985 *Journal of Molecular Biology* **181** 423 – 447 ISSN 0022-2836
[6] Janežič D and Brooks B R 1995 *Journal of Computational Chemistry* **16** 1543–1553 ISSN 1096-987X
[7] Tirion M M 1996 *Phys. Rev. Lett.* **77** 1905–1908
[8] Lu M and Ma J 2005 *Biophysical Journal* **89** 2395 – 2401 ISSN 0006-3495
[9] Tama F, Gadea F X, Marques O and Sanejouand Y H 2000 *Proteins: Structure, Function, and Bioinformatics* **41** 1–7 ISSN 1097-0134
[10] Ahmed A and Gohlke H 2006 *Proteins: Structure, Function, and Bioinformatics* **63** 1038–1051 ISSN 1097-0134
[11] Tripathi S and Portman J J 2009 *PNAS* **106** 2104–2109
[12] Zhang Y, Jasnow D and M Z 2008 *PNAS* **104** 18043–18048
[13] Meyer D F, Mabuchi Y and Z Grabarek Z 1996 *The Journal of Biological Chemistry* **271** 11284–11290
[14] Cohen F E and Prusiner S B 1998 *Annual Review of Biochemistry* **67** 793–819
[15] Clarkson M W and Lee A L 2004 *Biochemistry* **43** 12448–12458 pMID: 15449934
[16] Tousignant A and Pelletier J N 2004 *Chemistry and Biology* **11** 1037–1042
[17] Ma J 2005 *Structure* **13** 373 – 380 ISSN 0969-2126
[18] Barkema G T and Mousseau N 1996 *Phys. Rev. Lett.* **77** 4358–4361
[19] Malek R and Mousseau N 2000 *Phys. Rev.* **62** 7723–7728
[20] St-Pierre J F, Mousseau N and Derreumaux P 2008 *J. Chem. Phys.* **128** 0145101–1
[21] Carr J M, Trygubenko S A and Wales D J 2005 *The Journal of Chemical Physics* **122** 234903
[22] Jacobs D J, Rader A J, Kuhn L A and Thorpe M F 2001 *Proteins: Structure, Function, and Bioinformatics* **44** 150–165 ISSN 1097-0134
[23] Mamonova T, Hespenheide B, Straub R, Thorpe M F and Kurnikova M 2005 *Physical Biology* **2** S137
[24] Mousseau N, Derreumaux P, Barkema G T and Malek R 2001 *Mol. Graph. and Modeling* **19** 78–86
[25] Wei G, Derreumaux P and Mousseau N 2004 *The Journal of Chemical Physics* **119** 6403–6406
[26] Santini S, Wei G, Mousseau N and Derreumaux P 2004 *Structure* **12** 1245–1255
[27] Yun M R, Mousseau N and Derreumaux P 2007 *The Journal of Chemical Physics* **126** 105101–105110
[28] Chu J W and Voth G A 2006 *Biophysical Journal* **90** 1572–1582
[29] Marinica M C, Willaime F and Mousseau N 2011 *Phys. Rev. B* **83** 094119
[30] Machado-Charry E, Caliste D, Genovese L, Mousseau N and Pochet P 2011 *J. Chem Phys.*
[31] Neria E, Fischer S and Karplus M 1996 *Journal of Chemical Physics* **105** 1902–1921
[32] Ponder 2010 *http://dasher.wustl.edu/tinker*

[33] Maupetit J, Tuffery P and Derreumaux P 2007 *Proteins* **69** 394–408
[34] Lazaridis T and Karplus M 1999 *PROTEINS: Structure, Function and Genetics* **35** 133–152
[35] Laghaei R, Mousseau N and Wei G 2010 *J. Phys. Chem.* **114** 7071–7077
[36] Baker D 2000 *Nature* **405** 39–43
[37] M Zhang T T and Ikura M 1995 *Nature structural biology* **2** 758–767
[38] Tripathi S and Portman J J 2008 *The Journal of Chemical Physics* **128** 205104
[39] Dupuis L and Mousseau N *in press, J. Chem. Phys.*

A Study of NetCDF as an Approach for High Performance Medical Image Storage

Marcone Magnus, Thiago Coelho Prado, Aldo von Wangenhein
Post-graduate Program in Computer Science,
Department of Informatics and Statistic,
Federal University of Santa Catarina,
Florianópolis, SC - Brasil

E-mail: marcone, coeio, awangenh@inf.ufsc.br

Douglas D J de Macedo, M A R Dantas
Post-graduate Program in Knowledge Engineering and Managemet,
Department of Knowledge Engineering,
Federal University of Santa Catarina,
Florianópolis, SC - Brasil

E-mail: macedo, mario@inf.ufsc.br

Abstract. The spread of telemedicine systems increases every day. The systems and PACS based on DICOM images has become common. This rise reflects the need to develop new storage systems, more efficient and with lower computational costs. With this in mind, this article discusses a study for application in NetCDF data format as the basic platform for storage of DICOM images. The study case comparison adopts an ordinary database, the HDF5 and the NetCDF to storage the medical images. Empirical results, using a real set of images, indicate that the time to retrieve images from the NetCDF for large scale images has a higher latency compared to the other two methods. In addition, the latency is proportional to the file size, which represents a drawback to a telemedicine system that is characterized by a large amount of large image files.

1. Introduction

The communication availability in recent years led the development and dissemination of applications aimed at meeting the needs of geographically dispersed users. One of these are the Telemedicine Systems, which have been applied as a way to provide access to the health of people or communities who are isolated or lack qualified medical treatment in one way or another. It's use in these cases can reduce transportation costs and maximize the use of the technology park installed in hospitals and medical clinics [1].

In this way, the spread of PACS (*Picture Archiving and Communication Systems*) along with systems to support medical diagnosis based on images have been spreading steadily. This approach was applied to the project called RCTM *Rede Catarinense de Telemedicina* [2], [3] . This project was supported by the State of Santa Catarina Government and developed by the Telemedicine Laboratory [4], a member of the Cyclops Project [5] from Federal University of Santa Catarina, to support the lives of people around the state.

Currently, the RCTM is present in 286 cities of Santa Catarina, generating a monthly flow around 1.5 Terabytes of data, which are represented by more than 50000 medical exams performed by the population. The expectation for the coming years is to surpass the 150000 medical exams per month which will generate around 5 Terabytes per month of medical data. This reality leads to serious problems of data storage, that, for legal reasons, should be stored for a minimum period of 20 years.

Today the RCTM's data is stored in a relational database, which has brought high storage costs with unsatisfactory performance. Thus, the development of new methods and techniques for storing data efficiently and with lower cost are being developed. This follows from the data replication to remote sites [6], [7], new data compression techniques and new ways of storing other data formats using [8].

In this article we will discuss a new format for storing data based on the data format NetCDF (*Network Common Data Format*) applied to telemedicine systems. The motivation for the development of this work is based on positive experience cited in the work developed by *Macedo et. al.* [8] that used data format HDF5 (*Hierarchical Data Format 5*) as the basic platform of data storage for the telemedicine system.

This article is structured in 6 sections, this being the first. Section 2 delineates a background of the research, seeking to base the fundamental concepts for understanding of the proposal. In this section are discussed telemedicine systems, DICOM Standard, HDF5 and NetCDF file formats. Section 3 exposes work-related research. Section 4 is the approach of the research. Section 5 exposes the exprimental results, and the final section is the conclusion and future works.

2. Background

2.1. DICOM

In the mid 70's, with the introduction of digital images from various modalities, manufacturers of medical equipment created their your own standard of communication and data storage. As this type of digital image became popular and more companies manufactured equipment that generated it, the need for a single standard for all manufacturers increased. It was precisely why in the early 80's that the ACR (American College of Radiology) together with the NEMA (National Electrical Manufacturers Association) created a single standard [9].

The first standard was published in 1985 and was called ACR-NEMA and was the first non-proprietary standard for storage and communication of medical data. In 1988 came the second version of the standard and in 1993 became what would be the third version that was called ACR-NEMA DICOM 3.0 [10]. Currently there are no plans for a new version but new features are added to the existing standard, the last being in 2011. DICOM, since then, is the basis for most of the PACS.

2.1.1. DICOM Information Object Definitions The DICOM objects are generated by medical equipment and generaly contains images and graphics. In addition the DICOM objects supports different types of data encapsulation such as waveforms, medical reports, PDFs, among others.

The components of the data structure of the DICOM described in the standard, that represents a study in the real world, are called IODs (Information Object Definitions). In the case of examinations which contains images such as CT scans and MRIs, there is information related to it, such as patient information, equipment and everything that is relevant for medical diagnosis. The IOD is structured in IE (Information Entities) that defines logical separations (the patient's information, characteristics of the medical equipment, etc.), which in turn contains IOMs (Information Objetct Modules) that contain data elements. The data element is the unit of DICOM information and are containers that represent the data. There are elements for time

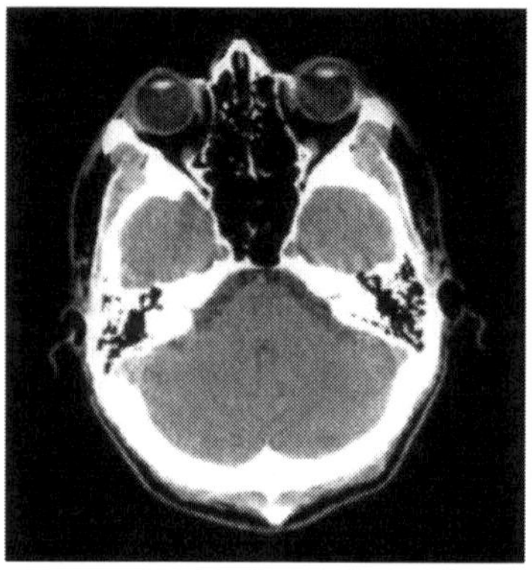

Figure 1. Computed Tomography

formats, date formats, UIDs, names, etc. In the standard are described for each IOD which IEs, IOMs and Data Elements are related [11].

2.1.2. DICOM Service Class In addition to the data format described in the previous section, there are parts in the standard that formalize the communication protocol and service classes that can be supported. For this work we used the class provider of C-STORE service which is responsible for reception of medical imaging equipment for storage and interpretation [12].

2.2. Hierarchical Data Format

The HDF5 is the current version of a hierarchical data format created at NCSA (National Center for Supercomputing Applications). It is a self describing data format to store high amout of data. It can be employed in many areas like simulations involving physics, storage of geospatial data acquired by satellites or even in medicine. There are API for using the HDF5 data model in various languages like C, C++, Fortran and Java.

It resembles the XML format but the HDF5 stores data in binary format and lets you access data in specific parts of a file without having to load the entire content over random access. In addition, it supports aggregate data types that are compound atomic data type. It was designed to be able to execute partial I/O (read/write chunks) and parallel I/O (through MPI).

In contrast to relational database, hierarchical relashionship between data are expressed in a natural way. Relational database provides an excellent support for query based search on a comparison of field, but does not have good performance for sequential processing of all records or subsets of the data based coordinate-style lookup [13], like the extraction of a region of interest from an image.

2.3. NetCDF

Network Common Data Form (NetCDF) is a set of array-oriented interface for data access and a free collection of data access libraries for various languages like C, Fortran, C++, Java, among others. The NetCDF libraries support a machine-independent format for representing scientific data. Together, the interfaces and libraries support the creation, access and sharing of scientific data [14].

The NetCDF was developed and is maintained by Unidata Program from UCAR (*University Corporation for Atmospheric Research*). NetCDF is very popular data format for storing results of astronomical and physical climate simulations.

The NetCDF tool is older than HDF, the latter arose when in fact one of the goals was to replace the former in many applications. Because of this, many studies comparing these two tools were made. Many innovations that did not exist or did not work well in NetCDF were implemented in the HDF. In both latest versions of NetCDF (NetCDF4) the main change was the integration with HDF5. In conclusion, NetCDF4 uses many new features of HDF without losing compatibility with older versions.

The data model NetCDF4 that was implemented using a storage tier based on HDF5, works in all these limitations. In this model of enhanced data, a file has an initial group unnamed. Each group can contain one or more named variables, dimensions, attributes, groups and types. A variable is still a multidimensional array whose elements are all the same type, each variable may have attributes, and each format variable is specified by its size, which can be shared. However, in this enhanced model, one or more dimensions should be unlimited in size, so the data can be efficiently added to the variables along with any of these dimensions [14].

3. Related Works

The NetCDF has been a very popular and well established model for many years. There are several projects in various scientific fields that have based their systems or support for NetCDF. In this section we will see some of the projects and business entities that use or develop systems based on this tool.

The HALO (*Halogen Occultation Experiment*) was launched in atmospheric research satellite (SRVAS) in 1991, and ,after a period of adjustment, it began scientific observations. The experiment uses solar occultation to measure vertical profiles of O3, HCl, HF, CH4, H2O, NO, NO2, and temperature versus pressure with an instantaneous vertical field of view of 1.6 km on Earth. The latitudinal coverage is 80 degrees S and 80 degrees N, over a year and includes extensive observations of the Antarctic region during spring.

The main objectives of this project are to study the dynamics of atmospheres of various, mainly polar, regions, study on the importance of the relationship between natural and anthropogenic sources of chlorine, measurement of atmospheric gases and aerosols and detailed analysis of development and recovery of the hole in the ozone layer in Antarctica. Data from the experiments of halo are organized in NetCDF format. The *HALOE Data Viewer* makes every data type with a menu-driven interface to assist in locating files based on date, time, type and version of the data [15].

The LDEO (*Lamont-Doherty Earth Observatory*) is a research laboratory, at Columbia University, that complies with the geological and climatic behavior of the Earth. The observatory studies earthquakes, volcanoes and climate change in order to understand and predict the behavior of the Earth and how it affects and will affect society. Scientists observes Earth on a global scale, from its deepest interior to the outer layers of the atmosphere, on every continent and every ocean. They decipher the long history of the past, monitor the present and try to predict the future of Earth. The research institute converted all lasted geophysical data collected in 40 years by the scientists for the NetCDF data format [16].

The CHAMMP (*Climate Change Prediction Program*) is a program of the Department of Energy that conducts research in climate prediction. An important component of the program is that it unites CHAMMP emerging technologies in High Performance Computing for the development of computationally efficient models for precise numerical prediction of climate. Sponsored by the Institute of Biological and Environmental Research (OBER), the program involves an effort to develop computational methods capable of simulating future general circulation model of atmosphere and ocean. These programs are the core of advanced forecasting models that can be used to study climate change and they use NetCDF format as storage for their computer data [17].

The CSIRO (*Commonwealth Scientific and Industrial Research Organisation*) is a national scientific agency in Australia and one of the largest and most diverse in the world. Among it's areas of research are climate adaptation, preventive health, transformed energy, sustainable agriculture, and other ocean riches. The division of atmospheric research and the division for data coming from the ocean models uses for storage the NetCDF [18].

4. The Proposal

The main test to be performed with the proposed model of this work is a performance test comparing the results of the model that is used with HDF5. The chosen version was NetCDF4, which supports the hierarchical model of the HDF and fixes several bugs that the old versions of NetCDF had.

The DICOM data will be stored in a single file just like the HDF model does for hierarchical data representation. One can understand the model of this study through an analogy to the UNIX file system with its pattern of folders and subfolders. We define a folder structure based on the fixed table structure of the of DICOM database currently used in RCTM. Each DICOM image will be inserted and organized according to this pre-defined structure.

This structure is divided into four levels. They were chosen to be implemented similarly to the way in [19], from where this model was chosen considering the information contained in images in DICOM format, as can be seen in Figure 2. The model enabled a satisfactory arrangement for the experiment, allowing a good view, an excellent performance and ease of access to data [19].

Originally the model had a hierarchical level that represented the 'Hospital' because it was considered a national scenario application, which does not correspond to reality today. In this work the level of the 'Hospital' was removed considering the fact that a patient can take exams in more than one health institution.

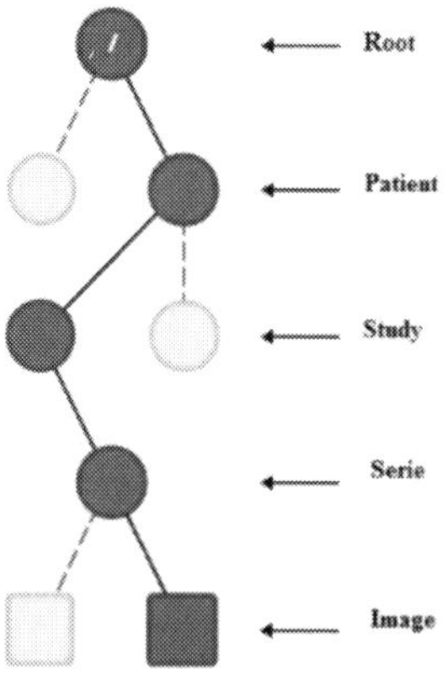

Figure 2. Hierarchical data model used in NetCDF.

As in the UNIX file system, the top of the hierarchy always starts with the "/" level and it is the only group in that level. At the second level we have the name of the patient. In this model there is still abstracted situations such as two patients with the same name, we know that this is possible and often but it is not the focus of this paper. The third level is the study and the fourth and final level is the series.

When we insert an image using DICOM, the application checks if there are already levels related to the image, if it does, the corresponding directory is open and have the data stored in an organized manner. If there are no directories, it is the first time that a patient is inserted

Journal of Physics: Conference Series **341** (2012) 012016 doi:10.1088/1742-6596/341/1/012016

in an examination system, a new study or series, the respective directories to the new data elements are created.

5. Experimental Results

According to the chosen model, the medical images are stored in a single NetCDF file and within that file they are organized and stored in accordance with the determined hierarchy. It was noted that as the file sizes increased some operations were getting slower.

Figure 3 each bar represents the time spent for each image stored. Note that the last image took almost 30 times more than the first to be stored, although both have roughly the same size. This test was performed on a set of 100 images, all with approximately equal size.

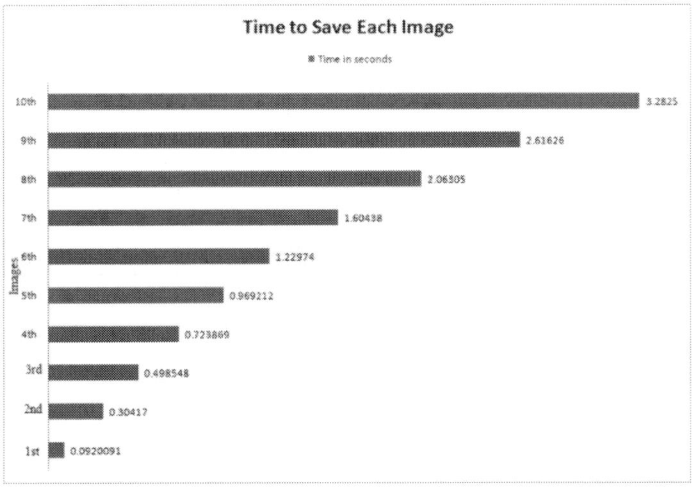

Figure 3. Behavior of image storage in NetCDF.

Isolating each image store procedure, we notice that not all of them were directly related to the increasing time. We performed the same test but this time we measured the main methods runtime. The results show that although the opening operation time increased, the wrinting method presented a linear behavior, spending about the same time running on all images.

6. Conclusions and Future Works

The results from the test showed a behavior that was not expected in this work. The model implemented is similar to the one in [19], both data formats share a common layer that handles the hierarchical portion of the tool. Despite the HDF5 promissing behavior, the NetCDF does not meet the expectations, since the storage time increases as the file grows.

Using a single file to store all the information is not a viable alternative to the problem faced in the RCTM. Though this means that the tools is not suitable for working with this model, it does not mean that it is not suitable for working in this enviroment.

We contacted the NetCDF development team and we were informed that in order to accelerate access to data recorded the entire NetCDF file is loaded into memory, in HDF5 this does not happen, it only loads the requested data. Thus we assured the unsuitable use of this tool for this model.

A new model is being studied to replace the hierarchical model storage of medical images with NetCDF. This new model should consider the tool limitations detected in this work, as well the advantages of the old format that is not hierarchical based.

7. References

[1] McNeill K M, Weinstein R S and Holcomb M J 2002 *Journal of American Medical Informatics Association* URL `http://www.pubmedcentral.gov/articlerender.fcgi?artid=61325`

[2] Maia R, von Wangenheim A and Nobre L 2006 *19th IEEE International Symposium on Computer-Based Medical Systems, 2006. CBMS 2006.* pp 495–500

[3] Wallauer J, Macedo D D J, Andrade R and von Wangenheim A 2008 *IT Professional* **10** 12–17 ISSN 1520-9202

[4] Laboratório de telemedicina Available in: http://www.telemedicina.ufsc.br. Acessed in: 07/14/2011. URL `http://www.telemedicina.ufsc.br`

[5] The cyclops project Available in: http://www.cyclops.ufsc.br. Acessed in: 07/14/2011 URL `http://www.cyclops.ufsc.br`

[6] Macedo D D J, Perantunes H W G, Andrade R, von Wangenheim A and Dantas M A R 2008 *21th IEEE International Symposium on Computer-Based Medical Systems, 2008. CBMS 2008.* pp 495–500

[7] Macedo D D J, Perantunes H W G, Comunello E, von Wangenheim A and Dantas M A R 2008 *IEEE Symposium on Computers and Communications – ISCC, 2008*

[8] Macedo D, Von Wangenheim A, Dantas M and Perantunes H 2009 *International Journal of High Performance Systems Architecture* **2** 99–106

[9] NEMA 2009 Digital imaging and communications in medicine (dicom) part 1: Introduction and overview Tech. rep. National Electrical Manufactures Association.

[10] Mildenberger P, Eichelberg M and Martin E 2002 *European radiology* **12** 920

[11] NEMA 2009 Digital imaging and communications in medicine (dicom) part 3: Information object definitions Tech. rep. National Electrical Manufactures Association.

[12] NEMA 2009 Digital imaging and communications in medicine (dicom) part 4: Service class specifications Tech. rep. National Electrical Manufactures Association.

[13] group H 2010 Hierarchical data format Acessado em: 28 abr. 2010 URL `http://www.hdfgroup.org/`

[14] UNIDATA 2010 Network common data format Acessed in: 28 abr. 2010 URL `http://www.unidata.ucar.edu/software/netcdf/`

[15] NASA 2010 Haloe Acessed in: 14/07/2011 URL `http://haloe.gats-inc.com/about/index.php`

[16] LDEO 2010 Lamont-doherty earth observatory Acessed in: 14/07/2011 URL `http://www.ldeo.columbia.edu/`

[17] CHAMMP 2010 Climate change prediction program Acessed in: 14/07/2011 URL `http://www.csm.ornl.gov/chammp/chammp.html`

[18] CSIRO 2010 Commonwealth scientific and industrial research organisation Acessed in: 14/07/2011 URL `http://www.its.csiro.au/csiro/channel/`

[19] de Macedo D, Comunello E, von Wangenheim A and Dantas M 2008 *Proceedings of the 14th Brazilian Symposium on Multimedia and the Web* (ACM) pp 20–27

High Performance Computing Symposium 2011 IOP Publishing
Journal of Physics: Conference Series **341** (2012) 012017 doi:10.1088/1742-6596/341/1/012017

Load Balancing Issues with Constructing Phylogenetic Trees using Neighbour-Joining Algorithm

S.M. Al Mamun
Department of Computer Science
University of Manitoba
Winnipeg, MB, Canada

E-mail: mamun@cs.umanitoba.ca

Abstract. Phylogenetic tree construction is one of the most important and interesting problems in bioinformatics. Constructing an efficient phylogenetic tree has always been a research issue. It needs to consider both the correctness and the speed of the tree construction. In this paper, we implemented the neighbour-joining algorithm, using Message Passing Interface (MPI) for constructing the phylogenetic tree. Performance is efficacious, comparing to the best sequential algorithm. From this paper, it would be clear to the researchers that how load balance can make a great effect for constructing phylogenetic trees using neighbour-joining algorithm.

1. Introduction

A phylogenetic tree shows the evolutionary relationship among the biological species. There are several well-established strategies to build a phylogenetic tree. Neighbour-Joining algorithm is one of them, which is frequently used to reconstruct a phylogenetic tree based on the distance between each pair of taxa, where taxa is group of organisms. Parallel implementation of this algorithm can make computationally efficient tree for large amount of data. Constructing the phylogenetic tree with high performance has become one of the major problems in bioinformatics. As the sequence becomes larger, huge amount of time is needed to calculate. For example, an input with 50,000x50,000 may not be even computed within several days. If we use the same input for parallel implementation, it would be done by a single day. The tree construction might be finished by several hours even if the code is highly optimized.

2. Background and Related Work

Saitou et al. [1] first designed neighbour-joining approach, which is an iterative algorithm. Several multiple sequence alignment package use neighbour-joining algorithm. Thompson et al. [2] used neighbour-joining algorithm in ClustalW, which is a widely used tool for aligning multiple sequences. Li [3] developed a parallel version of ClustalW, using message passing interface. RapidNJ [4] and NINJA [5] are another two implementations of neighbour-joining algorithm. Stamatakis et al. [6] developed another package named RaxML for phylogenetic tree construction using neighbour-joining algorithm. Sheneman et al. [7] designed a faster version, named as Clearcut which is based on relaxed neighbour-joining algorithm. Liu et al.[8]

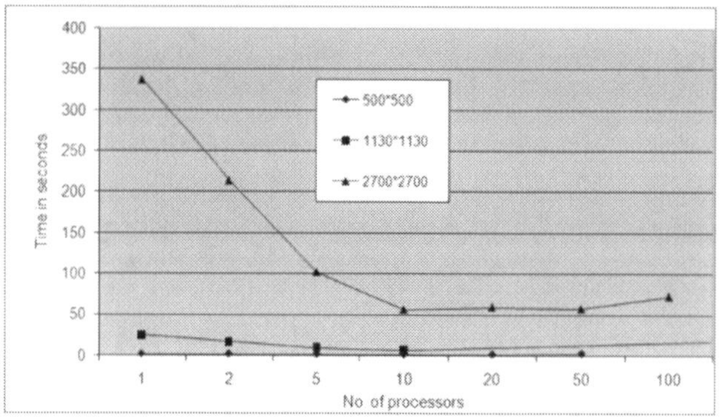

Figure 1. Processors Vs Time(in seconds)

implemented a new approach with neighbour-joining using CUDA. A lot of works are already done with phylogenetic trees using parallel neighbour-joining approaches, but to the best of my knowledge this is the first approach where load balancing is specifically focused.

3. The Neighbour-Joining Algorithm

Neighbouring algorithm is a hierarchical clustering algorithm. It takes a distance matrix, D as input where d_{ij} is the distance between two clusters named i, j. Neighbour-joining algorithm repeatedly joins pairs of closest clusters. This repeating process continues until the tree is completely resolved.

In some of the previous works, researchers used biological sequences as input. From the sequences, they calculated the distance matrix. On the other hand, designers directly provided distance matrix in some other implementations. In this paper, we emphasized on the neighbour-joining algorithm only and considered distance matrix as direct input.

4. Solution Strategy and Implementation

We implemented the parallel niejghbour-joining algorithm in C language using Message passing interface (MPI). At the A NxN distance matrix is used as input where each value is a distance between two species. The root processor reads the size of matrix and broadcasts it to other processors. The matrix was distributed by column-wise to the processors. Here, we faced a peculiar complexity when the datasets were not equally divisible by the number of processors. As a result, we used MPI_Alltoall instead of using MPI_Alltoallv. There might be curiosity why we did not use other libraries like, OpenMP. Because OpenMP performs well only in case of for loops. On the other hand, MPI is an excellent choice as it can handle distributing and gathering data, utilize the perfect flavor of a parallel programming. However, each processor then computes the local sum (which is denoted by r_j) and exchanges the sums with MPI all-to-all. Again, all the processors search a minimum distance, $D_{i,j}$ and exchange their value with another all-to-all MPI call. A global minimum distance is calculated to construct the tree. Each processor computes the node, comparing with all of it's distances and the received minimum distance. At the end, root processor draws the tree as described in the previous section.

5. Experimental Framework

Programs were run in the Helium machine, which is a cluster of Linux servers. It consists of one head node(Sun Fire X4200 machine) and five other computing nodes(Sun Fire X4600 machines), operating system is CentOS 5. Here, head node is a dual core 2.6GHz AMD Opteron processor

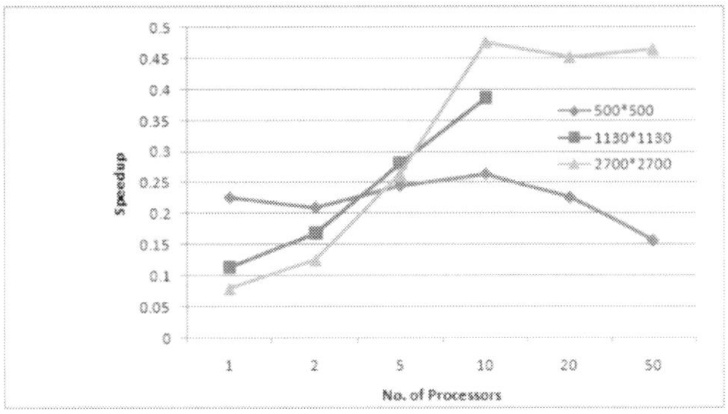

Figure 2. Processors Vs Speedup

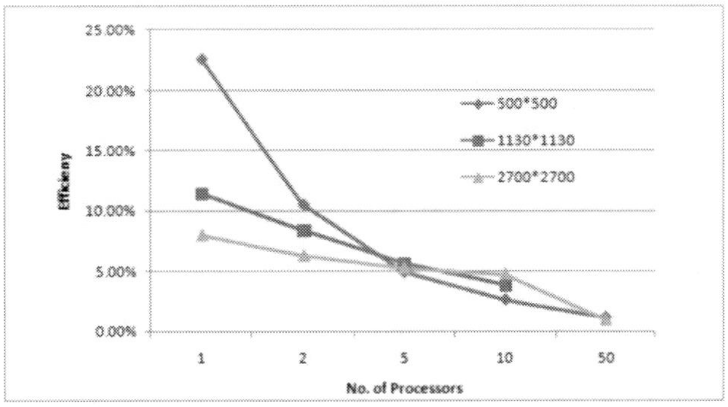

Figure 3. Processors Vs Efficiency

with 2GB memory and each computing node features 8-processor dual-core cache-coherent Non-Uniform Memory Access Symmetric Multiprocessing (ccNUMA SMP) computing model with a dual core 2.6GHz processor and 32GB memory .

6. Results

Several distance matrices were used as input. We used 3 matrices of size 500x500, 1130x1130 and 2700x2700 from `http://www.daimi.au.dk/~cstorm/courses/AiBTaS_e05/project2/`. For these three datasets, the runtime for different parameters was quite similar. That's why we did not put any graph that would show the mean runtime for the three datasets.

When the load per processor is too high, the performance is poor. As the load decreases, performance gets better. Again, when the load per processor is too low, the performance becomes poor. That means, an average load gives the best performance. In figure 1, different timings are plotted for all these 3 datasets and also for the sequential program. For sequential program, we used quicktree [9] as it can build the tree amazingly faster than the other existing sequential implementations. For the same three datasets, quicktree gives the execution time of 0.55 second, 2.91 seconds and 27.11 seconds.

Figure 2 and 3 show the speedup and efficiency for various number of processors. In figure 2, the speedup is highest for 10 nodes. Then, speedup starts decreasing. In figure 3, efficiency also decreases with the number of increasing nodes. But it is reasonable up to 10 nodes. After that, it becomes worse. As, it is highly iterative algorithm, the speedup and efficiency may be a bit low. In figure 4, the cost for different datasets is shown. The same case happens here as like efficiency, up to 10 nodes costs are reasonable. For more than 10 nodes, costs have been

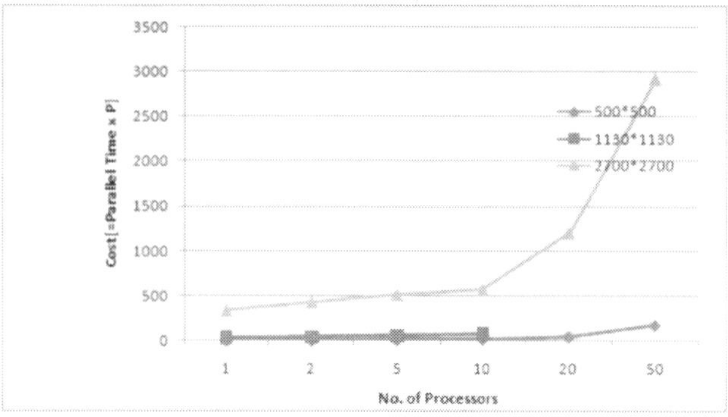

Figure 4. No. of Processors Vs Cost

increased drastically. It means that for neighbour-joining algorithm whatever the matrix size is, we will get the best performance for a certain number of processors.

7. Conclusion and Future Work

We implemented the simplest version of neighbour-joining algorithm. From the experiment and comparison with the performance of quicktree, we can see in figure 2 that when each processor is responsible for a reasonable amount of workloads, it gives the best performance. Only then, the highest possible speedup can be gained. In future, we will be working on it so that it can handle when data size can not be divided by the number of processors. We will also apply the hybrid programming techniques to examine the load balancing issues with constructing phylogenetic trees using neighbour-joining algorithm.

8. Acknowledgments

I would like to thank Dr. Parimala Thulasiraman for all of her helpful guidelines and feedbacks starting from the beginning through the end of this implementation. I would also like to thank Dr. Michael Domaratzki who encouraged me to face the challenge and suggested various guidelines during the implementation.

References

[1] Saitou N and Nei M 1987 *Mol Biol Evol.* **4** 406–425
[2] Thompson J D, Higgins D G and Gibson T J 1994 *Nucleic Acids Research* **22** 4673–4680
[3] Li K B 2003 *BIOINFORMATICS* **19** 1585–1586
[4] Simonsen M, Mailund T and Pedersen C 2002 *Algorithms in Bioinformatics* vol 5251 pp 113–122
[5] Wheeler T 2009 *Proceedings of the 9th Workshop on Algorithms in Bioinformatics (WABI 2009)* pp 375–389
[6] Stamatakis A 2006 *BIOINFORMATICS* **22** 2688–2690
[7] Sheneman L, Evans J and Foster J A 2006 *BIOINFORMATICS* **22** 2823–2824
[8] Liu Y, Schmidt B and Maskell D L 2009 *IEEE International Workshop on High Performance Computational Biology (HiCOMB 2009)*
[9] Howe K, Bateman A and Durbin R 2002 *BIOINFORMATICS* **18** 1546–1547

Journal of Physics: Conference Series **341** (2012) 012018 doi:10.1088/1742-6596/341/1/012018

Scalability of Self-organizing Maps on a GPU cluster using OpenCL and CUDA

Sabine McConnell, Robert Sturgeon, Gregory Henry, Andrew Mayne, and Richard Hurley

Department of Computing and Information Systems, Trent University, Peterborough, On, Canada

E-mail: `sabinemcconnell@trentu.ca`

Abstract. We evaluate a novel implementation of a Self-Organizing Map (SOM) on a Graphics Processing Unit (GPU) cluster. Using various combinations of OpenCL, CUDA, and two different graphics cards, we demonstrate the scalability of the SOM implementation on one to eight GPUs. Results indicate that while the algorithm scales well with the number of training samples and the map size, the benefits from using the data-parallel approaches offered by the GPU are severely limited when combined with the Message Passing Interface (MPI) in this setting, and comparable to speedups of GPU-based implementations as compared to optimized sequential code. Speedups achieved range from 3 to 32, for various map and training data sizes. We also observed a performance penalty for the OpenCL implementation as compared to CUDA.

Since the introduction of NVIDIA's release of the Compute Unified Device Architecture (CUDA) in November 2006, which enabled programmers to bypass traditional graphics Application Programming Interfaces (APIs) such as OpenGL for General-Purpose GPU computing, the suitability of graphics cards for computationally intensive data-parallel applications has been demonstrated by a wide range of applications. Recent examples include the calculation of powder diffraction patterns [5] and real-time stream scheduling that utilizes both GPU and CPU [14].

Early on, the medical community recognized the potential for the use of Graphic Processing Units in, for example, organ segmentation [6]. With the advent of CUDA and the removal of the need for substantial background knowledge in graphics, the number of applications increased steadily where today, the use of GPUs has a strong base in medical computation. In addition to, but separate from the use of GPUs, the medical community has also taken advantage of data mining, which is typically defined as the extraction of novel information from large dataset. Successful applications of this include an early evaluation of neural networks and genetic programming for classification and generalization on multiple diagnosis problems by Brameir and Banzhaf [3], as well as the diagnosis of cardiopulmonary disease [4], breast cancer [1], and diabetes [2].

With the exponential growth of datasets over the past decade and the increased use of medical data-mining apppliations, the data-mining community has moved into high-performance settings including accelerators which are characterized as hardware that performs certain computations faster than the CPU. Examples for such accelerators include Field Programmable Gate Arrays, the Cell Broadband Engine Architecture (CBEA), and GPUs. Implementations of data-mining techniques on GPUs include a neural network for bankruptcy prediction [12], Expectation Maximization for Gaussian Mixture Models [8], and Support Vector Machines [10]. Liao et al. [10] present a parallel implementation evaluated on six different datasets, achieving a maximum speedup of 212 for the kernel alone, and a speedup of up to 94 as compared to the CPU version when taking memory transfers into account. To the best of our knowledge, no implementations of data-mining techniques exist to date for clusters of GPUs. In this paper, we evaluate the scalability of Self-Organizing Maps, a particular data-mining technique, on the SHARCNET cluster Angel, which supports both Open Computing Language (OpenCL) and CUDA. We investigate the scalability of the implementations with respect to the size of the input dataset, the size of the map, and the effects of the different programming environments.

The remainder of this paper is organized as follows. In Section 1, we introduce the concepts of data mining and self-organizing maps. We describe the implementation along with the experimental setup and environment in Section 2, followed by a discussion of the results in Section 3. Section 4 summarizes our findings and points to future work.

1. Data Mining and Self-Organizing Maps

1.1. Data Mining

Over the past two decades, the field of data mining as a tool to extract novel information from typically large datasets has matured to a point where it is used to answer a wide variety of questions in a large range of application areas. In order to do so, a range of techniques that can be categorized into *predictive* and *descriptive* techniques, are utilized. Predictive techniques include *Decision Trees*, *Rule-Based Approaches* and *Support Vector Machines* and develop a model from data containing predefined classes and class labels. The model is then used to predict class labels for previously unclassified data. In contrast, descriptive techniques such as *k-means clustering* and *Expectation Maximization* detect structure inherent in the data without the use of classes and labels.

One particular challenge in data mining (other than the need to analyze data sets on the order of Tera- and Petabytes) is its interactive nature, together with the wide range of available techniques. Typically, the model-building stage that uses a single particular technique follows a preprocessing step in which the data is transformed to the format expected by the selected technique (for example, numerical values only for Support Vector Machines). This means that if the learning technique is changed, the preprocessing step has to be modified and repeated. In addition, the data set itself is often refined iteratively. For all but small data sets, this interactive process results in prohibitively large run times. For that reason, the data-mining community has turned towards parallel implementations using a variety of environments, focusing most recently on accelerator architectures including GPUs and the CBEA.

1.2. Self-Organizing Maps

Self-organizing maps, originally proposed by Kohonen [7] are variants of Artificial Neural Networks, which themselves date back to the proposal by McCulloch and Pitts [11] that used artificial neurons as a basic building block in modeling a complex structure. A SOM is a typically two- or three-dimensional map of neurons that consist of weights (one for each dimension of the input data), which represents a higher-dimensional data space in fewer dimensions in order to employ human pattern recognition capabilities. During the training stage, the input data is presented to the map repeatedly and the winning neuron (the closest map neuron in terms of Euclidean distance) is determined, then updated along with its neighbourhood to more closely represent the input data. Both sequential and batch training algorithms are available. In a sequential approach, the map updates are calculated based on the distance between the winning node and the input sample, according to the rule $\mathbf{m}_i(t+1) = \mathbf{m}_i(t) + \alpha h_{ci}(t)[\mathbf{x}(t) - \mathbf{m}_i(t)]$, where $\mathbf{m}_i(t)$ denotes the m_{th} neuron at iteration t, α the learning rate, h_{ci} the neighborhood function, and $\mathbf{x}$ the input vector. Over time, both the neighborhood function and the learning rate decrease. Convergence of the map occurs often over a substantial number of iterations through map updates calculated from the complete input data set. In contrast, the batch training algorithm replaces the node updates that are based on Euclidean distances with weighted averages of the input samples that are assigned to the map nodes. In both cases, the neighborhood function is often reduced to updating only the winning node. The computationally significant portion of the learning algorithm is the determination of the Euclidean distance between each input sample and all map nodes, while the required synchronization of map updates is the limiting factor for any distributed-memory implementation.

2. Experimental Setup and Environments

In this section, we describe the implementation of the SOMs and the environments under which it was evaluated. The data-parallel version using the GPU performs the calculations of the distance and map updates using the streaming processors, while the CPU is responsible for kernel calls and memory transfers. The map is accessed frequently, and, as a result, resides in shared memory of the GPU in our implementation. Each thread is responsible for calculation of the distance of one particular input datum to the map, determining the winning map node for its assigned input datum. The updates to the map therefore need to be atomic, which results in stalls for threads if two input data share a winning node. If more than one GPU is used, then the training dataset is distributed over the processors and partial map updates are collected and processed at the root, followed by a broadcast of the updated map at each iteration.

We used two different APIs that are currently candidates for a developing standard for General Purpose GPU programming: CUDA and OpenCL. CUDA has undergone nine releases as of the time of this writing and was designed by NVIDIA specifically for their GPUs. In contrast, OpenCL is designed for use across heterogeneous platforms, and is being developed by the Kronos Group, a not-for-profit industry consortium. Both use the concept of a kernel to represent a unit of code used for general purpose computation. The memory hierarchy model is conceptually similar in both, only differing slightly in the names of hierarchical levels. Allocation and management of device memory is similar in concept as well.

OpenCL mainly differs from CUDA in its approach to managing and executing instructions on the GPU. OpenCL employs a model which abstracts the program flow, logic and synchronization required for computation from device and vendor specific API calls used to carry out the computation. As such, OpenCL programs are required at run time to explicitly manage and determine which devices will be used. The choice of devices then dictates how the underlying interactions between the device and host will be carried out. Another major difference is the way in which OpenCL programs typically use a run-time compiler to produce kernel binaries, for a specified device, from raw strings of code. In contrast, CUDA kernels are compiled along with the main program for a specific target device/system. Compiled OpenCL kernels can be stored to disk and used in subsequent instances of the program. However, depending on the OpenCL implementation the kernel may be stored in an intermediate language (device specific) instead of executable binary.

We used two different environments for our evaluation: a Zotac GeForce GT 220 with 1GB DDR2 and the Sharcnet cluster Angel. The former is hosted by an AMD Athlon 64 X2 Dual Core Processor 5400+ with 2GB memory and a clock speed of 2.8 GHz. Angel consists of 11 NVIDIA Tesla S1070 GPU servers; each server contains 4 GPUs and 16GB global memory, with a total peak performance of over 345 GFlops (double precision). The GPU servers are connected via a DDR Infiniband and GigE copper network in a standard star topology with a single switch.

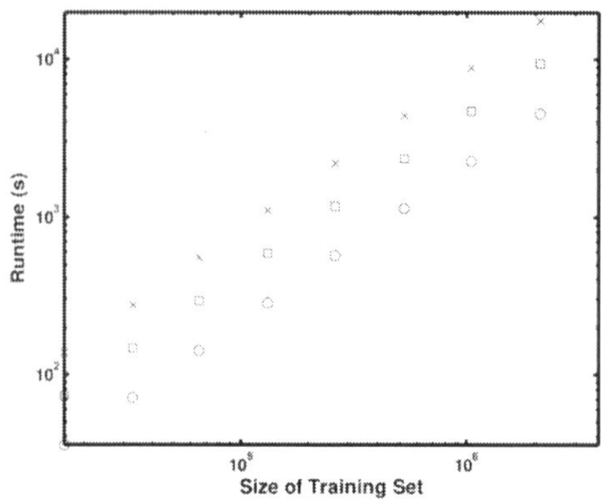

Figure 1. Runtime as a function of varying training set size on a single CPU. Shown are results for a map size of 64 nodes with 4 (o) and 8 (□) attributes) and 256 nodes and 4 attributes (x).

3. Results

In order to determine the scalability of our implementation, we present the following results. Section 3.1 shows how our implementation scales on a single GPU, using two different graphics cards in combination with OpenCL and CUDA. For comparison, we include results obtained from implementation on a CPU only. Section 3.2 discusses the run times and speedups for the SOM using MPI only, and up to eight processors. Finally, in Section 3.3, we combine MPI with CUDA to determine the scalability of the hybrid implementation, using up to 8 GPUs. For each section, the data sets range in size from 16 384 to 2 097 152 samples of simulated data, with 4 or 8 attributes, and a map size of either 64 or 256 nodes. All results are averaged over five trials, for 500 iterations.

3.1. Scalability on a Single GPU

Figure 1 shows the scalability of the SOM implementation on a single CPU, for varying training set sizes and map sizes, on a log-log plot. Results indicate that the implementation scales well with the number of samples and map size, but that the run times are prohibitive for interactive data mining beyond 10 000 samples. In addition, the confidence intervals are too small to show on the figure.

Next, we evaluate the scalability of our implementation using varying numbers of training samples and map sizes, as well as two graphics cards in combination with OpenCL and CUDA. Shown are results for a map size of 64 nodes with 4 (o) and 8 (□) attributes and 256 nodes and 4 attributes (x), for Figures 2 through 4.

Figure 2 shows a log-log plot of the run times for the CUDA implementation on a single GPU using the GT 220, for three different map sizes and varying size of training data up to 2 Million samples. In general, the algorithm scales close to linear with the training set size. For example, increasing the number of training samples by a factor of 128 resulted in a runtime increase of a factor of 125 for a map with 64 nodes, and eight attributes. Results for other training set sizes are similar, given a fixed map size. In addition, the implementation scales well with the map size (increasing the number of nodes in the map by a factor or four results in less than threefold increase in runtime), but less with the number of attributes of the training data and the map. That is, doubling the number of attributes for a given size of the map resulted in an increase by roughly a factor of three, for all training set sizes. We believe that the increase in memory transfers for updates of the larger map is responsible for this effect.

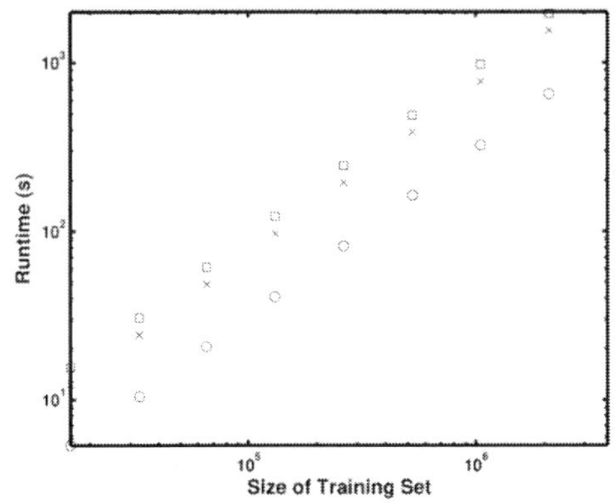

Figure 2. Runtime as a function of varying training set size using CUDA and the GT 220.

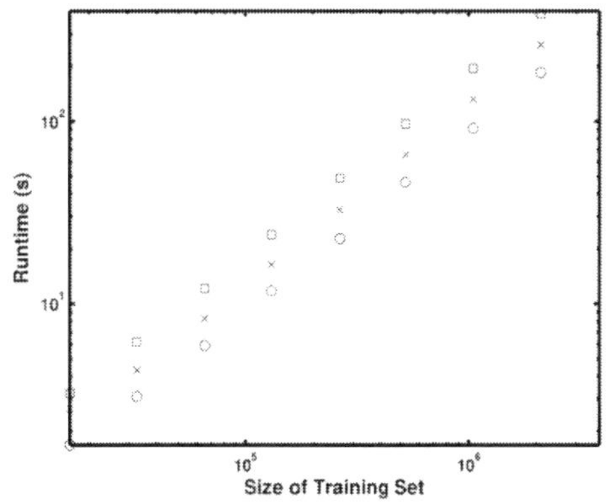

Figure 3. Runtime as a function of varying training set size using CUDA and NVIDIA's Tesla 1070.

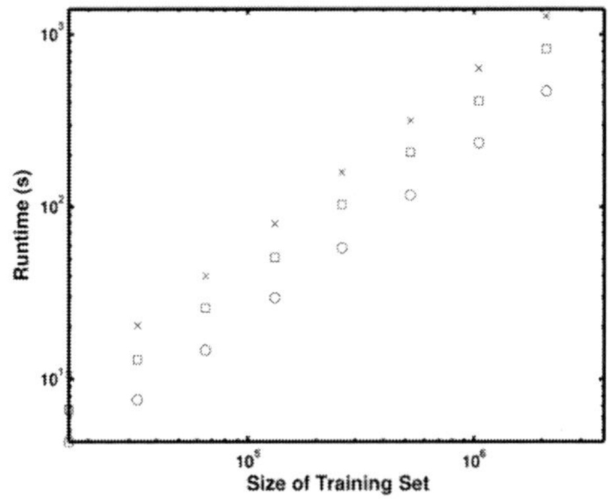

Figure 4. Runtime as a function of varying training set size using OpenCL and NVIDIA's Tesla 1070.

The speedups achieved using CUDA on the GT 220 vary for the different map sizes and increase slightly as the training size is increased, for a fixed map size. For example, the speedup increases from 11.14 to 11.39, for the largest map, as the size of the training data increases by a factor of 2^7. The average speedups for the two smaller maps are 6.90 (4 attributes) and 4.79 (8 attributes), both with small standard deviations.

Figure 3 shows a log-log plot of the run times for the CUDA implementation on a single GPU on Angel, for three different map sizes and varying size of training data, up to 2 Million samples. While trends similar to those discussed for Figure 2 are visible, of particular interest is the fact that there is a substantial decrease in runtime in general, for all training set and map sizes, as compared to the results obtained by the GT 220. For example, in the smallest map, runtime reductions fall between 3.20 and 3.54, while for the largest map the runtime decrease by a factor that ranges from 4.67 to 5.91. The reductions in runtime for the CUDA implementation on Angel increase slightly with the size of the training data and are summarized in Tables 1 through 3.

Figure 4 shows a log-log plot of the run times for the OpenCL implementation on a single GPU on Angel, for three different map sizes and varying size of training data up to 2 Million samples. While the implementation again scales with the size of the training data, we observed substantial runtime penalties as compared to the implementation using CUDA, which were most prominent for the largest map. While the speedup is roughly reduced by a factor of two for the smaller maps, there is a four to five fold drop in performance, depending on the size of the training set, for the largest map.

Table 1. Speedups for OpenCL and CUDA using two GPUs as compared to the sequential implementation, for a map with 64 nodes (4 attributes) and varying sizes of training data

Training Data	GT 220(CUDA)	Tesla(CUDA)	Tesla(OpenCL)
16 384	6.62	21.21	8.17
32 768	6.80	23.01	9.37
65 536	6.88	24.06	9.69
131 072	6.95	24.12	9.60
262 144	6.97	25.03	9.85
524 288	6.98	24.53	9.75
1 048 576	6.98	24.76	9.74
2 097 152	6.99	24.71	9.73

Table 2. Speedups for OpenCL and CUDA using two GPUs as compared to the sequential implementation, for a map with 64 nodes (8 attributes) and varying sizes of training data

Training Data	GT 220(CUDA)	Tesla(CUDA)	Tesla(OpenCL)
16 384	4.72	22.77	11.05
32 768	4.77	23.83	11.36
65 536	4.80	24.24	11.37
131 072	4.80	24.61	11.58
262 144	4.81	24.10	11.45
524 288	4.82	24.27	11.41
1 048 576	4.81	24.23	11.44
2 097 152	4.82	24.38	11.46

Table 3. Speedups for OpenCL and CUDA using two GPUs as compared to the sequential implementation, for a map with 256 nodes (4 attributes) and varying sizes of training data

Training Data	GT 220(CUDA)	Tesla(CUDA)	Tesla(OpenCL)
16 384	11.14	52.01	12.91
32 768	11.25	63.83	13.46
65 536	11.33	66.16	13.87
131 072	11.36	66.87	13.87
262 144	11.38	67.40	13.91
524 288	11.38	67.06	13.90
1 048 576	11.39	67.05	13.87
2 097 152	11.39	67.37	13.86

3.2. MPI only

In this section, we discuss the results for the implementation using MPI only. We evaluate the implementation of the SOM using a single map size (256 nodes, 4 attributes) with varying numbers of training samples, up to 2 Million samples. Using Angel, we vary the number of processors from 1 to 8 over 500 iterations through the training data.

Figures 5 and 6 show the runtimes for the four smallest and largest datasets, respectively. The data depicted in the figures show that there is a trend towards better performance for larger

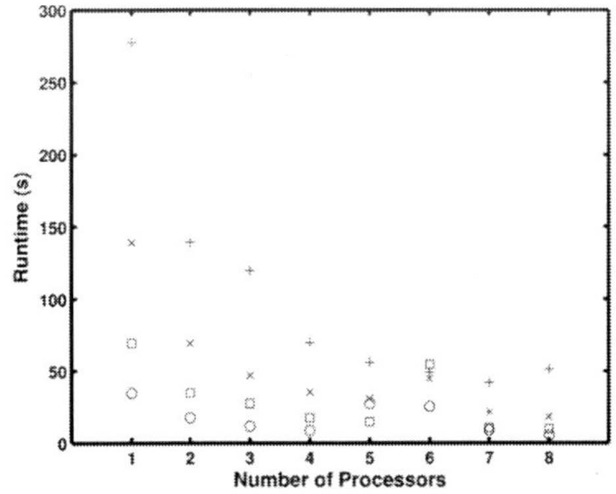

Figure 5. Runtimes for the Four Smallest Datasets: 16 384 (o), 32 768 (□), 65 536 (x), and 131 072 (+) samples.

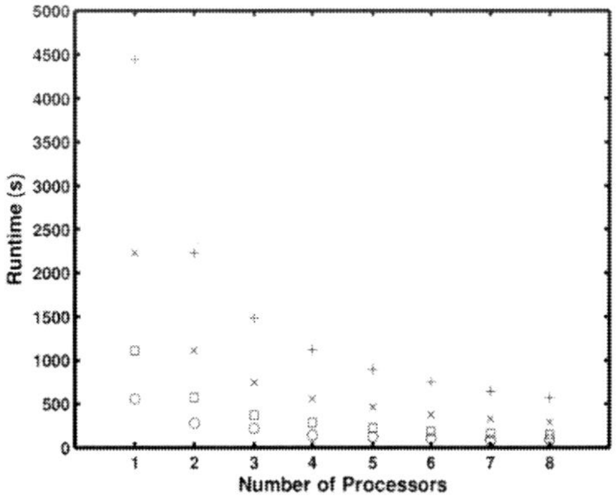

Figure 6. Runtimes for the Four Largest Datasets: 262 144(o), 524 288 (□), 1 048 576 (x), and 2 097 152 (+) samples.

datasets, as expected, where the costs for initial data distribution and communication of partial updates after each iteration are better amortized. The smaller datasets also show more variation in the runtimes for each of the trials than the larger datasets, which is caused by the network characteristics. In all cases, the confidence intervals are too small to show on the figures.

Table 4. Speedups for Increasing Training Data Size and Number of Processors, for a map size of 256 nodes (4 attributes) and 16 384 to 2 097 152 samples (shown in rows), using up to 8 processors (shown in columns).

Training Data	2	3	4	5	6	7	8
16 384	1.99	2.99	3.97	1.27	1.38	3.91	6.45
32 768	2.00	2.53	3.97	4.69	1.27	6.66	7.11
65 536	2.00	2.98	3.97	4.53	3.11	6.45	7.62
131 072	1.99	2.32	3.97	4.94	5.65	6.60	5.41
262 144	1.99	2.51	3.90	4.47	5.61	6.58	6.30
524 288	1.94	2.98	3.89	4.94	5.93	6.81	7.52
1 048 576	1.99	2.99	3.97	4.76	5.93	6.78	7.51
2 097 152	2.00	2.99	3.97	4.95	5.91	6.90	7.83

Table 4 shows the attained speedups for the MPI-only version of the Self-Organizing Map. The maximum speedup is achieved by the largest data set, with a speedup of 7.83 for eight processors. With few exceptions, we observe that, as expected, the speedup grows consistently with the training set size, for a given number of processors. For a fixed-size dataset, the speedup

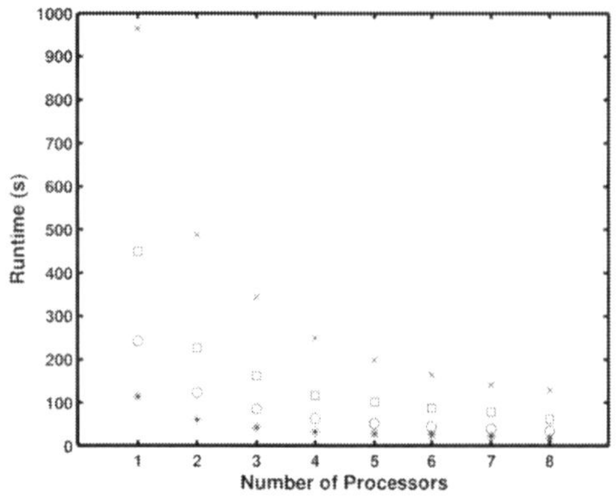

Figure 7. Runtime as function of the number of processors, 64 map nodes, 4 attributes.

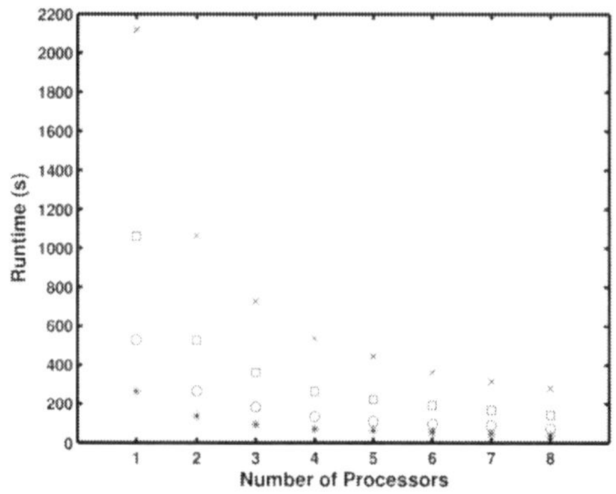

Figure 8. Runtime as function of the number of processors, 64 map nodes, 8 attributes.

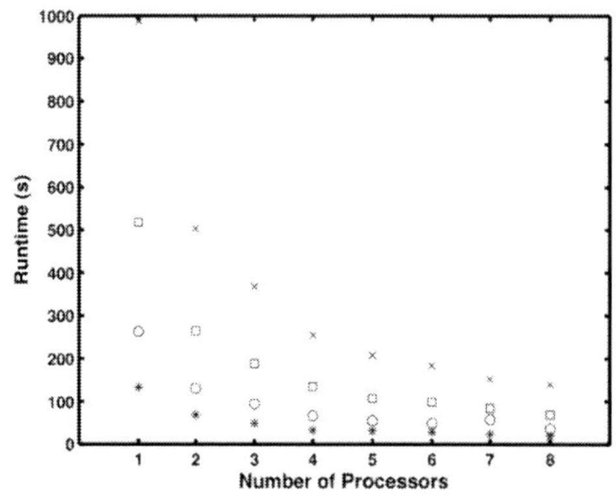

Figure 9. Runtime as function of the number of processors, 256 map nodes, 4 attributes.

grows close to linear with the number of processors, which is especially evident for the larger datasets.

3.3. Multiple GPUs

Finally, we evaluated the performance of the implementation using CUDA in combination with MPI. Figures 7 through 9 show the run times of the hybrid code, for various map sizes and 500 iterations. Shown are the results for 262 144 (∗), 524 288 (o), 1 048 576 (□), and 2 097 152 (x) samples. We observe the following trends: For the larger datasets, our implementation scales fairly well with the number of processors when combining MPI with CUDA, for all map sizes, as compared to the results obtained with a single GPU. The efficiencies are highest for two GPUs, but then drop off consistently. However, for the three largest datasets, the speedup as compared to using a single GPU does not drop below for a value of 7, for eight processors. The overhead in our implementation stems in part from the initial distribution of the data, but, more importantly, also from the required synchronization of the map updates.

The achieved speedup increases with the size of the training set, for fixed number of processors and map size; the largest performance improvement when combining multiple GPUs was 7.55

and occurred for the largest training set and map size. For all combinations of number of processors and size of training data, the speedup increases when increasing the dimensionality of the data from 4 to 8 attributes. In general, the increase in speedup when increasing the number of attributes is less than 10%, with a few exceptions. The largest increase in speedup when increasing the dimensionality of the data was observed for 32 768 samples and 7 processors, with a performance improvement of 40%. In addition, there is no general trend towards either a performance drop or improvement when increasing map size from 64 to 256 nodes. Overall, the speedup achieved as compared to the sequential version when combining CUDA with MPI ranges from 3.30 (single GPU, smallest dataset) to 32.45 (8 GPUs, second-largest data set).

For comparison, Table 5 shows the relative speedup when adding CUDA to the MPI implementation of the Self-Organizing Map, for different training set sizes and number of processors. The low speedups are similar to those found by Lee et al. [9], who compared GPU implementations to optimized sequential code. With the exception of the second largest data set for 7 processors, all performance gains are below a factor of 5. This relatively poor performance of the combination of MPI and CUDA stems from the relatively low computationally requirements of the learning algorithm at each iteration, combined with the overhead at each iteration.

Table 5. Relative Speedups for Distributed Version With and Without CUDA for Increasing Training Data Size and Number of Processors, for a map size of 256 nodes (4 attributes) and 16 384 to 2 097 152 samples (shown in rows), using up to 8 processors (shown in columns).

Training Data	1	2	3	4	5	6	7	8
16 384	3.30	2.66	2.45	2.07	6.82	6.48	2.35	1.40
32 768	3.70	3.23	2.49	2.69	2.65	10.90	2.18	2.15
65 536	4.79	3.31	3.18	3.10	2.74	3.60	1.79	2.40
131 072	4.18	3.99	4.45	3.88	2.73	3.30	2.59	4.60
262 144	4.16	4.06	4.57	4.27	3.91	3.43	3.55	4.43
524 288	4.22	4.36	3.99	4.30	4.09	3.84	2.84	4.01
1 048 576	4.30	4.21	3.94	4.13	4.36	3.82	3.89	4.32
2 097 152	4.50	4.43	4.03	4.39	4.31	4.09	4.20	4.06

4. Conclusions and Future Work

We implemented and evaluated a Self-Organizing Map on the Sharcnet GPU cluster Angel. An MPI-only version scaled well with the number of processors. As well, the GPU-only version, implemented using CUDA and OpenCL and evaluated using two different grapics cards, achieved speedups of up to 67 on the Tesla GPU server. However, the hybrid MPI/CUDA version showed an increase in speedup of roughly a factor of four only, caused mainly by the synchronization requirements of the algorithm. One possible avenue for future work would be the use of GPUDirect using CUDA 4, which allows memory access to remote GPUs from within kernels but was not supported by the cluster we used. We also observed a performance penalty for the OpenCL implementation as compared to CUDA, and therefore prefer the use of the latter. Moreover, OpenCL is designed for a more heterogeneous set of devices, as opposed to CUDA's GPU-specific approach. Future work will include the implementation of additional data-mining algorithms on the GPU cluster, as well as the evaluation of pre-processing techniques required by various learning techniques, with the overall goal of providing a data-mining toolkit for GPU clusters based on both CUDA and OpenCL.

Acknowledgments
Acknowledgments. This work was made possible by a Fellowship Award from the Shared Hierarchical Academic Research Computing Network (SHARCNET:www.sharcnet.ca).

References
[1] Abdelaal, M., and Farouq, M., *Using data mining for assessing diagnosis of breast cancer*Proceedings of the International Multiconference on Computer Science and Information Technology pp. 1117
[2] Bose, I., *Data Mining in Diabetes Diagnosis and Detection*, Data Warehousing and Mining: Concepts, Methodologies, Tools, and Applications, J. Wang (Ed.), pp. 1817-1824, 2008
[3] Brameier, M. and Banzhaf, W., *A comparison of linear genetic programming and neural networks in medical data mining*, IEEE Transactions on Evolutionary Computation, 5(1), pp. 17-26. Feb 2001
[4] Flietstra, B., *A Data Mining Approach for Acoustic Diagnosis of Cardiopulmonary Disease*, MSc Thesis, Massachusetts Institute of Technology, June 2008
[5] Geliso L., Azanza Ricardo C. L., Leoni, M. and Scardi, P., *Real-space calculation of powder diffraction patterns on graphicsprocessing units*, Journal of Applied Crystallography 43:647-653, 2010.
[6] Grady, L., Schiwietz, T., and Ahron, S., *Random walks for interactive organ segmentation in two and three dimensions: Implementation and validation*, Proceedings of MICCAI 2005 II, 2005
[7] Kohonen, T.: The Self-Organizing Map. Proceedings of the IEEE, Vol. 78(9), pp. 1551-1558.
[8] Kumar, N., Satoor, S. and Buck, I., *Fast Parallel Expectation Maximization for Gaussian Mixture Models on GPUs Using CUDA*, 11th IEEE International Conference on High Performance Computing and Communications (HPCC 09), pp. 103-109, 25-27 June 2009
[9] Lee, V. et al., *Debunking the 100X GPU vs. CPU Myth: An Evaluation of Throughput Computing on CPU and GPU*, in ISCA10:Proceedings of the 37th Annual International Symposium on Computer Architecture. New York, NY, USA: ACM, 2010, pp. 451460.
[10] Liao, Q., Wang, J., Webster, Y., and Watson, IA., Journal of Chemical Information and Modeling, 2009 Dec; 49(12):2718-25.
[11] McCulloch, W. and Pitts, W., *A logical calculus of the ideas immanent in nervous activity* Bulletin of Mathematical Biophysics 5:115-133, 1943
[12] Ribeiro, B., Lopes, N., and Silva, C., *High-Performance Bankruptcy Prediction Model using Graphics Processing Units*,IEEEWCCI 2010 IEEE World Congress on Computational Intelligence July, 18-23, 2010 - CCIB, Barcelona, Spain
[13] Singh, J. and Aruni, I., *Accelerating Power Flow studies on Graphics Processing Unit*, Proceedings of the Annual IEEE India Conference 2010 (INDICON), pp 1-5, Dec. 2010
[14] Verner U., Schuster A., and Silberstein, M., *Processing data streams with hard real-time constraints on heterogeneous systems*, ICS11, to appear

High Performance Computing Symposium 2011 IOP Publishing
Journal of Physics: Conference Series **341** (2012) 012019 doi:10.1088/1742-6596/341/1/012019

Predicting Alzheimer's disease by classifying 3D-Brain MRI images using SVM and other well-defined classifiers

S. Matoug*[1], A. Abdel-Dayem, K. Passi, W. Gross and M. Alqarni
Department of Mathematics and Computer Science
Laurentian University
935 Ramsey Lake Road Sudbury ON P3E 2C6 CANADA

E-mail: [1]sx_matoug@laurentian.ca

Abstract. Alzheimer's disease (AD) is the most common form of dementia affecting seniors age 65 and over. When AD is suspected, the diagnosis is usually confirmed with behavioural assessments and cognitive tests, often followed by a brain scan. Advanced medical imaging and pattern recognition techniques are good tools to create a learning database in the first step and to predict the class label of incoming data in order to assess the development of the disease, i.e., the conversion from prodromal stages (mild cognitive impairment) to Alzheimer's disease, which is the most critical brain disease for the senior population.

Advanced medical imaging such as the volumetric MRI can detect changes in the size of brain regions due to the loss of the brain tissues. Measuring regions that atrophy during the progress of Alzheimer's disease can help neurologists in detecting and staging the disease.

In the present investigation, we present a pseudo-automatic scheme that reads volumetric MRI, extracts the middle slices of the brain region, performs segmentation in order to detect the region of brain's ventricle, generates a feature vector that characterizes this region, creates an SQL database that contains the generated data, and finally classifies the images based on the extracted features. For our results, we have used the MRI data sets from the Alzheimer's Disease Neuroimaging Initiative (ADNI) database[2].

1. Introduction

AD causes nerve cell death and tissue loss throughout the brain, resulting to brain tissue shrinking and larger ventricles (chambers within the brain that contain cerebrospinal fluid). When AD is suspected, the diagnosis is first confirmed with behavioural assessments and cognitive tests and often followed by

* To whom any correspondence should be made
[2] http://adni.loni.ucla.edu/

a brain scan [1]. Advanced medical imaging with computed tomography (CT) or magnetic resonance imaging (MRI), and with single photon emission computed tomography (SPECT) or positron emission tomography (PET) can be used to help exclude other cerebral pathology or subtypes of dementia [1]. Moreover, it may predict conversion from prodromal stages (mild cognitive impairment) to Alzheimer's disease [1].

The following report describes the whole process of pattern recognition where several steps are performed such as access ADNI database, describe the medical data, read the volumetric MRI, extract the middle slice of the brain region, perform segmentation methods in order to detect the region of brain's ventricle, generate a vector of attributes that characterizes this region, create an SQL database that contains the generated data, perform clustering to get the class labels and finally perform some classification methods based on the clustering results.

Section 2 describes briefly the organizational schema of the system implementation. Section 3 describes some of the tools used to access the medical ADNI database, gives an overview of the type of medical data files and discusses the problems encountered during the first step of accessing data. Section 4 includes the segmentation methods used to extract the regions of interest and shows the way they were used during the implementation process. Section 5 defines the vectors of attributes and shows their comprehensive statistical analysis. Section 6 introduces briefly the classification methods and shows the results of the used database in addition to classical databases results. Finally, in the conclusion, we summarize the different steps, assess the overall work and bring the future projection to the work.

2. Organizational schema of the system implementation

The proposed system is to perform 2D image segmentation of the original 3D MRI neuroimage brain data followed by attribute extraction and classification/prediction in order to assess, in the future, whether the patient is developing the AD and at which stage the disease stands. There are many existing segmentation techniques applied for medical image segmentation, including statistical methods, thresholding, edge detection, region-based techniques and more recently multi-resolution (using wavelets, ridgelets, etc.) techniques (figure 1, [18]). The choice of one method will depend on the type and quality of the image and the statistics of the extracted regions. The choice of attributes depends closely on the object's shape and statistics such as the statistical moments of order n, the textures, the geometric measures, etc. Once, we extract the attributes, we create a data base that is well defined and large enough to contain both training and testing data. A statistical analysis is made on the database to extract meaningful parameters and distributions that might assist the learning and classification processes. The first step of classification is to proceed to the learning process in order to produce the classes' categorical labels, and then we perform the classification. A step of accuracy measure of the classifier might be added to assess the accuracy of the employed classifier. The classification process is shown in figure 1.

3. Data access

The data we are using comes from Alzheimer's Disease Neuroimaging Initiative database (ADNI). ADNI is a multisite longitudinal clinical/imaging/genetic/biospecimen/biomarker study, whose goal is to determine the characteristics of AD as the pathology evolves from normal ageing to mild symptoms, to MCI, to dementia. It is a generally accessible data repository, which describes longitudinal changes in brain structure and metabolism.

ADNI uses several medical file formats such as the classical Analyse Format (hdr/img) that contains a header file and a separate *3D* image and the next generation of medical images based on the Analyze Format, called NIFTI which is an *nii* structure containing both the header file and the *3D* image.

The headers contain the information about the data such as the patient sex and age, the type of radiography, the view, size of voxels, etc, all of them stored into an info structure and the data itself as a 3D matrix usually of single type. All the nifti medical image files in ADNI database have the same standard which is: ADNI_pppp_S_ssss_Sequence_Sxxxx_Iyyyyy.nii where *pppp* is the patient ID, *ssss* is the site ID, *Sequence* is the Sequence and processing steps, *Sxxxx* is LONIUID and *yyyyy* is the Image ID.

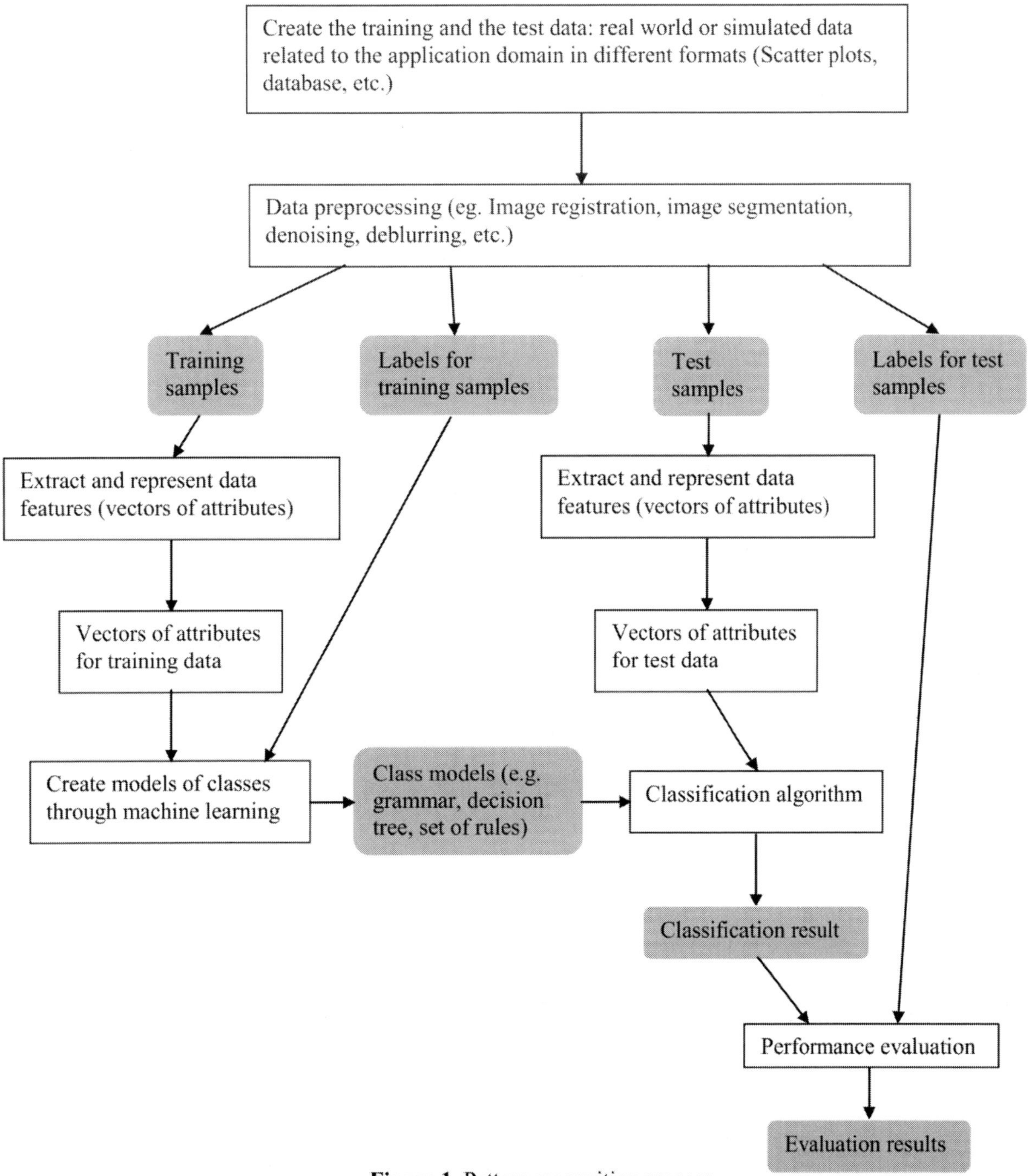

Figure 1. Pattern recognition process

In our work, we used Analyze and Nifti medical image formats without loss of generality, since both header files contain the same information even though the architecture of the structure is different

Journal of Physics: Conference Series **341** (2012) 012019 doi:10.1088/1742-6596/341/1/012019

from one format to another. The data is first read using a matlab gui called readData3D[3], which allows the user to open medical 3D files. It supports the following formats : Dicom Files (.dcm , .dicom), V3D Philips Scanner (.v3d), GIPL Guys Image Processing Lab (.gipl), HDR/IMG Analyze (.hdr), ISI Files (.isi) and NifTi Files (.nii). etc.
Using the Matlab statement: *[D, info] = ReadData3D;*
Where D corresponds to the 3D MRI data and info is a structure containing the header information of the data such as the age, the sex, the view, etc. In case of Analyze 7.5 Format which uses radiological orientation (LAS), the data should be flipped for correct image display in MATLAB and reoriented for easier interpretation (stand patient up).
We also used other packages that allow *3D* view and extraction of statistical information such as *Twfu_bpm* toolbox[4] for multimodal image analysis called biological parametric mapping (BPM), based on a voxel-wise use of the general linear model. It has a high degree of integration with the SPM[5] (statistical parametric mapping) software relying on it for visualization and statistical inference.
The medical data is accessed through the different packages and stored into *3D* arrays where each element corresponds to a voxel with the three space coordinates and the intensity value. Depending on the chosen view, we might flip or interchange the three dimensions of each medical data. The next step is to extract meaningful information from the data using segmentation methods on the middle slice by taking the middle 2D array (or image) from the transversal view of the *3D* data.

4. Segmentation
The segmentation is a well known tool that has the purpose of partitioning the image into regions of interest in order to extract 'meaningful' objects; several methods were developed to segment images based on thresholds, feature extraction and Clustering techniques.

4.1. Thresholding
When images contain different contrasting objects, thresholding provides effective means for obtaining segmented images. Thresholding techniques are based on partitioning the intensities using global or local threshold calculations techniques such as Otsu [3] and Niblack methods [12], where each threshold classifies the voxels in *3D* images (or pixels in 2D images) into different modes using a clustering criterion. Otsu method [13] is a clustering technique that tends to produce two tight clusters by minimizing their overlap (misclassified pixels). The threshold is adjusted dynamically by increasing the spread of one cluster and decreasing the spread of the other one. The goal then is to select the threshold that minimizes the combined spread.

4.2. Edge detection
Other segmentation methods are based on edge detection techniques such as Canny [4] , active contours or snakes using the technique of matching a deformable model to an image by means of energy minimization [5], [11]. The active contour, or snake algorithm was introduced by Kass [11]. Given an approximation of the boundary of an object in an image, an active contour model deforms the initial boundary to lock onto characteristic features within the region of interest. The contour is deformed iteratively until it matches the boundary of the region of interest by looking for the minimum of energy of a given problem. The energy functional is a weighted combination of internal and external forces depending on the shape of the snake and location within the image.

---
[3] http://www.mathworks.com/matlabcentral/fileexchange/29344
[4] http://www.fmri.wfubmc.edu/cms/software
[5] http://www.fil.ion.ucl.ac.uk/spm/

4.3. *Region-based*

Region-based segmentation uses different techniques such as seeded region-growing [2], split-and-merge [8], watershed [10] and Wavelet-based segmentation [9] which is based on mathematical concepts such as quadrature mirror filtering, sub-band coding and pyramidal image processing.
The accessed data, as explained in the previous subsection, shows slices of images that can be visualized into different views (coronal, transversal or longitudinal). Figure 2 shows a transversal view of the middle slice. Since the cost is very important, we noticed that some of the upfront or down-front slices, spatially low positions (below brain mass), show very low intensity values which leads to the choice of thresholding values that remove those slices in order de decrease the cost of the algorithms. Before segmenting the reduced slices, we noticed that the area of interest that corresponds to the most visible ventricle is located in the middle, which corresponds to the middle transversal slice. To do that, we used the Otsu global thresholding using custom GUI tool to select best threshold level values where the starting intensities correspond to the user chosen value. This will lead to ignore low level intensities (back-ground air, CSF and other soft tissues) which correspond to the intensity 21 in our case. Using the same Matlab gui, we ignored the high levels (skull and other hard Tissues) which correspond to the intensities above 1995. For the images presenting a big area around the ventricle chambers (figure 2), we used the active contour techniques to extract the ventricle chamber's region (figure 3). But for smaller areas, the latter algorithm was unsuccessful compared with the DRLSE segmentation which extracts the active contour using the Distance Regularized Level Set Evolution (DRLSE) formulation.

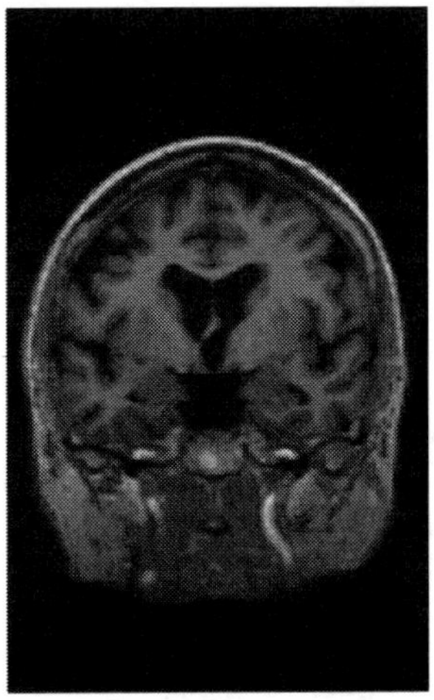

Figure 2. Transversal view of the middle slice

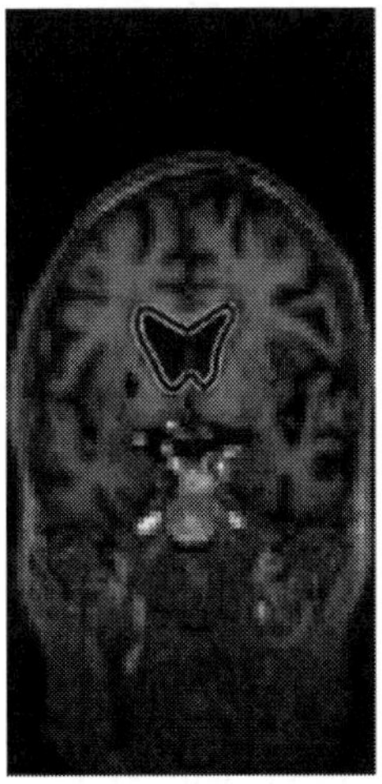

Figure 3. Active contour of a slightly reoriented and resized middle slice.

5. Attribute selection

We could assess the AD disease based on the shape of the ventricle chamber's area. Since AD causes the loss of brain mass due to molecules created in this area and being spread all over the brain, we assumed in this paper that we can assess the whole brain based on this specific area in order to decrease the cost of calculation. The more the chambers are big the more the AD is spread, which leads to the characterization of this area based on the shape of the chambers using statistical and geometrical attributes. The resulting attributes are respectively the surface area of the extracted region (Surf), the perimeter (Per), the first and second statistical moments (Mean, Std), 28 horizontal distances (W01, W02, …, W28), the height (Height) and the coordinates of the center of gravity of the region (Gx, Gy). The attributes are normalized into the range [0 1]. In order to make a first assessment of the behavioral trend of each attribute, we made a statistical analysis of the 35 attributes using stacked columns where each attribute is summarized by a column bar containing five key data points (also called five-number summary) named max, third quartile (Q3), median, first quartile (Q1) and min. The median is the middle point of the classified set of data. The first and third quartile values are respectively the 25th and 75th percentiles. The first four attributes (surface, perimeter, first and second statistical moments) and the last three attributes (height and coordinates of the centre of gravity) tend to vary more clearly and the widths shown from the 17th o the 32nd bar carry very little information and tend to have similar values which clearly is not a good tool to classify between the different medical image data base (figure 4).

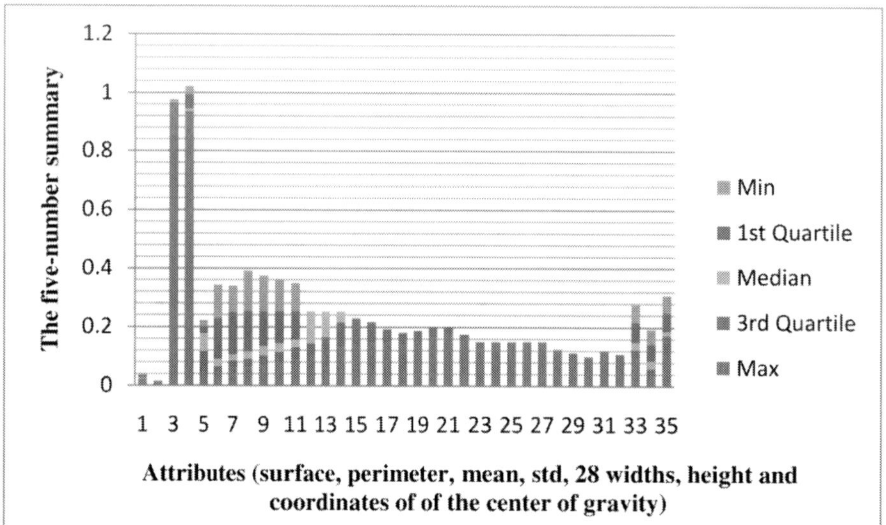

Figure 4. Statistical trend of 105 medical image data base based on the five-number summary (Min, 1st Quartile, Median, 3rd Quartile and Max)

6. Classification

The last step of the general framework is to proceed to the classification process starting by the K-Nearest Neighbors (KNN) clustering technique in order to label the classes and then creating a similarity criterion that classifies each data into the proper class name or label using the support vector machine (SVM) classification method. The SVM is a classifier that attempts to create a linear vector that segments the classes equally. In the event that is not linearly inseparable, the SVM algorithm lifts the space into a higher dimensional plane, until, on that plane, a vector can separate the classes. It is a particularly good classifier, albeit a slow one [6], [16], [17]. It supports various mathematical formulations. We used the C-Support Vector Classification or C-SVC [20], [21] that solves the following optimization problem

$$\begin{aligned} \min_{\omega,b,\xi} \quad & \frac{1}{2}\omega^T \omega + C\sum_{k=0}^{n} \xi_i \qquad (1) \\ \text{subject to} \quad & y_i\left(\omega^T \phi(x_i) + b\right) \geq 1 - \xi_i \\ & \xi_i \geq 0, i = 1, \ldots, l, \end{aligned}$$

given training vectors $x_i \in R^n$, i=1, …,l, and an indicator vector $y \in R^l$
where $\phi(x_i)$ maps x_i into a higher-dimensional space and C > 0 is the regularization parameter.

Clustering is a procedure which aims to categorize objects. In particular, the objective is to categorize objects in groups that do not have class labels. Popular methods of clustering include; K-Nearest Neighbors, K-Means and K-Medoids [15], [7].
The results of clustering are summarized in tables 1 and 2 for the number of clusters k = 2, 5 and 7 respectively, showing the clusters' centroids of each attribute. Since our data contains 105 samples, If we choose $k = 2$ for two classes, based on the results of the KNN, we will get a vector of 105 labels of 1's and 2's for the 105 samples as shown in table 3.

Table 1: KNN results showing the clusters' centroids of each attribute for k=2 and 5

	k=2		**k=5**				
Attributes	**Class 1**	**Class 2**	**Class 1**	**Class 2**	**Class 3**	**Class 4**	**Class 5**
Surf	0.0098	0.0211	0.0059	0.0091	0.0219	0.0208	0.0043
Per	0.0053	0.0084	0.0039	0.0048	0.0081	0.0093	0.0024
Mean	0.0221	0.0118	0.0079	0.0089	0.0128	0.0094	0.9738
Std	0.0706	0.0626	0.0510	0.0613	0.0694	0.0535	0.9942
W01	0.0877	0.0853	0.1153	0.0679	0.0374	0.1346	0.0222
W02	0.1404	0.1708	0.1201	0.1446	0.1669	0.1723	0.0889
W03	0.1477	0.1926	0.1112	0.1614	0.2055	0.1736	0.0889
W04	0.1363	0.2023	0.0898	0.1542	0.2210	0.1715	0.0889
W05	0.1221	0.1999	0.0701	0.1431	0.2229	0.1612	0.0667
W06	0.1053	0.1920	0.0407	0.1298	0.2182	0.1505	0.0667
W07	0.0910	0.1820	0.0340	0.1124	0.2087	0.1403	0
W08	0.0589	0.1706	0.0053	0.0742	0.1965	0.1288	0
W09	0.0380	0.1576	0.0013	0.0453	0.1822	0.1144	0
W10	0.0196	0.1437	0.0013	0.0180	0.1693	0.0988	0
W11	0.0107	0.1260	0.0007	0.0073	0.1578	0.0769	0
W12	0.0073	0.1080	0	0.0057	0.1428	0.0570	0
W13	0.0038	0.0877	0	0.0005	0.1165	0.0469	0
W14	0.0035	0.0667	0	0	0.0852	0.0403	0
W15	0.0055	0.0539	0	0	0.0654	0.0421	0
W16	0.0082	0.0436	0	0	0.0538	0.0435	0
W17	0.0082	0.0381	0	0	0.0471	0.0408	0
W18	0.0077	0.0299	0	0	0.0376	0.0348	0
W19	0.0092	0.0212	0	0	0.0296	0.0329	0
W20	0.0098	0.0184	0	0	0.0262	0.0329	0
W21	0.0094	0.0181	0	0	0.0174	0.0375	0
W22	0.0078	0.0145	0	0	0.0054	0.0367	0
W23	0.0071	0.0159	0	0	0.0020	0.0390	0
W24	0.0065	0.0164	0	0	0.0007	0.0392	0
W25	0.0049	0.0164	0	0	0.0007	0.0348	0
W26	0.0046	0.0136	0	0	0.0013	0.0297	0
W27	0.0033	0.0111	0	0	0.0013	0.0229	0
W28	0.0030	0.0111	0	0	0.0013	0.0220	0
Height	0.1042	0.0762	0.0922	0.1147	0.0819	0.0749	0.0667
Gx	0.0903	0.0579	0.0813	0.0985	0.0577	0.0631	0.0667
Gy	0.0785	0.1393	0.0528	0.0705	0.1411	0.1489	0.0469

Table 2: KNN results showing the clusters' centroids of each attribute for k=7

Attributes	Class 1	Class 2	Class 3	Class 4	Class 5	Class 6	Class 7
Surf	0.0167	0.0043	0.0055	0.0228	0.0095	0.0081	0.0237
Per	0.0071	0.0024	0.0037	0.0083	0.0049	0.0047	0.0110
Mean	0.0124	0.9738	0.0088	0.0097	0.0089	0.0082	0.0094
Std	0.0624	0.9942	0.0520	0.0661	0.0623	0.0579	0.0498
W01	0.1468	0.0222	0.1082	0.0287	0.0178	0.1226	0.1197
W02	0.1871	0.0889	0.1140	0.1629	0.1327	0.1544	0.1584
W03	0.1928	0.0889	0.1066	0.2032	0.1598	0.1574	0.1572
W04	0.1921	0.0889	0.0825	0.2207	0.1536	0.1482	0.1540
W05	0.1856	0.0667	0.0597	0.2237	0.1417	0.1385	0.1400
W06	0.1735	0.0667	0.0285	0.2200	0.1266	0.1265	0.1297
W07	0.1620	0	0.0212	0.2116	0.1173	0.1014	0.1206
W08	0.1506	0	0.0063	0.1987	0.1093	0.0206	0.1094
W09	0.1385	0	0.0016	0.1843	0.0639	0.0116	0.0925
W10	0.1177	0	0.0016	0.1729	0.0229	0.0034	0.0792
W11	0.0840	0	0.0008	0.1646	0.0107	0.0034	0.0598
W12	0.0575	0	0	0.1517	0.0082	0.0029	0.0499
W13	0.0369	0	0	0.1297	0.0010	0	0.0474
W14	0.0249	0	0	0.0946	0	0	0.0497
W15	0.0214	0	0	0.0709	0	0	0.0596
W16	0.0149	0	0	0.0587	0	0	0.0698
W17	0.0114	0	0	0.0519	0	0	0.0680
W18	0.0049	0	0	0.0429	0	0	0.0625
W19	0.0028	0	0	0.0338	0	0	0.0616
W20	0.0021	0	0	0.0299	0	0	0.0624
W21	0	0	0	0.0203	0	0	0.0742
W22	0	0	0	0.0067	0	0	0.0725
W23	0	0	0	0.0036	0	0	0.0764
W24	0	0	0	0.0049	0	0	0.0733
W25	0	0	0	0.0070	0	0	0.0620
W26	0	0	0	0.0071	0	0	0.0526
W27	0	0	0	0.0064	0	0	0.0398
W28	0	0	0	0.0064	0	0	0.0380
Height	0.0626	0.0667	0.0903	0.0834	0.1053	0.1215	0.0982
Gx	0.0513	0.0667	0.0780	0.0585	0.0885	0.1095	0.0810
Gy	0.1315	0.0469	0.0512	0.1364	0.0778	0.0607	0.1654

Table 3. Class labels corresponding to the 105 samples (S1, S2, ..., S105) for k=2 (2 classes) using KNN clustering technique

S1	**S2**	**S3**	**S4**	**S5**	**S6**	**S7**	**S8**	**S9**	**S10**	**S11**	**S12**
1	2	2	2	1	1	2	2	2	2	2	2
S13	**S14**	**S15**	**S16**	**S17**	**S18**	**S19**	**S20**	**S21**	**S22**	**S23**	**S24**
2	2	2	2	2	2	1	2	2	2	2	2
S25	**S26**	**S27**	**S28**	**S29**	**S30**	**S31**	**S32**	**S33**	**S34**	**S35**	**S36**
2	1	2	2	2	2	2	2	2	1	1	1
S37	**S38**	**S39**	**S40**	**S41**	**S42**	**S43**	**S44**	**S45**	**S46**	**S47**	**S48**
1	2	2	2	2	2	1	2	2	1	1	1
S49	**S50**	**S51**	**S52**	**S53**	**S54**	**S55**	**S56**	**S57**	**S58**	**S59**	**S60**
2	2	2	1	2	2	2	2	2	1	2	1
S61	**S62**	**S63**	**S64**	**S65**	**S66**	**S67**	**S68**	**S69**	**S70**	**S71**	**S72**
1	2	2	1	1	2	2	2	2	1	1	1
S73	**S74**	**S75**	**S76**	**S77**	**S78**	**S79**	**S80**	**S81**	**S82**	**S83**	**S84**
1	1	2	2	2	2	2	2	2	2	1	2
S85	**S86**	**S87**	**S88**	**S89**	**S90**	**S91**	**S92**	**S93**	**S94**	**S95**	**S96**
2	1	2	1	2	2	2	2	1	1	1	2
S97	**S98**	**S99**	**S100**	**S101**	**S102**	**S103**	**S104**	**S105**			
2	1	2	1	2	2	1	1	1			

For the SVM, we could not interpret the results since they represent 35 dimensions, however we can examine the results visually by plotting only 2-D vector of attributes. We set the parameter C of class i to *weight*C* in C-SVC in equation 1.

Using the first two attributes that correspond to the surface and perimeter of the objects and by forcing a linearly separable solution, we got a classification accuracy of 0% for 7 support vectors (SV). The weights are respectively -767.36739 and 1074.7484, the bias $b_o = 4.6073$ the margin = 0.0015145. The classification results are plotted in figure 5 showing a predominant *class 2* for the first two attributes.

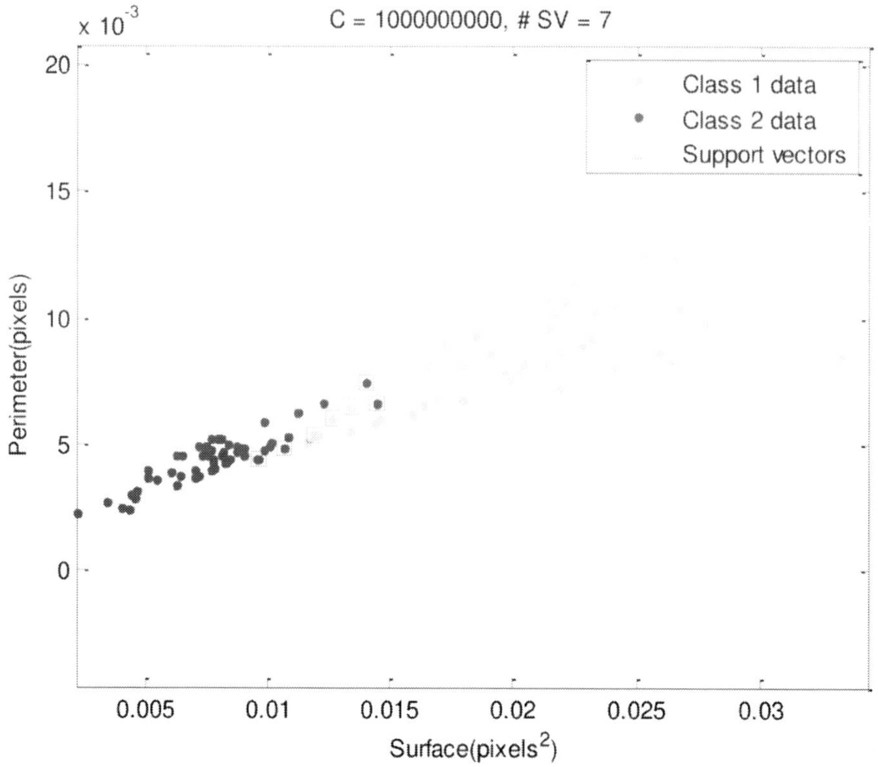

Figure 5. SVM classification results for attributes 1 and 2 (Perimeter vs. surface) with linearly separable solution

The same work was performed for other pairs of attributes. Table 4 summarizes the SVM classification results corresponding to 7 pairs of attributes by forcing a linearly separable solution. The results correspond to the number of support vectors, the weights, the parameter b_o, the margin and a visual classification assessment (good, fair, poor).

Table 4. SVM classification results for a pair of attributes with linearly separable solution

Attribute x	Attribute y	SV	weights	b_o	Margin	Visual assessment
Mean	Std	87	-6.8384 .8242	1.6761	0.2923	Poor
Width01	Width02	59	1.0811 6.4423	0.74054	0.30617	Poor
Width03	Width10	17	-6.7621 -17.4341	3.6638	0.10695	Good
Width10	Height	4	-14.6321 -2.5301	3.2781	0.13469	Good
Gx	Gy	36	7.5521 -6.5821	1.7852	0.19964	Fair
Surface	Height	14	95.2414 -7.4722	0.83633	0.020935	Good
Perimeter	Width11	13	131.9472 11.7253	0.19204	0.015098	Good

The results shown in table 4, imply a good classification in the case of geometric attributes (e.g the widths and the surface) and fair to poor classification in the case of statistical attributes (mean, standard deviation and center of gravity).
Other techniques of classification have been used but did not produce greater results due to the choice of the vector of attributes and the small size of the data base.

The first one is the Bayesian Network which is a decision making model. This model makes decisions based on the dependencies of the previous conditions using probability theory. It takes a set of particular inputs and based on the probability of the occurrence of the inputs in the training set, it makes a decision of where the new input is supposed to be classed. The highest probability calculated for a particular output based on the inputs is the decision that is finally chosen as the output [3], [6], [19].
The second method is the neural network which is a learning system that creates interconnecting artificial neurons. It receives an input through its input layer and produces an output that depends on the architecture of the neural network, the values of the initial weights and the number of hidden layers. The hidden layer contains a set of neurons that have various weights to produce and output from each neuron. Once each neuron has a result, those results are aggregated with another set of weights to produce the final output [3].

7. Conclusion

Medical image processing and machine learning tools can help neurologists in assessing whether a subject is developing the Alzheimer disease. A machine learning system has been developed in order to extract meaningful information from the ADNI database, where the ventricle chambers are extracted using a segmentation method based on the statistical features of the region of interest. As a first impression, we chose this region to see if it corresponds to a good marker, however, the shape and size of ventricle did not have a big impact on the final result and we noticed that some other regions might be more affected. New features should be generated in order to characterize the new region. Due to the time factor, we could not classify the regions using different methods as expected, however we created different tools and simulated the whole system to work for the real image database. The developed algorithms have been tested on real medical data from ADNI database.
What was successfully completed is the creation of a framework to continue researching the use of SVMs towards mining medical images. This framework has potential to make potent predictions based exclusively on the properties of a patient's hippocampus based on the strength of the SVM's ability to classify objects. With further research, a system could be developed, coupled with a better feature set to produce an early detection for patients.
Ongoing research is focusing on the enhancement of the classification algorithms, adding more data to the system, enhancing the segmentation tool and assessing the results of the classification.

References

[1] Schroeter ML, Stein T, Maslowski N, Neumann J 2009 Neural correlates of Alzheimer's disease and mild cognitive impairment A meta-analysis including 1351 patients. *NeuroImage* **47** 1196-1206.
[2] Adams, R., and Bischof, L. 1994 Seeded region growing. IEEE Transactions on Pattern Analysis and Machine Intelligence **16** 641-647.
[3] Bishop, C. M. 1995 Neural Networks for Pattern Recognition. (New York, Oxford University Press, Inc.).
[4] Canny, J.1986 A computational approach to edge detection. *IEEE Transactions on Pattern Analysis and Machine Intelligence PAMI*-**8**, 6.
[5] Caselles, V., Kimmel, R., and Sapiro, G. 1997 Geodesic active contours. Int. J. Comput. Vision **22**, 61-69.
[6] Duda, R. O., Hart, P. E., and Stork, D. G. 2001 Pattern Classification, (New York, 2. ed. Wiley).
[7] Filippone, M., Camastra, F., Masulli, F., and Rovetta, S. 2008 A survey of kernel and spectral methods for clustering. Pattern Recogn. **41** 176-190.
[8] Horowitz, S., and Pavlidis, T. 1977 Picture segmentation by a directed split and merge procedure. CMetImAly77, 101-111.

[9] Mallat, S. G. 1989 A theory for multi-resolution signal decomposition: The wavelet representation. IEEE Trans. Pattern Anal. Mach. Intell. **11** 674-693.
[10] Meyer, F. 2001 An overview of morphological segmentation. IJPRAI **15** 1089-1118.
[11] Kass M, Witkin A and Terzopoulos D, 1988 Snakes: Active contour models, Int. J. Comp. Vision **1** 321-331.
[12] Niblack, W. 1986 An Introduction to Digital Image Processing. (New Jersey, Prentice Hall)
[13] Otsu, N. 1979 Threshold selection method from gray-level histograms. *IEEE Trans. Systems, Man, and Cybernetics* **9** 62-66.
[14] Sharma N, Ray AK, Sharma S, Shukla KK, Pradhan S, Aggarwal LM. 2009 Segmentation of medical images using simulated annealing based fuzzy C Means algorithm. *Int J Biomed Engg Technol.* **2** 260-278.
[15] Taylor, C. 1997 Classification and kernel density estimation. Vistas in Astronomy **41** 411-417.
[16] Vapnik, V. N. 1995 The nature of statistical learning theory.(New York, Springer-Verlag, Inc.).
[17] Weston, J., Mukherjee, S., Chapelle, O., Pontil, M., Poggio, T., and Vapnik, V. 2000 Feature selection for svms. *Advances in Neural Information Processing Systems* (MIT Press) **13** 668-674.
[18] Zaidi H, Ruest T, Schoenahl F, Montandon ML 2006 Comparative assessment of statistical brain MR image segmentation algorithms and their impact on partial volume correction in PET. *Neuroimage.* **32** 1591-1607.
[19] Zhang, H., and Su, J. 2008 Naive bayes for optimal ranking. *J. Exp. Theor. Artif. Intell.* **20** 79-93.
[20] Boser, B. E., Guyon, I. , and Vapnik, V. 1992. A training algorithm for optimal margin classifiers. Proceedings of the Fifth Annual Workshop on Computational Learning Theory, 144-152.
[21] Cortes, C., and Vapnik, V. 1995. Support-vector network. Machine Learning, **20** 273-297

Journal of Physics: Conference Series **341** (2012) 012020 doi:10.1088/1742-6596/341/1/012020

Implementation of Segmentation Methods for the Diagnosis and Prognosis of Mild Cognitive Impairment and Alzheimer Disease

S. Matoug*[1], A. Abdel-Dayem[2]
Department of Mathematics and Computer Science
Laurentian University
935 Ramsey Lake Road Sudbury ON P3E 2C6 CANADA
E-mail: [1]sx_matoug@laurentian.ca, [2]aabdeldayem@laurentian.ca

Abstract. Alzheimer's disease (AD) is the most common form of dementia affecting seniors age 65 and over. When AD is suspected, the diagnosis is usually confirmed with behavioural assessments and cognitive tests, often followed by a brain scan. Advanced medical imaging is a good tool to predict conversion from prodromal stages (mild cognitive impairment) to Alzheimer's disease. Since volumetric MRI can detect changes in the size of brain regions, measuring those regions that atrophy during the progress of Alzheimer's disease can help the neurologist in his diagnostic. In the present investigation, we present an automatic tool that reads volumetric MRI and performs 2-dimensional (volume slices) and volumetric segmentation methods in order to segment gray matter, white matter and cerebrospinal fluid (CSF). We used the MRI data sets database from the Open Access Series of Imaging Studies (OASIS).

1. Introduction

AD leads to nerve cell death and tissue loss throughout the brain, resulting in brain tissue shrinking and larger ventricles (chambers within the brain that contain cerebrospinal fluid). When AD is suspected, the diagnosis is usually confirmed with behavioral assessments and cognitive tests, often followed by a brain scan [1]. Advanced medical imaging with computed tomography (CT) or magnetic resonance imaging (MRI), and with single photon emission computed tomography (SPECT) or positron emission tomography (PET) can be used to help exclude other cerebral pathology or subtypes of dementia [1]. Moreover, it may predict conversion from prodromal stages (mild cognitive impairment) to Alzheimer's disease [2].

In this paper, we describe succinctly image segmentation methods that segment the slice brain's neuroimages in order to extract the ventricles' region, as a way to get a first visual assessment.

* To whom any correspondence should be made

Section 2 describes briefly the theory behind the image segmentation. Section 3 describes and discusses the system implementation including the data access, visualization and different segmentation methods while incorporating different algorithms.

2. Theory behind the system implementation

The proposed system is to perform 3D image segmentation of original 3D MRI neuroimage brain data. The theory about 2D segmentation is easily transported into 3D, however the cost of the algorithms is theoretically squared O (3D) = O $(2D)^2$. The 3D segmentation transforms the original voxels in 3D images into 3D regions where each region, identified by a different label, represents meaningful physical behaviors defined by a vector of attributes (average, standard deviation, etc.). There are many existing segmentation techniques applied for medical image segmentation, including statistical methods, thresholding, edge detection, region-based techniques and more recently multi-resolution (using wavelets, ridgelets, etc.) techniques [3], [4]. The choice of the method depends on the type and quality of the image.

2.1. Thresholding

When images contain different contrasting objects, thresholding provides effective means for obtaining segmented images. Thresholding techniques are based on partitioning the intensities using global or local threshold calculations techniques such as Otsu [5] and Niblack methods [6], where each threshold classifies the voxels (or pixels) into different modes using a clustering criterion.

2.1.1. The Otsu Method

The Otsu method [5] is a clustering technique that tends to produce two tight clusters by minimizing their overlap (misclassified pixels). The threshold is adjusted dynamically by increasing the spread of one cluster and decreasing the spread of the other one. The goal then is to select the threshold that minimizes the combined spread. We define the within-class variance as the weighted sum of the variances of each cluster:

$$\sigma^2_{within} = n_B(T)\sigma^2_B(T) + n_O(T)\sigma^2_O(T) \tag{1}$$

$$\sigma^2_{between} = n_B(T)n_O(T)\left(\mu^2_B(T) + \mu^2_O(T)\right) \tag{2}$$

Where:

$n_B(T) = \sum_{i=0}^{T-1} p(i)$: the number of pixels in the first cluster

$n_O(T) = \sum_{i=T}^{N-1} p(i)$: the number of pixels in the second cluster

$\sigma^2_B(T)$: the variance of the pixels in the background (below threshold)

$\sigma^2_O(T)$: the variance of the pixels in the foreground (above threshold)

$\mu_B(T)$: the mean of all pixels less than the threshold

$\mu_O(T)$: the mean of all pixels greater than the threshold

[0, N-1]: is the range of intensity levels.

Developed algorithm

The optimal threshold is the one that maximizes the between-class variance (or, conversely, minimizes the within-class variance). So, for each potential threshold *T* we:

1. Separate the pixels into two clusters according to the threshold.
2. Find the mean of each cluster.
3. Square the difference between the means.
4. Multiply by the number of pixels in one cluster times the number in the other.

2.1.2. Niblack method
Niblack's algorithm [6] calculates a local threshold T for each pixel. The threshold T is computed by using the mean μ and standard deviation σ of all the pixels in the pixel neighborhood, and is denoted as: $T = \mu + k*\sigma$, where the parameter k is a constant, which determines how much of the total object is extracted, and is usually chosen between 0 and 1. The value of k and the size of the neighborhood influence the result of thresholding.

2.2. Edge detection
Other segmentation methods are based on edge detection techniques such as Canny [7], active contours or snakes using the technique of matching a deformable model to an image by means of energy minimization [8], [9].

2.2.1. Canny edge detection
The Canny edge detection algorithm [7] aims to the following optimal properties:

- Good detection: the algorithm should detect as many real edges in the image as possible.
- Good localization: the edges marked should be as close as possible to the edge in the real image.
- Minimal response: a given edge in the image should only be marked once, and where possible, image noise should not create false edges.

Canny's algorithm is based on finding an optimal function as the first derivative of a Gaussian, originally described by the sum of four exponential terms. The effectiveness and cost of the algorithm depends on the size of the Gaussian filter and the hysteresis thresholds.

2.2.2. Active contours
The active contour [8], [9] is also sometimes called snake algorithm. Given an approximation of the boundary of an object in an image, an active contour model deforms the initial boundary to lock onto characteristic features within the region of interest. The contour is deformed iteratively until it matches the boundary of the region of interest by looking for the minimum of energy of a given problem. The energy functional is a weighted combination of internal and external forces depending on the shape of the snake and location within the image. It is defined by the following algorithm:

The integral energy functional to be minimized is given by:

$$E^*_{snake} = \int_0^1 E_{snake}(v(s))ds \tag{3}$$

$$= \int_0^1 [E_{int}(v(s)) + E_{image}(v(s)) + E_{con}(v(s))]ds$$

Where $E_{int} = \alpha(s)\left|\frac{dv}{ds}\right|^2 + \left|\frac{d^2v}{ds^2}\right|^2$ is the internal spline energy.
$\alpha(s)$ and $\beta(s)$ are the elasticity and stiffness of the snake respectively,
E_{image} is derived from the image data over which the snake lies and it is modeled as a weighted combination of different functional, and
E_{con} comes from external constraints that force the snake toward or away from particular features.
The effectiveness of the algorithm depends on the initial choice of the approximate shape and starting position. A priori information is then used to move the snake toward an optimal solution.

2.3. Region-based segmentation
Region-based segmentation uses different techniques such as seeded region-growing [10], split-and-merge [11], watershed [12] and Wavelet-based segmentation [13] which is based on mathematical concepts such as quadrature mirror filtering, sub-band coding, and pyramidal image processing.

2.3.1. Region-growing segmentation:
Region-growing segmentation [10] starts with initial seed points chosen from the target region or without a priori knowledge, taken from the picks of the histogram. It checks the neighborhood pixels and adds them to the region if they are similar to the chosen seeds using a similarity criteria (homogeneity predicates) based on a vector of characteristics (attributes) in the image such as the average, standard deviation, texture, etc.

2.3.2. Split-and-merge segmentation:
Split-and-merge segmentation [11] consists of two different parts. The split process keeps dividing the image into smaller regions that do not respect a criterion of similarity. In the merge process, neighboring regions, resulting from the split process that respects a similarity criterion, using a vector of predicates, are merged into bigger regions.

3. System implementation and discussion

3.1. Data access
Medical images database consist of headers containing the information such as the patient sex and age, the type of radiography, the view and the size of voxels. The data itself is stored as a 3D matrix usually of single type. Each type of data has a different info structure. Matlab was used to access the data.

3.2. Segmentation
The resulting *3D* data shows slices of images that can be visualized into different views (coronal, transversal or longitudinal). Since the cost is very important, we noticed that some of the upfront or down-front slices, spatially low positioned (below brain mass), show very low intensity values which leads to the choice of thresholding values that remove those slices in order to decrease the cost of the algorithms. Before segmenting the reduced slices, we noticed that the area of interest that corresponds to the most visible ventricle is located in the middle slices, which correspond to the transversal slices numbers beyond 40 and less than 200; figure 1, shows the grey view of the slices numbered from 121 to 135. In order to get the initial threshold value that segments the slices into different regions, we generate the histograms of the slices (grey level images). The histograms of the different slices (figure 2) show that the origin that corresponds to the extreme dark regions has a peak (many data) however; we ignored low level intensities because it is due to the background air, cerebrospinal fluid (CSF) and other soft tissues which correspond to the intensities 0 to 21. We also ignored the high level intensities that correspond to the skull and other hard tissues which correspond to the intensities above 1995 (table 1). The resulting preprocessed images are shown in figure 3 after ignoring the low and high intensities.

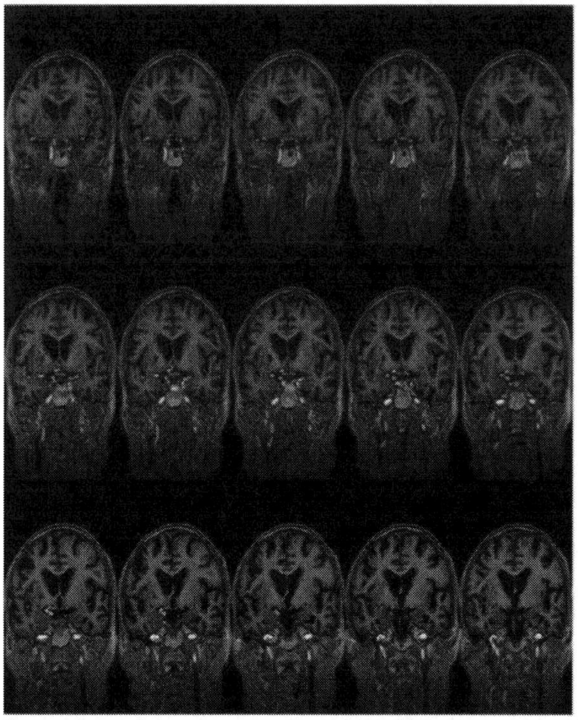

Figure 1. Grey view of the transversal middle slices (numbers 121 to 135)

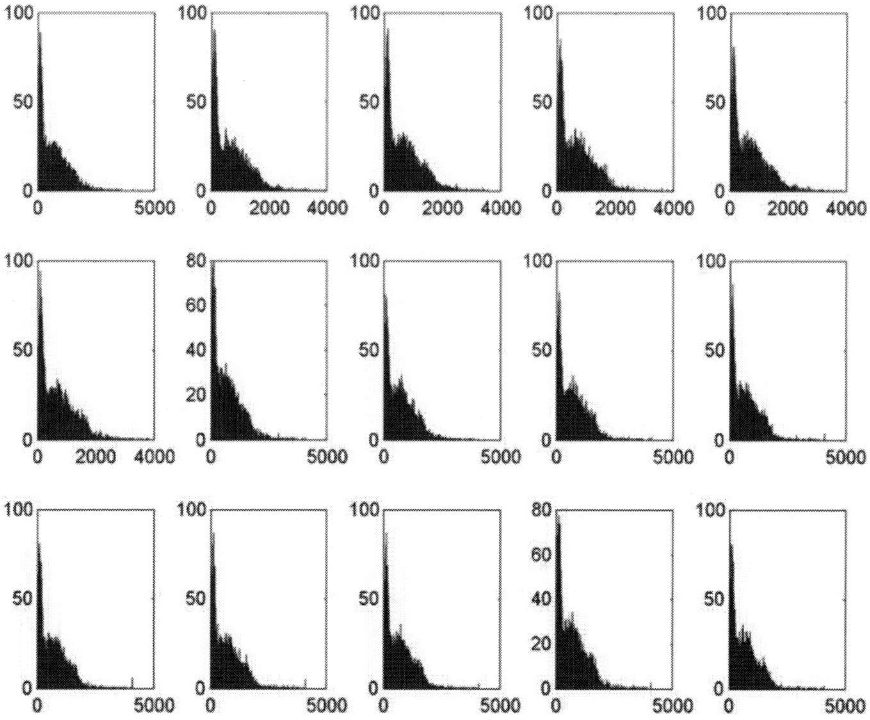

Figure 2. Histograms of the transversal middle slices (numbers 121 to 135)

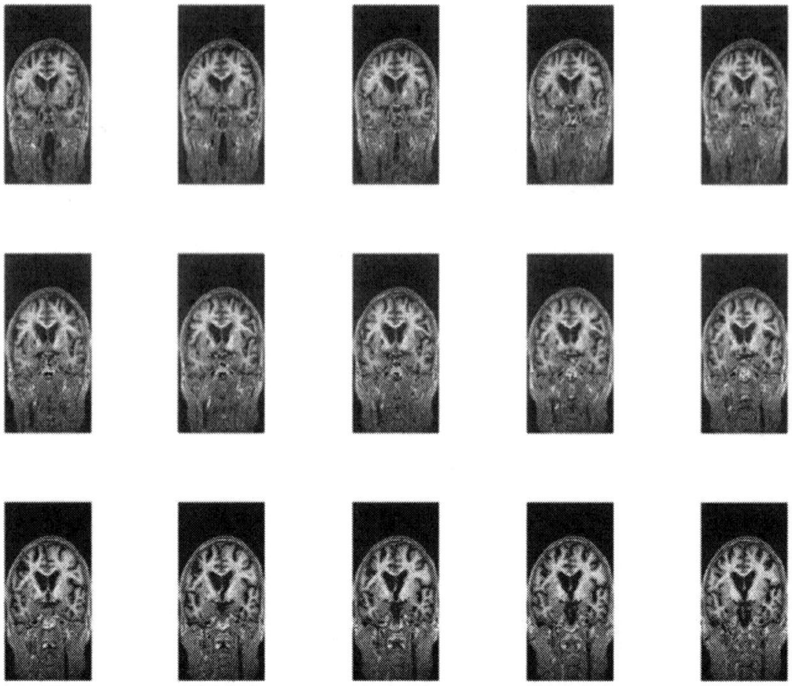

Figure 3. Results of preprocessing after removing low and high intensities of the transversal middle slices (numbers 121 to 135)

The first performed segmentation method was the Otsu global thresholding by selecting initial threshold level values using the histograms of figure 2 and performing an adaptive thresholding based on the iterative threshold and the in-between variances. The resulting image gives the entire area of the brain based on the intensity value (figure 4). The result could be used for selecting some markers as initial points to more sophisticated segmentation methods but is insufficient by itself. The Niblack adaptive (local) thresholding depends on the choice of the parameter k and the size of the filter. Increased size values ($N \geq 11$) showed better results by using smoothing filters (figure 5).
The edge detection using Canny gave correct output edges with less noise compared to Marr-and-Hildreth or Sobel, the ventricle region is well detected and stable however some noise is introduced outside the region itself that must be thresholded a second time (figure 6). The active contour gave much better results (figure 7) than Canny's, however the initialization is not automatic and is based on the central slice. The cost has been reduced by initializing the edge to the same contour for every slice.
One of the best results has been given by the region growing algorithm, however the choice of the initial seed has to be more automatic and less prone to errors, the cost is 10 times bigger since the seed is chosen for every slide; figure 8 shows the resulting segmented image for a *distance*=0.01 and an initial seed point of coordinates x=80 and y=85. Finally the watershed method extracted the ventricle region but over-segmented the live tissue of the brain (figure 9).
To decrease the cost, we calculated the singular value decomposition (SVD) of each slice, with n (= rank(Slice)-5) as the number of decomposed images, the image is well reconstructed (figure 10) with an average error of 10^{-2}.

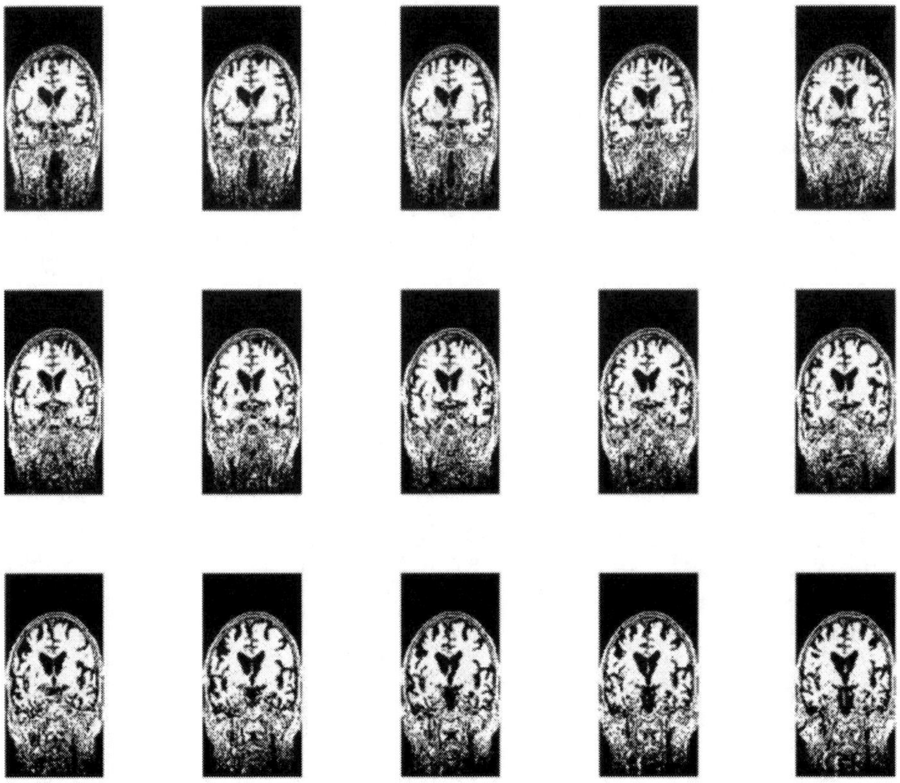

Figure 4. Results of using global Otsu thresholding of the transversal middle slices (numbers 121 to 135)

Table 1. Cost (number of pixels) of removing low (less than the 21 intensity value) and high (above the 1995 intensity value) level intensities.

Slice number	lower intensities	higher intensities
121	228	404
122	227	369
123	209	381
124	223	336
125	229	371
126	203	398
127	234	407
128	210	434
129	229	357
130	229	308
131	217	285
132	193	276
133	212	286
134	241	275
135	219	243

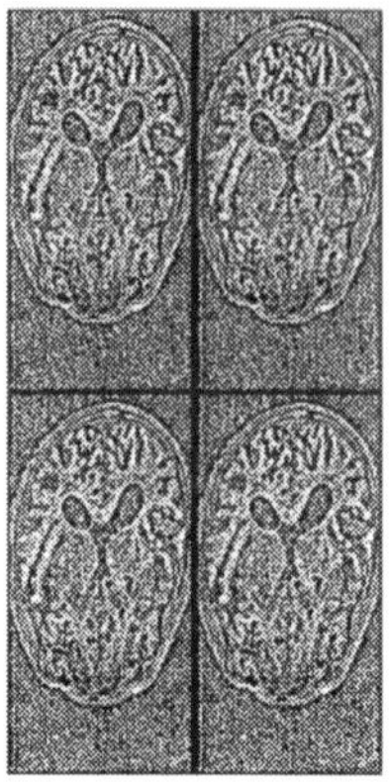

(a) ULk=0001, URk=0002, BLk= 0003, BRk=0004, N=5.

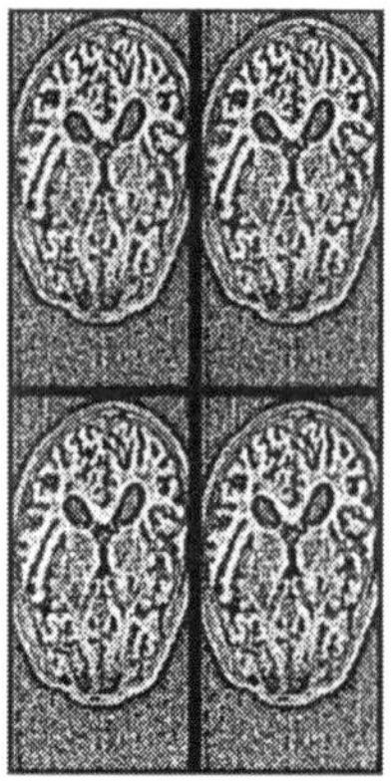

(b) ULk=0001, URk=0002, BLk=0003, BRk=0004, N=9

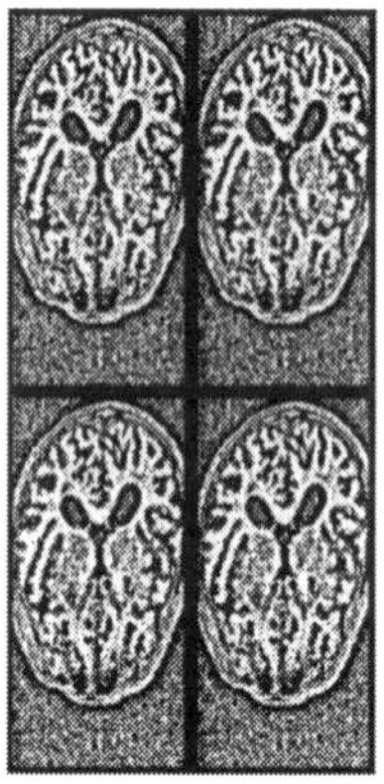

(c) ULk=0001, URk=0002, BLk=0003, BRk=0004, N=11

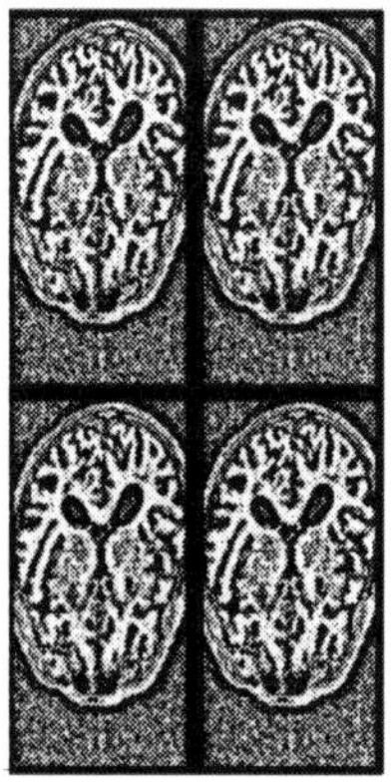

(d) ULk=0001, URk=0002, BLk=0003, BRk=0004, N=13.

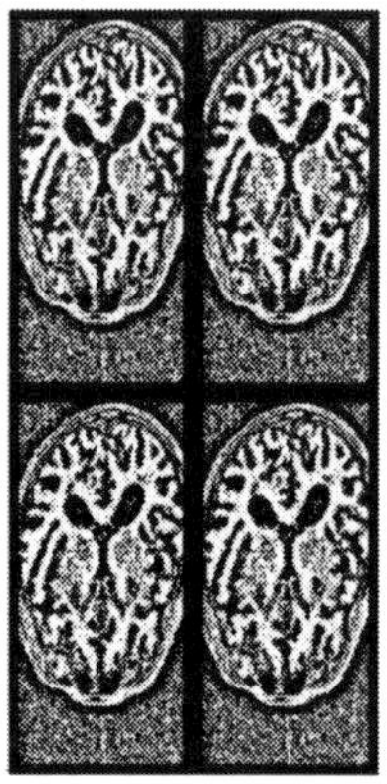

(e) ULk=0001, URk=0002, BLk=0003, BRk=0004, N=15.

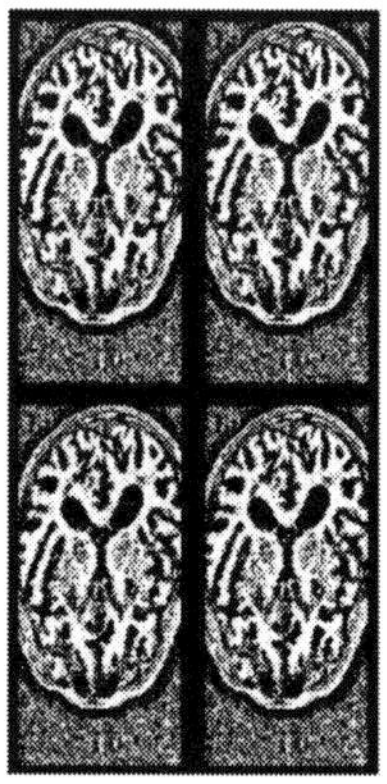

(f) ULk=0001, URk=0002, BLk=0003, BRk=0004, N=17.

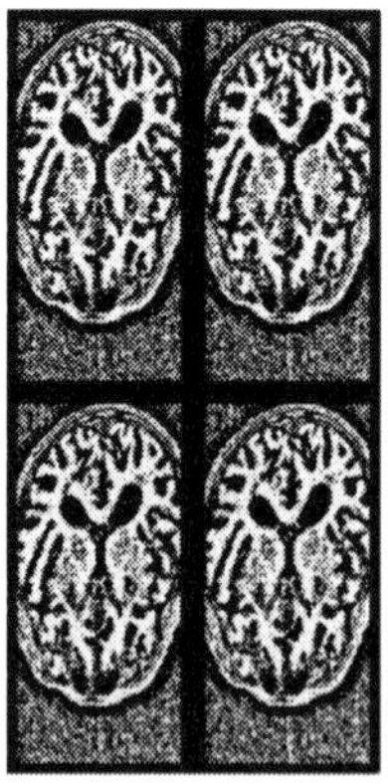

(g) ULk=001, URk=002, BLk=003, BRk= 004, N=17.

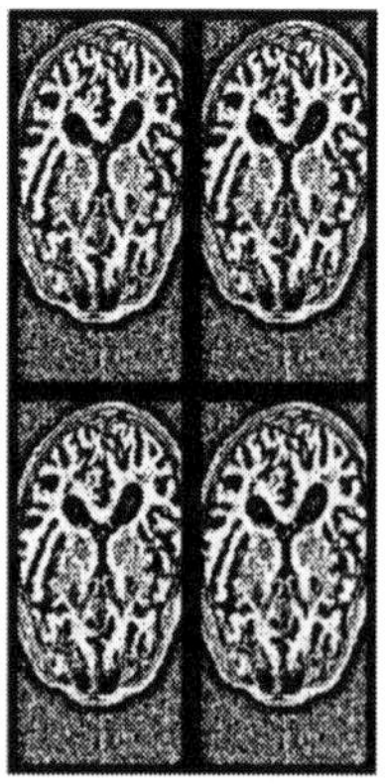

(h) ULk=-001, URk=-002, BLk=-003, BRk=-004, N=17.

Figure 5. Results of using Niblack local thresholding on the middle slice with different values of the coefficient *k* and the size of the filter N, where ULk, URk, BLk and BRk correspond respectively to the k value respectively for the upper left, upper right, bottom left and bottom right images.

Journal of Physics: Conference Series **341** (2012) 012020 doi:10.1088/1742-6596/341/1/012020

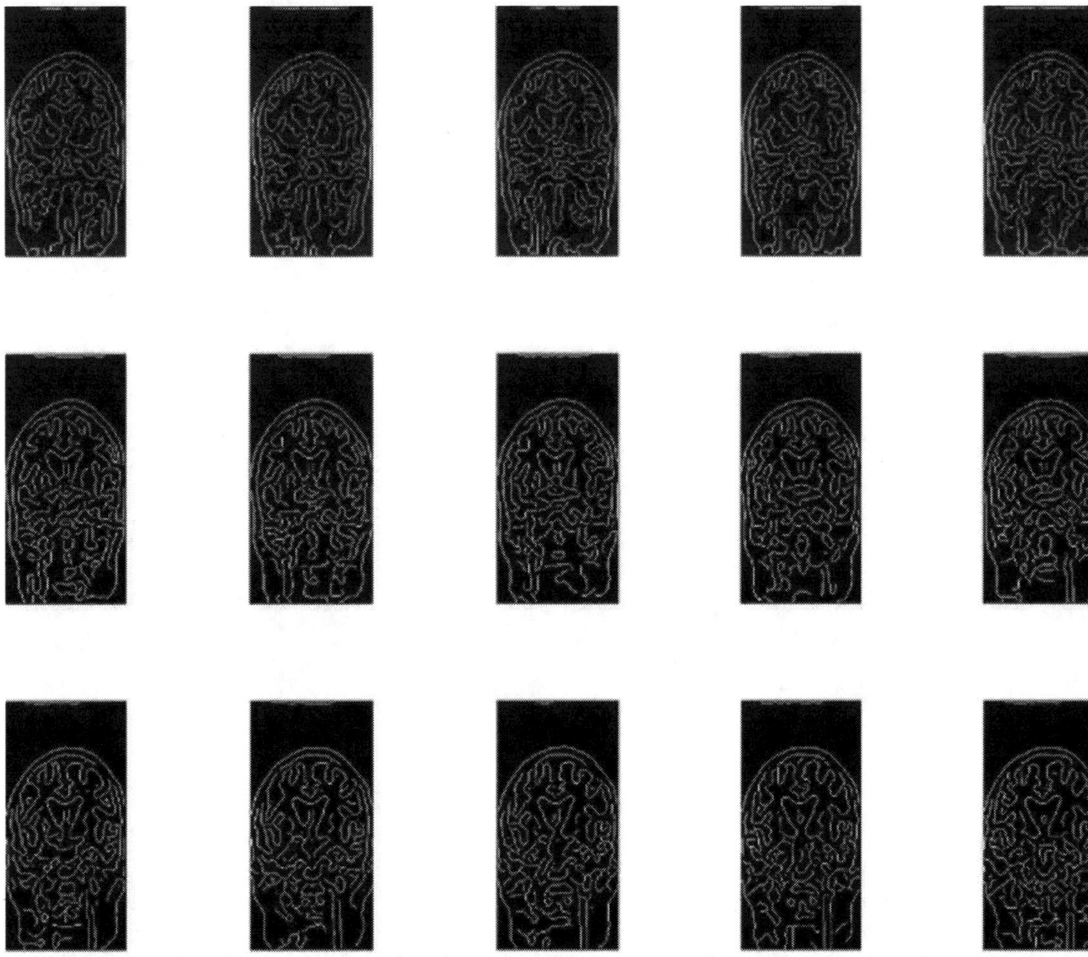

Figure 6. Results of using Canny edge detector of the transversal middle slices (numbers 121 to 135)

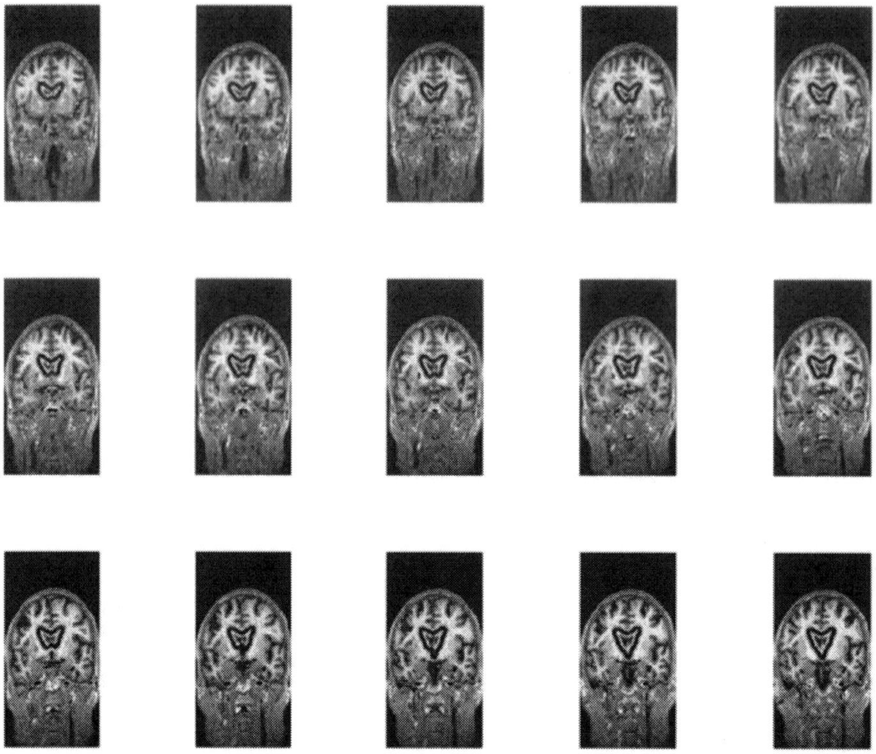

Figure 7. Results of using the active contour of the transversal middle slices (numbers 121 to 135)

Figure 8. Segmentation of the middle slice using region growing method

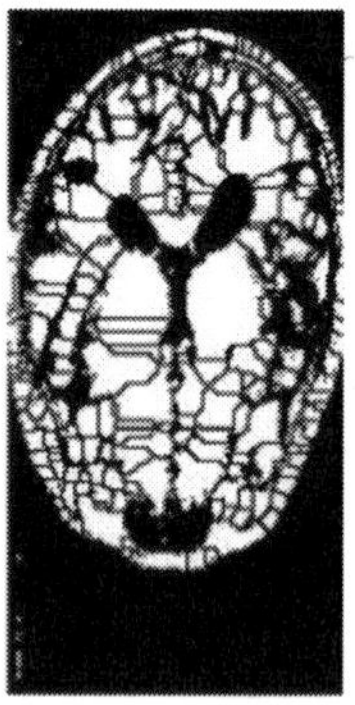

Figure 9. Segmentation of the middle slice using Watershed method.

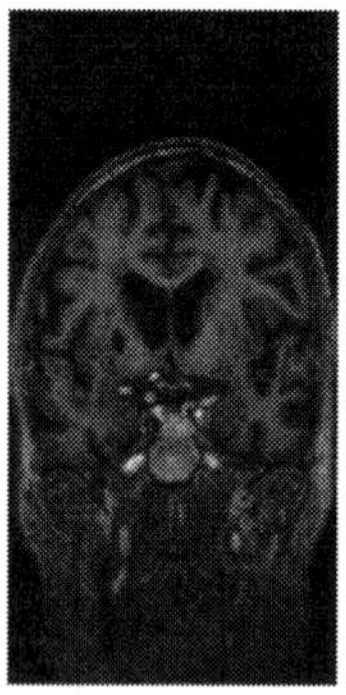

Figure 10. singular value decomposition (SVD) of the middle slice

4. Conclusion

Image segmentation is very useful for MRI medical images, to detect regions of interest for radiotherapy planning and brain damage diagnostic. Several segmentation methods have been adapted and applied on *2D* and *3D* medical images. The developed algorithms have been tested on real medical data from OASIS database. The system was used to extract the ventricles from each slice, and these results were visually inspected and compared to the real images. Thresholding was applied to remove unnecessary regions and edge detection techniques in addition to region segmentations were used to extract the ventricles. Ongoing research is focusing on reducing the cost and improving the precision of the algorithms by using more intelligent algorithms including the SVD decomposition and more adaptive seeds initialization.

References

[1] National Institute for Health and Clinical Excellence, Dementia: Quick reference guide. London, (UK), 2nd Edition, November 2006. http://www.nice.org.uk/nicemedia/pdf/
[2] Schroeter ML, Stein T, Maslowski N, Neumann J 2009 Neural correlates of Alzheimer's disease and mild cognitive impairment A meta-analysis including 1351 patients. *NeuroImage* **47** 1196-1206.
[3] Sharma N, Ray AK, Sharma S, Shukla KK, Pradhan S, Aggarwal LM. 2009 Segmentation of medical images using simulated annealing based fuzzy C Means algorithm. *Int J Biomed Engg Technol*. 2:26078.
[4] Zaidi H, Ruest T, Schoenahl F, Montandon ML 2006 Comparative assessment of statistical brain MR image segmentation algorithms and their impact on partial volume correction in PET. *Neuroimage*. **32** 15911607.
[5] Otsu, N 1979 A threshold selection method from gray-level histograms. *IEEE Trans. Systems, Man, and Cybernetics,* **9** 62-66.
[6] Niblack W 1986 An Introduction to Digital Image Processing.(New Jersey: Prentice Hall,)
[7] Canny J 1986 A computational approach to edge detection*IEEE Transactions on, PAMI* **8** 679698.
[8] Caselles V, Kimmel R and Sapiro G 1997 Geodesic Active Contours, Int. J. Comput. Vision, **22** . 61-79.
[9] Kass M, Witkin A and Terzopoulos D, Snakes 1988 Active contour models, Int. J. Comp. Vision **1**321-331.
[10] Adams R and Bischof L 1994 Seeded Region Growing, *IEEE Trans. Pattern Anal. Mach. Intell.*, **16**,. 641-647.
[11] Horowitz SL and Pavlidi, T 1977Picture Segmentation by a Directed Split and Merge Procedure , CMetImAly77, 101-11.
[12] Fernand Meyer, F 2001 An Overview of Morphological Segmentation, *IJPRAI*, **15** 1089-1118,.
[13] Mallat, S. G. 1989 A Theory for Multiresolution Signal Decomposition: The Wavelet Representation, *IEEE Trans. Pattern Anal. Mach. Intell.*, **11** 674-693.

Journal of Physics: Conference Series **341** (2012) 012021 doi:10.1088/1742-6596/341/1/012021

An Approach Using Parallel Architecture to Storage DICOM Images in Distributed File System

Tiago S Soares, Thiago C Prado, M A R Dantas
Informatics and Statistic Department
Federal University of Santa Catarina
Florianpolis, Brazil
E-mail: steinmetz@telemedicina.ufsc.br, coeio@inf.ufsc.br, mario@inf.ufsc.br

Douglas D J de Macedo
Post-Graduate Program of Knowledge Engineering and Management
Federal University of Santa Catarina
Florianpolis, Brazil
E-mail: macedo@inf.ufsc.br

Michael A Bauer
Department of Computer Science
University of Western Ontario, UWO
London, Canada
E-mail: bauer@uwo.ca

Abstract. Telemedicine is a very important area in medical field that is expanding daily motivated by many researchers interested in improving medical applications. In Brazil was started in 2005, in the State of Santa Catarina has a developed server called the CyclopsDCMServer, which the purpose to embrace the HDF for the manipulation of medical images (DICOM) using a distributed file system. Since then, many researches were initiated in order to seek better performance. Our approach for this server represents an additional parallel implementation in I/O operations since HDF version 5 has an essential feature for our work which supports parallel I/O, based upon the MPI paradigm. Early experiments using four parallel nodes, provide good performance when compare to the serial HDF implemented in the CyclopsDCMServer.

1. Introduction

Since the HDF release many scientific and general purpose applications facing difficulties when storing large amounts of data started to use it as an efficient alternative to store and access large files. HDF is used from clinical applications for managing large collections of X-ray images to oil companies when storing and accessing large 3D seismic soundings. Developed by HDF Group's of University of Illinois, the primary goal of HDF project was accessibility over multi-platforms , later other goals like access ability, efficient storing of large objects and data viewing tools [1].

Published under licence by IOP Publishing Ltd

In Brazil, the Federal University Of Santa Catarina's Telemedicine Laboratory (LABTELEMED) [4] have two main projects, the first is called CyclopsDCMServer which is a DICOM medical image facility developed by The Cyclops Group [6] to provide DICOM image storage and access by using wide area networks (WAN) and the second one is a large-scale telemedicine network called Santa Catarina Telemedicine Network (RCTM - Acronyms for "Rede Catarinense de Telemedicina" in Portuguese language) [2]. RCTM connects different health institutions such as hospitals and basic care instalations encompassing 286 municipalities in the state of Santa Catarina. This network provide access for more than 10 DICOM modalities like electrocardiograms (ECG), magnetic resonance (MRI), computed tomography (CT) and computed radiography (CR) [3].

The RCTM network uses a Picture Archiving and Communication Systems (PACS) proper developed to aquire all patient's clinical information such as DICOM images and personal data to the CyclopsDCMServer. Figure 1 show the PACS distribution in Santa Catarina state. A DICOM based PACS consist in hardware and software designed for medical use and is supported by major medical imaging equipment manufacturers. It embraces digital image acquisition devices, digital image archives and workstations. Its basic function starts when the image is aquired by the proper equipment (e.g. Tomograph) the DICOM server provides segmentation service over the incoming information, processing and storing the images in a centralized database.

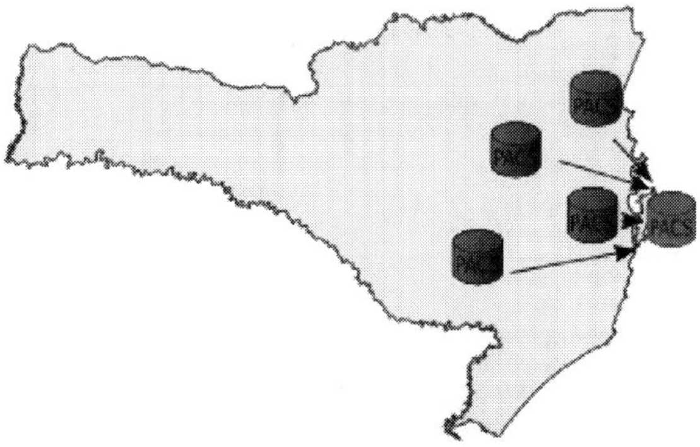

Figure 1. PACS Network.

Currently the CyclopsDCMServer stores all PACS information in a PostgreSQL object-relational database with an aproximate amount of data of 50 terabytes [3]. Such magnitude leads to problems like scalability, information distribution, operational cost and ability to use high performance system techniques. To overcome these issues a new server architecture based on PVFS and HDF5 was conceived. Parallel virtual le system (PVFS) is a distributed le systems designed to scale petabytes of data and provide high access rates [5], and Hierarchical Data Format 5 is a data model to handle high volume and complex data. This paper is structured in background with a summary of the CyclopsDCMServer application.The next two sections introduces the DICOM standard and the HDF5 definitions while section 4 describe our methodology follow by obtained results, our conclusions and future work.

2. Background

The CyclosDCMServer was developed for both scenarios, medical image storage and workstation manipulation, currently it suports DICOM based PACS equipments distributed over hospitals and radiology clinics and communication between the server and medical equipment is performed through TCP/IP. The server aims at storage and retrieval of indexed DICOM files which are store in a relational PostgreSQL database. In additional for CyclopsDCMServer, a new research started in 2008, to explore new architecture for DICOM images using a distributed file systems. The goal is to address issues of telemedicine environments based on relational database systems such as scalability, information distribution, ability to use high performance system techniques and operational costs. Among the usual procedures to avoid the scalability issue, the project design to use high performance distributed systems, like clusters and grids [3]. The first approach distinguish from the usual system is to store all information hierarchically, such as organize and store in HDF5 data format. The second step was to use a distributed file system to keep the HDF file distributes across the nodes.

Another important contribution is the HDF5 Wrapper Library (H5WL). It was created to supply the scarcity of DICOM applications, once usually only support drivers for standard DBMSs. The library contains a wrapper object which is used to create, locate, collect and store information related to DICOM images on HDF5 files. The hierarchical data structured is organized is based on DICOM information hierarchy [7] and an example of the structure is shown in Figure 2, where the data is organized around six layers: root, hospital, patient, study, series and image.

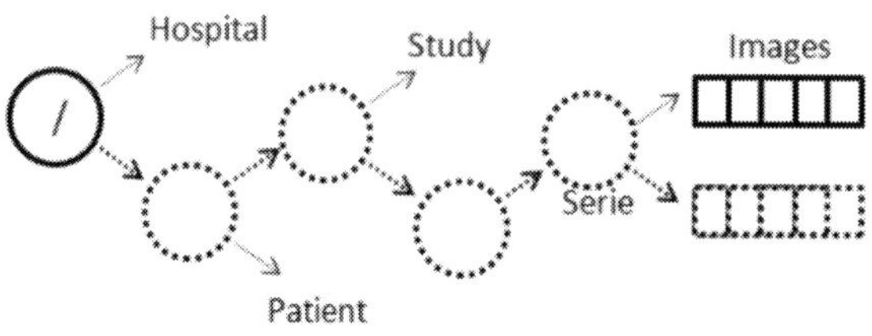

Figure 2. HDF structure

When CyclopsDCMServer receive a DICOM file, the hierarchy is created through H5WL methods, creating groups that provide information to identify an image inside the file. It is important to note that image (p. e. tomography picture) represents almost the full size of a DICOM file and normally represents as the last hierachical structure. Basically, the image structure is represented in binary and can easily been converted to usual image like JPEG [7] and representing the compress DICOM image.

Based on 25 experiences with the system, it showed an average improvement in storage of about 16% when compared with the usual system using DBMSs. However, in term of retrieval operations, there was a drop in performance, with average decrease in performance of around 21%. This was due to mechanisms used to provide similar behavior to that of standard DBMSs when retrieving information. This paper addresses this problem by proposing an extension to the architecture to take more advantage of a parallel environment. This architecture is detailed in Section VI.

2.1. DICOM

The main proposed of Digital Imaging and Communications in Medicine (DICOM 3.0) is to standard the medical digital images as visualization, storage and impression. Previously was known as ACR_NEMA Standards Publication PS3 and created in 1983 by the American College of Radiology, conceiving a unique standard for medical devices and facilitating the expansion of digital images [7]. Before this standard was established, each manufacture created their own solution for visualization, storage and impression of digital images, resulting in incompatibility between medical devices.

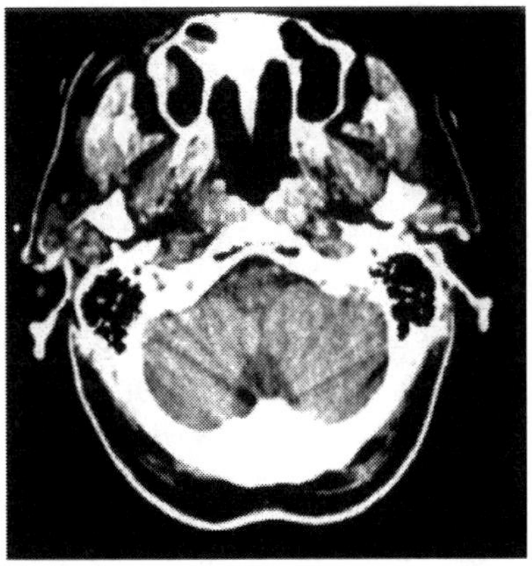

Figure 3. Example of DICOM imag

Completed in September 1992, DICOM third version came with major revision, supplying increasing variety of digital devices and their communications protocols. This version was called DICOM 3.0, as it followed two earlier ACRNEMA editions Taking many important features from earlier and other standards, the early versions of focused on the improvement and correction of some issues, and where those publications provided specifications related to hardware interfaces, it introduced a set of data format and commands for software packages. Nowadays the standard is reviewed annually and updated with new supplements if necessary [7]. Our current CyclopsDCMServer supports eight DICOM modalities: computed radiography; computed tomography; magnetic resonance; nuclear medicine; ultrasound; X-ray angiography; electrocardiograms; DICOM structured reporting [8].

2.2. HDF5

The HDF group at the University of Illinois, initially in the 90s, developed a new product aiming to work with large scientific data, including high performance data manipulation supporting random access, number encoding in native format, data compression, individual data set encryption, and storage strategies for parallel I/O and multidimensional data structures; the current version is HDF5. There is two important feature of HDF. First is that files can contain binary data as multi-dimensional arrays and allow direct access to parts of the file without first parsing the entire contents [1].

Second feature is supporting for standard parallel I/O interfaces. Known as The Parallel Hierarchical Data Format 5 (Parallel HDF5), for work with it is required MPI/IO interface, which is supported through MPICH ROMIO [9]. Now days, ROMIO has support to most common file system, but only offer bracing to PVFS as distributed file system. The purpose of Parallel HDF5 is to make it easy for users to use the library, providing compatibility with serial HDF5 file. One approach is to read and write data by hyperslab [1], i.e., a multidimensional array that can be spread by rows, columns, patterns and chunks, and a hyperslab selection could be a logically contiguous collection of points, or it can be a regular pattern of points or blocks, depending on the type used.

When works with parallel I/O, the properties of communication that will realize the I/O operation its important to synchronise the nodes. The Parallel HDF5 library has available two types of properties (collective and independent data access) and four hyperslab model (Contiguous Hyperslab, Regularly Spaced Data, Pattern and Chunk) [1]. In HDF5 there are two essential structures which form the base for the library: dataset and group. Dataset is a multi-dimensional array of datatype; HDF stores and organize all kinds of data from atomic to composed types, similar to the C structure construct. Special array operations, such as chunks, compression and extendibility are available through the HDF library and can be applied to a dataset. Other important structure is the dataspace. Through a dataspace is possible to require components of dataset or even an attributes are defined, as well as array ranks, sizes and types. The group is similar to UNIX directories, though cycles are allowed. Every file is started with a root group, represented as /, and could be followed by the name of another group or a dataset as show in figure 2.

3. Related work

The works related are suchlike to our work, concerning with the use of parallel I/O as a solution for I/O bottlenecks access for large amounts of stored data. The first related work presented in Nikhil Laghave [10], is very similar to our work. This work is focused on the use of a parallel I/O library for scalability issues involving fermion dynamics for nuclear structure (MFDn). This work used the HDF5 parallel version for parallel I/O, testing with collective and independent models. As result once file sizes increasing above 20 GB, parallel HDF5 becomes more cost-eective than sequential I/O for sufficiently large datasets than sequential binary I/O.

The work of A. Adelmann [11] focused on using parallel I/O for particle-based accelerator simulation which involved vast quantities of data and dimensional arrays. He used parallel I/O performance for MPI code as well parallel HD5 by developed API call H5part. H5part is a portable high performance parallel data interface for particle simulation. He compared read and write performance in simulations between H5part, mpi-io and one file per process. HDF5 showed good performance in writing, though mpi-io showed better results.

Finally, H. Yu [12] even not using parallel HDF as solution, he demonstrated an interesting work which dealt with large earthquake simulations. He faced scalability issue and I/o bottleneck, due earthquake simulation required large file to storage. To solve his problem, he developed an application using parallel I/O strategies through MPI I/O to address his needs. The results was considerable to remove the I/O bottleneck and also hide pre-processing costs.

4. The parallel architecture

Our work comes to propose a new architecture for Macedos approach, in order to avoid I/O bottlenecks and get better performance using parallel data access to HDF files stored in the PVFS distributed file system. To accomplish our work, we configured MPI environment to work with Parallel HDF5. As cited in session 2.3, the Parallel HDF5 library requires a parallel MPI/IO interface through ROMIO and when working with MPI, it is necessary to design it to running in a cluster environment. It is important to note the requirement to use the mpirun shell script to run any MPI application, which attempts to hide the differences in starting jobs for various devices from the user [13]. We created an additional procedure to work with the CyclopsDCMServer. This procedure should be called every time when is required to retrieve or store some medical information.

For now, we build an application concerning with I/O access, which our application just responsible to direct access a file for reading and writing a dataset. Figure 4 illustrates how the architecture works. It was crucial to modify the H5WL functionality. Instead of having the H5WL responsible for reading and writing the binary information created by PACS, it will be treated as a new parallel application. When some client requests to store a DICOM files, it necessary to initiate the parallel application by calling mpirun shell script. After, all communication between them will be made by socket connections.

The communication is done by the master process (represented by MPI process zero) and H5WL, which the messages is represent function parameters, like the location that is the target of an operation (group path), the image buffer and the number of MPI processes. In write functions, per example, the H5WL will first receive the DICOM file, create a new hierarchy

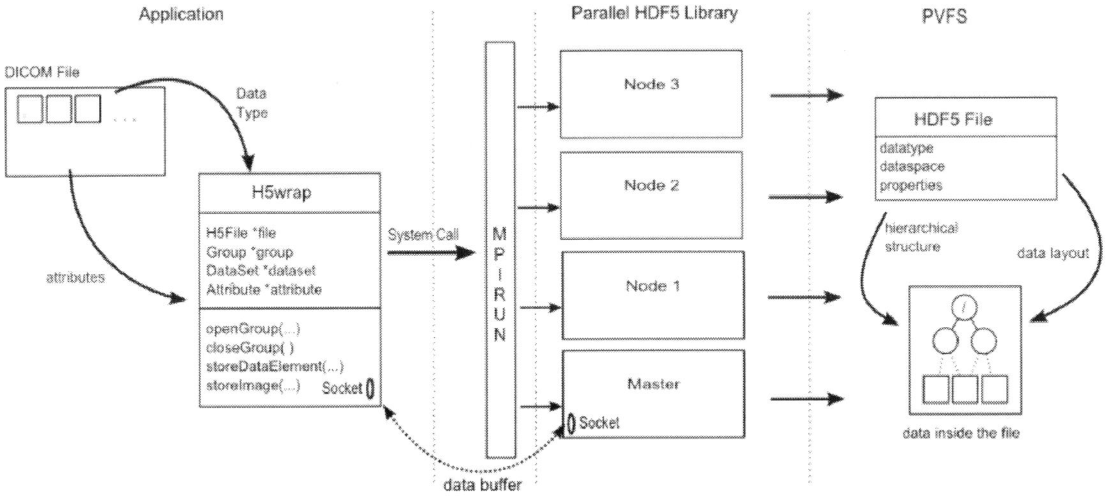

Figure 4. Parallel architecture

of the image based on DICOM file layers, get the path location for new image (JPEG image) and then call mpirun procedure to start the MPI application. The master process will delegate the communication with H5WL to retrieve the function to perform, get the location (group) of image in the HDF5 structure, the arguments for the job and the stream of images to be stored. Receiving the stream, the master node has to distribute the memory buffer to each process in small buffer equal to number of nodes. Finally, once the jobs have executed, the main process will return to the wrapper the status of reading or writing the buffer.

5. Experiments

Our experiments are based on the Parallel HDF5 architecture, adapted to use PVFS, MPI and sequential CyclopsDCMServer. The parallel environment consists in four node cluster, as specified in Table 1 and one dedicated computer for DICOM server. Unfortunately, the environment is non-dedicated cluster and belongs to Telemedicine Laboratory and share a connection network 100 Mbs Ethernet. The operating system installed on all nodes is CentOS with kernel 2.6.32, and there only one metadata PVFS node. Each node has a PVFS client for access to the PVFS file system and MPI compiled. It is important to note that our results do not take into consideration external factors, like computer users, using the same network.

Table 1.

Name	CPU	Memory	HD
Node1	AMD athlon x2 2.1 GHz	2 Gb	20 Gb
Node2	AMD athlon x2 2.8 GHz	3 Gb	20 Gb
Node3	Intel PentiumR Dual 1.80 GHz	1 Gb	20 Gb
Node4	Intel Core i5 3.2 GHz	3 Gb	20 Gb

All the experiments were conducted with CyclopsDCMServer sequential and parallel architecture and involve only comparison of storing a new DICOM file in HDF file, discarding all delay from communication between server and master node; future comparisons will compare file retrieval. This experiment measures the time spent write an image buffer into a HDF file repeating 25 times with different DICOM files for each test. The selection of the files used is the same files were used for the both experiments. The time collected is the time required to write an image, ignoring sequential information like patient name, hospital and etc. The reason that this measure was chosen is because an image represents most of a DICOM file, nearly 50 Mb. All these 25 images were created by CT equipment that generates monochromatic images with 512x512 pixels with 16 bits per pixel.

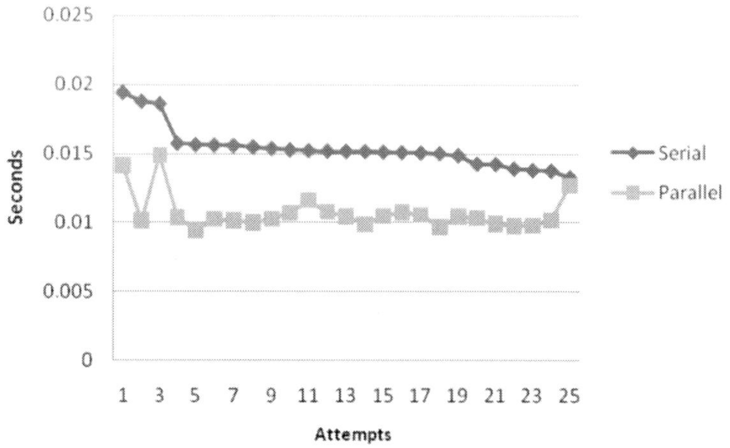

Figure 5. Time of writing operation

The figure 5 compares the performance between serial CyclosDCMServer and our integrated parallel architecture. Collected 25 stored times, our architecture shows better performance than the serial with the average for parallel method being 0.0107176 seconds, while the average for serial was 0.01539956 seconds, representing an improvement of around 30 percent.

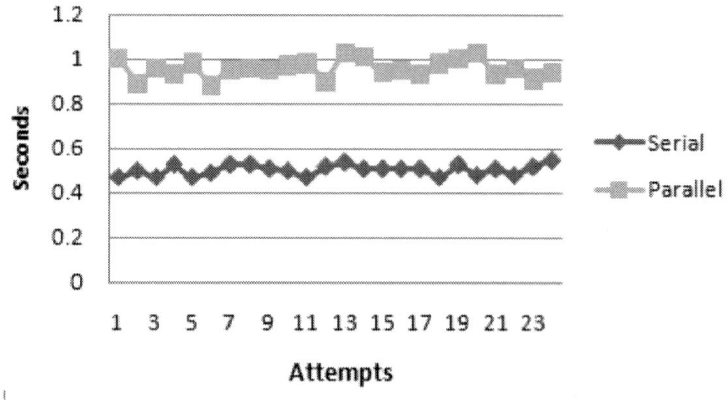

Figure 6. Time of server process

However, the figure 6 shows the server performance in all process, collecting time since receiving a DICOM image, to storage in HDF file. The experience demonstrates the difference between the architectures, in specific, the leakage of using system call for initiate MPI process. As we can note, the performance of CyclopsDCMServer with additional of parallel architecture, has a delay about twice time than serial server.

Journal of Physics: Conference Series **341** (2012) 012021 doi:10.1088/1742-6596/341/1/012021

6. Conclusion and future work

The objective of this paper was to introduce a new parallel architecture designed to reduce the bottleneck I/O issues in the serial architecture provide by HDF. With the goal to get better performance in I/O access in HDF file at PVFS, we introduced an extension to the architecture for the CyclopsDCMServer previously introduced by Macedo et al. [14].

In related work, we shown many studies with similar issues in bottleneck which have used parallel I/O to avoid the problems to obtain great results.

Although our experiments involved only the writing function to compare serial and parallel architecture in the same environment, we shown that the parallel architecture result in thirty percent improvement on I/O operations, but still having performance issues from system call.

The retrieval time of a DICOM image collected from a HDF file using parallel I/O is our priority work although the aproach presented in this work doesn't inclued the retrieval operation since the server application was not completly done.

Future work includes analyzing the significance MPI nodes numbers to measure the variance of I/O performance and minimize the communication between H5WL and master MPI node since causes performance decrease. Running performance experiments in a virtual cluster machines comparing the network performance in a virtualization enviroment over standard cluster is another goal.

7. Reference

[1] HDFGroup 2011 Access: January URL `http://www.hdfgroup.org`

[2] Wallauer J, von Wangenheim A, Andrade R and De Macedo D 2008 *21st IEEE International Symposium on Computer-Based Medical Systems* (IEEE) pp 105–107

[3] De Macedo D, Von Wangenheim A, Dantas M and Perantunes H 2009 *International Journal of High Performance Systems Architecture* **2** 99–106 ISSN 1751-6528

[4] Labtelemed Laboratorio de telemedicina Access: January. 2011 URL `http://www.telemedicina.ufsc.br`

[5] PVFS Parallel virtual file system, version 2 Access: January. 2011 URL `http://www.pvfs.org`

[6] Romio: A high-performance, portable mpi-io implementation Access: December. 2010 URL `http://cyclops.telemedicina.ufsc.br`

[7] Pianykh O 2008 *Digital Imaging and Communications in Medicine (DICOM): A practical introduction and survival guide* (Springer Verlag) ISBN 354074570X

[8] Cyclops 2011 Cyclops project Access: January URL `http://www.cyclops.ufsc.br`

[9] Division M Romio: A high-performance, portable mpi-io implementation Access: March. 2011 URL `http://www.mcs.anl.gov/research/projects/romio/`

[10] Laghave N, Sosonkina M, Maris P and Vary J 2009 *Computational Science–ICCS 2009* 84–93

[11] Adelmann A, Ryne R, Shalf J and Siegerist C 2006 *Particle Accelerator Conference, 2005. PAC 2005. Proceedings of the* (IEEE) pp 4129–4131 ISBN 0780388593

[12] Yu H, Ma K and Welling J 2005 *Supercomputing, 2004. Proceedings of the ACM/IEEE SC2004 Conference* (IEEE) p 49 ISBN 0769521533

[13] Division M Mpich2: High-performance and widely portable mpi Access: December. 2010 URL `http://www.mcs.anl.gov/research/projects/mpich2`

[14] de Macedo D, Perantunes H, Maia L, Comunello E, von Wangenheim A and Dantas M 2008 *Computers and Communications, 2008. ISCC 2008. IEEE Symposium on* pp 658 –663 ISSN 1530-1346

Solving Bivariate Polynomial Systems on a GPU

Marc Moreno Maza and Wei Pan
University of Western Ontario and Intel Corporation
E-mail: moreno@csd.uwo.ca; wei.pan@intel.com

Abstract. We present a CUDA implementation of dense multivariate polynomial arithmetic based on Fast Fourier Transforms over finite fields. Our core routine computes on the device (GPU) the subresultant chain of two polynomials with respect to a given variable. This subresultant chain is encoded by values on a FFT grid and is manipulated from the host (CPU) in higher-level procedures.

We have realized a bivariate polynomial system solver supported by our GPU code. Our experimental results (including detailed profiling information and benchmarks against a serial polynomial system solver implementing the same algorithm) demonstrate that our strategy is well suited for GPU implementation and provides large speedup factors with respect to pure CPU code.

1. Introduction

Graphics processing units (GPUs) are becoming very attractive for scientific computing applications that are targeting high performance. Many application software have gained massive speedup factors by integrating GPU support, in areas like Monte Carlo simulation, weather forecasting, climate research, molecular modeling, quantum mechanical physics and astrophysics. See for instance the research areas in the website of `GPGPU.org` [2] or the application section in the Wikipedia page for GPGPU [1], where many references and links are provided. Symbolic computation is also entering the area of many-core computing, but few reports have been published so far [4, 8, 10, 17].

In this paper, and up to our knowledge, we report on the first GPU implementation of a polynomial system solver over finite fields. A GPU implementation of a bivariate solver over the real numbers is reported in [3]. Solving polynomial systems is a driving subject in symbolic computation with many successful results from theoretical to practical aspects. Adapting this knowledge and acquired experience is, however, very challenging.

The difficulty starts at the level of dense multivariate polynomial arithmetic. For instance, techniques that have appeared to be very effective for multicore implementation [18, 19] do not apply to GPU implementation and had to be revisited [17], see ??.

Efficient multicore implementation relies on a good usage of the different levels of cache memories (L1, L2, L3) and in minimizing the impact of performance bottlenecks due to phenomena like true sharing, false sharing, memory contention, etc. On the other hand, concurrency platforms (such as OpenMP, Intel Cilk Plus, KAAPI) for multicore architectures are generally based on the *fork-join parallelism model* and provide tools for automatic scheduling; this helps programmers focusing on exposing parallelism for their targeted applications.

The GPU programming model is based on *data-parallelism* and a large part of the scheduling is done statically by the programmer through the decomposition of the application into GPU kernels. Efficient GPU implementation relies on optimizing memory usage for maximum bandwidth, maximizing occupancy of the device in order to hide memory transfer latency and optimizing instruction usage for maximum throughput.

The motivation of our work is to support polynomial system solvers based on the notion of a *regular chains* [11] where the core algorithms [16] often rely on the theory of polynomial subresultants [9]. In [12], we have shown that the dominant cost of those algorithms can be essentially reduced to that of subresultant chain computations.

In Section 2, following an idea proposed by Collins in [6], we explain how we compute a subresultant chain by evaluation-interpolation, based on FFT techniques. Once we obtain a so-called FFT grid on which the input polynomials are sampled, we employ the Brown's Algorithm to compute subresultant chains of univariate polynomials. All these computations are performed on the device for which we have designed a specific implementation of Brown's Algorithm, see Section 3.

This sampled subresultant chain can then be exploited by a high-level algorithm running on the host. A simple case is the resultant computation described in Section 3. A more advanced one is that of the bivariate solver algorithm of [13] that we report in Section 4. Actually, we implemented this algorithm with and without GPU code support. Our pure CPU implementation is serial C code. The specifications of the GPU cards used in our experimentation are listed in Appendix A while profiling information for our GPU kernels appear in Appendix C.

Our experimental results show that, for resultant computation our GPU-supported code outperform our pure CPU code by a speedup factor of 34 (resp. 69) on sufficiently large input bivariate (resp. trivariate) polynomials. For the largest examples we tested our GPU-supported bivariate solver outperform its CPU counterpart by a factor of 7.5. For this latter experimentation, we should insist on the fact that a significant part of the computation (univariate polynomial GCDs) are still performed by CPU code. Removing this bottleneck on the critical path of the whole application is work in progress and could be an opportunity to use two GPU cards concurrently.

One may ask whether the algorithms implemented in this work could also lead to a successful multicore implementation. We actually tried and the answer is no for sufficiently large input data. Indeed, our most powerful GPU card can efficiently handle a sampled subresultant chain with a size in the order of 1 GB. Since the construction of the sampled subresultant chain essentially consists of many successive traversals of this data structure, a multicore implementation will suffer from high rate of cache misses due to the fact that L2 to L3 caches are today in the order of several MB.

2. Subresultant computations

This section is devoted to a high level description of polynomial subresultant computation, by means of an FFT-based modular algorithm. Let $\mathbb{K}$ be a field and $n \geq 1$. In the sequel, we consider $P, Q \in \mathbb{K}[x_1, \ldots, x_{n+1}]$ two non-constant polynomials with the same main variable $y := x_{n+1}$. We define $p := \deg(P, y) \geq q := \deg(Q, y)$. We denote by $\mathbb{B}$ the ring $\mathbb{K}[x_1, \ldots, x_n]$ and by S_j the j-th subresultant of P, Q in $\mathbb{B}[y]$, for $0 \leq j < q$.

FFT grid. Given n positive integers $m_1, \ldots, m_n$, we call the following finite set Θ a *grid* in $\mathbb{K}^n$

$$\begin{aligned}\Theta &= \Theta_1 \times \Theta_2 \times \cdots \times \Theta_n \\ &= \{(u_1, \ldots, u_n) \mid u_i \in \Theta_i \text{ for each } i\}\end{aligned} \tag{1}$$

where Θ_i is a finite subset of $\mathbb{K}$ with size m_i, for each i. We say that the grid Θ is *valid* for a polynomial $f \in \mathbb{K}[x_1, \ldots, x_n][y]$, if the leading coefficient of f in y does not vanish at any point of Θ. Assume furthermore that, for each i, the integer m_i is a power of 2 and that there exists $\omega_i \in \mathbb{K}$ which is a m_i-th primitive root of unity. If Θ_i is chosen as follows:

$$\Theta_i = \{\omega_i^j \mid j = 0 \cdots m_i - 1\},$$

then we say that Θ a *FFT grid* in $\mathbb{K}^n$. The *format* of Θ is $(m_1, \ldots, m_n)$ and its *size* is the product $m_1 \cdots m_n$. The *evaluation-interpolation method* for computing subresultants S_j, for $0 \leq j < q$, proposed by Collins in [6], proceeds as follows.

($\mathbf{S}_1$) Compute an upper bound for the degree $\deg(S_0, x_i)$ and set m_i to it, for $i = 1 \cdots n$.

($\mathbf{S}_2$) Construct a grid Θ of format $(m_1, \ldots, m_n)$, valid for both P and Q.

($\mathbf{S}_3$) Evaluation: compute $P(u, y)$ and $Q(u, y)$ for each $u \in \Theta$.

($\mathbf{S}_4$) For all $u \in \Theta$ and all $0 \leq j < q$, compute the subresultant $S_j(u)$ of $P(u, y)$, $Q(u, y)$ in y.

($\mathbf{S}_5$) Interpolation: for each $0 \leq j < q$, construct S_j from the image set $\{S_j(u) \mid u \in \Theta\}$.

For Step ($\mathbf{S}_1$), a well-known degree bound can be derived from the Sylvester matrix of P and Q, see the paper of Monagan [15] for detailed discussions. For Step ($\mathbf{S}_2$), let h be the product of the leading coefficients of P and Q in y, which is a nonzero polynomial in $\mathbb{K}[x_1, \ldots, x_n]$. Constructing a valid grid for h deterministically is difficult. In practice, one can generate a grid at random and check whether the grid is valid or not. Step ($\mathbf{S}_4$) is equivalent to computing the subresultant chains of m univariate polynomial pairs. A standard tool is Brown's subresultant algorithm [5].

Step ($\mathbf{S}_3$) and Step ($\mathbf{S}_5$) are instances of the so-called *multipoint evaluation* and *multipoint interpolation* problems, respectively. In general, these operations can be performed by means of subproduct tree techniques [9]. We do not analyze this point of view further since our focus is on FFT grids. If Θ is a FFT grid, then FFT-based multipoint evaluation and interpolation run in $O(m \log(m))$ operations in $\mathbb{K}$.

The approach used in our implementation is summarized by Figure 1, where all the computations are converted into the bivariate case, by means of Kronecker's substitutions. This turns multivariate FFT computations into univariate ones, which simplifies both the analysis and the implementation. In practice, it is wiser to conduct large univariate FFTs (or large bivariate FFTs, see [18] for details) rather than multivariate FFTs with small sizes along one or more dimensions. Another feature of our implementation is the use of random translations of the form $\phi_a : x \longmapsto x + a$ (applied to the polynomials $F', G' \in \mathbb{Z}_c[x, y]$, see Figure 1) in order to find a valid FFT grid. Details, including a probability analysis can be found in [20].

3. Brown's algorithm on the GPU

One of the most important algebraic tools for solving polynomial system symbolically is the polynomial subresultants. We refer to [7, 21] for the theory of polynomial subresultants. In this section, after briefly reviewing some properties of polynomial subresultants, we present our CUDA FFT-based implementation of Brown's subresultant algorithm for bivariate (or trivariate) polynomials over a prime field $\mathbb{K} = \mathbb{Z}_c$, of characteristic $c > 2$.

Let $\mathbb{D}$ be an integral domain and let $P, Q \in \mathbb{D}[y]$ be non-constant polynomials of degree p, q respectively. We assume $p \geq q$. The polynomials computed by the subresultant algorithm form a sequence, called the *subresultant chain* and denoted by $S = \mathrm{src}(P, Q, y)$. This sequence consists of $q+1$ polynomials $(S_q, \ldots, S_0)$ in $\mathbb{D}[y]$, starting at $S_q = \mathrm{lc}(Q)^{p-q}Q$ and ending at the resultant $S_0 = \mathsf{res}(P, Q, y)$. The polynomial S_j is called the subresultant of index j. For each index $0 \leq j \leq q$, if S_j is not zero, its degree is at most j and S_j is said regular if $\deg(S_j, y) = j$. Hence, S_q is a regular subresultant by definition. See [12] for a complete and formal definition of a subresultant chain.

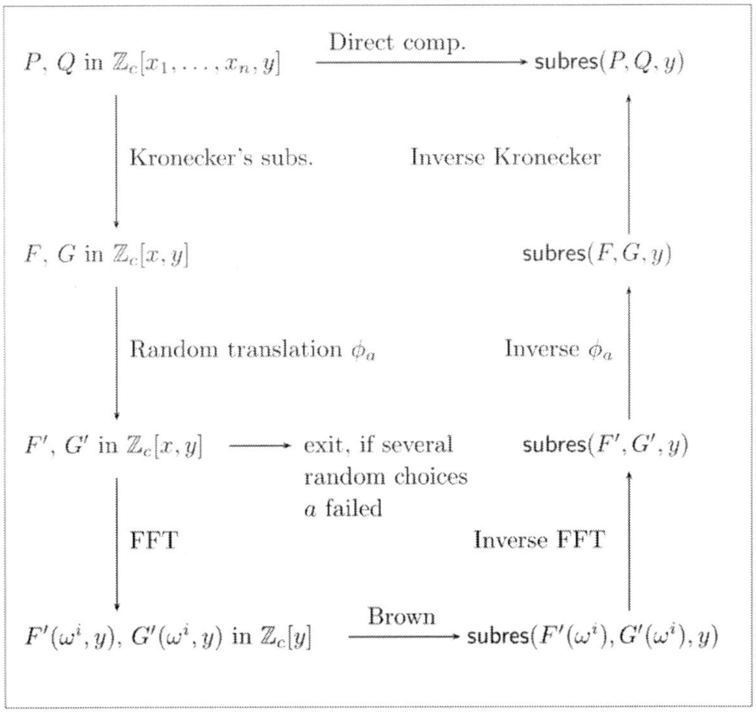

Figure 1. Compute the subresultant chain via an FFT based modular algorithm

3.1. Brown's algorithm

Algorithm 1: Brown's subresultant algorithm

Input : polynomials $F, G \in \mathbb{K}[x]$ such that $\deg(F) \geq \deg(G) > 0$
Output : the subresultant chain of F and G

```
S_i ← 0 for 0 ≤ i < deg(G);
B ← prem(F, −G, x), A ← G, α ← deg(F) − deg(G);
while B ≠ 0 do
    d ← deg(A), e ← deg(B), δ ← d − e;
    S_{d−1} ← B;
    S_e ← lc(A)^{α(1−δ)} lc(B)^{δ−1} B;
    if e = 0 then break;
    B ← lc(A)^{−αδ−1} prem(A, −B, x), A ← S_e, α ← 1;
return S_i for 0 ≤ i < deg(G);
```

The main subroutine involved in Brown's Algorithm is *pseudo-remainder* computation. Recall that given $f, g \in \mathbb{K}[x]$, the *pseudo-division with remainder* of f by g computes $q, r \in \mathbb{K}[x]$ with

$$\mathrm{lc}(g)^{1+\deg(f)-\deg(g)} f = qg + r, \quad \deg(r) < \deg(g), \tag{2}$$

assuming $g \neq 0$. The polynomials q and r are uniquely determined by Equation (2) and computed by Algorithm 2. These polynomials are called respectively the *pseudo-quotient* and *pseudo-remainder* of $f(x)$ w.r.t. $g(x)$. We also denote by $\mathrm{prem}(f, g, x)$ the pseudo-remainder r.

Algorithm 2: The pseudo-remainder algorithm

Input : polynomials $f, g \in \mathbb{K}[x]$ such that $\deg(f) \geq \deg(g) > 0$
Output : the pseudo-remainder of f by g in x

```
r ← f;
for i ← deg(f) − deg(g) down to 0 do
    r ← lc(g)r − lc(f)x^i g;
return r;
```

We present our CUDA implementation for constructing FFT-based subresultant chain of bivariate polynomials. The first step is to evaluate the input polynomials $F, G \in \mathbb{Z}_c[x, y]$. For our implementation, univariate polynomials are all dense, encoded as a vector of coefficients. Multivariate dense polynomials with n variables are encoded recursively as a univariate polynomial with $(n-1)$-variable polynomials as coefficients. For example, $F = 1 + 2x + 3xy + 4x^2 + 5y^2$ can encoded as a vector $[1, 2, 4, 0, 3, 0, 5, 0, 0]$, read as

$$F = (1 + 2x + 4x^2) + (0 + 3x + 0x^2)y + (5 + 0x + 0x^2)y^2.$$

Our subroutine `list_fft_univariate` computes a list of univariate n-point FFTs on a list polynomials of the same size q. For $n = 2^k$ and $q \geq 1$, the Stockham FFT factorization implies

$$I_q \otimes \mathrm{DFT}_n = \prod_{i=0}^{k-1} (I_q \otimes \mathrm{DFT}_2 \otimes I_{2^{k-1}})(I_q \otimes D_{2,2^{k-i-1}} \otimes I_{2^i}) (I_q \otimes L_2^{2^{k-i}} \otimes I_{2^i}), \quad (3)$$

which can be efficiently realized from three types of CUDA kernels, similar to the Stockham FFT, as presented in [17]. More details appear in our article [17] reporting on a GPU implementation of FFT over finite fields.

Example 1 *Let $F = a(x) + b(x)y + c(x)y^2 + d(x)y^3$ be a bivariate polynomial where a, b, c, d, have degree less than 8. Let ω be a 8-th primitive root of unity. After evaluating $F(x, y)$ at $x = (1, w, w^2, w^3, w^4, w^5, w^6, w^7)$, the layout of the evaluation data is*

$$M = \begin{bmatrix} a(1) & a(\omega) & a(\omega^2) & \cdots & a(\omega^6) & a(\omega^7) \\ b(1) & b(\omega) & b(\omega^2) & \cdots & b(\omega^6) & b(\omega^7) \\ c(1) & c(\omega) & c(\omega^2) & \cdots & c(\omega^6) & c(\omega^7) \\ d(1) & d(\omega) & d(\omega^2) & \cdots & d(\omega^6) & d(\omega^7) \end{bmatrix}$$

The i-th column corresponds the univariate polynomial $F(\omega^i, y)$ in y for $i = 0 \cdots 7$. Transposing matrix M gives

$$M^t = \begin{bmatrix} a(1) & b(1) & c(1) & d(1) \\ a(\omega) & b(\omega) & c(\omega) & d(\omega) \\ a(\omega^2) & b(\omega^2) & c(\omega^2) & d(\omega^2) \\ a(\omega^3) & b(\omega^3) & c(\omega^3) & d(\omega^3) \\ a(\omega^4) & b(\omega^4) & c(\omega^4) & d(\omega^4) \\ a(\omega^5) & b(\omega^5) & c(\omega^5) & d(\omega^5) \\ a(\omega^6) & b(\omega^6) & c(\omega^6) & d(\omega^6) \\ a(\omega^7) & b(\omega^7) & c(\omega^7) & d(\omega^7) \end{bmatrix}.$$

Assuming that the leading coefficient $d(x)$ in y of $F(x, y)$ does not vanish at any power of ω. Then each row of M^t can be regarded as a univariate polynomial of degree 3.

As illustrated in the above example, the second step is to transpose the evaluated F and G for preparing the subresultant constructions. The major cost is to compute the subresultant chains at all evaluation points. To accelerate the overall performance, we implemented Brown's subresultant algorithm in CUDA for computing a sequence of subresultant chains in a highly parallel manner and we present two different approaches for this task.

Coarse-grained approach. The most direct way is to let each CUDA thread run a univariate Brown's subresultant algorithm. This approach works all the time and for practical problems where the number of threads can reach a big value, say 2^{12}, this can still bring a reasonable amount of speedup. However, this approach incurs a large amount of memory accesses to the GPU global memory within a thread block, which limits the peak performance of the implementation.

Fine-grained approach. The second approach is to break a list of univariate Brown's subresultant computations into a sequence of lists of univariate polynomial pseudo-divisions. Provided that at each step all A_j's have the same degree and all B_j's have the same degree [1], Algorithm 1 can be turned into the "list version", namely Algorithm 3. Initially, all F_j's have the same degree according to the choice of ω, (as all G_j's do).

For two univariate polynomials $P(x)$ and $Q(x)$, let $S_{k_1}, \ldots, S_{k_\ell}$ be all the regular subresultants of P and G with $0 \le k_1 < \cdots < k_\ell < \deg(Q)$. Then $(k_1, \ldots, k_\ell)$ is called the degree sequence of P and Q. In fact, they are the degrees of all remainders in the Euclidean algorithm with input P, Q. Therefore, the above generic assumption says: the degree sequence of the subresultant chain of $F(\omega^i, y), G(\omega^j, y)$ is independent of j.

Algorithm 3: List of subresultant chains

Input : A list of pairs of polynomials $F_j, G_j \in \mathbb{K}[x]$ for $0 \le j < m$ such that $\deg(F_j) \ge \deg(G_j) > 0$, all F_j having the same degree, and all G_j having the same degree.

Output: the subresultant chain of F_j and G_j, for $0 \le j < m$.

- **for** $0 \le i < \deg(G_j),\ 0 \le j < m$ **do**
  - $S_i^j \leftarrow 0$;
- **for** $0 \le j < m$ **do**
  - $B_j \leftarrow \mathrm{prem}(F_j, -G_j, x)$;
  - $A_j \leftarrow G_j$;
- $\alpha \leftarrow \deg(F_j) - \deg(G_j)$;
- **for** $0 \le j < m$ **do**
  - **while** $B_j \ne 0$ **do**
    - $d \leftarrow \deg(A_j),\ e \leftarrow \deg(B_j),\ \delta \leftarrow d - e$;
    - $S_{d-1}^j \leftarrow B_j$;
    - $S_e^j \leftarrow \mathrm{lc}(A_j)^{\alpha(1-\delta)} \mathrm{lc}(B_j)^{\delta-1} B_j$;
    - **if** $e = 0$ **then break**;
    - $B_j \leftarrow \mathrm{lc}(A_j)^{-\alpha\delta-1} \mathrm{prem}(A_j, -B_j, x),\ A_j \leftarrow S_e^j,\ \alpha \leftarrow 1$;
- **for** $0 \le j < m$ **do**
  - **return** $\{S_i^j \mid 0 \le i < \deg(G_j)\}$;

In Algorithm 3, the generic assumption implies that

[1] A nonzero constant has degree 0, and 0 has degree -1.

- at Line 8, if $B_j = 0$ holds for some $0 \le j < m$, then all B_j are zero;
- at Line 9, all A_j have the same degree d, and all B_j have the same degree e;
- at Line 11, all S_e^j can be computed in a *Single Instruction Multiple Data (SIMD)* fashion.
- Similarly, at Line 13, all pseudo-divisions can be performed in a SIMD fashion.

Consequently, each of the lines 9 to 13 in the **while** loop can be executed in a SIMD fashion. Moreover, several threads can cooperate in computing one S_e^j or one B_j. In particular, our core subroutine performs a list of pseudo-divisions in a fine-grained way.

Example 2 *Let* $f = a_3x^3 + a_2x^2 + a_1x + a_0$ *and* $g = b_2x^2 + b_1x + b_0$. *To obtain the pseudo-remainder* $\mathrm{prem}(f, -g, x)$ *of* f *and* g, *we compute*

(1) $h_2 = -b_2 f + a_3 x g = c_2x^2 + c_1x + c_0$,

(2) $h_1 = -b_2h_2 + c_2 g = d_1x + b_0$.

Alternatively, the pseudo-remainder $\mathrm{prem}(f, -g, x)$ *can be computed in two steps:*

$$(S_1) \quad c_2 = \begin{vmatrix} a_3 & a_2 \\ b_2 & b_1 \end{vmatrix}, \quad c_1 = \begin{vmatrix} a_3 & a_1 \\ b_2 & 0 \end{vmatrix}, \quad c_0 = \begin{vmatrix} a_3 & a_0 \\ b_2 & 0 \end{vmatrix};$$

$$(S_2) \quad d_1 = \begin{vmatrix} c_2 & c_1 \\ b_2 & b_1 \end{vmatrix}, \quad d_0 = \begin{vmatrix} c_2 & c_0 \\ b_2 & b_0 \end{vmatrix}.$$

As illustrated in the above example, the basic unit is to perform a single reduction step, called `list_reduce`, which takes two lists of univariate polynomials LF and LG as input, and computes

$$h_i = \mathrm{lc}(g_i) f_i - \mathrm{lc}(f_i) x^{\deg(f_i) - \deg(g_i)} g_i,$$

where f_i is the i-th polynomial in LF and g_i is the i-th polynomial in LG. We assume that polynomials in LG have the same degree dG, and polynomials in LF have the same degree dF. The result LH consists of the computed h_i's. The source code of this kernel is included in Appendix B. In terms of the kernel `list_reduce`, a list of pseudo-remainders can be computed as in Algorithm 4 using a double-buffer method, where operation `swap` only exchanges two pointers not their contents.

Algorithm 4: List of pseudo-remainders

Input : Two lists LF, LG of polynomials such that $\mathrm{LF}[i] = F_i$, $\mathrm{LG}[i] = G_i$ for $0 \le i < m$, $\deg(F_i) \ge \deg(G_i) > 0$, all F_i having the same degree dF, and all G_i having the same degree dG.

Output : The list LH of polynomials such that $\mathrm{LH}[i] = H_i = \mathrm{prem}(F_i, -G_i, x)$ for $0 \le i < m$.

```
list_reduce(LX, dF, LF, dG, LG, p);
for d ← dF − 1 down to dG do
    list_reduce(LY, d, LX, dG, LG, p);
    swap(LX, LY);
Copy out the result from LX to LH;
```

---

d	t_0	t_1	t_1/t_0
30	0.23	0.29	1.3
40	0.23	0.43	1.9
50	0.27	1.14	4.2
60	0.27	1.53	5.7
70	0.31	3.95	12.7
80	0.32	4.88	15.3
90	0.35	5.95	17.0
100	0.50	19.10	38.2
110	0.53	17.89	33.8
120	0.58	19.72	34.0

Figure 2. Computing resultants for bivariate dense polynomials in seconds

3.2. Computing resultants

The data computed from Brown's Algorithm, either with the coarse-grained method or with the fine-grained method, is called a *subresultant chain cube* or simply an *scube*. It consists of the images of all the subresultants of two polynomials on a FFT grid. Any subresultant or any coefficient of a subresultant can be interpolated from the scube, by means an inverse FFT computation inside the GPU.

For two bivariate polynomials of partial degree d in both x and y, the size of their scube is in the order of d^4. For instance, this size is approximately 632MB for $d = 100$. For applications like computing the resultant or several subresultants (as in our bivariate solver), it is natural to keep the scube within the GPU memory and copy only the desired result (say a subresultant) back to the main main, whenever necessary.

Throughout this paper, our code is compiled with gcc 4.2.4 and nvcc 2.2. All the benchmarks were conducted on a desktop with the processor Intel Core 2 Quad CPU Q9400 @ 2.66GHz and 6 GB main memory. In Appendix A, we briefly list the specifications of two NVIDIA's graphics cards used in our benchmarking: Geforce GTX 285 and Telsa C2050. Without any modifications, our code constructs a scube up to 2 times faster on the latter GPU, mainly due to the better support double-precision floating point calculations. Please see Appendix B for detailed description on integer multiplications modulo a prime number using floats.

In Figure 2, we report the timing for computing resultant $\mathsf{res}(F_1, F_2, y)$ with bivariate random dense polynomials $F_1, F_2 \in \mathbb{Z}_c[x, y]$ such that $c = 469762049$ and $d = \deg(F_i, x) = \deg(F_i, y)$ for $i = 1, 2$. In the figure, the first column, labeled by d, shows the partial degree d. The second one, labeled by t_0, is the timing for GPU implementation of the FFT-based scube method, which includes the time for moving result back to the main memory. The third column, labeled by t_1, shows the results for the CPU implementation of the FFT-based scube serial C code in the `modpn` library [14]. The last column reports the ratio between the two implementations. Note that all the resultants in these experiments are computed with the fine-grained method. The maximal speedup we achieve is approximately 38.

In addition to the one-dimensional Stockham FFT, we have also implemented a two-dimensional Stockham FFT for handling trivariate polynomials. This can be applied to compute the scube for a trivariate input. Figure 3 lists our experimental results for computing resultants for trivariate random dense polynomials in $\mathbb{K}[x, y, z]$. The first column shows the common partial degree d in x, y and z. The other three column have the same meaning as in the bivariate case. Note that all but $d = 15$ and $d = 18$ are based on the fine-grained approach. When the coarse-grained method is forced to be used, the speedup it achieves drops significantly. This confirms experimentally the merit of the fine-grained approach.

We observe that the GPU-based implementation achieves a much larger speedup factor in the

d	t_0	t_1	t_1/t_0
7	0.22	0.16	0.7
8	0.23	0.76	3.3
9	0.24	0.85	3.5
10	0.25	0.98	3.9
11	0.24	1.10	4.6
12	0.30	4.96	16.5
13	0.31	5.52	17.8
14	0.32	6.07	19.0
15	0.78	8.95	11.5
16	0.65	31.65	48.7
17	0.66	34.55	52.3
18	3.46	47.54	13.7
19	0.73	51.04	69.9
20	0.75	43.12	57.5

Figure 3. Computing resultants for trivariate dense polynomials in seconds

trivariate case (approximately 70 for the best cases) than in the bivariate case (approximately 38). We are currently investigating the phenomenon.

4. Bivariate Solver

In this section, we discuss the GPU acceleration towards solving bivariate polynomials systems over finite fields by means of triangular decompositions. This work extends the papers [12, 14]. Throughout this section, we consider two bivariate polynomials F_1 and F_2 are in $\mathbb{K}[x, y]$, with ordered variables $x < y$ and with coefficients in a field $\mathbb{K}$.

4.1. Preliminaries

The key idea behind solving bivariate polynomials follows from the fact that the common roots of F_1 and F_2 are likely to be described by the common roots of a single triangular set:

$$\begin{cases} A(x) & = 0 \\ B(x, y) & = 0 \end{cases}$$

Generically, the univariate polynomial $A(x)$ describes the projection of $V(F_1, F_2)$ on the x-axis, whereas the bivariate polynomial $B(x, y)$ is expected to have partial degree 1 in y. We usually obtain $A(x)$ from the resultant of F_1 and F_2 in y, and compute $B(x, y)$ in a polynomial GCD computation.

Example 3 ([14]) *Consider* $F_1 = x^2 + y + 1$ *and* $F_2 = x + y^2 + 1$. *We have* $\mathsf{res}(F_1, F_2, y) = x^4 + 2x^2 + x + 2$, *which is a squarefree polynomial when* $\mathbb{K} = \mathbb{Q}$ *or* $\mathbb{K} = \mathbb{Z}_c$ *with* $c \neq 3, 7$. *It is not hard to show that* $A(x) = x^4 + 2x^2 + x + 2$ *and* $B(x, y) = y + x^2 + 1$ *are a desired pair of polynomials. That is, the common solutions of* $F_1(x)$ *and* $F_2(x)$ *are exactly those of* $A(x)$ *and* $B(x, y)$. *In fact, the polynomial* $B(x, y)$ *is a gcd of* F_1 *and* F_2 *modulo* $A(x)$ *in a technical sense introduced in [16] and that we illustrate here. For a formal definition and algorithms, please refer to [12].*

Observe that $A(x)$ writes $(x^2+x+1)(x^2-x+2)$, thus we have

$$\begin{cases} F_1 = y - x, \\ F_2 = (y-x)(y+x), \end{cases} \quad \text{if } x^2+x+1=0,$$
$$\begin{cases} F_1 = y + x - 1, \\ F_2 = (y-x+1)(y+x-1), \end{cases} \quad \text{if } x^2-x+2=0.$$

Hence, using the Chinese Remaindering Theorem, we obtain

$$B(x,y) = \gcd(F_1, F_2) \bmod A(x) = y + x^2 + 1.$$

4.2. Algorithm

Algorithm 7 is an improved version of the bivariate solver routine presented in [13]. It can also be viewed as a specialized version of the general regular gcd algorithm presented in [12]. The improvement is two folds. First Algorithm 7 makes no genericity assumptions on the input polynomials F_1, F_2, such as the resultant $\mathsf{res}(F_1, F_2, y)$ is not zero. Secondly, in some non-generic situations, Algorithm 7 avoids unnecessary computations. However, and in order to avoid technical details irrelevant to this paper's objectives, Algorithm 7 computes a decomposition into triangular sets, rather than regular chains. Regular chains have additional algebraic properties which are often desirable in practice. Deriving a decomposition into regular chains from a decomposition into triangular sets is easy and we refer to [16] for related procedures.

Observe that Algorithm 7 takes three polynomials as input. Indeed, in addition to F_1 and F_2, it takes a third polynomial g, which is univariate in x and which plays the role of a "book keeper". Algorithm 7 computes the common zeros of these three polynomials via a triangular decompositions into triangular sets. Observes also that Algorithm 7 is essentially a wrapper function for Algorithm 6. This second algorithm takes as input two bivariate (non-univariate, non-constant) polynomials F_1, F_2 and two univariate polynomials g, h. Algorithm 6 computes the common zeros of F_1, F_2, g that do not can cancel h. It can be seen as the core procedure of our bivariate solver. This is precisely in this procedure that the subresultant chain of F_1, F_2 is computed and then manipulated to obtain the necessary subresultant polynomials. Once calculated, the subresultant chain of F_1, F_2 is traversed in a bottom-up manner as in the algorithm of [12].

Observe that Algorithm 6 starts by forwarding to Algorithm 5 the case where the resultant of F_1, F_2 w.r.t. y is zero. After Line 7, the polynomial R is a squarefree factor of the resultant. In the **while** loop, we maintain a level counter i, initialized to 1. During the loop execution, we have the following invariant: R is a non-constant univariate polynomial and R divides the coefficient in y^v of S_v for all $0 < v < i$. The inner **while** loop from Line 12 to Line 16 searches for the first regular subresultant S_j such that R does not divide $\mathrm{lc}(S_j, y)$. This search skips all the non-regular subresultants. The leading coefficient of S_j splits the squarefree polynomial R into two parts $G = \gcd(R, \mathrm{lc}(S_j, y))$ and $R' = R/G$. The pair (R', S_j) is added to the output, and the algorithm continues with G to the level $j+1$.

Finally, Algorithm 5 can be seen as another wrapper function which handles the case where the input polynomials F_1 and F_2 have a non-trivial factor, and thus a zero resultant.

4.3. Complexity analysis

We analyze the algebraic complexity of Algorithm 6 for the generic dense bivariate input. To further simplify the analysis, we assume $\deg(F_1, x) = \deg(F_1, y) = \deg(F_2, x) = \deg(F_2, y) = d$. That is, both F_1 and F_2 are random dense square polynomials with partial degrees d. We also assume $g = 0$ and $h = 1$.

The degree of the resultant is bounded by $B = 2d^2 + 1$, which can be read from the Sylvester matrix of F_1 and F_2. Hence the FFT size n is the smallest power of 2 such that $n = 2^e \geq B$.

Algorithm 5: ModularGenericSolve2ZeroResultant(F_1, F_2, g, h)

Input : $F_1, F_2 \in \mathbb{K}[x,y]$ and $g, h \in \mathbb{K}[x]$ such that $h(x) \neq 0$ implies $\gcd(\mathrm{lc}(F_1, y), \mathrm{lc}(F_2, y))(x) \neq 0$ and such that F_1, F_2 have positive degree in y and such that F_1, F_2 have a zero resultant in y

Output : a set of triangular sets $\{(A_1, B_1), \ldots, (A_e, B_e)\}$ such that

$$V(F_1, F_2, g) \setminus V(h) = \bigcup_{i=1}^{e} V(A_i, B_i).$$

1 $G \leftarrow \gcd(F_1, F_2)$;
2 $result \leftarrow \{(g, G)\}$;
3 $F_1 \leftarrow F_1$ quo G;
4 $F_2 \leftarrow F_2$ quo G;
5 **if** $F_1 \in \mathbb{K}[x]$ **then**
6 $\quad F_1 \leftarrow$ squarefreePart(F_1);
7 $\quad F_1 \leftarrow F_1$ quo $\gcd(F_1, h)$;
8 $\quad F_1 \leftarrow \gcd(g, F_1)$;
9 **if** $F_1 \in \mathbb{K}$ **then return** $\emptyset$;
10 **if** $F_2 \in \mathbb{K}[x]$ **then**
11 $\quad F_2 \leftarrow$ squarefreePart(F_2);
12 $\quad F_2 \leftarrow F_2$ quo $\gcd(F_2, h)$;
13 $\quad F_2 \leftarrow \gcd(g, F_2)$;
14 **if** $F_2 \in \mathbb{K}$ **then return** $\emptyset$;
15 **if** $F_1, F_2 \in \mathbb{K}[x]$ **then**
16 $\quad g \leftarrow \gcd(F_1, F_2)$;
17 $\quad$**if** $g \notin \mathbb{K}$ **then return** $result \cup \{(g, 0)\}$;
18 **if** $F_1 \in \mathbb{K}[x]$ **then return** $result \cup \{(F_1, F_2)\}$;
19 **if** $F_2 \in \mathbb{K}[x]$ **then return** $result \cup \{(F_2, F_1)\}$;
20 **return** $result \cup$ ModularGenericSolve2(F_1, F_2, g, h)

Let ω be a proper n-th primitive root of unity, any power of which does not cancel the leading coefficients of F_1 and F_2 in y. Then for each $0 \leq i < n$, evaluating F_1 and F_2 over ω^i gives a pair of univariate polynomials of degree d. The subresultant algorithm on each pair costs $\Theta(d^2)$ operations in $\mathbb{Z}_c$, from which we derive the cost to construct the FFT based scube $\Theta(d^4)$.

For generic input F_1, F_2, we expect that the loop from Line 7 to Line 11 find $j = 1$. That is, the expected gcd of F_1 and F_2 modulo their resultant in y is the subresultant of index 1. This also implies that $G = 1$ at Line 19. In this case, the return value of Algorithm 6 is the pair (A, B) where A is the squarefree part of the resultant and B is the subresultant of index 1. Hence, the major steps are

- subresultant chain construction at Line 5,
- computation of the squarefree part at Line 7,
- interpolation of the subresultant S_1 at Line 13,
- remainder computation at Line 14,
- gcd computation at Line 19.

The cost to interpolate the subresultant S_1 is $O^{\sim}(d^2)$, which is the cost to interpolate

Algorithm 6: ModularGenericSolve2(F_1, F_2, g, h)

Input : $F_1, F_2 \in \mathbb{K}[x, y]$ and $g, h \in \mathbb{K}[x]$ such that $h(x) \neq 0$ implies that $\gcd(\mathrm{lc}(F_1, y), \mathrm{lc}(F_2, y))(x) \neq 0$ and such that F_1, F_2 have positive degree in y

Output : a set of triangular sets $\{(A_1, B_1), \ldots, (A_e, B_e)\}$ such that

$$V(F_1, F_2, g) \setminus V(h) = \bigcup_{i=1}^{e} V(A_i, B_i).$$

1 $h \leftarrow$ squarefreePart(h);
2 $g \leftarrow$ squarefreePart(g);
3 $g \leftarrow g$ quo $\gcd(g, h)$;
4 **if** $g \in \mathbb{K}$ **then return** $\emptyset$;
5 Compute the subresultant chain S of F_1, F_2 in y;
6 **if** $S_0 = 0$ **then return** ModularGenericSolve2ZeroResultant(F_1, F_2, g, h) ;
7 $R \leftarrow$ squarefreePart(S_0);
8 $R \leftarrow R$ quo h;
9 $R \leftarrow \gcd(R, g)$;
10 $result \leftarrow \emptyset$, $i \leftarrow 1$;
11 **while** $R \notin \mathbb{K}$ **and** $i \leq \deg(F_2, y)$ **do**
12   **while** $i \leq \deg(F_2, y)$ **do**
13   Let S_j be the regular subresultant with $i \leq j \leq \deg(F_2, y)$ with j minimum;
14   **if** R *divides* $\mathrm{lc}(S_j, y)$ **then** $i \leftarrow i + 1$;
15   **else**
16   **break**;
17   **if** $i > \deg(F_2, y)$ **then**
18   **return** $result \cup \{(R, F_1)\}$;
19   $G \leftarrow \gcd(R, \mathrm{lc}(S_j, y))$;
20   **if** $G \in \mathbb{K}$ **then return** $result \cup \{(R, S_j)\}$;
21   $result \leftarrow result \cup \{(R \text{ quo } G, S_j)\}$;
22   $R \leftarrow G$, $i \leftarrow j + 1$;
23 **return** $result$;

two coefficients of S_1 with inverse FFTs. The classic algorithm to compute the gcd of two univariate polynomial of degree d^2 costs $O(d^4)$ field operations. However, by means of the half-gcd technique, this can be reduced to $O^{\sim}(d^2)$. Hence, the squarefree part at Line 2 and the gcd computation at Line 14 both cost $O^{\sim}(d^2)$ operations in $\mathbb{Z}_c$. By means of Newton iteration, the fast remainder computation (Ch. 9 of [9]) costs $O^{\sim}(d^2)$ field operations at Line 14. In our implementation, the subresultant construction and polynomial interpolations, of cost $\Theta(d^4)$, are performed in parallel inside the GPU; the remaining portion costs $O^{\sim}(d^2)$ operations in $\mathbb{Z}_c$.

4.4. Experimentation

In this section, we report our benchmarking results for solving bivariate systems. The dense polynomials F_1 and F_2 are taken from $\mathbb{Z}_c[x, y]$ randomly, with $c = 469762049$. In Figure 4, the partial degrees of F_1 and F_2 are all d, indicated in the first column. The following four columns indicate timing as follows:

t_0: GPU implementation of FFT-based scube construction,

Algorithm 7: ModularSolve2(F_1, F_2, g)

Input : $F_1, F_2 \in \mathbb{K}[x, y]$ and $g \in \mathbb{K}[x]$
Output : a set of triangular sets $\{(A_1, B_1), \ldots, (A_e, B_e)\}$ such that

$$V(F_1, F_2, g) = \bigcup_{i=1}^{e} V(A_i, B_i).$$

if $g = F_1 = F_2 = 0$ **then return** $\{(0, 0)\}$;
if $g, F_1, F_2 \in \mathbb{K}$ **then return** $\emptyset$;
if $F_1 \in \mathbb{K}[x]$ **and** $F_2 \in \mathbb{K}[x]$ **then**
  $A \leftarrow \gcd(F_1, F_2, g)$;
  if $A \in \mathbb{K}$ **then return** $\emptyset$;
  return $\{(A, 0)\}$;
if $F_1 \in \mathbb{K}[x]$ **then**
  return ModularSolve2($F_1 + F_2, F_2, g$);
if $F_2 \in \mathbb{K}[x]$ **then**
  return ModularSolve2($F_1, F_1 + F_2, g$);
$h \leftarrow \gcd(\mathrm{lc}(F_1, y), \mathrm{lc}(F_2, y))$;
$G \leftarrow$ ModularGenericSolve2(F_1, F_2, g, h)
if $h \in \mathbb{K}$ **then return** G;
$F_1 \leftarrow \mathrm{reductum}(F_1, y)$, $F_2 \leftarrow \mathrm{reductum}(F_2, y)$;
return $G \cup$ ModularSolve2($F_1, F_2, \gcd(g, h)$);

d	t_0	t_1	t_2	t_3	t_2/t_0	t_3/t_1
25	0.23	0.30	0.10	0.17	0.4	0.6
30	0.25	0.35	0.14	0.25	0.6	0.7
35	0.23	0.42	0.34	0.51	1.5	1.2
40	0.25	0.46	0.42	0.64	1.7	1.4
45	0.24	0.51	0.48	0.77	2.0	1.5
50	0.28	0.67	1.14	1.56	4.1	2.3
55	0.27	0.76	1.33	1.84	4.9	2.4
60	0.29	0.88	1.54	2.20	5.3	2.5
65	0.30	1.13	3.50	4.44	11.7	3.9
70	0.31	1.20	3.94	4.94	12.7	4.1
75	0.31	1.34	4.42	5.56	14.3	4.2
80	0.32	1.42	4.84	6.06	15.1	4.3
85	0.34	1.74	5.40	6.95	15.9	4.0
90	0.33	1.80	5.94	7.54	18.0	4.2
95	0.45	2.47	13.09	15.44	29.0	6.3
100	0.48	2.56	14.23	16.66	29.7	6.5
105	0.49	2.74	15.53	18.14	31.7	6.6
110	0.52	2.93	16.78	19.58	32.1	6.7
115	0.54	3.35	18.05	21.26	33.4	6.3
120	0.55	3.80	24.41	28.60	44.4	7.5

Figure 4. Solving bivariate systems in seconds

t_1: total time for solving with GPU support,
t_2: CPU implementation of FFT-based scube construction,
t_3: total time for solving without any GPU support.

The fifth column shows the speedup for the subresultant chain construction and the last column shows the speedup for solving bivariate random dense systems. Figure 5 shows the comparison of the running time with and without GPU acceleration. The largest speedup we achieve is approximately 7.5.

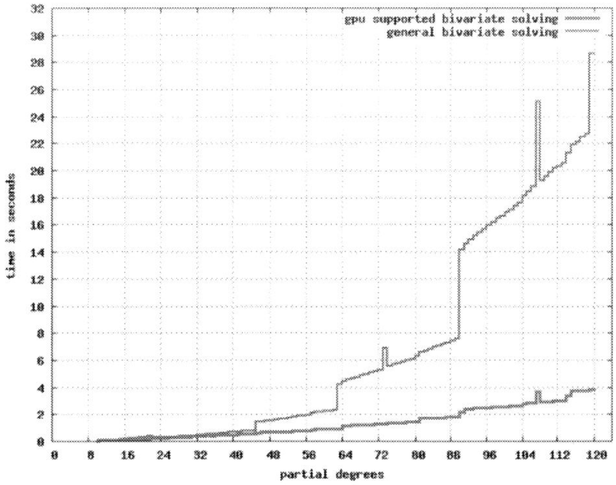

Figure 5. Solving bivariate random dense polynomial systems over a finite field

When asymptotically fast arithmetic is used for univariate polynomials, in terms of algebraic complexity, the dominate part of solving generic bivariate system is the subresultant construction. This could be observed from Figure 6. In the histogram, the left column (red) indicates the time spent for subresultant chain construction, and the right column (green) indicates the total time. After plugging the GPU support in the FFT-based subresultant chain computations into the bivariate solver, the univariate polynomial arithmetic (for large polynomials) dominates the total running time as shown in Figure 7. This also limits our speedup factor, since univariate polynomials GCDs are currently all computed serially by the CPU.

5. Conclusions and remarks

In this paper, we have reported a GPU implementation of dense multivariate polynomial arithmetic focusing on the computation of subresultant chains. With respect to a pure CPU code, we obtain very large speedup factors for this task. We have also reported on the implementation of a bivariate polynomial system solver, based on our GPU code. By speeding up subresultant computations, our GPU-supported solver has substantially increased its performance w.r.t. its pure CPU code counterpart.

The subresultant construction also illustrates the following pattern, which is frequent in symbolic computation and which is well-suitable for GPU computing. More precisely, the total amount of work (in terms of number of coefficient operations) is essentially proportional (may be up to logarithmic factors) to the size of the output, This type of algorithms seem to have sufficient parallelism for current multicore architectures. However, due the output data size,

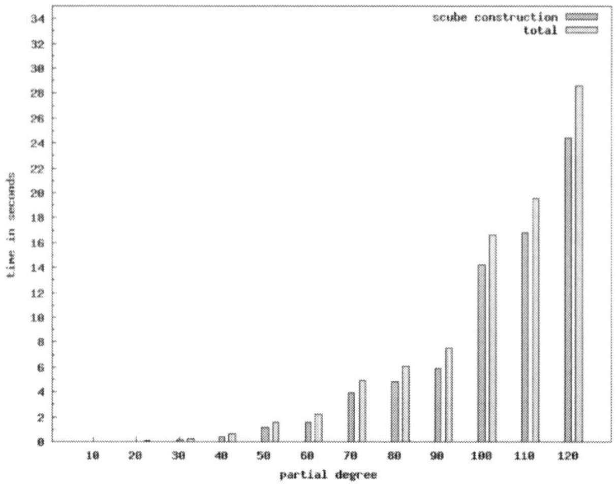

Figure 6. Subresultant computations inside bivariate dense solver, C code only

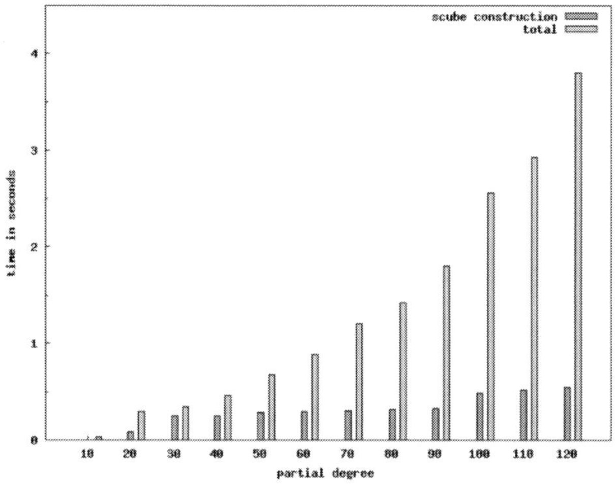

Figure 7. Subresultant computations inside bivariate dense solver, C with GPU code

which is often much larger than than the L2 or L3 cache size, performance can be limited by the memory bandwidth between the L2 (or L3) cache and the main memory.

As shown in Section 4.4, the speedup we can achieve in the bivariate solver is largely limited by the computation of univariate polynomial gcds of large degree. In the future work, we would like to improve this part and in turn fully utilize the speedup from our GPU scube construction.

References

[1] *Wikipedia page for* GPGPU. http://en.wikipedia.org/wiki/GPGPU.

[2] *GPGPU.org.* http://gpgpu.org/category/research.

[3] E. Berberich, P. Emeliyanenko, and M. Sagraloff. An elimination method for solving bivariate polynomial systems: Eliminating the usual drawbacks. Technical report, Max-Planck-Institut für Informatik, Saarbrücken, Germany, 2011.

[4] B. Boyer, J.-G. Dumas, and P. Giorgi. Exact sparse matrix-vector multiplication on gpu's and multicore architectures. In *Proceedings of the 4th International Workshop on Parallel and Symbolic Computation*, PASCO '10, pages 80–88, New York, NY, USA, 2010. ACM.

[5] W. Brown. The subresultant PRS algorithm. *Transaction on Mathematical Software*, 4:237–249, 1978.
[6] G. Collins. *The Calculation of Multivariate Polynomial Resultants*, volume 18, pages 515–532. Journal of the ACM, 1971.
[7] M. El Kahoui. An elementary approach to subresultants theory. *J. Symb. Comp.*, 35:281–292, 2003.
[8] P. Emeliyanenko. A complete modular resultant algorithm targeted for realization on graphics hardware. In *Proceedings of the 4th International Workshop on Parallel and Symbolic Computation*, PASCO '10, pages 35–43. ACM, 2010.
[9] J. Gathen and J. Gerhard. *Modern Computer Algebra*. Cambridge University Press, 1999.
[10] L. Jacquin, V. Roca, J.-L. Roch, and M. Al Ali. Parallel arithmetic encryption for high-bandwidth communications on multicore/gpgpu platforms. In *Proceedings of the 4th International Workshop on Parallel and Symbolic Computation*, PASCO '10, pages 73–79, New York, NY, USA, 2010. ACM.
[11] M. Kalkbrener. *A generalized Euclidean algorithm for computing triangular representations of algebraic varieties*, volume 15, pages 143–167. J. Symb. Comp., 1993.
[12] X. Li, M. Moreno Maza, and W. Pan. Computations modulo regular chains. In *Proc. ISSAC'09*, pages 239–246, New York, NY, USA, 2009. ACM Press.
[13] X. Li, M. Moreno Maza, R. Rasheed, and É. Schost. The modpn library: Bringing fast polynomial arithmetic into maple. In *MICA'08*, pages 73–80, 2008.
[14] X. Li, M. Moreno Maza, R. Rasheed, and É. Schost. The modpn library: Bringing fast polynomial arithmetic into maple, 2008. To appear in the J. of Symbolic Computation.
[15] M. Monagan. Probabilistic algorithms for computing resultants. In *ISSAC '05: Proceedings of the 2005 international symposium on Symbolic and algebraic computation*, pages 245–252, New York, NY, USA, 2005. ACM.
[16] M. Moreno Maza. On triangular decompositions of algebraic varieties. Technical Report TR 4/99, NAG Ltd, Oxford, UK, 1999. http://www.csd.uwo.ca/~moreno.
[17] M. Moreno Maza and W. Pan. *Fast polynomial arithmetic on a GPU*, volume 256. J. of Physics: Conference Series, 2010.
[18] M. Moreno Maza and Y. Xie. Balanced dense polynomial multiplication on multicores. In *Proc. of PDCAT'09*. IEEE Computer Society, 2009.
[19] M. Moreno Maza and Y. Xie. FFT-based dense polynomial arithmetic on multi-cores. In D. Mewhort, editor, *Proc. HPCS 2009*, volume 5976 of *LNCS*, Heidelberg, 2010. Springer-Verlag Berlin.
[20] W. Pan. *Algorithmic Contributions to the Theory of Regular Chains*. PhD thesis, University of Western Ontario, 2010.
[21] C. Yap. *Fundamental Problems in Algorithmic Algebra*. Princeton University Press, 1993.

Appendix A. Specifications of our GPU cards

Our CUDA code has been run for the two NVIDIA GPUs and we briefly list their specifications below.

GPU	GTX 285	Tesla C2050
Capability	1.3	2.0
Multiprocessors	30	14
Cores	240	448
Clock Rate	1.15G GHz	1.15 GHz
Memory Bandwidth	159 GB/sec	144 GB/sec
Double Floating Point	partially	fully
Global Memory	1GB	3GB
Shared Memory	16KB	48KB or 16KB
L1 Cache	none	48KB or 16KB
L2 Cache	none	768KB
Concurrent Kernels	no	up to 16

The first card belongs to the GeForce 200 series, which starts to support double-precision floating point numbers in hardware, but only one double-precision floating point unit (FPU) per multiprocessor (a group of 8 or 32 cores). The second card belongs to the Fermi series released in April 2010, in which each core is integrated with a double-precision FPU.

Appendix B. Source code

The following CUDA kernel performs a list of eliminations over a finite field $\mathbb{Z}_c$, which takes two lists of univariate polynomials LF and LG, and computes

$$h_i = \mathrm{lc}(g_i) f_i - \mathrm{lc}(f_i) x^{\deg(f_i) - \deg(g_i)} g_i,$$

where f_i is the i-th polynomial in LF and g_i is the i-th polynomial in LG. It requires that polynomials in LG have the same degree dG, and polynomials in LF have the same degree dF. The result LH consists of h_i's computed.

```
__global__ void list_reduce(int *LH, int dF, int *LF,
  int dG, int *LG, int p)
{
  int bid = blockIdx.x;
  int tid = bid * blockDim.x + threadIdx.x;
  int qtid = tid / dF;    // pair index
  int rtid = tid % dF;

  int *F = LF + qtid * (dF + 1); // first poly
  int *G = LG + qtid * (dG + 1); // second poly
  int *H = LH + qtid * dF;       // output poly

  // The configuration is the following
  //          u ...... a
  //    v ...... b
  // where a is the current coefficient to eliminate,
  // b is the current leading coefficient,
  // and u, v are cofficients to be adjusted.
  // For each pair (u, v), compute
  //    a * v - u * b mod p
  // and store it to H[rtid];

  int dgap = dF - dG;
  int a = F[dF];  // the leading coefficient of F
  int b = G[dG];  // the leading coefficient of G
  int u = F[rtid];
  int v = ((rtid >= dgap) ? G[rtid - dgap] : 0);

  int t1 = mul_mod(a, v, p);  // t1 = a * v mod p
  int t2 = mul_mod(b, u, p);  // t2 = b * u mod p
  H[rtid] = sub_mod(t1, t2, p);
}
```

The modular multiplication mul_mod uses the following method in our code, assuming that the inverse $pinv = 1/p$ of p has been pre-computed.

```
int mul_mod(int a, int b, int p, double pinv) {
  int q = (int)((((double)a) * ((double)b)) * pinv);
  int result = a * b - q * p;
  if (result < 0) result += p;
  return result;
}
```

The above code requires two double-precision floating point multiplications. Up to our knowledge, this is the simplest and best method to do the modular multiplications, even for those GPUs which only partially support double-precision floats, like Nvidia GTX 285.

Appendix C. Profiling results

To visualize the performance of our implementations, we use the Nvidia's visual profiler `cudaprof` to analyze CUDA kernel calls. It is very helpful to find out the bottlenecks of an implementation.

Figure C1 shows the kernel statistics of the Stockham FFT. In this figure, the x-axis shows the kernel call indices in chronological order and the y-axis is proportional to the GPU time for each kernel. The first kernel copies the coefficient vector of the univariate polynomial into the GPU global memory, followed by a sequence of kernel calls to `double_expand_ker` to compute powers $\{1, \omega, \omega^2, \ldots, \omega^{n/2-1}\}$. The last kernel copies the result back to the main memory. In the middle, it has basic butterflies, twiddling and data transpositions. Each of those kernel names uses _a or _b as a suffix, since we implement the same algorithm for handling input data in different ranges. This enables the kernels to efficiently utilize the shared memory space of the GPU, whenever necessary.

Figure C1 was generated using our first GPU GTX 285 and the FFT size is $n = 2^{26}$. The memory bandwidth will be the only bottleneck if the goal is to speedup a single univariate FFT. It is advisable to keep the data inside GPU global memory as long as possible, as we did for computing the subresultant chains.

Figure C2 was generated using our second GPU Telsa C2050 for constructing the scube of two random dense bivariate polynomials of partial degree 100 in each variable. There are three major components: FFT evaluations of a list of univariate polynomials (the region in a rectangle shape), initialization of the output (the highest single column), and the fine-grained Brown's algorithm (the region in a triangle shape). The last one dominates the overall computation, i.e. the kernel `list_reduce`. We gain approximately an extra speedup factor of 2 by switching from the GTX 285 to Telsa C2050, due to the better support of double-precision floating point calculations.

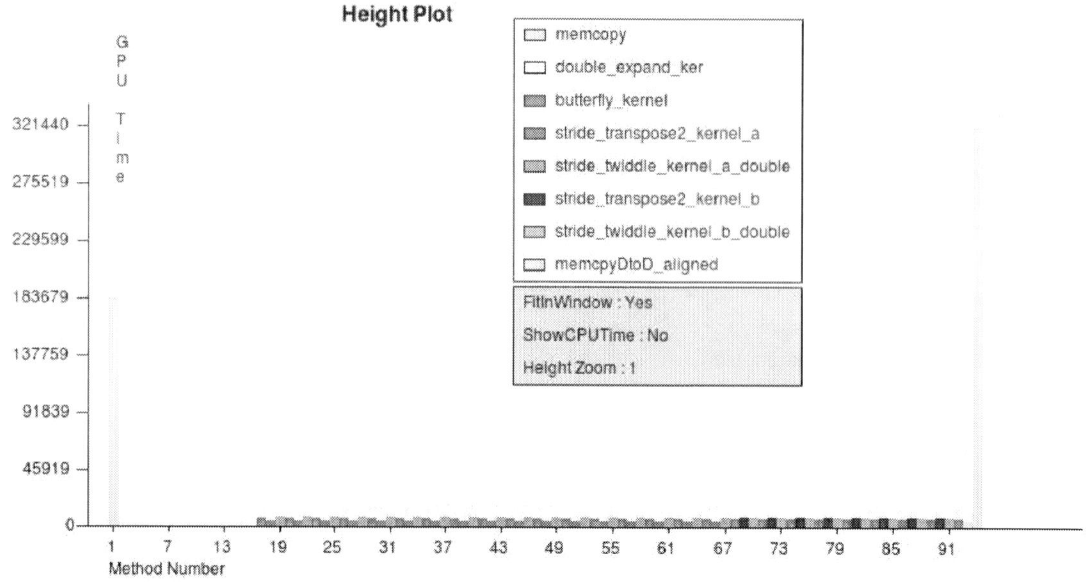

Figure C1. Kernel statistics of the Stockham FFT

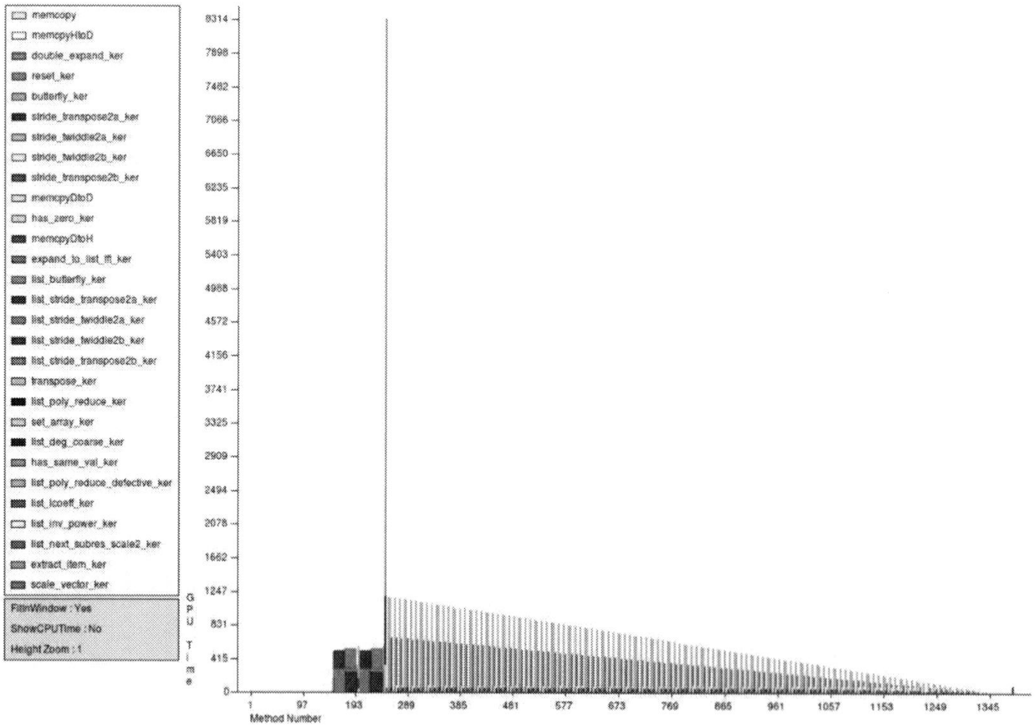

Figure C2. Kernel statistics of a scube construction

Complex lens design: searching for a needle in a haystack

Sergey Mashchenko

Department of Physics and Astronomy, McMaster University, Hamilton, ON, L8S 4M1, Canada

E-mail: syam@physics.mcmaster.ca

Abstract. Optical design of complex (multi-element) lenses is traditionally considered to be part science and part art, primarily because of the enormous complexity of the problem. Recent advances in high performance computing (HPC) made it feasible to adopt a purely scientific approach in discovering new lens designs. In this paper, I formulate the task of finding a new lens design that satisfies a given set of constraints as a search for the global minimum of a function of unknown and very large ($\sim 30-100$) number of dimensions. I address the significant complication that only a tiny fraction of the volume of the free parameters space is physically accessible. I propose a smart lens drafting algorithm which circumvents this difficulty. I present my numerical code which can be used to discover novel complex lens designs in a fully automatic fashion. I discuss the HPC aspects of the problem of searching for minima of high dimensionality functions.

1. Introduction

A purpose of a lens is to produce an image. A single-element lens produces a low quality image, which is usually not good enough for many applications, including photography, astronomy, medicine and so on. Compound, or complex lenses (consisting of more than one element) can produce significantly better quality images. Unfortunately, there is no general theory for complex lenses. As a result, the field of optical lens design is considered to be partly art, partly science [1].

Modern optical lens design consists of two relatively independent steps:

(i) **"Art"**. Choosing a good starting point: a draft design which would already be able to produce an image (not necessarily of good quality), with the main lens parameters (focal length, aperture, type of projection etc.) been not too far from the target values. The following approaches are used [2]:
- A mental guess.
- A previously designed lens in the company files.
- Purchase of a competing lens and analysis of its structure.
- A search through the patent files.

(ii) **"Science"**. Lens optimization. Starting from the draft design, lens parameters are gradually optimized in an attempt to achieve the target lens parameters (optical,

Journal of Physics: Conference Series **341** (2012) 012023 doi:10.1088/1742-6596/341/1/012023

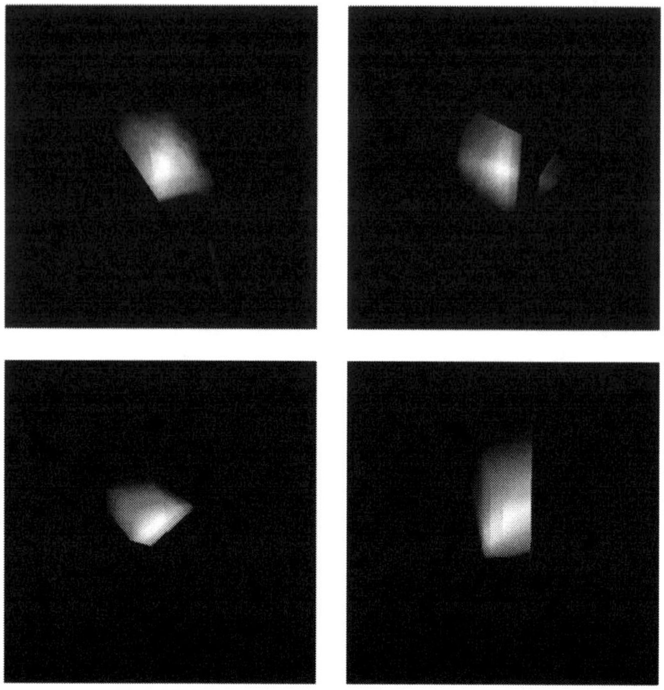

Figure 1. Four randomly oriented two-dimensional slices through the neighborhood of a local minimum of the merit function for a complex lens consisting of five elements (all surfaces are conic) with an aperture stop. The local minimum is at the centre. Number of dimensions is 39. The square size is one in scale-free units. Black areas are physically inaccessible. I use logarithmic scale to plot the merit function ($\sim$ 30 orders of magnitude from white to almost black color).

mechanical, cost etc.). It is usually accomplished by minimizing a specially designed multi-dimensional merit function. Both approximate (aberration theory) and direct (ray tracing) methods can be used in designing the merit function.

2. Automatic lens design

The current two-step approach to complex lens design makes it impossible to design complex lenses in a fully automatic fashion. I address this issue by reformulating the complex lens design problem as a single-step procedure of finding the global (or a very good local) minimum of the multi-dimensional merit function.

Monte Carlo minimization is the method of choice for this type of problem. Specifically, a random initial point is generated, and a multi-dimensional minimizer is used to descend to the nearest local minimum. The process is repeated many times, until the global (or sufficiently good local) minimum is found.

To avoid the unknown biases introduced by approximate theories, and to make the whole multi-dimensional space of the lens parameters fully explorable, in my algorithm I use the direct method (tracing of hundreds and thousands of rays through the system) to design the merit function.

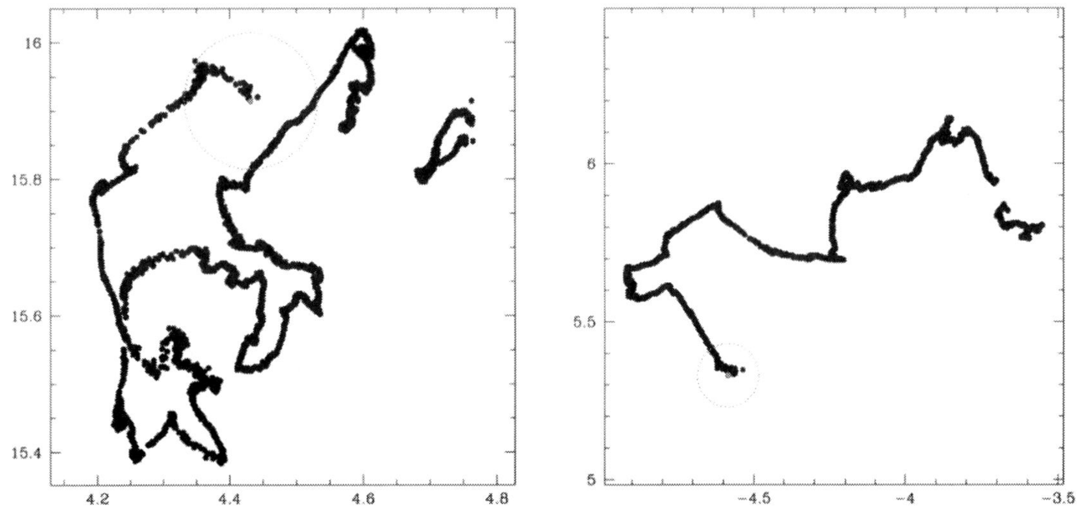

Figure 2. Projection of the 39-dimensional path taken by the optimization code for the lens design from Figure 1 on two randomly oriented planes. The coordinates are in scale-free units. The initial point (produced by my smart drafting module, § 2.2) is in red; the red circle shows the initial search radius for the optimizer. The path consists of approximately half a million merit function measurements, and multiple (~ 30) optimization hops (see § 2.4).

2.1. Properties of the global merit function

My merit function is computed by (1) tracing of hundreds to thousands of rays of different wavelengths, coming from sources located at different distances from the lens and from the optical axis of the system, (2) finding the optimal focal plane, where these rays produce the best quality image, (3) computing the χ^2 deviation of the rays in the focal plane from the ideal (theoretical) positions, and (4) computing and applying a large number of punishment coefficients (multipliers to χ^2). Rays about to be lost (approaching the edges of lens surfaces, the aperture stop, and the rear lens flange) are artificially dimmed (their weight in χ^2 computation approaches zero as they are approaching the edge); this ensures that the merit function stays smooth even if some rays disappear or re-appear.

The most important punishment coefficients are for vignetting, focal length (to drive the system toward the target focal length value), glass weight (to minimize the weight of the lens), back focal length (to ensure that it is larger than a given minimum value), medium (both glass and air) thickness (to keep it within the allowed range), deviation from a nearest real glass (see § 2.3), and distortion (deviation from a given mapping function, which could be gnomonic, stereographic, orthographic etc.).

A serious complication is that the merit function is mostly "vacuum": only a tiny fraction of the total hyper-volume of the lens parameter space is physically accessible. This can be seen in Figure 1, which shows the neighbourhood of a local minimum of 39-dimensional merit function (for a five elements lens design). In one dimension, the extent of the physically accessible space (white blob) is ~ 0.3 of scale-free units (the size of the square window is 1). In 39 dimensions this will result in the volume filling factor of $0.3^{39} \simeq 4 \times 10^{-21}$. The probability to detect this physically accessible blob by using a straightforward Monte Carlo approach (generating random initial points within this hyper-cube) is hence negligibly small. The solution is to design a

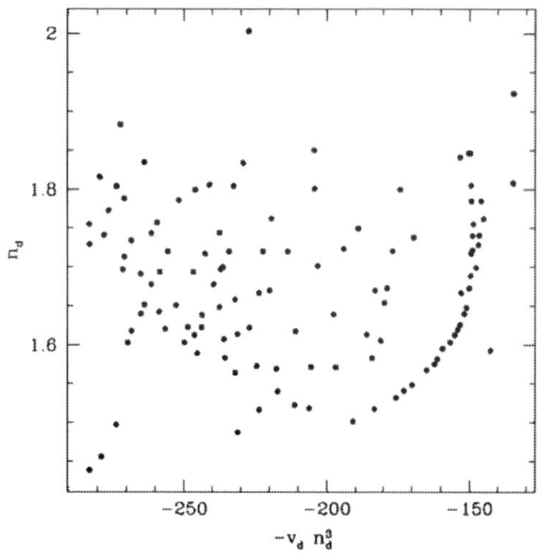

Figure 3. Plot of the optical glass from the online catalog of Ohara Corporation in $(-v_d n_d^3, n_d)$ coordinates.

"smart" algorithm for generating random draft lens designs. I describe my drafting algorithm in § 2.2.

Tiny local minimum islands (like the one in Figure 1) is not the only type of structure exhibiting by the global merit function. As one can see in Figure 2, local minima are connected by filaments. The fact that these filaments are not visible in Figure 1 suggests that they are of a lower dimensionality. In my serial farm runs for 5–8 elements designs I found that the local minima blobs have sizes ~ 0.3 of scale-free units, whereas the distance between the initial, draft design point and the minimum is much larger: $\sim 2 - 10$ (typically ~ 5). The fact that in my experiments initial random points never land directly inside local minima islands strongly suggests that filaments occupy much larger volume than local minima blobs. The existence of lower-dimensionality filaments can also be deduced from theoretical arguments, specifically from the presence of degeneracies in optical designs (e.g. changing both the surface curvature and the refractive index in concert can keep a lens element's power constant).

2.2. Smart lens drafting

My algorithm for generating random draft lens designs (used as initial points in the global Monte Carlo minimization of the merit function) is based on ray tracing, and is very fast: it usually takes seconds to generate one good random design, on one CPU. Draft designs produced by the algorithm are capable of creating an image, with the focal length, aperture, back focal length, rear flange opening and other lens parameters reasonably close to the target values. There is some control over the amount of optical aberrations, distortions, and vignetting present in draft designs. The algorithm also assigns random glass to the lens elements, using my optical glass model (see the next section).

2.3. Glass model

Multi-dimensional optimizers are not designed to handle non-continuous (discrete) parameters. Most of the lens parameters (surface curvatures, vertex coordinates etc.) are naturally continuous, but some are not. To make the whole multi-dimensional space of lens parameters explorable and optimizable, one has to reduce the number of discrete lens parameters to an absolute minimum. Ideally, only the number of individual lens elements should be left as a discrete parameter; all other discrete quantities should be converted to pseudo-continuous ones.

Optical glass is one of such discrete parameters. Though it is physically feasible to produce

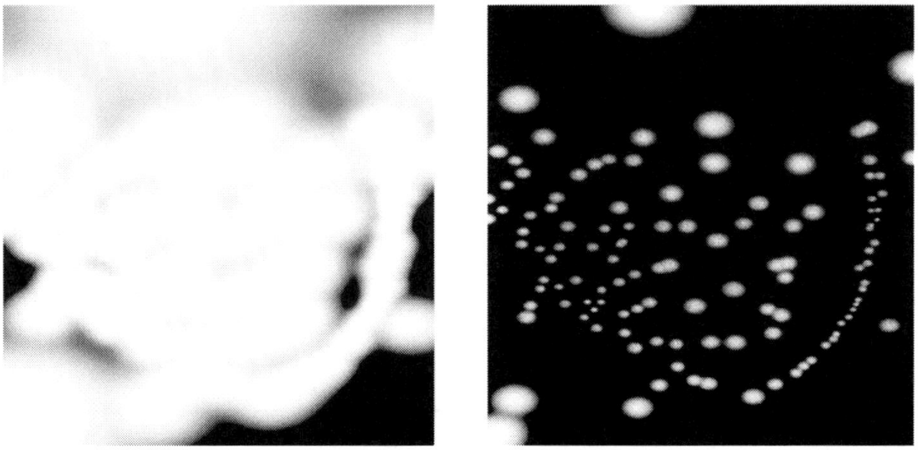

Figure 4. Glass punishment function (the same coordinates as in Figure 3). The range is from no punishment (white) to severe punishment (black). Left panel: at the beginning of the optimization process, when the image quality is poor. Right panel: at the end of the optimization, when the image quality is high. The punishment function is computed separately for each element of the lens. The merit function is multiplied by all of the individual punishment functions.

optical glass with smoothly varying properties, lens design and manufacturing is based on a discrete set of ~ 100 commercially available types of optical glass.

The commercially available optical glass is reasonable uniformly scattered in the plane with $x \equiv v_d n_d^3$ and $y \equiv n_d$ coordinates (see Figure 3; here v_d is the Abbe number and n_d is the refractive index for d-line). I found that the following equation is a good fit for the refractive index n as a function of the wavelength λ (in nm) and n_d and v_d parameters:

$$n(\lambda, n_d, v_d) = (((-1.1286752 \times 10^{-8}\Delta\lambda + 4.6105463 \times 10^{-6})\Delta\lambda - 1.3883940 \times 10^{-3})\Delta\lambda + \\ +5.2671310 \times 10^{-1})\Delta\lambda(1 - n_d)/(109.87v_d) + 0.99864003n_d + 1.5353551 \times 10^{-3}. \quad (1)$$

Here $\Delta\lambda = \lambda - 596$ nm. For the 109 glass types from the Ohara Corporation online catalog[1] and seven spectral lines (A', r, C, He–Ne, e, F, g), the standard deviation of equation (1) from the real glass is 1.8×10^{-4}.

At the draft design stage, I generate random glass by choosing a random point in the (x, y) plane. Such glass is physically plausible, but it is not manufactured.

At the optimization stage, the glass properties of each lens element are described by the two continuous parameters, x and y. Refractive indices for all wavelengths are computed using the equation (1). Initially, x and y are allowed to vary freely (within the physically plausible region); as the quality of the lens is improving, the system is driven towards commercially available glass types by utilizing a punishment function (see Figure 4).

At the end, the optimized model glass is quite close to real types of glass. The final optimized lens design is produced by switching to the nearest real glass types and re-optimization of the lens.

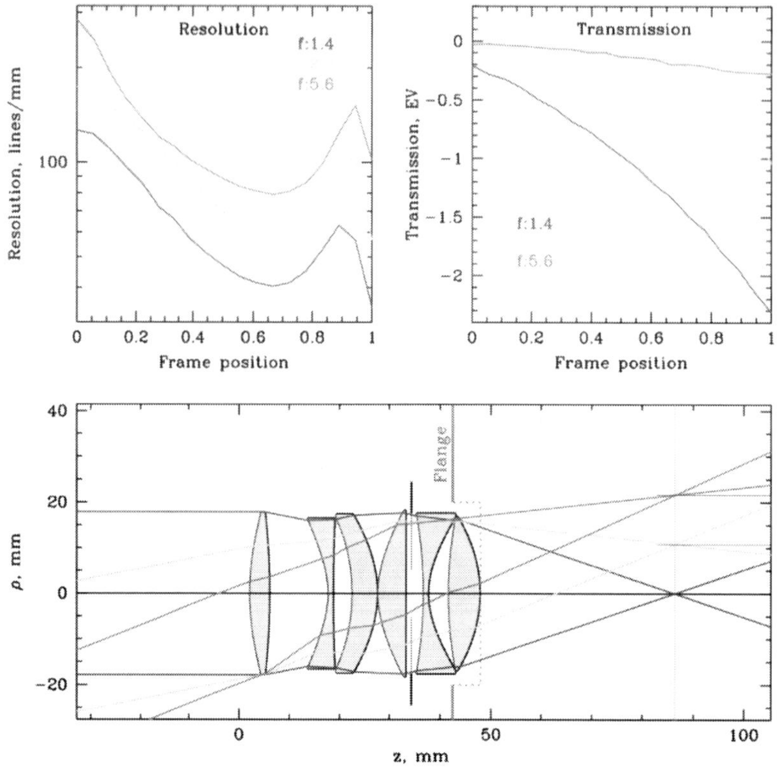

Figure 5. Optimized design for a 5 elements SLR lens with the focal length 50 mm and the f-number 1.4 (bottom panel). All surfaces are conic. Optimization included the location and diameter of the aperture stop, and the constraints imposed by the camera (minimum back focal length, location and diameter of the rear lens flange). The resolution curves (top left panel) and transmission curves (top right panel) are shown for three different values of f-number (1.4, 2.8, 5.6).

2.4. The code

I have written a code which implements the algorithm described in the previous sections. It is written in C and has more than 10,000 lines of code. The code consists of draft and optimization modules, and requires no interaction. As it may take weeks, months or even years to discover a high quality design on a single computer, the code should be run in parallel as a serial farm on a cluster.

The code makes a heavy use of GSL[2] (GNU Scientific Library), specifically the random number generation (see § 4) and multi-dimensional minimization routines. As my merit function is based on ray tracing, its partial derivatives are not readily available. This narrowed down my choice for minimizer to the Nelder-Mead simplex algorithm [3]. I modified the GSL's routine for Nelder-Mead simplex minimization to make it suitable for parallel (serial farming) applications.

I found that for lenses consisting of more than 2 − 3 elements the simplex algorithm does

[1] http://www.oharacorp.com/catalog.html
[2] http://www.gnu.org/software/gsl

not directly converge to a true minimum. Instead, the code converges to a complex saddle point. To address this I implemented a "multi-hop" approach: the code performs a sequence of re-optimizations ("hops"), each re-optimization starting from the previous best result, with gradually shrinking initial search radius. Once the distance traveled during a few consecutive hops becomes very small, a convergence to a true local minimum is achieved. Typically between 20 and 50 hops are required to get to a local minimum.

At this point, my code only uses the simplest global optimization approach – searching for the local minimum in the neighbourhood of randomly generated initial (draft design) points. My observation that almost all good designs are among the best $\sim 5\%$ designs at the end of the first optimization hop allowed me to accelerate the search by a factor of 10 by using the following strategy: after the first serial farm run, when only first hop is performed, I choose the best 5% of designs for the second stage (multi-hop) run, where true local minima convergence is achieved. The only acceptance criterion for each initial random point run is a convergence to a true local minimum. Only a small fraction of found local minima correspond to high quality designs. More complex strategies (like simulated annealing) may be tried in the future, but it is not clear if they will result in a more efficient code.

This is still a work in progress. I am currently working on implementing such lens features as internal focusing and zoom.

3. Preliminary results

Figure 5 shows one of the preliminary results obtained using my code – an optical design for a standard (focal length 50 mm) fast (f-number 1.4) photographic SLR lens. Only these two lens parameters (focal length and f-number) plus the camera mount mechanical constraints were initially specified; all the rest (number of lens elements, glass types, surface parameters, parameters of the aperture stop) were discovered by the code (ran in parallel as a serial farm on a cluster) fully automatically, without any human intervention.

4. HPC aspects

Monte Carlo based numerical codes require a high quality pseudo-random number generation. GSL implements many different random generators, with the default one, MT19937, being of very high quality: its Mersenne prime period is $\sim 10^{6000}$, and it is equi-distributed in 623 dimensions [4].

Unfortunately, any pseudo-random number generator is as good as its initialization procedure, and usually employed initialization methods may not be adequate for Monte Carlo based codes executed many thousands of times in a parallel (serial farming) setting.

GSL provides an important feature which makes it possible to design an almost perfect parallel Monte Carlo setup: in addition to using seed numbers, it also allows to read and write the full state of the random number generator – a small binary file – which uniquely defines the exact position of the generator in the Mersenne prime period.

My serial farming random number generation setup consists of the following elements:

- Only once, a large number N of unique random seed numbers is generated. (N should be larger than the maximum number of jobs utilizing this algorithm one would ever run in parallel on a cluster or clusters with a shared file system). These seeds are used to initialize the random number generator, and the corresponding generator states are written to the shared file system as N binary state files.
- Any time a code utilizing this algorithm is run, the first thing it does is to borrow an available state file. The file is used to initialize the random number generator, and is made unavailable to any other job for the duration of the current job. Periodic checkpointing (updating the state file with the current state) is performed. At the end of the run (either

normal, or initiated by the KILL signal), the file is updated with the final state of the generator, and then released back to the pool of available state files.

Given the extremely large Mersenne prime period of the MT19937 algorithm, the above setup essentially guarantees that pseudo-random number sequences generated in a parallel (serial farming) setting will always be unique.

5. Conclusions

The ongoing evolution of HPC hardware has reached the point when fully automatic complex lens design became feasible. In this paper, I presented my numerical code which can be used to discover new lens designs when run as a serial farm on HPC clusters. The code is based on the algorithm for smart lens drafting, optical glass model, and parallel pseudo random number generation setup I have developed.

Acknowledgments

This work was supported by the Shared Hierarchical Academic Research Computing Network (SHARCNET).

References

[1] Smith W J 2008 *Modern Optical Engineering: The Design of Optical Systems*, Fourth Edition (New York: McGraw Hill) p 409
[2] Kingslake R and Johnson R B 2010 *Lens Design Fundamentals*, Second Edition (Burlington, MA: Elsevier) p 9
[3] Nelder J A and Mead R 1965 *Computer Journal* **7** 308
[4] Matsumoto M and Nishimura T 1998 *ACM Transactions on Modeling and Computer Simulation* **8** no 1, 3

Journal of Physics: Conference Series **341** (2012) 012024 doi:10.1088/1742-6596/341/1/012024

A Parallel Supercomputer Implementation of a Biological Inspired Neural Network and its use for Pattern Recognition

Vincent de Ladurantaye, Jean Lavoie, Jocelyn Bergeron, Maxime Parenteau, Huizhong Lu, Ramin Pichevar, Jean Rouat [1]

NECOTIS, GEGI, Univ. de Sherbrooke Qué., CANADA. http://www.gel.usherbrooke.ca/necotis

Abstract. A parallel implementation of a large spiking neural network is proposed and evaluated. The neural network implements the *binding* by synchrony process using the Oscillatory Dynamic Link Matcher (ODLM). Scalability, speed and performance are compared for 2 implementations: Message Passing Interface (MPI) and Compute Unified Device Architecture (CUDA) running on clusters of multicore supercomputers and NVIDIA graphical processing units respectively. A global spiking list that represents at each instant the state of the neural network is described. This list indexes each neuron that fires during the current simulation time so that the influence of their spikes are simultaneously processed on all computing units. Our implementation shows a good scalability for very large networks. A complex and large spiking neural network has been implemented in parallel with success, thus paving the road towards real-life applications based on networks of spiking neurons. MPI offers a better scalability than CUDA, while the CUDA implementation on a GeForce GTX 285 gives the best cost to performance ratio. When running the neural network on the GTX 285, the processing speed is comparable to the MPI implementation on RQCHP's Mammouth parallel with 64 notes (128 cores).

1. Introduction

One of the greatest engineering challenges faced in the 21st century is to understand and reproduce the functioning of the brain. Binding by synchrony is an important function that explains a great number of behaviors like perception of objects and polysensoriality [1, 2]. With the arrival of parallel computing, simulation of large artificial neural networks is becoming feasible. It is now realistic to implement the binding function for use in artificial intelligence or in the area of computational neuroscience. The challenge behind such work is the great number of interconnections making each unit dependent on many others and thus limiting data independence, which is crucial for effective parallel computing. A parallel implementation of the binding function with spiking neural network (SNN), as proposed by Pichevar *et al.* [3], is described. Also, the scalability for large networks and performance comparison between multi-core computers (MPI) and graphical processing units (GPUs : CUDA) implementations is evaluated.

[1] CUDA conception and experiments by V. de Ladurantaye; MPI experiments by V. de Ladurantaye and M. Parenteau; MPI architecture conceived or realized by J. Lavoie, J. Bergeron, H. Lu, M. Parenteau, R. Pichevar and J. Rouat; V. de Ladurantaye and J. Rouat are primary authors.

2. Parallel Implementations of Spiking Neural Networks

Biologically accurate models of neurons like the Hodgkin-Huxley model [4] require the integration of 4 non-linear equations which is very computationally intensive. On the other hand, the integrate and fire model [5] is less biologically realistic but requires a single linear equation. Izhikevich proposed a 2-equation model that is a good compromise between speed and biological accuracy [6]. To use spiking neurons on parallel supercomputers, SNN simulators such as NEURON, PGENESIS and NEST have been developed [7]. Those simulators are based on MPI and have been reported to have a good scalability for large network simulations [8]. However, with the recent arrival of GPU technology, much interest is now on graphical processor implementations. The Izhikevich [9, 10, 11] and the integrate and fire models [12] have been ported to GPU using the CUDA plateform. Pallipuram *et al.* [13] have compared implementations on different GPU programming models and architectures. Also, Han and Tarek have proposed a mixed architecture using an MPI implementation running on clusters of GPUs [14]. These implementations are usually done on networks of relatively simple complexity, with the maximum number of synaptic connections never exceeding one million. In this work, we implement a large scale SNN using integrate-and-fire neurons in a network which is almost fully connected with a number of synaptic connexions as high as several billions. Furthermore, we simulate and use the binding function in an image comparison task. To our knowledge there is no other parallel implementation of image processing applications based on the binding occurring in the brain.

3. Model Description and application in image comparison

3.1. Neuronal model

The implemented Oscillatory Dynamic Link Matcher (ODLM) network [3] uses leaky integrate-and-fire neurons. The membrane potential $v(t)$ of an isolated neuron evolves through time according to:

$$v(t) = \frac{I_0}{C}(1 - e^{-\frac{t}{\tau}}) \tag{1}$$

When $v(t)$ becomes greater or equal to a fixed threshold θ, at time t_{spike}, then the neuron emits a spike $\delta(t - t_{spike})$ and $v(t)$ is reset to 0. τ is a time constant that controls the amount of leaky current and thus, the speed. C is a constant and I_0 is a constant input current.

3.2. The architecture of the Oscillatory Dynamic Link Matcher for image comparisons

For the image comparison task at hand, the SNN comprises two layers with local connectivity within each layer and full connectivity between different layers. A neural layer is attributed to each image and each pixel is associated with a spiking neuron as illustrated in figure 1. Synaptic connexions are symmetrical and bidirectionnal (Fig. 1). For a given neuron (i, j), the state equation of the system is given by

$$\frac{dv_{i,j}}{dt} = -\frac{1}{\tau}v_{i,j}(t) + \frac{1}{C}I_0 + S_{i,j}(t) \tag{2}$$

with

$$S_{i,j}(t) = \sum_{\mathbf{k},\mathbf{m} \in N^{ext}(i,j)} \{w^{ext}_{i,j;\mathbf{k},\mathbf{m}}\delta(t - t_{spike_{\mathbf{k},\mathbf{m}}})\} + \sum_{\mathbf{k},\mathbf{m} \in N^{int}(i,j)} \{w^{int}_{i,j;\mathbf{k},\mathbf{m}}\delta(t - t_{spike_{\mathbf{k},\mathbf{m}}})\} - \eta G(t) \tag{3}$$

$S_{i,j}(t)$ is the total contribution that a neuron (i, j) receives from all neurons $(\mathbf{k}, \mathbf{m})$ it is connected to; whether they are on the same layer (int) or on a different layer (ext). $N^{ext}(i, j)$ and $N^{int}(i, j)$ are the sets of firing neurons $(\mathbf{k}, \mathbf{m})$ to which neuron (i, j) is connected. $G(t)$ is a global inhibitor that regulates the neural network activity. η is a constant that controls the influence of the global inhibitor.

The connecting weights have the respective expressions

$$w^{int}_{i,j;\mathbf{k},\mathbf{m}} = \frac{w^{int}_{max}}{Card\{N^{int}(i,j)\}} e^{-\lambda|p_{i,j}-p_{\mathbf{k},\mathbf{m}}|} \text{ inside a layer} \quad (4)$$

$$w^{ext}_{i,j;\mathbf{k},\mathbf{m}} = \frac{w^{ext}_{max} f(\text{SegCompactness, SegAveragedgrey})}{Card\{N^{ext}(i,j)\}} e^{-\lambda|p_{i,j}-p_{\mathbf{k},\mathbf{m}}|} \text{ between layers} \quad (5)$$

In this simple application $p_{\mathbf{k},\mathbf{m}}$ and $p_{i,j}$ are pixel values associated respectively to neurons $(\mathbf{k},\mathbf{m})$ and (i,j). $Card\{N^{int}(i,j)\}$ and $Card\{N^{ext}(i,j)\}$ are normalization factors corresponding to the number of active neurons $(\mathbf{k},\mathbf{m})$ that are connected to neurons (i,j) whether they are inside the same layer (int) or in a different layer (ext). Two processing phases are defined: the segmentation and the matching. During the **segmentation** phase, the connections between layers are not activated ($w^{ext}_{i,j;\mathbf{k},\mathbf{m}} = 0$). Once the images have been independently segmented, the two layers are connected with $w^{ext}_{i,j;\mathbf{k},\mathbf{m}}$ derived from equation 5, thus creating the **matching** phase. f(SegCompactness, SegAveragedgrey) is a weighting function that modulates the external weight connections depending on the averaged grey level and of the compactness (segment squared perimeter over the surface) of the segment to which neuron (i,j) belongs.

3.3. Binding by synchronization and the image comparison experiment

Two images shown in figure 2(a) are respectively presented to each layer. Neurons with strong connections tend to synchronize with each other, thus forming segments of synchronized neurons representing a similar part of the image (fig. 2(b)). Instead of using image features like contours, orientations, etc., pixel grey values are inputs to the computation of the connexion weights (equations 4, 5). This is not an ideal situation as recognition scores are lower with pixel values than with image features. However, this naive input setting yields a much more complex network with the number of connexions growing exponentially with the size of the images. With two 200p x 200p images, the network comprises 80 000 neurons with more than 1.6 x 10^9 connections. In this work our main interest is in fact the implementation of a complex network. At this stage, we are not interested in simplification and reduction of the network size. The matching phase begins once the segmentation is completed. It also uses the information from the segments to connect neurons between the two layers according to equation 5. Neurons within segments with similar grey scale values and compactness are strongly connected and thus the matching phase tends to synchronize similar parts of the two images as shown in figure 2(c).

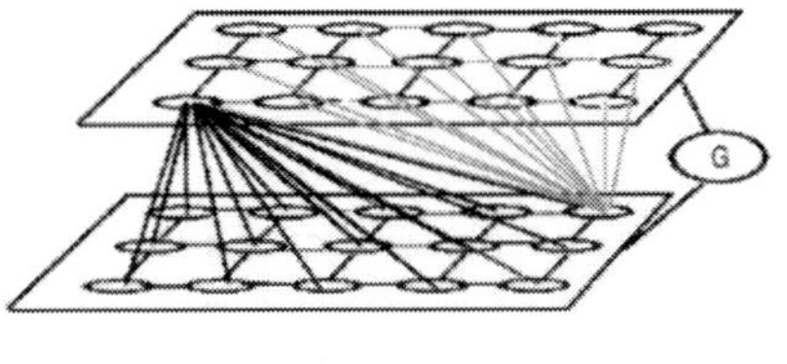

Figure 1. Architecture of the 2 layer neural network: neurons are locally coupled inside a layer and fully connected between layers. G is a global regulation unit that reduces the activity of the neural network (with inhibition) when the NN is too active.

4. Network parallelization

4.1. Distribution on processing units and event-driven processing

Neurons are uniformly distributed on different computing units and since we have a great number of neurons, this allows for a good level of scalability. However, the problem we face is that neurons are not independent of each other. A neuron can be affected by a spike coming from any processing unit at any

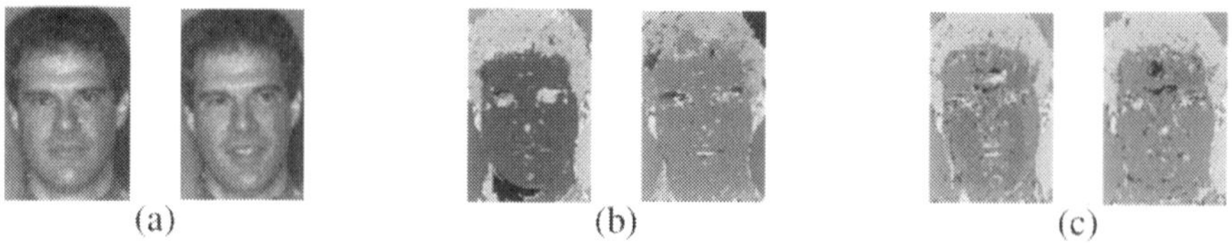

Figure 2. (a) Images used as input. (b) Segments formed after the segmentation phase. The color encodes the time at which neurons emit spikes. (c) Results of the matching phase between the two images.

time. In a network that is very strongly connected, this yields a data dependency that is very problematic for massively parallel architectures. An event-driven time management is therefore implemented as described in [7] to avoid calculating uneventful time steps between neuron spikes. By finding the neuron with the highest membrane potential in the network, one can advance the time to the moment at which the neuron will fire.

Since neurons are distributed on different processing units, they do not have access to neurons spiking on a different thread (threads can be associated to a different processing units). Thus, a global mechanism must be used that regroups the spiking neurons in a list that is available to all computing units. This is illustrated in figure 3. Individual neurons are then able to independently iterate through that list and react accordingly if they are connected to a neuron that spikes.

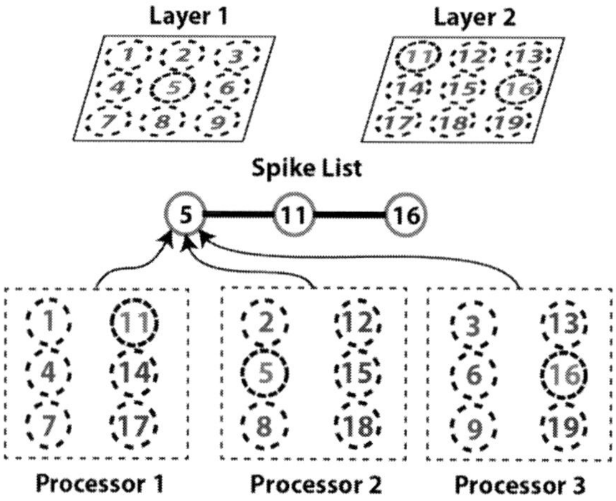

Figure 3. Example of parallel distribution of neurons and interaction with the global spike list. Neurons from layer 1 and 2 are distributed evenly on the 3 processors. This example shows the network in a state where there are 3 neurons spiking simultaneously (in red). All spiking neurons are put into the spiking list which is available to all processors. All processors iterate in parallel through the spiking list and process the spikes on their local neurons.

4.2. Differences between parallel and serial implementations

Memory management is different in the serial (conventional) and the parallel implementations. In the sequential version, all synaptic weights are stored in memory. However, for the parallel version, because of limited local memory size, synaptic weights for all neurons cannot be stored. Instead, the synaptic weight between two neurons is computed (with equations 4 and 5) when the receiving neuron fetches the spiking neuron information in the global spiking list. As such, the spiking list contains the ID of the spiking neurons along with the information needed by other neurons to compute the synaptic weight linking them.

5. Platform implementation

Implementation of the model was realized on two different architectures: Message Passing Interface (MPI) and Compute Unified Device Architecture (CUDA). The MPI implementation is used for platforms such as multi-core PCs or clustered computers whereas CUDA is for Nvidia's GPUs. Each architecture brings a different level of parallelization, but also different constraints. This is described in the following subsections.

5.1. MPI Implementation

Computations are distributed as processes which can be scattered on different computing nodes. Each node is independent and has its own local memory. MPI is an interface that allows messages to be passed between the different processes to exchange data and synchronize computation. However, communications and data exchange are costly and need to be kept to a minimum by trying to constrain dependent data to a single process. Once computations run within the scope of a single process, it is equivalent as doing computation on a CPU.

For the neural network implementation, each process is in charge of a group of neurons that resides in local memory for the duration of the simulation. Neurons are distributed on available processes with no particular order as the connections are too numerous to restrict communications within individual nodes. MPI collective operations are used to gather data distributed on the different processes, and general broadcast messages are used to instruct the processes of the computations they need to do on their neurons. The general program flow of the MPI implementation is shown in figure 4 (a).

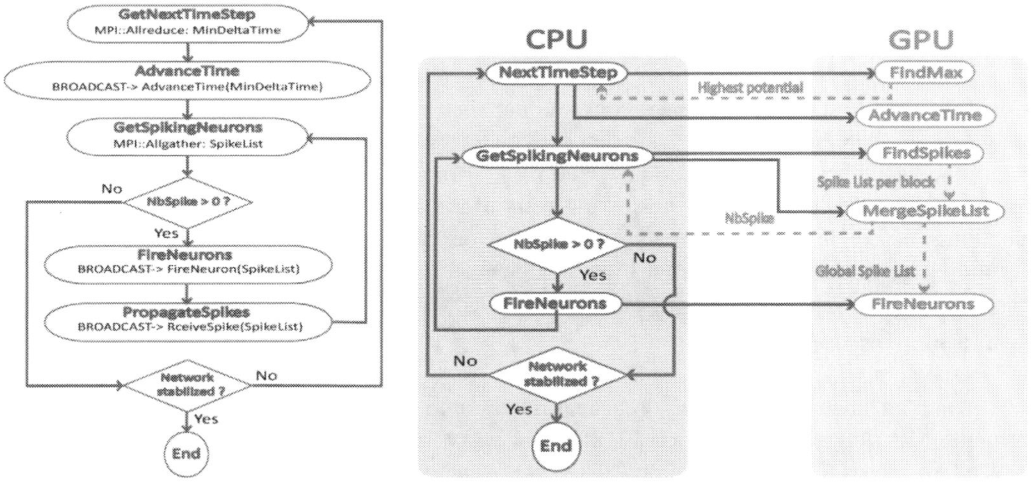

(a) (b)

Figure 4. Execution flow of the program for both MPI (a) and CUDA (b) implementations. Both implementations follow the same general algorithm. It starts by finding the neuron with the highest potential. Then simulation time is advanced to when it fires. A list of spiking neurons is collected and spikes are finally computed. This is done iteratively untill there are no more spikes left. This iterative process represents one simulation cycle. Once a cycle is done, network stability is verified (the network is stable when the spike patterns are the same over multiple cycles).

Each process finds the neuron with the highest potential within its assigned neurons and computes the time needed for it to spike. Using the collective operation MPI_AllReduce, which allows to find the minimal value of a variable distributed among all processes, the smallest time (*MinDeltaTime*) is isolated. It is then broadcasted in the system within a *AdvanceTime* message, that instructs the processes to advance the simulation time on their neurons. The global spiking list is assembled using the collective operation MPI_Allgather, which allows the gathering of data that is distributed on all processes. The *FireNeurons* and *PropagateSpikes* messages are then broadcasted with the spike list, instructing the processes to check if they are responsible for neurons affected by the spike list. The *FireNeurons* message instructs the processes to handle the computations of the spiking neurons themselves, that is resetting their internal potential. The *PropagateSpikes* message on the other hand, handles the computations on the neurons receiving a spike from a neuron in the spiking list.

5.2. Cuda Implementation

Computations are launched from the PC as kernels running on the GPU. A kernel consists of CUDA code that is distributed as blocks on the GPU. A block is a group of threads running on the graphic card processors, and collaborating together on the same data. Nvidia's GPUs have global memory, which is available to all threads, and also local memory which is very fast and shared within a block. Local memory is very limited, and thus, CUDA code needs to be designed to maximize its use, and also to minimize exchanges with the PC which are very costly.

For the neural network implementation, the CPU controls the flow of the program as shown in figure 4 (b). When the neurons need to be manipulated, a CUDA kernel is called to launch the calculations on the GPU. A different kernel is used for each specific task because local resources usage changes depending on the computations. Kernels are individually optimized to use available resources as efficiently as possible, so the number of neurons handled per block varies depending on the kernel. Typically, each kernel is separated in blocks handling several hundreds of neurons at a time. All neuron data is kept in global memory. Each time a kernel is called, it loads the data of its assigned neurons in local memory, executes the required computations and returns the result in global memory.

The *FindMax* and *AdvanceTime* kernels find the next neuron to spike and advance the simulation time respectively. The highest neuron potential is sent to the CPU so it can keep track of the simulation time. The *FindSpike* kernel finds the neurons that have a potential above the firing threshold and creates a spike list for each CUDA block. Since blocks do not share memory, the *MergeSpikeList* kernel is used to merge all the spike list and create a single list. The total number of spiking neuron is sent to the CPU to determine which kernel needs to be launched next. If the list is not empty, the *FireNeurons* kernel is launched to compute the impact of the spiking neuron on the network. The *FireNeurons* kernel works, as mentioned before, by making all neurons in the network iterate through the global spiking list, and by performing the appropriate computations if a local neuron is affected by a spike.

6. Results and Discussion

6.1. Experimental setup

The MPI implementation was tested on a dual quad core (AMD Opteron 2.2Ghz), and on RQCHP's Mammouth MP (2 CPUs/nodes (3.6GHz), 8GB/nodes, switch-infiniband). The CUDA implementation was tested on a Tesla C870 and a GeForce GTX285 from Nvidia.

For each implementation, the same 2 input images (Fig. 2) are presented to the neural network. Various network sizes are obtained by modifying the image resolutions. As the initialization values are randomized, the total number of iterations can vary between simulations. To ensure equivalent workload, instead of waiting for the network to stabilize, the number of iterations is arbitrarily fixed : 250 iterations for the segmentation of each image and 500 matching iterations for a total of 1000 cycles. To validate the different implementations, an image comparison benchmark is run and we verify that results are the same for all platforms.

6.2. Results

Table 1 gives the simulation times for different network sizes on different platforms. As can be seen, the MPI implementation running on the dual quad core is slower than the sequential implementation of the model. For Mammouth MP, performance is similar to the sequential version using only 2 nodes, which is

4 CPUs. As shown in figure 5, the execution time is greatly reduced as the number of CPUs is increased. Therefore, more than 8 CPUs on the dual quad core,would be faster than the sequential version.

Platform	Network size in nb. of neurons			
	4.1k	16.6k	66.3k	265.2k
Sequential CPU	3	59	-	-
Dual Quad Core	5	74	1394	41383
Mammouth MP (2 Nodes)	4	60	1838	-
Mammouth MP (32 Nodes)	2	4	43	684
Mammouth MP (64 Nodes)	3	4	37	327
Tesla C870	2	8	75	2345
GeForce GTX 285	2	4	19	467

Table 1. Execution time in seconds for the simulation of different network sizes on different platforms. Sequential CPU simulations for network bigger than 16.6k neurons were not run because the memory requirement would be too high.

Figure 6 shows the scalability of our MPI implementation on Mammouth MP from 8 nodes up to 256 nodes. The speedup scales well with the number of processing nodes. The bigger the neural network, the better the speedup. When the network size gets relatively small (thousands of neurons for each processing core), the speedup function is no longer linear (Fig. 6).

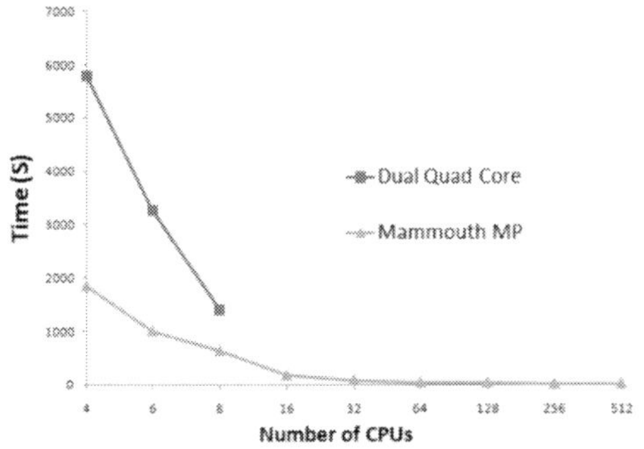

Figure 5. Execution time of the Dual Quad Core versus Mammouth MP

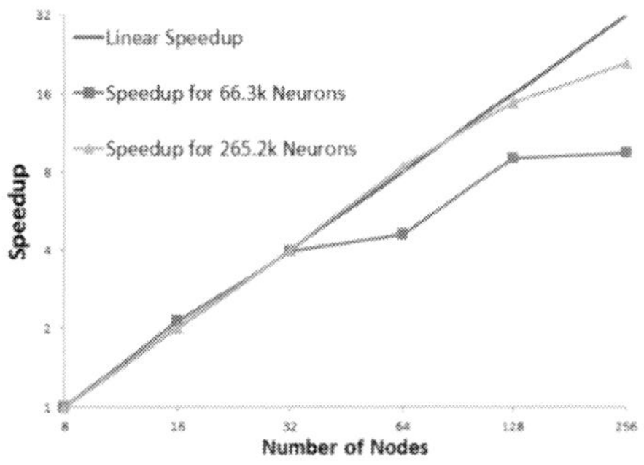

Figure 6. Speedup on Mammouth MP for the number of node used

As for the CUDA implementation, (Table 1) the C870 is much slower than the GTX 285. This is mainly explained by the higher bandwidth on the GTX 285 rather than the greater number of processing units. Neurons are loaded in memory much faster, an operation that has to be done at each kernel execution. With 64 nodes, Mammouth Parallel (MP) has roughly equivalent performance to the GTX 285. It is important to note however, that the MPI implementation scales better with increasing network size than the CUDA implementation. This is shown more explicitly in figure 7, where a slowdown is observed when increasing the network size. The slope of the graph is greater for both GPUs than for MP. In fact, the CUDA implementation has to reload the neuron's data from global memory for every kernel execution while on the MPI implementation, the neuron's data never moves from the process to which they are initially assigned. The bigger the network, the more data transfer have to be done in the CUDA implementation whereas there is simply no transfer needed for MPI.

7. Conclusion

Parallel implementations of a strongly connected large scale spiking neural network were presented for both MPI and CUDA architectures. On Mammouth MP, scaling of the MPI implementation with the number of available cores was shown to be very good, following an almost linear speedup with very

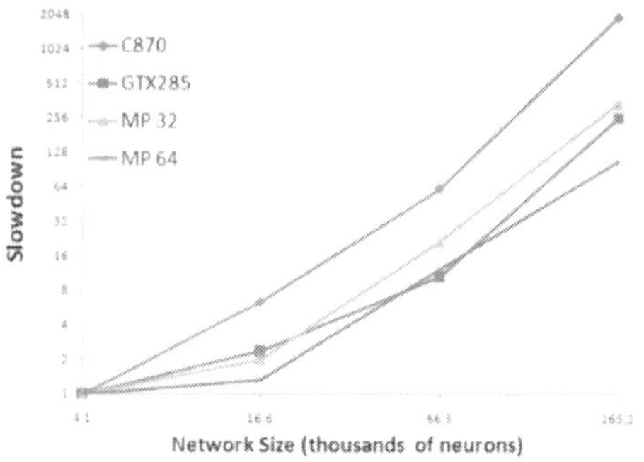

Figure 7. Slowdown observed on different platforms with increasing network size

large neural networks. The CUDA implementation presented similar performance using a GTX 285, compared with Mamouth MP with 64 nodes. However, the MPI version was shown to scale better with network size than the CUDA version.

Considering the of cost to performance ratio for parallel implementation of spiking neural networks, GPUs show the best trade off. Although, the current CUDA implementation should be redesigned to avoid reloading the neurons' data, which doesn't necessarily change from a kernel to the other. Local data persisting through kernel execution would be a great feature for this application (though it may not be technically feasible).

For real image processing applications, it is not necessary to let the network iterate as much as in the presented experiments. Also, suitable initial conditions accelerate the convergence speed of the neural network. So care should be taken as not to extrapolate the current processing speed to real applications. The goal of this paper was to demonstrate the parallel implementation effectiveness of a spiking neural network implementing the binding operation between two layers of neurons. The findings can be generalized to any application other than image processing as long as the topology of the neural network is comparable (local connections inside layers and full connections between layers) and that weights between neurons can be computed online.

Acknowledgment
CRSNG, Univ. de Sherbrooke, OPAL-RT for funding the research. Vincent Tremblay Boucher, Julien Beaumier-Éthier, Alexandre Provencher and Jonathan Verdier for CUDA programming. Many thanks to Sean Wood for correcting English.

References

[1] Milner P 1974 *Psychological Review* **81** 521–535
[2] Molotchnikoff S and Rouat J 2011 *Frontiers in Bioscience*
[3] Pichevar R, Rouat J and Tai L 2006 *Neurocomputing* **69** 1837–1849
[4] Hodgkin A L and Huxley A F 1952 *J Physiol* **117** 500–544
[5] Gerstner W 1998 *Pulsed Neural Networks* (The MIT Press)
[6] Izhikevich E 2003 *IEEE Transactions on Neural Networks* **14** 1569–1572
[7] Brette R and *et al* 2007 *Journal of Computational Neuroscience* **23** 349–398
[8] Plesser H E, Eppler J M, Morrison A, Diesmann M and Gewaltig M-O 2007 *Euro. Conf. on Parallel Processing* 672–681
[9] Nageswaran J M, Dutt N, Krichmar J L, Nicolau A and Veidenbaum A 2009 *IJCNN 2009* 3201–3208
[10] Fidjeland A and Shanahan M 2010 *IJCNN 2010*
[11] Yudanov D, Shaaban M, Melton R and Reznik L 2010 *IJCNN 2010*
[12] Nageswaran J, Dutt N, Wang Y and Delbrueck T 2009 *2009 IEEE Int. Symp. on Circuits and Systems* 1917 – 20
[13] Pallipuram V K, Bhuiyan M and Smith M C 2011 *Journal of Supercomputing* 1 – 46
[14] Han B and Taha T 2010 *IJCNN 2010*

Journal of Physics: Conference Series **341** (2012) 012025 doi:10.1088/1742-6596/341/1/012025

High-Performance Parallel and Stream Processing of X-ray Microdiffraction Data on Multicores

Michael A. Bauer[1], **Alain Biem**[2], **Stewart McIntyre**[1], **Nobumichi Tamura**[3] **and Yuzhen Xie**[1]

[1]University of Western Ontario, London, Canada
[2]IBM Research, Yorktown Heights, United States
[3]Advanced Light Source, Lawrence Berkeley National Lab, United States

E-mail: bauer@uwo.ca; biem@us.ibm.com; smcintyr@uwo.ca; ntamura@lbl.gov; yxie@csd.uwo.ca

Abstract. We present the design and implementation of a high-performance system for processing synchrotron X-ray microdiffraction (XRD) data in IBM InfoSphere Streams on multicore processors. We report on the parallel and stream processing techniques that we use to harvest the power of clusters of multicores to analyze hundreds of gigabytes of synchrotron XRD data in order to reveal the microtexture of polycrystalline materials. The timing to process one XRD image using one pipeline is about ten times faster than the best C program at present. With the support of InfoSphere Streams platform, our software is able to be scaled up to operate on clusters of multi-cores for processing multiple images concurrently. This system provides a high-performance processing kernel to achieve near real-time data analysis of image data from synchrotron experiments.

1. Introduction

High-brilliance synchrotron X-ray microbeams combined with fast large-area two-dimensional detectors can produce large blocks of X-ray microdiffraction (XRD) image data at unprecedented rates. In principle, up to 1 to 10 images per second of 4MB size can be produced. This data enables scientists to study crystalline texture in polycrystalline materials using Polychromatic X-ray Microscopy (PXM), and has led to refinements that permit the mapping of crystallographic orientation and deviatoric strain of such materials [7, 11, 12]. Even at this early stage, such measurements involve the accumulation of data sets as large as 60,000 of 4MB images. Using even advanced desktop computers it would take several days to index the diffraction patterns produced using software such as XMAS (X-ray Microdiffraction Analysis Software) developed at the Advanced Light Source [2] and the 3D X-ray Microdiffraction Analysis Software Package at the Advanced Photon Source [3]. A recent work [8] attempted to use cell processors to accelerate the processes.

Funded by CANARIE [1], the Active Network Interchange for Scientific Experimentation (ANISE) project [15] aims at developing a high speed network and processing systems to achieve near real-time data analysis for synchrotron experiments. Being able to process this data in near

[1] http://www.canarie.ca/

real-time would enable users to make sound decisions about issues on time-sensitive materials under study. In this paper we report on the development of a software system for analyzing *synchrotron white-beam Laue diffractions* using parallel and stream processing on clusters of off-the-shelf multi-core processors: today's mainstream architectures. This is the result of an ANISE sub-project, serving as one of the high-performance data processing kernels for synchrotron experimental data analysis.

Stream processing is a data-centric programming model. Given a set of data which can be structured as a stream of tuples, a series of operations (processing elements) are applied to each tuple or some set of tuples in the stream. Processing elements are usually pipelined, much like an assembly line. Pipelining does not reduce the time for individual instruction execution, but instead, it improves instruction throughput. Our implementation platform is the *IBM InfoSphere Streams* [6]. InfoSphere Streams enables the development and execution of applications that process information in data streams. It provides built-in operators as well as a programming interface for end-users to create user-defined operators to operate on data streams. The runtime system and the scheduler of InfoSphere Streams help to seamlessly deploy stream applications to clusters of multi-core processors.

We consider a sequence of procedures for analyzing synchrotron white-beam Laue diffractions for the mapping of crystallographic orientation and deviatoric strain of polycrystalline materials. As reported in [7, 11, 12], the main process for analyzing Laue patterns consists of the following procedures: Image parsing, Background filtering (optional), Blob searching, Peak fitting, Indexing and Strain refinement. Following the resources of the three software packages mentioned above, we review the crystallographic principles of Laue diffraction and its mathematical model for the purpose of parallel and stream processing, which is presented in Section 2. In Section 3, we introduce our techniques for adapting the algorithms and restructure their memory access patterns for efficient parallel and stream processing. We also apply high-performance software engineering to the sequential procedures for processing images individually in order to optimize the code and improve memory usage. Our implementation in IBM InfoSphere Streams platform and benchmark results are summarized at the end of Section 3.5.

2. Review of Laue Microdiffraction Methods for Parallel and Stream Processing

Laue (white-beam) diffraction is a standard crystallographic method for determining crystal orientation without rotation of the sample. It is advantageous for micron-sized samples since current high quality diffractometers have a sphere of confusion of about 10μm and any rotation would move the sample out of the micro-beam. Synchrotron-based micro-beam combined with modern charge-coupled device (CCD) detector has made Laue microdiffraction a promising method for the study of material properties with submicron spatial resolution. In such measurement systems the sample remains fixed except for translation motion in the plane of the sample for scanning an area of interest.

Figure 1 is a schematic view of the local geometry around the sample presented in [1]. Most micro-beam Laue experiments such as [1, 13] use the following setup: The sample is inclined at approximately 45 degrees to the white X-ray beam. The Laue diffraction pattern is collected by the X-ray CCD camera located above the sample at $2\theta = 90$ degrees (angle between the incident beam and the vector from the sample to the detector centre). In general, the lower limit of the photon energy bandpass lies between 5 and 10 keV, and the upper limit between 22 and 35 keV.

When a broad-bandpass micro-beam intercepts a sample and illuminates a small number of crystal grains, the wave from the white radiation with particular wavelength that satisfies the Bragg law, $n\lambda = 2d \sin\theta$, is diffracted. In the previous formula, the parameter λ is the wavelengh of a wave, d is the spacing of the planes of crystal atoms, θ is the incident angle, and n is an integer. The diffracted beams produce Laue diffraction patterns, which are recorded by the CCD camera as reflection spots (i.e. energy-intensities and pixel positions) in a 2D image.

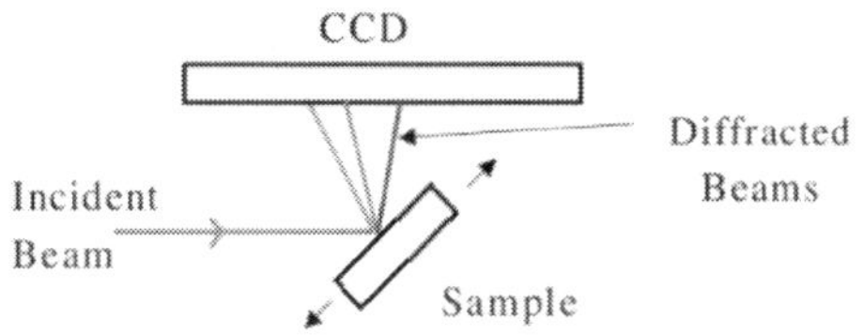

Figure 1: A schematic of X-ray microdiffraction geometry

Mathematical methods and software to obtain accurate strain information from synchrotron-based Laue patterns were first developed by a group at the Oak Ridge National Laboratory [3, 7, 12], and further improved at the Advanced Light Source [1, 2, 11]. To analyze synchrotron white-beam Laue diffractions for the mapping of crystallographic orientation and deviatoric strain of a single crystal or polycrystalline materials, the following sequence of processes takes place.

Image parsing: to convert a CCD image C in binary format (SPE, Pilatus TIFF or Mar TIFF) into a matrix E_0 of energy intensity values (integer or double). The coordinates of a pixel in C is encoded by the indices of the corresponding element in the matrix E_0.

Background filtering: to estimate the background B of E_0 (typically coming from elastic or inelastic scattering processes, fluorescence that do not contribute to the Bragg reflection peaks) and subtract B from E_0, generating an image E. This process is optional. However for some samples where the background is too high, it is necessary to filter out the background in order to accurately locate the reflection peaks.

Blob searching: to find the regions (or blobs) in E (or E_0 if no background filtering is applied) where pixel values are above a certain threshold (i.e. significant reflections) and then identify the potential peak positions of local maxima. Usually 8-way connectivity check and two-dimensional local maxima for peak searching are sufficient.

Peak fitting: to characterize the two-dimensional profile of each peak (by fitting to either Gaussian, Lorentzian or Pearson VII function) so as to find the correct positions of the reflection spots. Peak fitting is important because the accuracy of the fitted peak positions directly influence the results of indexing and strain refinement.

Indexing: to determine the Miller indices (hkl) of each reflection and the number of reflecting grains. To do so, the fitted peaks are sorted by their integrated intensities and a subset N of the most intense reflections is first considered. The value of N can be carefully chosen to allow the algorithm to index a set of overlapping diffraction patterns coming from different illuminated grains.
First, a calibration procedure is used to obtain the instrument geometrical parameters (i.e. the position of the detector), including the pixel coordinates of the center channel, the distance of the detector to the sample, angular positions and angular tilts with respect to the incoming beam direction. Given the direction of the incident beam $\vec{k}_{in}$, the position of the detector and the characteristics of the detector (i.e. the size of the active area both in number of pixels and in mm), for each reflection in N with a precise position on the CCD, its corresponding reflected-beam direction $\vec{k}_{out}$ can be computed. Assuming elastic scattering, the direction of the experimental scattering vector can thus be calculated as $\vec{q}_{exp} = \vec{k}_{out} - \vec{k}_{in}$.
The angles between each pair of these scattering vectors corresponding to the N reflections are then computed and compared with a "reference" list of angles between each pair of the theoretical scattering vectors determined from the crystalline structure of interest and the energy bandpass of the polychromatic beam. Next, the process looks for angular matches

between triplets of reflections within an adjustable angular tolerance. For each matched triplet and their corresponding (hkl) indices, a trial orientation matrix is computed as well as a complete list of reflections that should be visible on the CCD for the considered energy bandpass, CCD dimensions and geometry. The best match is the triplet that is able to index the largest number of measured reflections.

Strain refinement: to obtain only the deviatoric strain tensor (unit-cell distortion) and the stress tensor for the illuminated grains, since white-beam patterns only allow the measurement of unit-cell distortions, not changes in the unit-cell volume (the dilatational or hydrostatic strain). Therefore, only the lattice parameters a, b, c can be refined. The transformation matrix t_{ij}, which defines the transformation from the unstrained crystal vectors to the distorted unit-cell vectors, can then be derived. By eliminating the rotational component from t_{ij}, the deviatoric strain tensor ε'_{ij} within the X-ray illuminated volume can be obtained. With the knowledge of the absolute lattice value of a single reflection, the complete strain tensor ε_{ij} (i.e. the sum of the deviatoric tensor and the dilatational tensor) can in principle be determined from a single white-beam diffraction image. The stress tensor can thus be derived from the strain tensor and the literature values of stiffness constants C_{ijkl} as $\delta_{ij} = C_{ijkl}\,\varepsilon_{kl}$.

3. Parallel and Stream Processing of Laue Microdiffractions

In this section, we present the development of a system in IBM InfoSphere Streams for parallel and stream processing of synchrotron-based Laue microdiffraction images. Starting from the CCD image files, we organize the data involved in the Laue microdiffraction analysis as data streams and implement in IBM Streams a set of user defined operators (UDOPs) to work on the data streams and perform the functionalities of the processes introduced in Section 2. These user defined operators can be used to construct pipelines for processing multiple images in streaming mode. In the Appendix, Figure 4 shows a Streams live graph captured by the Streams studio when we use one pipeline with four processing elements for parallel peak fitting for processing images. Figure 5 illustrates four XRD processing pipelines running concurrently. In the following we report on our key implementation techniques.

3.1. Efficient Source Operator for Parsing Image Files

Synchrotron X-ray experiments usually generate thousands of image files. The large volume of files may have to be stored on external disks. Thus, reading these files from disk and parsing them may be a bottleneck. For reading an image file, we first find the position of the data strip and the byte count of the data and then read a whole block to save on disk seeking time. We then use bit-wise operations to convert a two-dimensional XRD image in binary format to a matrix of integer or double-precision energy intensity values. This makes our Source operator in Streams about ten times faster than the present SPE and TIFF image readers [9, 14].

3.2. Cache-efficient Background Filter

Except for the fine-pipelined parabolic background filter reported in [10], we implement in Streams an operator for estimating the background of CCD images based on Bruckner's method [4] and the Fortran procedure in XMAS. In this method, the image data needs to be traversed for many times. To obtain a cache-efficient implementation, we arrange the 2D image data in row-major layout in one consecutive array. Between the two calls to the 1D Bruckner procedure, we transpose the image so that the data for 1D Bruckner to operate on will always be aligned consecutively in order to minimize cache misses. We use the cache-oblivious algorithm introduced in [5] for matrix transpositions. As a result, our operator is about ten times faster than the Fortran procedure in XMAS. This cache-efficient 2D Brückner Filter is presented in Algorithm 1.

Algorithm 1: Cache-efficient 2D Brückner Filter

```
Input: M, an array containing a matrix (r × c) of intensity values in row-major layout
Output: the background of M
S ← ResizeImage(M, r, c, m, n); // reduce the dimension of M to m × n
// repeat the procedure for q times
for k from 1 to q do
    for each row R_i in S do
        // replace each value in R_i by the average of its 2N neighbours, and iterate for t times
        BackgroundFitBruckner1D(R_i, N, t);
    // replace the value of each point in S by the median of its W neighbors
    MedianFilter2D(S, m, n, W);
    Rotate S clockwise for 90 degrees into S_r;
    for each row R_i in S_r do
        // replace each value in R_i by the average of its 2N neighbours, and iterate for t times
        BackgroundFitBruckner1D(R_i, N, t);
    // replace the value of each point in S_r by the median of its W neighbours
    MedianFilter2D(S_r, m, n, W);
    Rotate S_r counter-clockwise for 90 degrees into S;
S_b ← ResizeImage(S, m, n, r, c); // enlarge the dimension of S to r × c
return S_b;
```

3.3. Memory-efficient Parallel Peak Fitting

For each significant reflection identified during blob searching, a non-linear least square solver is used to fit to a two-dimensional Gaussian, Lorentzian or Pearson VII function, so as to estimate the correct peak position. This is a computationally intensive procedure. The number of reflections in an image varies from tens to hundreds, depending on the properties of the sample. To gain performance for this procedure, we apply a parallel scheme that utilizes a dynamic scheduler to distribute the blobs for balanced parallel peak fitting.

This scheme is demonstrated in Figure 2. In the example, the blob searching operator takes a stream u of image data and identifies interesting blobs and then arrange the data associated with each blob (i.e. the coordinates of the center, the pixel values of the blob region) as a tuple in a stream bl. To prepare to be distributed for peak fitting, each blob tuple is labelled with a group ID to classify the blob tuples into groups. The largest group ID is the number of distributed peak fitting elements (F) available. Next, a splitter is used to split the tuples in stream bl by Group ID, generating F number of sub-blob streams. Each sub-blob stream will be taken by a peak fitting process element to carry out the fitting operation and organize the result into a tuple in a sub-peak stream. All the sub-peak streams are bundled together to be sorted and then used in indexing process.

In order to avoid memory contention as well as to maintain good data locality, right after a blob with pixel b as its center is found, we copy an area R of pixel values from the original image, as shown in Figure 3. The area R that we define is twice as large as the area R_b (normally a 10×10 matrix) which has b in its center and is needed for peak fitting. We associate the data set R to the blob of b and distribute this blob for parallel peak fitting. In peak fitting, the necessary area R_b is determined from area R by coordinate mapping. A larger area like R is required because the center of a reflection can be different (i.e. slightly shifted) from b by peak fitting. The new center, say p, will define an area R_p of dimension 10×10 in the original image. The area R_p is used to compute an integrated intensity which will be used to classify (i.e. rank the priority of) a reflection for later indexing. We determine the area R_p from area R again by coordinate mapping, with no reference to the original image. In this way, we avoid the many times of references to the original image, a potential source of memory contention for concurrent execution. Good data locality is obtained by associating each peak fitting with a small working data set like R.

In addition, we align the data in the peak fitting results in one flat array, where the data associated with each peak is a record with the integrated intensity value as its key. We create two routines for sorting records of data. One implements insertion-sort algorithm for sorting small number of records and the other one uses merge-sort for sorting relatively large number of records. In these sorting routines specialized for records of data, a record is moved as a block to achieve efficient memory operations.

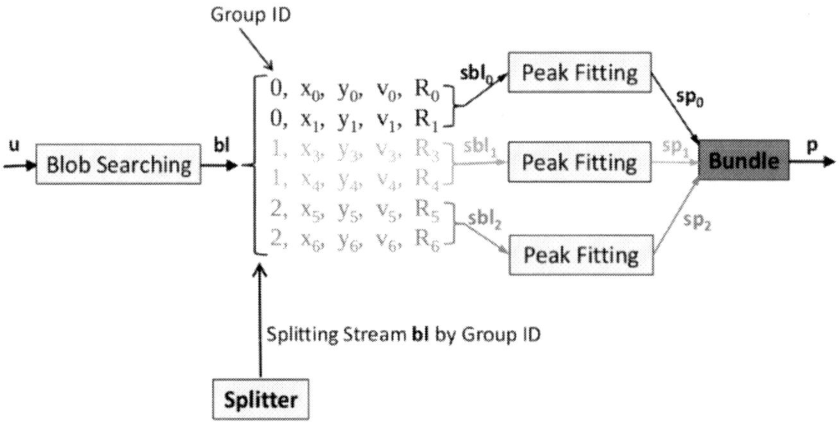

Figure 2: Dynamically scheduling blobs for parallel peak fitting

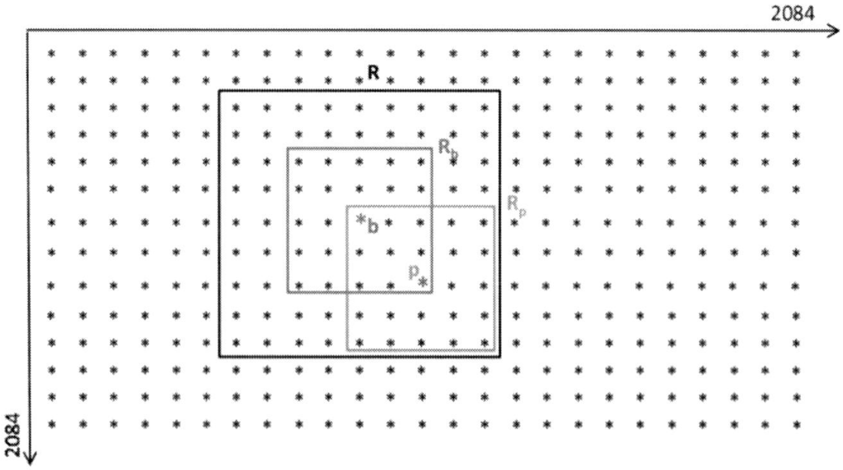

Figure 3: Efficient memory management for parallel peak fitting

3.4. Sharing Common Parameter Values in Indexing and Strain Analysis

For indexing and strain analysis, we eliminate the duplicate computations and data storage occurring in the sequential program. In InfoSphere Streams, we are able to generate streams of data which can later be shared by multiple processing elements. We take advantage of this feature and create a special operator to parse the experimental parameters, compute and then only keep the final derived parameter values necessary for indexing and strain analysis as a data stream.

More precisely, we design a Source operator for parsing the common parameter values (beamline, sample and detector geometrical parameters, unit-cell parameters, space group data and stiffness constants). Based on these inputs, three commonly used rotation matrices for conversion between different coordinate systems (beamline (lab), detector and sample) and two pairs of transformation matrices for conversion between lattice space and reciprocal space, between lattice space and direct (lab system) space are generated. Together with the space group data and stiffness constants, they form a stream which can be shared later for indexing and strain analysis for all the XRD images.

In addition, using this stream of parameter data, we create an operator for generating the theoretical reflection list which can be used for the indexing of all XRD images. We also observe that we can use matrix transposition instead of calculating the inverse for generating the inverse of a rotation matrix by Euler angles as composition of extrinsic rotations. For processing $10,000$ images, this can save 1.5 million double precision multiplications.

3.5. Benchmarks

Our benchmark system is an Intel Core2 Quad CPU Q9550 (2.83 GHz, 8 GB RAM and 6 MB L2 cache) running with Red Hat Enterprise Linux 5.4 operating system. It supports InfoSphere Streams Version 1.2. Our program can be configured to generate multiple pipelines to process multiple images concurrently in streaming mode. It takes about 2.5 seconds to process a single image of dimension size 2084×2084 in SPE format using one pipeline with four processing elements for parallel peak fitting, which is about 10 times faster than the present best C program. On this quad-core desktop, super-linear speedup is obtained for processing four images by four pipelines.

4. Concluding Remarks

We present a parallel and stream processing system on multicore processors to analyze X-ray microdiffraction images for the mapping of crystallographic orientation and deviatoric strains. It is the first streaming application in the field of XRD data analysis. Our approach and results show that stream processing is an effective model for efficiently using multicore processors for XRD image data analysis. With the support of IBM InfoSphere Streams platform, the set of user-defined operators that we have created for XRD image preprocessing and Laue pattern analysis could be generalized as part of a library of streaming operators for processing images derived from a variety of detectors. Our work-in-progress includes the configuration and deployment of this kernel to large systems with many cores for processing large set of XRD images efficiently.

5. Acknowledgments

We would like to acknowledge the support for this project from CANARIE and for the networking infrastructure that they provide to enable us to even begin to think about streaming image data. We would also like to thank all those involved in the Science Studio project and the ANISE project.

References

[1] A. A. MacDowell, R. S. Celestre1, N. Tamura, R. Spolenak, B. Valek, W. L. Brown, J. C. Bravman, H. A. Padmore, B. W. Batterman and J. R. Patel. Submicron X-ray Diffraction, http://escholarship.org/uc/item/2f47k34b, Lawrence Berkeley National Laboratory, USA, 2000.

[2] Advanced Light Source. X-ray Microdiffraction Analysis Software (XMAS), http://xraysweb.lbl.gov/microdif/user_resources.htm, Lawrence Berkeley National Laboratory, USA, 2010.

[3] Advanced Photon Source. 3D X-ray Microdiffraction Analysis Software Package, http://www.aps.anl.gov/Sectors/33_34/microdiff/downloads/, Argonne National Laboratory, USA, 2009.

[4] Sergio Brückner. Estimation of the Background in Powder Diffraction Patterns through a Robust Smoothing Procedure. *Journal of Applied Crystallography*, 33(3 Part 2):977–979, Jun 2000.

[5] M. Frigo, C. E. Leiserson, H. Prokop, and S. Ramachandran. Cache-Oblivious Algorithms. In *FOCS*, 1999.

[6] IBM. IBM InfoSphere Streams, http://www-01.ibm.com/software/data/infosphere/streams/, 2010.

[7] J. Chung and G. Ice. Automated Indexing for Texture and Strain Measurement with Broad-band Pass X-ray Micro-beams. In *Journal of Applied Physics*, number 9, Vol. 86, 1999.

[8] J. Qin and M. A. Bauer. Accelerated Synchrotron X-ray Diffraction Data Analysis on a Heterogeneous High Performance Computing System. In *J. Phys.: Conf. Ser.*, number 256 012007, 2010.

[9] Jon Tischler. WinView, Oak Ridge National Lab, TischlerJZ@ornl.gov, 2010.

[10] M. A. Bauer, A. Biem, S. McIntyre and Y. Xie. A Pipelining Implementation for Parsing X-ray Diffraction Source Data and Removing the Background Noise. In *J. Phys.: Conf. Ser.*, number 256 012017, 2010.

[11] N. Tamura and A. A. MacDowell and R. Spolenak and B. C. Valek and J. C. Bravman and W. L. Brown and R. S. Celestre and H. A. Padmore and B. W. Batterman and J. R. Patel. Scanning X-ray Microdiffraction with Submicrometer White Beam for Strain/Stress and Orientation Mapping in Thin Films. In *J. Synchrotron Rad.*, number 10, pages 137–143, 2003.

[12] N. Tamura and J.-S. Chung and G. E. Ice and B. C. Larson and J. D. Budai and J. Z. Tischler and M. Yoon. Strain and Texture in Al-Interconnect Wires Measured by X-ray Microbeam Diffraction. In *Mat. Res. Soc. Symp.*, number 563, pages 175–180, 1999.

[13] R. I. Barabash, G. E. Ice, N. Tamura, B. C. Valek, J. C. Bravman, R. Spolenak and J. R. Patel. Quantitative Characterization of Electromigration-induced Plastic Deformation in Al(0.5 wt%Cu) Interconnect. *Microelectron Engineering*, (75:1):24–30, 2004.

[14] Sam Leffler. libtiff 3.8.2, ftp://ftp.remotesensing.org/pub/libtiff/, Silicon Graphics, Inc., 2004.

[15] ANISE Team. Active Network Interchange for Scientific Experimentation Project, http://www.anise-project.com/, University of Western Ontario, London ON, Canada, 2010.

Appendix

Journal of Physics: Conference Series **341** (2012) 012025 doi:10.1088/1742-6596/341/1/012025

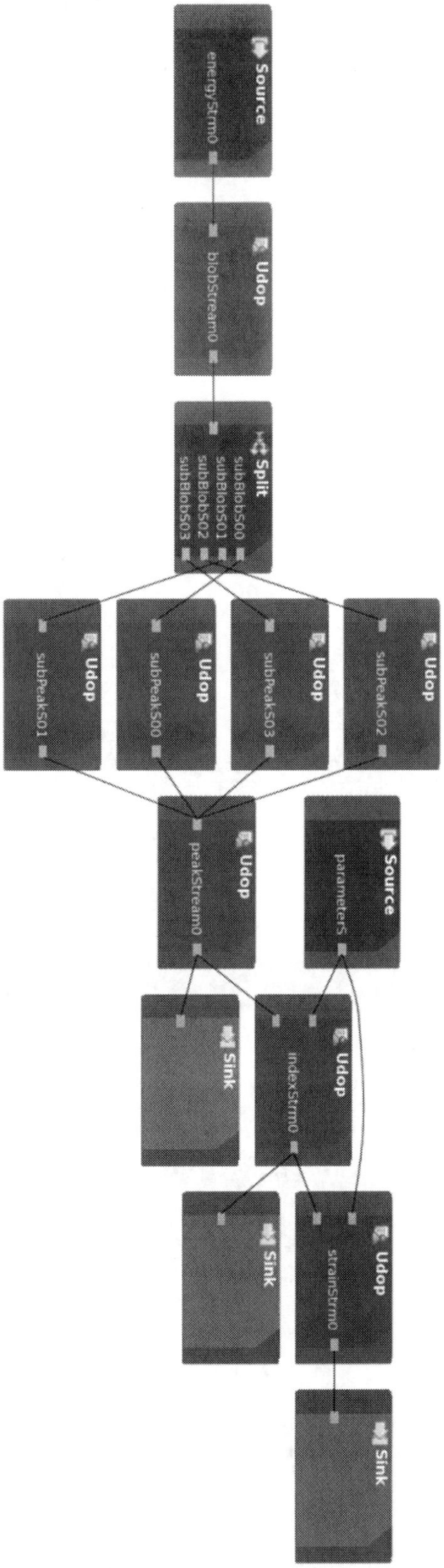

Figure 4: Streams live graph: one pipeline with 4 processing elements for parallel peak fitting

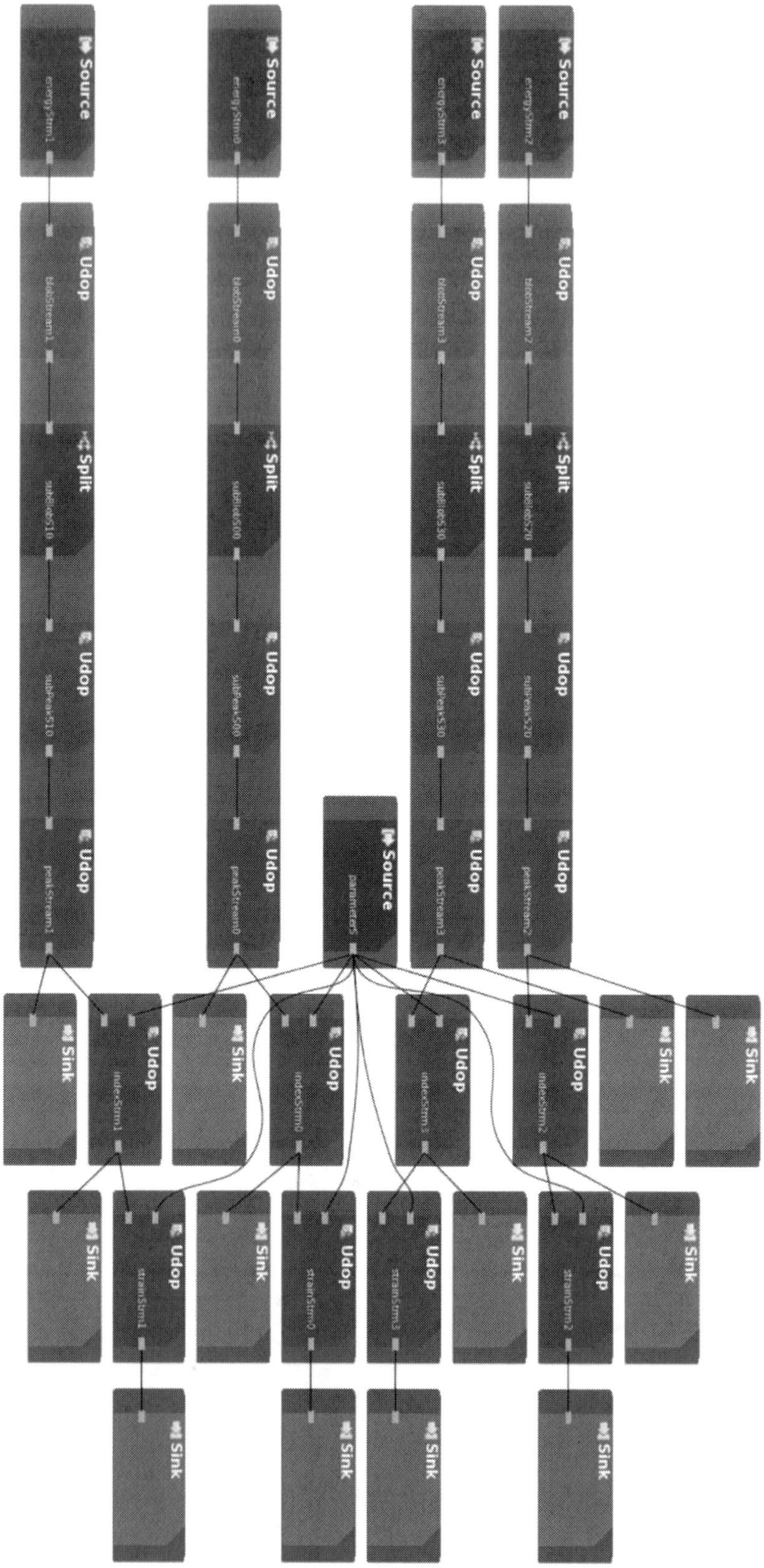

Figure 5: Four XRD processing pipelines running concurrently (the operators for blob searching and peak fitting are fused to minimize transfer latencies)

Journal of Physics: Conference Series **341** (2012) 012026 doi:10.1088/1742-6596/341/1/012026

Cache Complexity and Multicore Implementation for Univariate Real Root Isolation

Changbo Chen, Marc Moreno Maza and Yuzhen Xie
University of Western Ontario, London, Ontario, Canada
E-mail: cchen252,moreno,yxie@csd.uwo.ca

Abstract. We present parallel algorithms with optimal cache complexity for the kernel routine of many real root isolation algorithms, namely the *Taylor shift by 1*. We then report on multicore implementation for isolating the real roots of univariate polynomials with integer coefficients based on a classical algorithm due to Vincent, Collins and Akritas. For processing some well-known benchmark examples with sufficiently large size, our software tool reaches linear speedup on an 8-core machine. In addition, we show that our software is able to fully utilize the many cores and the memory space of a 32-core machine to tackle large problems that are out of reach for a desktop implementation.

1. Introduction

Isolating the real roots of a univariate polynomial is a driving subject in computer algebra. On the one hand, it is closely related to fundamental algorithms such as polynomial multiplication and *Taylor shift* [14]. On the other hand, it is one of the particular areas in scientific computing where computer algebra provides unmatchable tools, for example, counting rigorously the number of real roots of a polynomial. Many researchers have studied this problem under various angles from algebraic algorithms to implementation techniques. We refer the reader to the latest paper [13] for a survey of related works on this topic and the references therein.

Today, multicore processors with hierarchical memories have entered the mainstream of computer architectures. However, harvesting the power of multicores is challenging. Besides, understanding the implications of hierarchical memory has become an essential part of performance software engineering. In this work, we revisit a classical algorithm due to Vincent, Collins and Akritas [4] for isolating the real roots of univariate polynomials. We explore the parallelism of this algorithm and design fine-grained parallel algorithms with different strategies for the kernel routine, i.e., Taylor shift, aiming at attaining efficient implementations for polynomial real root isolation on multicores. This paper is an extension of the two-page abstract for our poster [2] presented in ISSAC 2010. Before this work, several other pioneer works [3, 5, 10, 11] have also been conducted on this topic with different techniques. The paper [3] reports on a first attempt on parallelizing real root isolation problem. Their work shows that the amount of parallelism is limited if only the traversal of the binary tree associated with the isolation problem is parallelized. Later works [5, 10, 11] focus on improving Taylor shift via high performance computing techniques. The paper [5] proposes new scheduling algorithms for the pyramid DAG associated with Taylor shifts and reduces the communication complexity overhead to linear. The paper [10, 11] improves the efficiency of Taylor shift computation and real root isolation based on a technique of register tiling.

This paper is organized as follows. Section 2 introduced the sequential algorithm of Vincent-Collins-Akritas and the Taylor shift operation. In Sections 3, 4 and 5, we present our strategies, *divide-and-conquer* and *blocking*, for a fine-grained parallelization of the Taylor shift procedure and a parallel algorithm for univariate real root isolation. We show that both can be optimal in terms of cache complexity. Their implementation face several challenges (scheduling, in place execution) that we address in Sections 6, 7, 8. In the experimentation, in Section 9, we report timings for processing well-known test examples as well as a real-world challenging problem, respectively on 8-core and 32-core machines. We conclude in Section 10 by discussing extensions of this study.

2. The sequential algorithm

Let p be a univariate squarefree polynomial with integer coefficients. To isolate the real roots of p, the Vincent-Collins-Akritas (VCA, for short) algorithm [4] computes a list of disjoint intervals with rational endpoints such that each real root of p belongs to a single interval and each interval contains only one real root of p. Clearly, the problem of isolating the real roots of p and that counting of them are essentially the same. Thus, one algorithm solving one of these problems can be adapted to solve the other. Since the output of the counting problem is simpler, we present, as Algorithm 1 below, the version of the VCA Algorithm solving this latter problem.

The main subroutine of Algorithm 1 is RootsInZeroOne, that is, Algorithm 2, in which the operation of dominant cost is performed at Lines 3 and 7. This computation, which substitutes x for $x + 1$ in p_1, is called the *Taylor shift by 1* (or simply Taylor shift) of the polynomial p_1.

Algorithm 1: RealRoots(p)

Input: a univariate squarefree polynomial p of degree d
Output: the number of real roots of p

1 **begin**
2 Let $k \geq 0$ be an integer such that the absolute value of all the real roots of p is less than or equal to 2^k;
3 **if** $x \mid p$ **then** $m := 1$ **else** $m := 0$;
4 $p_1 := p(2^k x)$;
5 $p_2 := p_1(-x)$;
6 $m' :=$ RootsInZeroOne(p_1);
7 $m := m +$ RootsInZeroOne(p_2);
8 return $m + m'$;
9 **end**

Algorithm 2: RootsInZeroOne(p)

Input: a univariate squarefree polynomial p of degree d
Output: the number of real roots of p in $(0, 1)$

1 **begin**
2 $p_1 := x^d p(1/x)$;
3 $p_2 := p_1(x + 1)$; //Taylor shift
4 Let v be the number of sign variations of the coefficients of p_2;
5 **if** $v \leq 1$ **then** return v;
6 $p_1 := 2^d p(x/2)$;
7 $p_2 := p_1(x + 1)$; //Taylor shift
8 **if** $x \mid p_2$ **then** $m := 1$ **else** $m := 0$;
9 $m' :=$ RootsInZeroOne(p_1);
10 $m := m +$ RootsInZeroOne(p_2);
11 return $m + m'$;
12 **end**

Executing Lines 4 and 5 in Algorithm 1 and executing Lines 2 and 6 in Algorithm 2 simply require to "traverse the data" and "update the coefficients", thus have a linear cost with respect to the data size. Apart from its recursive calls, the other computational costs of Algorithm 2 come from the Taylor shift which cost is more than linear (in fact quadratic, as we shall see later). Therefore, same for Algorithm 1, most of the work comes from this latter operation, which justifies the effort devoted to implementing it efficiently.

In order to better understand this operation, consider a polynomial $f(x) = a_d x^d + \cdots + a_1 x + a_0$ of degree d. The Taylor shift of f is the polynomial $f_0(x) = f(x+1)$ computed by the *Pascal's triangle* relation (or Horner's rule) $f_i(x) := f_{i+1}(x) \times (x+1) + a_i$ for i successively equal to $d-1, \ldots, 1, 0$ with $f_d(x) := a_d$, as illustrated for $d = 3$ in Example 1.

Example 1 *Let* $f(x) = a_3x^3 + a_2x^2 + a_1x + a_0$. *We construct the following Pascal's triangle.*

$$\begin{array}{ccccccccccc}
 & & 0 & & 0 & & 0 & & 0 & & \\
 & & \downarrow & & \downarrow & & \downarrow & & \downarrow & & \\
a_3 & \rightarrow & + & \rightarrow & + & \rightarrow & + & \rightarrow & + & \rightarrow & c_3 \\
 & & \downarrow & & \downarrow & & \downarrow & & & & \\
a_2 & \rightarrow & + & \rightarrow & + & \rightarrow & + & \rightarrow & c_2 & & \\
 & & \downarrow & & \downarrow & & & & & & \\
a_1 & \rightarrow & + & \rightarrow & + & \rightarrow & c_1 & & & & \\
 & & \downarrow & & & & & & & & \\
a_0 & \rightarrow & + & \rightarrow & c_0 & & & & & &
\end{array}$$

Performing the addition in the diagonal direction *is exactly the same as computing* $f(x+1)$ *in Horner's rule, which brings* $f(x+1) = c_3x^3 + c_2x^2 + c_1x + c_0$.

3. Parallelization of the Vincent-Collins-Akritas algorithm

A first opportunity for parallel execution appears in Algorithm 1. Indeed, the two calls to RootsInZeroOne can be performed concurrently. Similarly, in Algorithm 2, the two recursive calls can be performed concurrently. If the amount of work was always balanced among the two calls to RootsInZeroOne in Algorithm 1 as well as among the two recursive calls in Algorithm 2, then those concurrent calls would create significant parallelism. However, this situation will rarely happen in practice. Indeed, observe that one call to Algorithm 2 may quickly terminate at Line 5. In fact, Lines 4 and 5 implement the so-called *Descartes's Rule* (see [12] for detail) which is a criterion for detecting that a polynomial has no more than one root in the range $(0, 1)$. For this reason, that is, for the configuration of the real roots of p, the work is likely to be unbalanced among the recursive calls in Algorithm 2. Consequently, it is likely to be unbalanced among the two calls to RootsInZeroOne in Algorithm 1 as well. Therefore, one should look for other opportunities for parallel execution.

The only other possibility is the Taylor shift computations. There are clearly several ways of "filling the Pascal's triangle". Setting the origin $(0, 0)$ at the top left corner, one may proceed row by row, or column per column, or with the elements of coordinates (i, j) satisfying $i + j = k$ for k successively equal to $= 0, 1, \ldots, d$. We call this latter traversal *anti-diagonal*.

This observation suggests a potential for concurrency. As mentioned in the introduction, we discuss two parallelization schemes: one is a *divide-and-conquer* approach (d-n-c, for short) and the other is a *blocking strategy* (blocking, for short). They are discussed in detail in Sections 4 and 5, respectively.

4. A divide-and-conquer strategy for computing the Taylor shift in parallel

The elements of the Pascal's triangle, and thus the coefficients of $f(x+1)$, can be computed in a *divide-and-conquer* manner, demonstrated by the Subfigures (a) and (b) in Figure 1. Each triangular or square region is recursively divided into smaller regions until a base case is reached. Clearly, in both the triangular division and the square (or two-way tableau) division, opportunities for parallel execution are created. To be precise, the regions labeled by II can be computed concurrently.

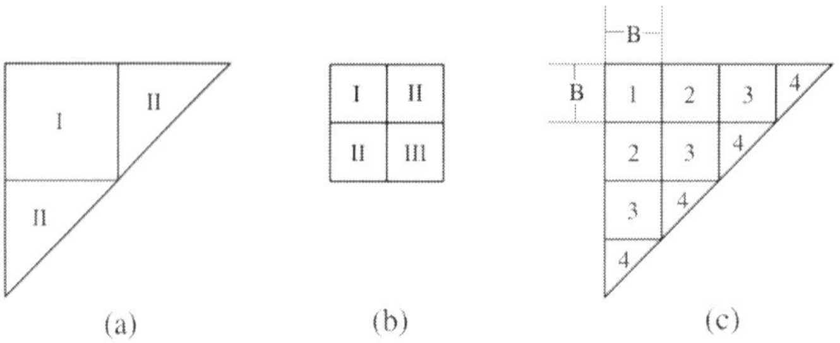

Figure 1. Illustration for the d-n-c and blocking strategies

In the rest of this paper, we assume that adding two integers is done within a constant amount of time. This property is true for machine integers. For multi-precision integers, this assumption is necessary in order to keep the analysis of the parallelization of Algorithms 1 and Algorithm 2 simple. For large input polynomials, relaxing this assumption is more realistic and estimating the cost of each addition in the construction of the Pascal's triangle lead to finer results and more opportunities for parallelism. The level of technical details, however, increase substantially and we will report this study in a forthcoming paper.

Since adding two integers depends linearly on their size, assuming that each integer addition is done in constant time implies that each integer fits within a constant number of bits. Let C be this number. Based on this assumption, we analyze the d-n-c approach for different complexity measures relevant to a multicore implementation, with an emphasis on cache complexity.

4.1. Work, span and parallelism estimates.

Our estimates follow the fork-join multithreaded programming model of the Cilk language [7]. For a two-way tableau with n as input data size, the work $U_1(n)$ required for filling this tableau satisfies $U_1(n) = 4U_1(n/2) + \Theta(1)$, for $n > 1$, and $U_1(1) \in \Theta(1)$, which implies $U_1(n) \in \Theta(n^2)$. (For simplicity, we may think of n as a power of 2.) For the same construction, the span (or critical path, or depth) satisfies $U_\infty(n) = 3U_\infty(n/2) + \Theta(1)$, for $n > 1$, and $U_\infty(1) \in \Theta(1)$, which leads to $U_\infty(n) \in \Theta(n^{log_2 3})$.

For a Pascal's triangle with n as input data size, the work $T_1(n)$ satisfies $T_1(n) = 2T_1(n/2) + U_1(n/2)$, for $n > 1$, and $T_1(1) \in \Theta(1)$, which gives $T_1(n) \in \Theta(n^2)$. For the same construction, the span $T_\infty(n)$ satisfies $T_\infty(n) = T_\infty(n/2) + U_\infty(n/2)$, for $n > 1$, and $T_\infty(1) \in \Theta(1)$, which implies that $T_\infty(n) = \Theta(n^{log_2 3})$. Therefore, for both constructions, the parallelism is $\Theta(n^{2-log_2 3})$. Since $2 - log_2 3$ is approximately 0.45, this indicates a relatively low parallelism, unless n is very large.

4.2. Space complexity estimate.

Consider first the sequential algorithm. One can observe that, at any point of the construction of the Pascal's triangle, the knowledge of $2n$, and only $2n$, integers (with $n = d + 1$) is needed in order to continue the construction until its completion. Moreover, the whole algorithm can be executed in place within the space allocated to the input $2n$ integers. At completion, the output n integers can be found in this allocated space, most likely in its first n slots.

The same property is easy to prove for the d-n-c approach. To do so, one first establishes by induction a similar property for the two-way tableau construction, see Subfigure (b) in Figure 1. Then, one proves by induction the desired property for the divide-and-conquer approach of the Pascal's triangle depicted by Subfigure (a) in Figure 1.

Finally, we conclude that the d-n-c approach can be executed in place in space $2\,n\,C$.

4.3. Cache complexity estimates.
We use the (Z, L) ideal cache model of [6], where Z is the number of words in the cache and L is the number of words per cache line. Let $Q_S(n)$ (resp. $Q_T(n)$) be the number of cache misses incurred by the d-n-c approach at the level of the two-way tableau (resp. Pascal's triangle) on input data of size n.

When n is small enough, all the data read and written by the algorithm fit in cache. In this case, the number of cache misses is proportional to the amount of read/written data. Thus, there exists a positive constant α such that for $n \leq \alpha Z$ we have $Q_S(n) = 2n/L + 2$. Here we use the fact that the whole algorithm is executed in place within $2n$ integer slots. The $+2$ in $2n/L + 2$ comes from the fact that scanning an array of m consecutive words in memory incurs m/L or $m/L + 1$, depending on alignment issues.

For $n > \alpha Z$, we have the obvious equality $Q_S(n) = 4Q_S(n/2) + \Theta(1)$. To summarize, we have:

$$Q_S(n) = \begin{cases} 2n/L + 2 & n \leq \alpha Z \\ 4Q_S(n/2) + \Theta(1) & \text{otherwise.} \end{cases}$$

Solving this recurrence equation, we obtain:

$$Q_S(n) = \Theta(n^2/(ZL)).$$

The cache complexity analysis of the d-n-c construction of the Pascal's triangle leads to a similar recurrence equation:

$$Q_T(n) = \begin{cases} 2n/L + 2 & n \leq \alpha Z \\ 2Q_T(n/2) + Q_S(n/2) + \Theta(1) & \text{otherwise,} \end{cases}$$

for a positive constant α which can be chosen equal to the previous one. Elementary calculations leads finally to:

$$Q_T(n) = \Theta(n^2/(ZL)). \quad (1)$$

Next we prove that such a cache complexity estimate result is optimal. Since the cache complexity of the Pascal's triangle is dominated by that of the Tableau construction, it is enough to show that $Q_S(n)$ is optimal. To this end, we rely on the lower bound of the I/O complexity on the product of two direct lines established by Hong and Kung [8]. As shown in Figure 7-1 of [8], a Tableau is exactly the product of two direct lines L_2. From Section 8 of [8], we have $Q = |V|/S$, where Q is the minimum I/O time, V is the sequential time and S is the number of red pebbles available. Translating those into the language of cache complexity, we have $|V| = n^2$, $S = Z$ and $Q/L = n^2/(ZL)$ is the minimum number of cache misses. Thus $Q_S(n)$ is cache optimal.

One may wonder how things could be worse. In fact, it is not hard to verify that, for the two-way Tableau a row-by-row construction or a column-by-column construction would lead to a cache complexity of $\Theta(n^2/L + n)$. Since, in practice, Z is quite large (typically a few Megabytes for a L2 cache), the result of Formula (1) is much more attractive.

5. A blocking strategy for computing the Taylor shift in parallel

As noted in Section 4, the d-n-c suffers from a relatively low parallelism. There is, in fact, a hint for increasing the parallelism. One can observe that, in the above triangle division, parts of the small triangle regions II can be evaluated before completing region I.

A first solution implementing this observation is to increase the number of "ways" in the above divide-and-conquer algorithm. However, this leads to very complex code. In addition, this increases the cache complexity,

A better solution is to adopt a *blocking strategy*. This is achieved by partitioning the entire Pascal's triangle into blocks of format $B \times B$, as shown in Subfigure (c) of Figure 1. Of course B should be tuned in order for a block to fit in cache. Then, the blocks are visited (and their elements are computed) following an anti-diagonal traversal.

5.1. *Work, span and parallelism estimates.*

The work for each block is $\Theta(B^2)$. Since we have $(n/B)^2$ blocks, we retrieve the work $T_1(n) \in \Theta(n^2)$. The span for each block is $\Theta(B^2)$. Indeed, each block is executed serially. Since there are n/B parallel steps in the anti-diagonal traversal, the span of the whole algorithm is $\Theta(Bn)$. So the parallelism is $\Theta(n^2/(Bn))$, which is $\Theta(n/B)$. This result is more attractive than the parallelism of the d-n-c approach. However, the d-n-c approach has the advantage of being cache-oblivious, that is, the knowledge of the characteristics of the cache is not required in order to minimize the number of cache misses in practice.

5.2. *Space complexity estimate.*

For each $B \times B$ block, the computation is sequential and is easily carried out in place within $2B$ integer slots. At the k-th parallel step k blocks are being computing while $2n/B - 2k$ ranges of B input integers have not been used yet. Therefore, in total $(2n/B - 2k)B + k\,2B = 2n$ integer slots are sufficient to run the whole algorithm.

5.3. *Cache complexity estimates.*

Let α be the constant introduced in the cache complexity analysis of the d-n-c approach. Assume that $B = \alpha Z$ holds. Then, the number of cache misses for each block is $2B/L + 1$ and the total number of cache misses is

$$Q(n) = \Theta((n/B)^2(2B/L + 1)) = \Theta(n^2/(BL)) = \Theta(n^2/(ZL)).$$

Therefore, provided that $B = \alpha Z$ holds, we retrieve the optimal cache complexity result established for the d-n-c approach.

6. Optimizing the multicore implementation

In this section, in Section 7 and in Section 8, we report on our implementation techniques for parallelization of Taylor shift computations targeting multicores. We show how to implement the d-n-c approach and the blocking strategy such that computing the Taylor shift of a univariate polynomial of degree d runs in-place using space $2Cn$ where $n = d + 1$. As before, C is the number of bytes for encoding a coefficient type (32-bit integer, 64-bit long integer or GMP multi-precision integer of bounded size). Recall that parallel execution can be achieved following either the blocking strategy or the the d-n-c approach, described respectively in Section 4 and Section 5. For both, the number of additions, and thus the work, is the same as that in the sequential construction of a Pascal triangle. However, the blocking strategy is theoretically more attractive in terms of parallelism whereas the d-n-c approach is cache-oblivious.

Given a polynomial of degree d, the number of blocks in the blocking strategy depends on the block size. Similarly, the number of division steps (or recursions) in the d-n-c approach depends on the *base case*. The best size of the base case shall be the one by which the program gives the best timing and speedup ratio, as a successful combination of many aspects, such as sufficient parallelism, low overhead of parallel constructs and function calls, low rate of cache misses, etc. In particular, the proper base case size may vary for polynomials with different coefficient size and on different computer architectures. Therefore, we include the size of the base case in our program as an adjustable (i.e. tunable) parameter. A user can first estimate the size of the base case based on the size of the coefficients of the input polynomial and the targeting computer architecture, and further fine-tune it by trial runs.

In the following two sections we describe our methods for in-place Taylor shift computation and for dynamically scheduling the tasks for concurrent execution according to the degree d of the input polynomial and the parameter B for the base case size. Section 7 covers the blocking strategy in detail while Section 8 offers an overview of the d-n-c strategy.

7. Optimizing the multicore implementation for the blocking strategy

Let $B > 0$ be the block size. For a polynomial f of degree d and $n = d+1 > B$, let q and r be the quotient and the remainder in the integer division of n by B. Thus we have $n = qB + r$, where $q \geq 1, r \geq 0$ and $r < B$. Depending on whether r is null or not, we classify the results of the blocking strategy into two cases: the *regular case* and *irregular case* corresponding respectively to $r = 0$ and $r > 0$. We illustrate them with examples in Figure 2.

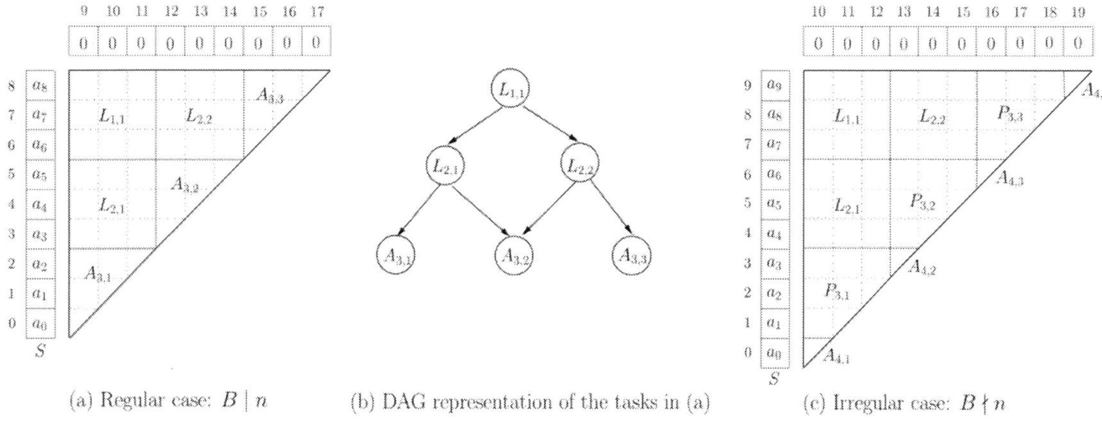

Figure 2. Case illustration for implementing the blocking strategy

7.1. The regular case: $B \mid n$

When $r = 0$, the entire computations in the Pascal triangle (let's name it PTR) are divided into small tableau of dimension $B \times B$ (for the intermediate computations) and small Pascal triangles of leg length B along the hypotenuse of PTR (near the end of the computation). For the example in Subfigure (a) of Figure 2 for which $d = 8$, $B = 3$ and hence $B \mid (n = 9)$, the blocking strategy divides the computations into three tableau of dimension 3×3 including $L_{1,1}$ on the first row and $L_{2,1}$ and $L_{2,2}$ on the second row, and three small Pascal triangles with leg length equal to 3 denoted by $A_{3,1}$, $A_{3,2}$ and $A_{3,3}$ on the last row.

In general we can deduce that there are $q-1$ rows of tableau of dimension $B \times B$ starting from the point of the right angle of PTR with each row parallel to the hypotenuse of PTR and only one row of small Pascal triangles of leg length B along the hypotenuse of PTR. Furthermore, if we index the $q-1$ rows of the small tableau by i where $i = 1, \cdots, q-1$, in the ith row there will be i number of tableaus with each dimension of $B \times B$. The number of small Pascal triangles along the hypotenuse of PTR is q.

Let us call the computation involved in each small tableau or a small Pascal triangle a *task*. The questions at this point are: How to schedule the tasks defined by the blocking strategy for parallel execution? What data structure can be used to minimize the synchronization cost for preserving the data coherency? How to store the intermediate data to obtain good locality?

According to the computational dependencies in the Taylor shift operation as shown in Example 1, the execution order of the tasks defined by the blocking method are deterministic. A task with indices (i, j) can not proceed until the one with indices $(i, j-1)$ and the one with indices $(i-1, j)$ are completed. In general, it can be represented by a Direct Acyclic Graph (DAG). In Figure 2, Subfigure (b) is a DAG representation of the example in Subfigure (a). However, the general DAG scheduling problem has been shown NP complete. In this work we design a parallel scheduling algorithm which is efficient in practice for computing Taylor shift on multicores. The basic idea is that the tasks with indices (i, j) for $j = 1, \cdots, i$ are executed in parallel. When all of them are completed, the tasks with indices $(i+1, j)$ for $j = 1, \cdots, i+1$ can start and run in parallel as well. This scheme is applied to all the tasks with indices (i, j) for $i = 1, \cdots, q$. In this way we exploit the parallelism offered by the many tasks on the same computational step while respecting the computational dependencies between two consecutive steps. The mapping between the tasks and CPUs are handled by the parallel platform Cilk++.

The entire Taylor shift operation is carried out in-place in an array S of size $2n$ with indices from 0 to $2n-1$. Initially the coefficients $a_0, a_1, \ldots, a_d$ of the input polynomial of degree d are copied to the first half space of S. The second half space of S are initialized to 0. An example is shown in Subfigure (a) of Figure 2. The working space of a task in S is scheduled so that it will not interleave with that of other tasks in parallel, to avoid from using locks for concurrent accessing to the same region of S. This is realized by properly arranging the space of S for storing the intermediate and the final results of the tasks in one parallel step, and its final results will be used as the input for some of the tasks in the next step.

More precisely, the working space of a task with indices (i, j), denoted by $S_{i,j}$, is the consecutive space between $S[n + (2j - i - 2)B]$ and $S[n + (2j - i)B]$ inclusively. Before the task starts, $S_{i,j}$ holds the input data (the results from some tasks in the $(i-1)$th step). The first half space of $S_{i,j}$ (with length B) holds the input data as the left side of a tableau or a triangle, while the second half holds the input data as the top side. The data accesses involved in the computation are carefully indexed so that all the computation can be done in-place. Further more, the final results are stored in the same space and are in the right order as the input for some of the tasks in the next step of computation. This way of memory management provides good data locality.

In the regular case, the in-place task processing is achieved by two procedures for the two different types of tasks, tableau and triangle, namely, TableauBaseInPlace(S, n, i, j, B), a sequential procedure to perform the computation for a task with indices (i, j) in the form of a tableau in work space $S_{i,j}$ and TriangleBaseInPlace(S, n, i, j, B), a sequential procedure to perform the computation for a task with indices (i, j) in the form of a Pascal triangle in work space $S_{i,j}$. The pseudo-code snippets for the two procedures are presented in Figure 3.

7.2. The irregular case: $B \nmid n$

An example of a case where $r > 0$ is given in Subfigure (b) of Figure 2. In the irregular case, we obtain also q rows of tasks. The tasks on the rows from 1 to $q-2$ are tableau with dimension of $B \times B$ and the tasks on the qth row (last row) are small Pascal triangles. However, the tasks on the $(q-1)$th row are neither tableau nor triangular. We call these tasks "polygon". It turns out that the scheduling algorithm for the regular case still works for the irregular case. We can also use a work space of size $2n$ to achieve in-place parallel computation. We only need to treat the "polygon" tasks by a special procedure for in-place processing, what we call PolygonBaseInPlace(S, n, i, j, B). Its pseudo-code snippet is shown in Figure 3.

```
TableauBaseInPlace(S, n, i, j, B){
  k = n + (2j − i − 2)B
  for u = B, ··· , 2B − 1 do
    S[k + u]+ = S[k + u − 1]
  for u = B − 2, ··· , 1 do
    for v = 0, ··· , B − 1 do
      S[k + u + v + 1] = S[k + u + v]
                        +S[k + u + v + 2]
  S[k]+ = S[k + 2]
  for u = 1, ··· , B − 1 do
    S[k + u] = S[k + u − 1]
              +S[k + u + 2]
  S[k + B] = S[k + B − 1]
}
```

```
TriangleBaseInPlace(S, n, i, j, B){
  k = n + (2j − i − 2)B
  for u = 0, ··· , B − 1 do
    S[k + B − 1] = S[k + B − 1]
                  +S[k + B + u]
    for v = B − 2, ··· , u do
      S[k + v] = S[k + v] + S[k + v + 1]
}
```

```
PolygonBaseInPlace(S, n, i, j, B){
  k = n + (2j − i − 2)B
  r = n mod B;  t = B − r − 1
  for u = B, ··· , 2B − 1 do
    S[k + u]+ = S[k + u − 1]
  for u = B − 2, ··· , t do
    for v = 0, ··· , B − 1 do
      S[k + u + v + 1] = S[k + u + v]
                        +S[k + u + v + 2]
  for u = t − 1, ··· , 1 do
    S[k + t + 1]+ = S[k + u]
    for v = t + 2, ··· , B + u do
      S[k + v]+ = S[k + v − 1]
  S[k]+ = S[k + t + 1]
  for u = 1, ··· , r do
    S[k + u] = S[k + u − 1]
              +S[k + u + t + 1]
}
```

Figure 3. Key sub-procedures for implementing the blocking strategy

8. Optimizing the multicore implementation for the d-n-c strategy

In section 4, we have described briefly the d-n-c algorithm. Here we provide some implementation details.

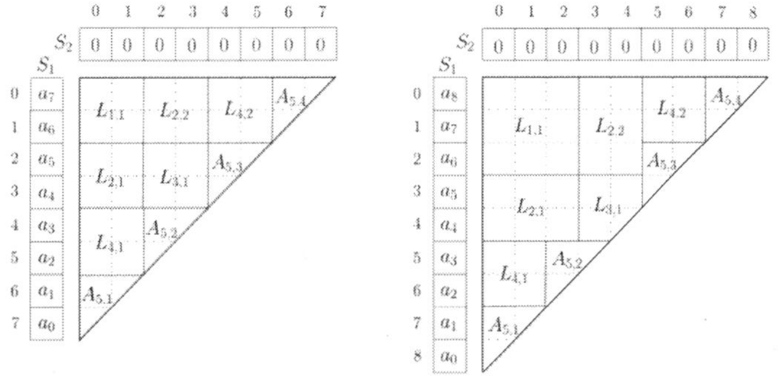

(a) Regular case: n is a power of 2 (b) Irregular case: n is not a power of 2

Figure 4. Case illustration for implementing the d-n-c strategy

When n is a power of 2, as shown in (a) of Figure 4, each block in the divided Pascal's triangle is either a square tableau or a triangle. We call this case the regular case. When n is not a power of 2, as shown in (b) of Figure 4, a tableau may not be square. Indeed, the lengths of two edges of a tableau may differ by one. This irregular case suggests us to implement the divide-and-conquer approach using two arrays of n elements rather than one array with $2n$ elements as for the blocking strategy.

For efficiency reason, there is usually a sequential base case for a parallel divide-and-conquer algorithm. In our implementation of d-n-c algorithm, there are two types of base cases: tableau base case and triangle base case. We provide here the in-place implementation of the tableau base case. The implementation of the triangle case is similar.

Journal of Physics: Conference Series **341** (2012) 012026 doi:10.1088/1742-6596/341/1/012026

Algorithm 3: TableauBase(p, s, q, t)

Input: a base block whose left edge is an array p of length s and whose top edge is an array q of length t

Output: the right edge of the block is stored in p and the bottom edge of the block is stored in q

```
begin
    for i = 0 to t − 1 do
        p[0] ← p[0] + q[i];
        for j = 1 to s − 1 do
            p[j] ← p[j] + p[j − 1];
        q[i] ← p[s − 1];
end
```

9. Experimentation

We have implemented in `Cilk++` [9] the algorithms presented in Sections 4 and 5 together with the corresponding parallel version of the VCA Algorithm for real root isolation. For integer arithmetic, we rely on the `GMP` library [1].

We have tested our software on well-known benchmark examples, including the Bnd and Cnd polynomials from [11], the Chebychev and Mignotte polynomials from [5], as well as randomly generated polynomials. All the benchmark runs are carried out on an 8-core machine with 8 GB memory and 6 MB of L2 cache. Each processor is an Intel Xeon X5460 @3.16 GHz.

The timing results for a set of test cases with different configurations are summarized in Tables 1, 2 and 3. In each case, $n-1$ and k denote the degree and the coefficient size (number of bits) of the input polynomial whereas B denotes both the *base case size* of d-n-c and the *block size* of blocking. The values of B reported herein are the ones which give the best timing after tuning.

The benchmark results show that our program reaches linear speedup on an 8-core machine for sufficiently large problems. We have also tested several other data traversal approaches for computing Taylor shifts and we have observed that our d-n-c and blocking outperform all of them. Overall, blocking demonstrates a slight win over d-n-c except for the case of Mignotte polynomials.

Table 1. Timings of ParTaylorShift (in seconds).

n	k	B	method	Bnd polynomial [11]			Cnd polynomial [11]			Random polynomial		
				8-core	1-core	Sp	8-core	1-core	Sp	8-core	1-core	Sp
5000	5000	50	blocking	1.33	6.49	4.9	0.92	2.26	2.5	1.33	6.50	4.9
5000	5000	8	d-n-c	1.45	6.63	4.6	0.94	2.33	2.5	1.45	6.63	4.6
10000	10000	50	blocking	7.70	50.76	6.6	4.40	17.46	4.0	7.79	50.78	6.5
10000	10000	8	d-n-c	8.54	51.66	6.0	4.22	17.64	4.2	8.51	51.65	6.1
25000	25000	50	blocking	103.9	778.9	7.5	42.6	261.2	6.1	103.9	778.7	7.5
25000	25000	8	d-n-c	110.1	789.6	7.2	41.5	261.7	6.3	110.2	789.7	7.2

In addition, we have used our software to isolate the real roots of a large polynomial coming from the study of limit cycles of dynamical systems on a 32-core machine. This polynomial has degree 426, and the bit size of its coefficients is 1900. It is available at `http://www.orcca.on.ca/~cchen/ammcs2011.txt`. The 32-core machine has 8 Quad Core AMD

Opteron 8354 @2.2 GHz connected by 8 sockets; each core has 64 KB L1 data cache and 512 KB L2 cache; every four cores share 2 MB of L3 cache; the total memory is 126 GB. In about 15 minutes, our software manages to isolate all the 78 real roots of this large polynomial using this 32-core machine. However, we could not process this polynomial on a standard desktop (Intel Core 2 Quad CPU @2.40 GHz and 3.0 GB memory).

Table 2. Timings of Parris for Chebychev and Mignotte polynomials (in seconds).

n	B	method	Chebychev polynomial [5]				Mignotte polynomial [5]			
			8-core	4-core	1-core	Sp	8-core	4-core	1-core	Sp
400	50	blocking	59.09	106.41	413.87	7.0	166.69	222.65	564.91	3.4
400	8	d-n-c	59.02	107.55	420.18	7.1	128.57	181.95	572.65	4.5
500	50	blocking	173.83	323.18	1269.61	7.3	not enough			
500	8	d-n-c	173.92	324.48	1279.05	7.4	memory			

Table 3. Timings of Parris for random polynomials (in seconds).

n	k	d-n-c				blocking			
		B	8-core	1-core	Speedup	B	8-core	1-core	Speedup
1000	1000	8	0.94	3.26	3.5	50	0.87	3.21	3.7
2000	2000	8	3.51	18.84	5.4	50	3.25	18.58	5.7
3000	3000	8	4.07	23.33	5.7	50	3.82	22.89	6.0
4000	4000	8	38.26	246.34	6.4	50	35.84	243.82	6.8
5000	5000	8	200.62	1372.70	6.8	50	184.07	1340.95	7.3

10. Concluding remarks

In this paper, we revisit the core routine of many univariate real root isolation algorithms, namely the Taylor shift operation. We discuss two strategies (divide-and-conquer and blocking) to implement this operation in parallel on multicores. Assuming that the cost of adding two integers is constant, we show that both strategies are cache optimal. For the second approach, however, this requires to choose the block size accurately.

On an 8-core machine, for benchmark examples of relatively large size, the speedup for both the Taylor shift computation and the real root isolation is nearly linear. We also demonstrate the effectiveness of our software by processing a large real-world polynomial on a 32-core machine with more than a hundred of gigabytes memory.

The assumption that the cost of adding two integers is constant is considerably strong for the Taylor shift operation. Consider, indeed, a univariate polynomial $p(x)$ of degree d, with integer coefficients. Assume that the *height* of $p(x)$ (that is, the bit size of the maximum absolute value of a coefficients in $p(x)$) is bounded by h. Then, the height of $p(x+1)$ is bounded by $h+d$ [14]. Therefore, to further improve the work presented here, we are preparing a new report where our cache complexity analysis of the Taylor shift operation takes into account this potential increase of height. Since two blocks of the same size in the Pascal's triangle can be quite different in terms of the work required to fill them, we would like to design a new blocking strategy with *dynamical granularity* aiming at balancing the work load among the blocks.

Journal of Physics: Conference Series **341** (2012) 012026 doi:10.1088/1742-6596/341/1/012026

References

[1] The GNU Multiple Precision Arithmetic Library. `http://gmplib.org`.
[2] C. Chen, M. Moreno Maza, and Y. Xie. Cache complexity and multicore implementation for univariate real root isolation. *ACM Commun. Comput. Algebra*, 44:97–98, January 2011.
[3] G. E. Collins, J. R. Johnson, and W. Küchlin. Parallel real root isolation using the coefficient sign variation method. *Lecture Notes in Computer Science*, (584):71–87, 1992.
[4] George E. Collins and Alkiviadis G. Akritas. Polynomial real root isolation using Descartes'rule of signs. In *proceedings of ISSAC'76*, pages 272–275, Yorktown Heights NY, 1976.
[5] T. Decker and W. Krandick. Parallel real root isolation using the descartes method. In *HIPC*, pages 261–268, 1999.
[6] M. Frigo, C. E. Leiserson, H. Prokop, and S. Ramachandran. Cache-oblivious algorithms. In *FOCS*, 1999.
[7] M. Frigo, C. E. Leiserson, and K. H. Randall. The implementation of the cilk-5 multithreaded language. In *ACM SIGPLAN*, 1998.
[8] Jia-Wei Hong and H. T. Kung. I/O complexity: the red-blue pebbling game. In *STOC*, pages 326–333, 1981.
[9] Intel Corporation. Cilk++. `http://www.cilk.com`.
[10] J. R. Johnson, W. Krandick, K. Lynch, D. G. Richardson, and A. D. Ruslanov. High-performance implementations of the descartes method. In *ISSAC '06*, pages 154–161, New York, NY, USA, 2006. ACM.
[11] J. R. Johnson, W. Krandick, and A. D. Ruslanov. Architecture-aware classical taylor shift by 1. In *ISSAC' 05*, pages 200–207, 2005.
[12] Maurice Mignotte. *Mathematics for computer algebra*. Springer, 1992.
[13] A. Strzebonski and E. Tsigaridas. Univariate real root isolation in an extension field. In *Proc. of ISSAC'11*, pages 321–328, New York, NY, USA, 2011. ACM.
[14] J. Von zur Gathen and J. Gerhard. Fast algorithms for taylor shifts and certain difference equations. In *ISSAC*, pages 40–47, 1997.

Journal of Physics: Conference Series **341** (2012) 012027 doi:10.1088/1742-6596/341/1/012027

High-throughput protein crystallization on the World Community Grid and the GPU

Yulia Kotseruba[1], **Christian A Cumbaa**[1] **and Igor Jurisica**[1,2]

[1]IBM Life Sciences Discovery Centre, Ontario Cancer Institute, University Health Network, 101 College Street, Toronto, Ontario M5G 1L7, Canada

[2]Department of Computer Science, University of Toronto, Toronto, Ontario, Canada

E-mail: `juris@ai.utoronto.ca`

Abstract.
We have developed CPU and GPU versions of an automated image analysis and classification system for protein crystallization trial images from the Hauptman Woodward Institute's High-Throughput Screening lab. The analysis step computes 12,375 numerical features per image. Using these features, we have trained a classifier that distinguishes 11 different crystallization outcomes, recognizing 80% of all crystals, 94% of clear drops, 94% of precipitates. The computing requirements for this analysis system are large. The complete HWI archive of 120 million images is being processed by the donated CPU cycles on World Community Grid, with a GPU phase launching in early 2012. The main computational burden of the analysis is the measure of textural (GLCM) features within the image at multiple neighbourhoods, distances, and at multiple greyscale intensity resolutions. CPU runtime averages 4,092 seconds (single threaded) on an Intel Xeon, but only 65 seconds on an NVIDIA Tesla C2050. We report on the process of adapting the C++ code to OpenCL, optimized for multiple platforms.

1. Introduction

X-ray crystallography is one of two principal methods of determining the 3-dimensional atomic structure of proteins. Structural genomics, the industrial-scale effort to solve protein structure, is a complex, expensive, and failure-prone process with several bottlenecks limiting the success rate. Protein crystallization is one. Proteins must be crystallized before X-ray diffraction can take place. The chemical conditions for successful protein crystallization are specific to each protein, and difficult to find. Crystallographers must search through chemical space for a *cocktail* (unique combination of buffers, precipitants, and salts) that will crystallize the target protein.

Robotic systems at the Hauptman-Woodward Institute (HWI) screen 1,536 cocktails in parallel against each protein. Each protein-cocktail trial mixes 200 nL of protein solution with 200 nL of cocktail in a conical plastic well. The resulting 400 nL droplet is photographed at multiple time-points, requiring visual assessment by a human expert to determine the outcome of the trial. The High-Throughput Screening lab at HWI[1] maintains and expands an archive of 120 million crystallization-trial images: 13,000 proteins × 1,536 cocktails per protein × multiple time-points. Fewer than 1% of these images have been scored by a human expert.

Our goals are to develop an automatic image-scoring system that replaces or substantially assists human experts, and to process the complete HWI image archive. A complete library of

[1] `http://www.hwi.buffalo.edu/faculty_research/crystallization.html`

scored crystallization results will accelerate the search phase in protein crystallization, and will improve throughput, consistency, and objectivity of the results. It will also enable the statistical optimization of the crystallization process, the rational design of crystallization screens, and the discovery of protein-crystallization principles and optimization strategies by mining a database of 20 million crystallization trials.

Section 2 describes our image analysis and classification system, its deployment on the World Community Grid to process the complete HWI archive, and some preliminary results on modeling protein crystallization. Section 3 describes our efforts to port our image analysis system to the GPU.

2. Automated analysis of protein crystallization-trial images

To interpret our images, we first transform each image from raw pixels into a set of numeric features measuring aspects of the image's texture and geometry. A separate classification stage interprets the image based on the features.

We compute 12,375 features per image in a complex, multi-layered, and CPU-intensive image-analysis step. An additional 2,533 features are derived from the primary 12,375, augmenting the feature-set and creating a final set of 14,908 features. All analysis concentrates on a 200-pixel-diameter region of interest centered on the circular droplet in the image. Several categories of image features are measured: basic image statistics, foreground-background topological measures, microcrystal-recognition filters, features measuring local grey-level intensity changes, statistics on the textural and geometric properties of foreground objects revealed by multiple filters (Sobel, Laplacian, and pseudo-Radon transforms), and Haralick grey-level co-occurrence matrix (GLCM) features [1]. Details of the feature computation are described in more detail in previous work [2].

To train and test our image-scoring system, we rely on two large databases of images hand-scored by crystallographers at HWI [3; 4]. In both sets, experts scored each image for the presence or absence of seven crystallization outcomes: *clear droplet*, *phase separation*, *precipitate*, *skin*, *crystal*, *junk*, and *unsure*. These outcomes can co-occur in the same image, with the exception of *clear*, which is exclusive. The first dataset contains 147,456 images, representing 1,536 trials of 96 different proteins, each photographed at a single time-point. Each image from this set was scored independently by at least three experts. The second database contains 17,895 images with the *crystal* outcome, selected from several hundred proteins. Images from this set were scored by a single expert. The two datasets were combined, and a random 10% reserved as a validation set.

2.1. *Image analysis on the World Community Grid*

The computing requirements for this analysis system are large. The main computational burden of the analysis is the set of GLCM features. GLCM features are computed for multiple neighbourhoods within each image, for multiple direction vectors and distances, and at multiple greyscale intensity resolutions. A total of 13 features $\times$ 68 million GLCMs are computed per image, requiring 4,091 seconds of CPU time on average on a modern PC workstation.

This expensive image analysis is largely being computed on the World Community Grid, a global philanthropic, scientific computing platform powered by the donated CPU-time of its members. Our *Help Conquer Cancer* project launched on the Grid in November 2007. Members have contributed over 75 CPU-millenia to the project as of July 5, 2011, at a rate of 55 CPU-years per day.

2.2. *Scoring the crystallization trials*

Using our training set of expert-scored images, and features computed from the World Community Grid, we have built a Random-Forest [5] image classifier that distinguishes 11 crystallization outcomes: *clear*, *precipitate*, *crystal*, *phase*, *precipitate & crystal*, *precipitate & skin*, *phase separation & crystal*, *phase separation & precipitate*, *skin*, *junk*, and *other*. Our classifier recognizes 80% of crystal-bearing images (aggregated), 94% of clears, and 94% of precipitates. Table 1 shows the accuracy of the classifier as measured against the independent validation set.

Table 1. Accuracy of the 11-way classifier.

outcome	precision	recall
clear	0.937	0.937
precipitate	0.94	0.717
crystal	0.637	0.622
phase separation	0.653	0.79
precip & crystal	0.422	0.568
precip & skin	0.511	0.796
phase & crystal	0.38	0.348
phase & precip	0.5	0.022
skin	0.545	0.58
junk	0.309	0.769
other	0.269	0.369

2.3. Modeling protein crystallization across time and chemical-space

We are developing a model of protein crystallization that incorporates the changing crystallization state of a single trial over time, and the network of probabilistic relationships linking individual crystallization trials to their neighbours in cocktail space and protein space. In a single crystallization trial, the crystallization state of a trial at time t is related probabilistically to the states at times $t-1$ and $t+1$. Thus, all images in a time series are best interpreted in tandem, effectively embedding the probabilistic outputs of the image classifier in a hidden Markov model. Depending on the starting conditions of the crystallization trial (e.g., supersaturated protein solution), several paths through the various crystallization outcomes are possible. These starting conditions then imply a mixture of HMMs.

Further, two crystallization trials of the same protein with similar cocktails, or similar proteins with the same cocktail, may have similar crystallization reactions. Thus, the outcome of one crystallization trial is probabilistically related to the outcomes of its neighbours in protein- and cocktail-space. Altogether, these layers of probabilistic interconnection form the probabilistic graphical model illustrated in Figure 1.

3. Image analysis on the GPU

Given recent successes elsewhere in applying GPU to multiple problems in image/video processing, bioinformatics, and medical imaging, etc., with reported run-time improvement factors of 10–1000×, we have ported our image analysis system to the GPU. We focussed our efforts on optimizing the GLCM features, which consume 98% of the run-time on the CPU version. We chose the OpenCL 1.0 language due to its support for multiple vendors and device types. We also implemented a CUDA version of our algorithm; however, with the NVIDIA drivers used (270.41.19), we observed no performance gain with CUDA on NVIDIA hardware, and thus kept only the OpenCL version.

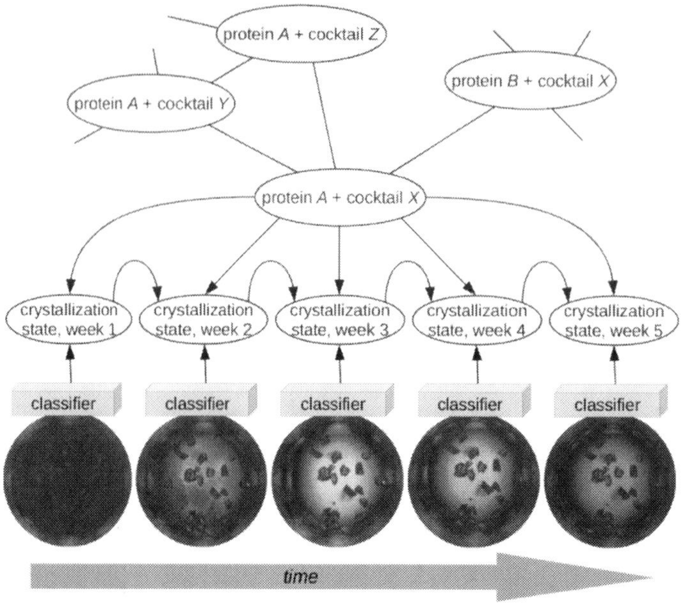

Figure 1. Scheme for a probabilistic model of protein crystallization.

3.1. Implementation Details

The first step in extracting GLCM features from the image is computing the GLCM itself. Every element (i, j) in this 2D histogram represents the co-occurrence frequency of grey levels i and j of pixel-pairs separated by a given (x, y) displacement vector. The width of the histogram is defined by greyscale-quantization level (64, 32, or 16 greys). The circular region of interest of each image contains 22,305 pixels. For each of these pixels, the GLCM is computed in a local window for 1,026 distance/direction combinations and 3 quantization levels, generating over 68 million histograms in total. Each matrix is then used to compute an array of 13 textural features. To reduce the amount of data retained, feature values are aggregated by displacement-vector magnitude for every quantization level. Then for every feature, the maximum range, minimum range, and mean value are stored.

Input image dimensions are 256×256 pixels, so we organize threads in a 256×256 grid with 64–256 threads per block, depending on the hardware. Each kernel computes all GLCMs for a single distance/direction pair, plus all associated features, resulting in 1,026 kernel calls in total. The kernel program is organized as follows: (1) construct the initial 64×64 GLCM from image data in global memory and reduce number of threads to 64; (2) compute the first set of Haralick features; (3) reduce the GLCM to 32×32 and reduce the number of threads to 32; (4) compute second set of features; (5) reduce the GLCM to 16×16 and reduce number of threads to 16; (6) compute the third set of features; (7) write results back to global memory.

This approach minimizes global memory usage: in each kernel execution, the image is read once from global memory for histogram construction, and in the end all features are written into the output array. It also reduces the amount of memory transferred between GPU and CPU: the input, composed of the image itself and several parameters, 65 kB total, and the output arrays, under 3 MB. As a result, the total memory transfer time accounts for less than 2% of the run-time.

3.2. Optimizations

The algorithm depends on statistics computed by summing across all rows or columns of the GLCM. Most GLCMs are sparse, with 10% or fewer non-zero values, usually tightly grouped along the main diagonal. It is not unusual also for a matrix to have a single non-zero value. We used sparseness and the symmetric property of GLCMs to limit iterations to non-empty rows and to simplify computation of some of the features.

Besides the GLCM, most Haralick features depend on one another, and on additional statistics. We cached these values to avoid expensive re-computations.

We limited the use of expensive operations such as `log()`, `exp()` and division. In places where avoiding `log` and `exp` was not possible, we used reduced-precision, hardware optimized functions `native_log()` and `native_exp()` instead. Division is not supported on GPUs directly and is normally implemented as reciprocal and multiplication. We were able to eliminate almost all division operations by precomputing reciprocals on CPU. These arithmetic optimizations alone made our algorithm run 13% faster. Conditional statements were replaced by `select()` function calls. Loops were unrolled where possible. Where applicable, we provided additional compiler options. The NVIDIA compiler tends to allocate more registers than needed, so we used `-cl-nv-maxrregcount=20` to limit register use per thread, and `-cl-mad-enable` for reduced-precision multiply-add operations, helping to speed up index calculations.

3.3. Problems

3.3.1. OpenCL implementations and hardware differences. We have dealt with NVIDIA and ATI implementations for Linux and Mac OS. Theoretically, a program written to the OpenCL API specifications will run correctly on all supporting devices. Obviously, this does not guarantee optimal performance due to the differences in hardware architectures. We intend to run our application on a variety of devices on all major platforms, so we wrote separate versions of the kernel code for CPU, NVIDIA and AMD cards. Note that for Intel CPUs on Mac OS 10.6, OpenCL always runs in single-threaded mode, and the ATI Stream SDK allows for up to 1,024 threads per block in a 1,024 $\times$ 1,024 $\times$ 1,024 grid on a CPU. We noticed that our application does not run efficiently on video cards with no dedicated memory, including almost all mobile GPUs. In some cases, the GPU version runs slower than CPU code. Further, some operating systems impose kernel run-time limits, usually 5–10 seconds. If a kernel exceeds this time, it is killed by the driver. To prevent this from happening, we run the OpenCL kernel on the CPU, which yields over 4-fold performance improvement compared to the original C++ program.

We encountered inconsistencies between NVIDIA and ATI implementations of OpenCL. For instance, memory barrier functions are treated differently. On ATI, every thread must hit the barrier before the kernel returns, whereas on NVIDIA every *active* thread must hit the barrier. So having a thread return before the `barrier()` on ATI will stop the execution indefinitely, while on NVIDIA it runs without any problems. Other minor differences between NVIDIA and ATI compilers changed across several driver updates.

3.3.2. Shared Memory Usage. One of the bottlenecks of this implementation is high usage of shared memory: almost 10 kB are required for each block to store GLCM arrays (8 kB), features, and intermediate results (2 kB). Most of the consumer cards and mobile GPUs have 16 kB of shared memory available. Some of the recent ATI mid-range cards have up to 32 kB and only top computing and gaming GPUs have 48 kB. As a result we could run at most 1–4 blocks of threads, while top cards usually allow for up to 8 blocks running concurrently. This essentially means that we were using only about half of the resources available (the NVIDIA profiler showed a maximum load of 66%). The only way to reduce the shared memory footprint was to compute GLCMs and features in two separate kernels, with the first kernel still requiring 8 kB, and the second much less. Since the GLCM computation consumes 50% of the kernel time,

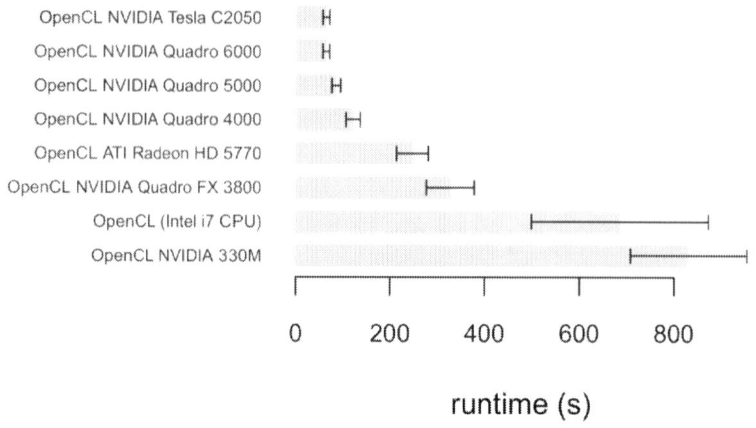

Figure 2. Image analysis run-times for OpenCL on NVIDIA (GPU), ATI (GPU), and Intel (CPU) platforms. The original, optimized C++ run-time is 4,092 seconds (not shown).

some performance gains were expected. In practice, no gains were realized, due to the costs of CPU↔GPU matrix-data transfers between kernel calls, and frequent access to the GLCMs in global memory during feature computation. Additionally, the GLCMs occupied 180 MB of space (22,305 pixels × 64 × 64 × 2 bytes), and most cards had a limit of 128 MB for a single global memory allocation.

3.4. Evaluation

Our main goal in this project was to improve performance without sacrificing precision. To evaluate our OpenCL implementation, we used as a standard the existing optimized C++ code, as runs on the World Community Grid. We ran our OpenCL implementation on several NVIDIA cards ranging from mobile devices to the newest scientific computing cards.

3.4.1. Performance Improvements. We were able to achieve a significant performance improvement for this problem. Figure 2 shows the time required to compute the Haralick features, averaged over 50 images randomly selected from our test set. Our final version of OpenCL code runs 6–60 times faster than the original optimized C++ code. It took over 3 days of CPU time to process our small 50 image test set, and under 2 hours on the NVIDIA Tesla C2050 GPU.

Kernel time is almost equally split between computing histograms and features. This ratio differs from the CPU version, which spends about 70% of time computing features. More detailed breakdown of GPU runtime is given in Figure 3.

We also measured the GFLOPS for our GPU implementation. On the Tesla C2050 we achieved 5.5 GFLOPS, which is only 1% of its maximum performance of 515 GFLOPS. Such low efficiency could be caused by several factors. None of the integer operations contributed to GFLOPS, so array-indexing and histogram-building was not taken into account. Additionally, the Haralick features require very complex memory access patterns and there were insufficient arithmetic operations to compensate for memory latencies and bank conflicts. Finally, as

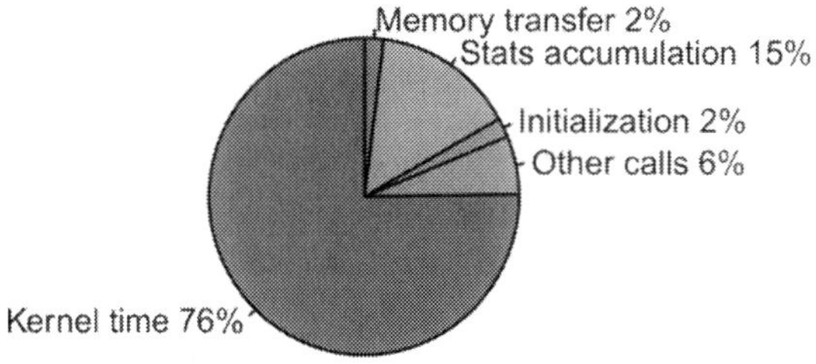

Figure 3. Breakdown of GPU run-time.

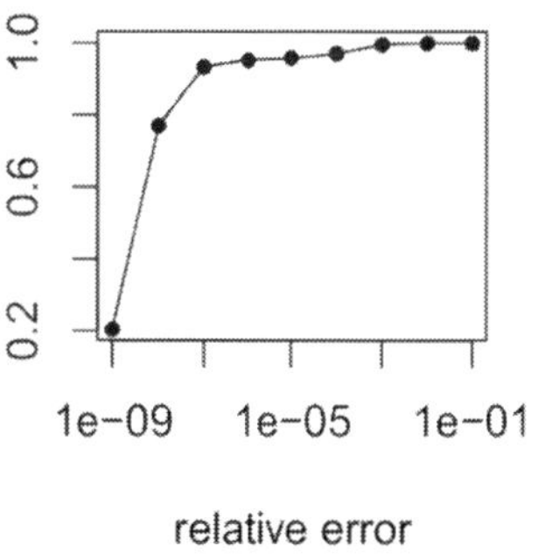

Figure 4. Cumulative distribution of feature values exceeding the relative error threshold.

described above, we were unable to saturate the GPU pipeline due to the high shared memory usage and gradual reduction of number of active threads.

3.4.2. Correctness. There is always a trade-off between precision and performance in GPU. Most current graphics hardware is not fully compliant with IEEE 754 standard for single-precision floating point operations. In particular, NVIDIA video cards flush denormals to zero and do not guarantee full precision for division and square root operations [6]. Since loss of precision is inherent in GPU computation we needed to measure its effects on computed results. We first computed relative error of OpenCL versus CPU results, defined as $\left|\frac{GPU-CPU}{CPU}\right|$. In the $CPU = 0$ case, we took an absolute error. It follows from the cumulative distribution plot on Figure 4 that almost 97% of values are correct up to four significant digits. More important is any possible effect on image classification. To verify that any shortcuts we took in the OpenCL version would not affect our classification, we tested the final code on a 4,500-image subset of our image-classifier's training/testing set. We then applied our classifier to both GPU- and CPU-computed image features and found only eight images that were assigned different classes. All eight of these images belonged to an ambiguous category (human experts disagreed on the image's true class), suggesting that discrepancies between GPU and CPU features pose little problem.

4. Conclusion

We have built and tested GPU-accelerated version of optimized C++ code. Given the complexity of the problem, we still achieved a 6–62-fold performance increase, depending on the hardware and OS. Despite some loss of precision, features computed on the GPU still give classification results consistent with the gold-standard CPU version. The GPU version therefore runs fast enough to allow same-day analysis of crystallization-trial images, and correctly enough to run alongside the CPU version on the World Community Grid. We expect GPU-enabled *Help Conquer Cancer* to enter beta testing by the end of 2011.

Acknowledgments

We wish to thank and acknowledge IBM and the World Community Grid volunteers for providing free computational power and technical support for this research through World Community Grid. For further information about World Community Grid, please view the following: `https://secure.worldcommunitygrid.org/index.jsp`. This work would not be possible without data generated at Hauptman-Woodward Medical Research Institute and insight from many discussions during our collaboration with Drs. George T. DeTitta, Edward H. Snell, Joseph R. Luft and their team, Geoff Franks, Thomas Grant, Dr. Stacey M. Gulde, Mary Koszelak-Rosenblum, Angela M. Lauricella, Dr. Michael G. Malkowski, Raymond Nagel, Stephen A. Potter, Meriem I. Said, Jennifer L. Smith, Max Thayer, and Christina K. Veatch. Dr. Jerry Heyman and Miso Cilimdzic from IBM assisted with testing, profiling, and tuning the CPU version of our image analysis application. Roy Bryant, Adin Scannell, and Olga Irzak at the University of Toronto developed an early CUDA prototype. We wish to thank Lenovo and NVIDIA for providing PC and GPU hardware for development and testing, and Richard Lu for assisting with computing resources and environment for post-Grid data processing. The authors gratefully acknowledge funding by NIH U54 GM074899, the Canada Foundation for Innovation (Grants #12301 and #203383), the Canada Research Chair Program, the Natural Science & Engineering Research Council of Canada (Grant #104105), the Ontario Research Fund (GL2-01-030), and IBM.

References

[1] Haralick R M, Shanmugan K and Dinstein I 1973 *IEEE Trans. Syst. Man Cybern.* SMC **3** 610

[2] Cumbaa C A and Jurisica I 2010 *J. Struct. Funct. Genomics* **11** 61–69

[3] Snell E H, Luft J R, Potter S A, Lauricella A M, Gulde S M, Malkowski M G, Koszelak-Rosenblum M, Said M I, Smith J L, Veatch C K, Collins R J, Franks G, Thayer M, Cumbaa C A, Jurisica I and Detitta G T 2008 *Acta Crystallogr.* D **64** 1123

[4] Snell E H, Lauricella A M, Potter S A, Luft J R, Gulde S M, Collins R J, Franks G, Malkowski M G, Cumbaa C A, Jurisica I and DeTitta G T 2008 *Acta Crystallogr.* D **64** 1131

[5] Breiman L 2001 *Mach. Learn.* **45**(1) 5–32

[6] Whitehead N and Fit-Florea A 2011 Precision & performance: Floating point and IEEE 754 compliance for NVIDIA GPUs Tech. rep. NVIDIA Corporation URL `http://developer.download.nvidia.com/assets/cuda/files/NVIDIA-CUDA-Floating-Point.pdf`

Monte Carlo dose calculation using a cell processor based PlayStation 3 system

James C L Chow[1,2], Phil Lam[1] and David A Jaffray[1,3]

[1] Department of Radiation Oncology, University of Toronto and Radiation Medicine Program, Princess Margaret Hospital, University Health Network, Toronto, Ontario M5G 2M9, Canada; [2] Department of Physics, Ryerson University, Toronto, Ontario, Canada; [3] Department of Medical Biophysics, University of Toronto, Ontario, Canada.

james.chow@rmp.uhn.on.ca

Abstract. This study investigates the performance of the EGSnrc computer code coupled with a Cell-based hardware in Monte Carlo simulation of radiation dose in radiotherapy. Performance evaluations of two processor-intensive functions namely, HOWNEAR and RANMAR_GET in the EGSnrc code were carried out basing on the 20-80 rule (Pareto principle). The execution speeds of the two functions were measured by the profiler *gprof* specifying the number of executions and total time spent on the functions. A testing architecture designed for Cell processor was implemented in the evaluation using a PlayStation3 (PS3) system. The evaluation results show that the algorithms examined are readily parallelizable on the Cell platform, provided that an architectural change of the EGSnrc was made. However, as the EGSnrc performance was limited by the PowerPC Processing Element in the PS3, PC coupled with graphics processing units or GPCPU may provide a more viable avenue for acceleration.

1. Introduction

To date, scientific computing is conducted by general-purpose computer processor (CPU) because of its relatively low cost. However, with the recent rapid development of non-traditional computer processor architectures driven mainly by the needs of the computer graphics and gaming industries, these architectures become popular. An example is the Cell processor used in the video game graphics and scientific computing.[1-4] The Cell Processor was designed to be a hybrid of conventional processors (*e.g.* Opteron, Core 2) and specialized processors such as graphics processing units (GPUs). Jointly developed by Sony Computer Entertainment, Toshiba and IBM (known collectively as STI) development started in 2001 and spanned 4 years, at a total cost of approximately $400 million. The first major commercial application of the Cell was the Playstation3 (PS3) game console. The low cost of the console has made it an interesting choice for scientific computing on a budget. A notable example is Gaurav Khanna's "PS3 Gravity Grid," (http://gravity.phy.umassd.edu/ps3.html) which is used to solve problems in astrophysics.

The EGSnrc Monte Carlo code[5,6] is used commonly to provide accurate dosimetric calculation of kV-MV photons and electrons for inhomogeneous medium in radiotherapy.[7-10] Since energy deposition or dose from each photon or electron particle is by definition statistically independent as well as governed by the same process,[11] such Monte Carlo simulations can be classified as "embarrassingly parallel" problems when implemented. The parallelizability of EGSnrc, coupled with the low cost of Cell-based hardware and the success of other efforts in similar fields suggested that an investigation into the possibility of porting EGSnrc to a Cell-based platform would be worthwhile. Therefore, this study is aimed at (1) to determine whether EGSnrc is a suitable candidate for parallelization on the Cell platform; and (2) to recommend software architecture most suited to accomplish and implement.

2. Methods

2.1 Tools

2.1.1 PS3

In the PS3, the Cell processor has 9 cores, 8 of which are homogeneous as shown in Figs. 1(a) and 1(b). The main processor is known as the PowerPC Processing Element, or PPE. The remaining 8 processors are known as Synergistic Processing Elements (SPEs). Each core is completely independent, containing a processing unit, local cache and DMA module to access the main memory, which is shared.[12] This is possible through the Element Interface Bus (EIB), which provides 204.8GB/s of bandwidth. The Cell processor in the PS3 contains one SPE that is disabled for yield purposes, and reserves another for OS use, leaving a total of 6 SPEs for program use.

The instruction set is a superset of the IBM PowerPC set. In fact, the PPE is binary-compatible with the PowerPC 970 architecture. This is a boon to developers, as it will allow any application compiled for this architecture to run natively on the Cell. It also allows for applications to be developed in steps. For instance, an algorithm can be written in the conventional fashion, then tested and benchmarked on the Cell processor without having to deal with the complexity of parallelization. Once the algorithm has been verified, it can be converted into a parallel algorithm taking advantage of the SPEs and other features of the Cell not present in the PPC970 architecture. This separates algorithm design, verification and acceleration into separate stages, simplifying the development process.

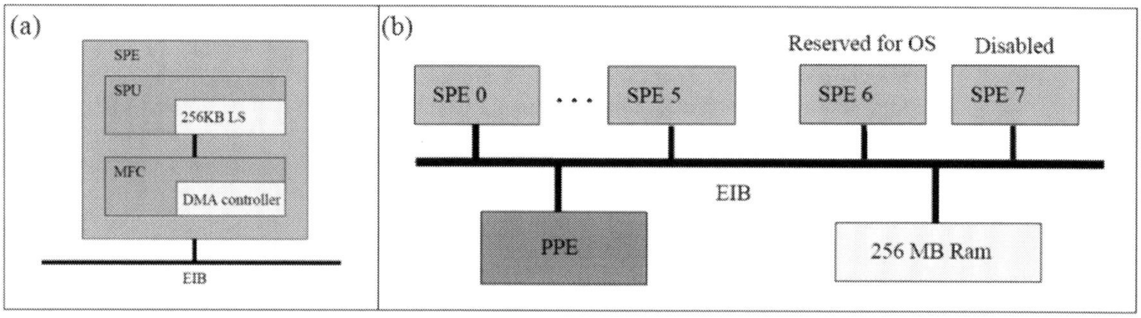

Figure 1. Block diagrams showing the architecture of the Cell processor: (a) details of the SPE, and (b) SPEs, PPE, EIB, DMA, local cache and main memory.

2.1.2 Linux system and compilers

In this study, Linux operation system called "Yellow Dog Linux", or YDL for short developed by Terra Soft Solutions, was used as the programming platform. It is based on CentOS, and is designed to support the PS3/Cell. YDL features kernel support for all of the hardware available on the Cell, including PPE and SPEs.

There are two sets of compilers available for the Cell processor running Linux. IBM distributes its XL series of compilers for Fortran and C. There are also Open Source GNU compilers available for Fortran and C/C++. Both of these produce separate executables for the PPE and SPEs, combining them into a single program image for execution. Traditional optimization technique is used, though the IBM XL compilers generally produce speedier executables.

2.2 Performance evaluation

Performance evaluation was carried out on small portions of the EGSnrc code. These portions were chosen following the informal 80-20 rule (Pareto principle) of large systems.[13] The rule states that a typical program spends 80% of its time executing 20% of the code. The reasoning was that the architecture code surrounding the essential algorithms could be adapted regardless, and it was only the core of the program that needed to be evaluated. Two processor-intensive functions, as shown in Fig. 2, named HOWNEAR and RANMAR_GET were selected in the performance test. These functions collectively account for over 20% of execution time, while comprising less than 1% of the total number of lines of code. However, it should be noted that such evaluation only tests if the whole EGSnrc code is worthwhile to parallelize. It is because a speed-up of 1.2 based on the Amdahl's law shows that only parallelizing these two functions is inadequate.[14] The law stated that the speed-up can be achieved by distributing a computer program over p processors cannot exceed p / $[1 + f \cdot (p - 1)]$, where f is the fraction of the work of the program that must be done in serial mode. The execution speed of these functions was measured in situ during execution of the original program using the profiler *gprof*, which specified the number of executions and the total time spent in the function, allowing for an estimate. The functions were then re-implemented in a testing architecture designed using the ideas discussed above. This proof-of-concept provided identical functionality as the original, serial functions, but did so in a parallel, accelerated way.

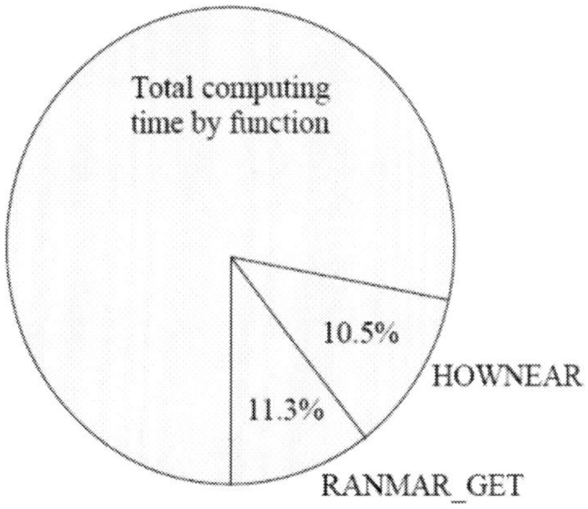

Figure 2. Pie chart showing the proportion of time each of the major functions takes as part of the whole with RANMAR_GET and HOWFAR highlighted.

HOWNEAR is employed to calculate the minimum distance of the current particle to a voxel boundary through a series of logical comparisons.[6] RANMAR, on the other hand, is the random number generator used in EGSnrc.[6] It is responsible for generating a reproducible, independent random number stream. These key algorithms were isolated from the EGSnrc code and re-implemented in a test framework. This framework involved dividing the program functionality in two - isolating branching code for execution on the PPE and allowing the SPEs to perform the requisite calculations. Execution began as a single thread on the PPE, which then spawned a number of child threads as shown in Figs. 3(a) and 3(b). Each thread made a series of decisions before queuing for access to an SPE. This was implemented in an asynchronous, time-slice based multithreaded model, with each PPE thread competing for an available SPE. Locks were used to synchronize SPE resources; a given PPE thread could queue its' task with a 'captured' SPE once it acquired its' corresponding lock by passing mailbox messages containing a pointer to instructions and data. The SPE, upon receiving this message, retrieved the data in question, executed the function and wrote the result back, informing the calling thread that the operation was completed. This process was self-managed (*i.e.* each thread would compete for available resources) via conventional locking mechanisms. For testing, the number of available SPEs and the size of the task were varied. Additionally, the "size" of the task performed by the SPEs was varied by changing the number of times the function was executed per request. This was represented using the variable TASK_SIZE; for TASK_SIZE=1, the function (RANMAR_GET or HOWNEAR) was executed only once per request made by a PPE thread. For TASK_SIZE=10, the function was looped ten times, the SPE returning the results from all ten executions. Tests were done for a variety of values for TASK_SIZE.

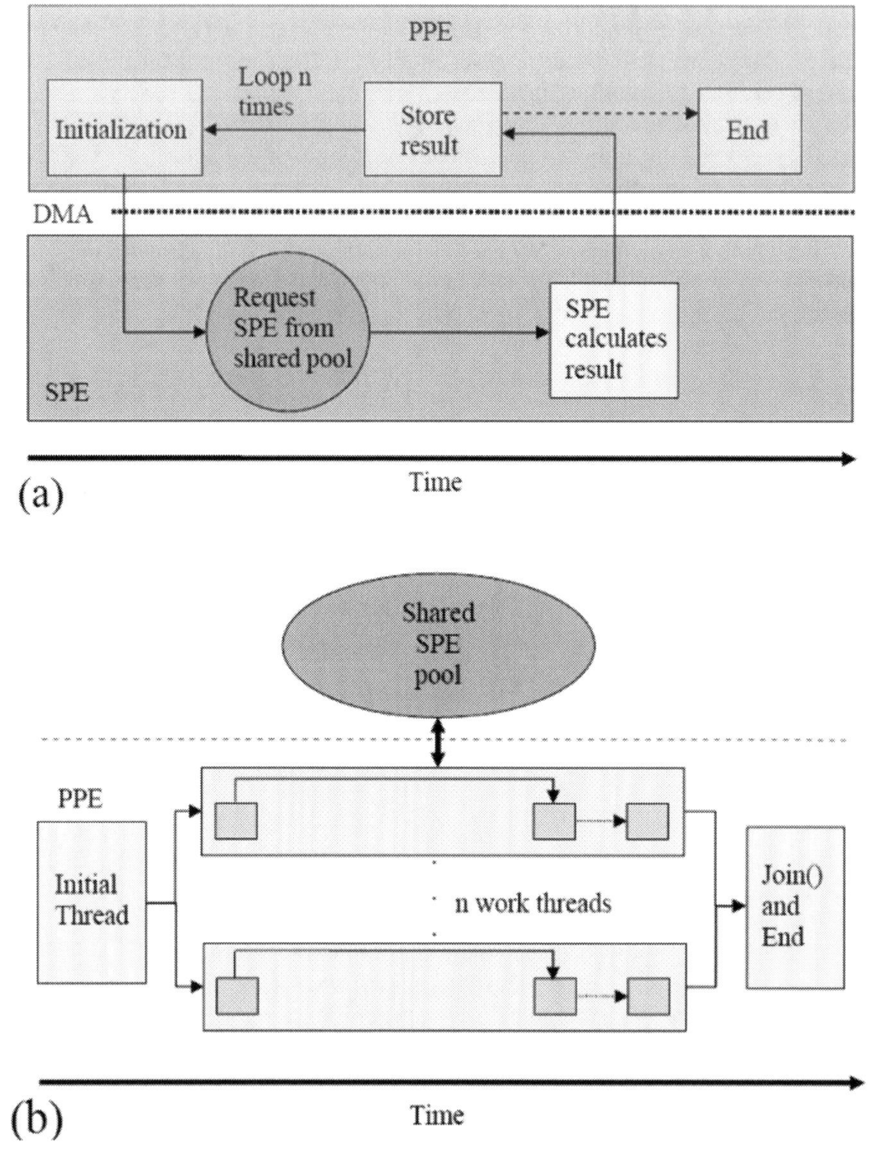

Figure 3. Diagrammatic overview of the test program for (a) single PPE thread operation, and (b) overall program operation. One PPE thread spawns multiple PPE threads, which make a few decisions before competing for available SPEs to calculate on.

3. Results and discussion

Figures 4(a) and 4(b) show the results obtained from execution of the test framework. Tests were performed varying the number of activated SPEs and the number of times each function was looped per PPE thread request. An ×86 baseline is shown on the graphs for comparison; this datum indicates the rate at which a desktop computer can execute the same function. The increase in speed due to the addition of a parallel processing element (*i.e.* SPE) can be seen as the slope of the curve for a given TASK_SIZE, and is known as the parallelizability. The higher this slope, the greater the benefit of adding more parallel elements; a highly parallelizable algorithm exhibits linear behavior when examined in this way. In Figs. 4(a) and 4(b), we found that there is a significant dependence of acceleration on TASK_SIZE. This is due to the relatively large cost of communication latency between PPE and SPE: a significant amount of time is wasted initializing transfers of data to and from the SPE. As the TASK_SIZE increases, the relative cost of latency decreases (processing and data transfer time increase) and an increase in execution speed is observed (this is most clearly seen when only one SPE is activated). Second, there is a dependency of parallelizability on TASK_SIZE. This is due to the overhead of executing the PPE threads to dispatch requests for function execution on the SPEs. As the TASK_SIZE increases, parallelizability increases because the number of PPE requests for a given number of executions of the test functions decreases. For HOWNEAR, the test framework is limited by the processing power of the PPE at TASK_SIZEs of 1 and 10; the PPE cannot dispatch tasks quickly enough to keep all SPEs in use. At TASK_SIZE = 100 or greater, the PPE is able to dispatch enough requests to provide the SPEs with work to do. This is evidenced by the positive slope of curve shown in Fig. 4(a). This is also the case with RANMAR_GET, though the function requires more cycles to complete than HOWNEAR and thus requires a smaller TASK_SIZE to become parallelizable. It can be seen that under optimal conditions both HOWNEAR and RANMAR_GET achieved acceleration of approximately one half order of magnitude. However, the functions have to be executed in relatively large batches to avoid overloading the PPE and avoiding penalties due to bus transfer latency.

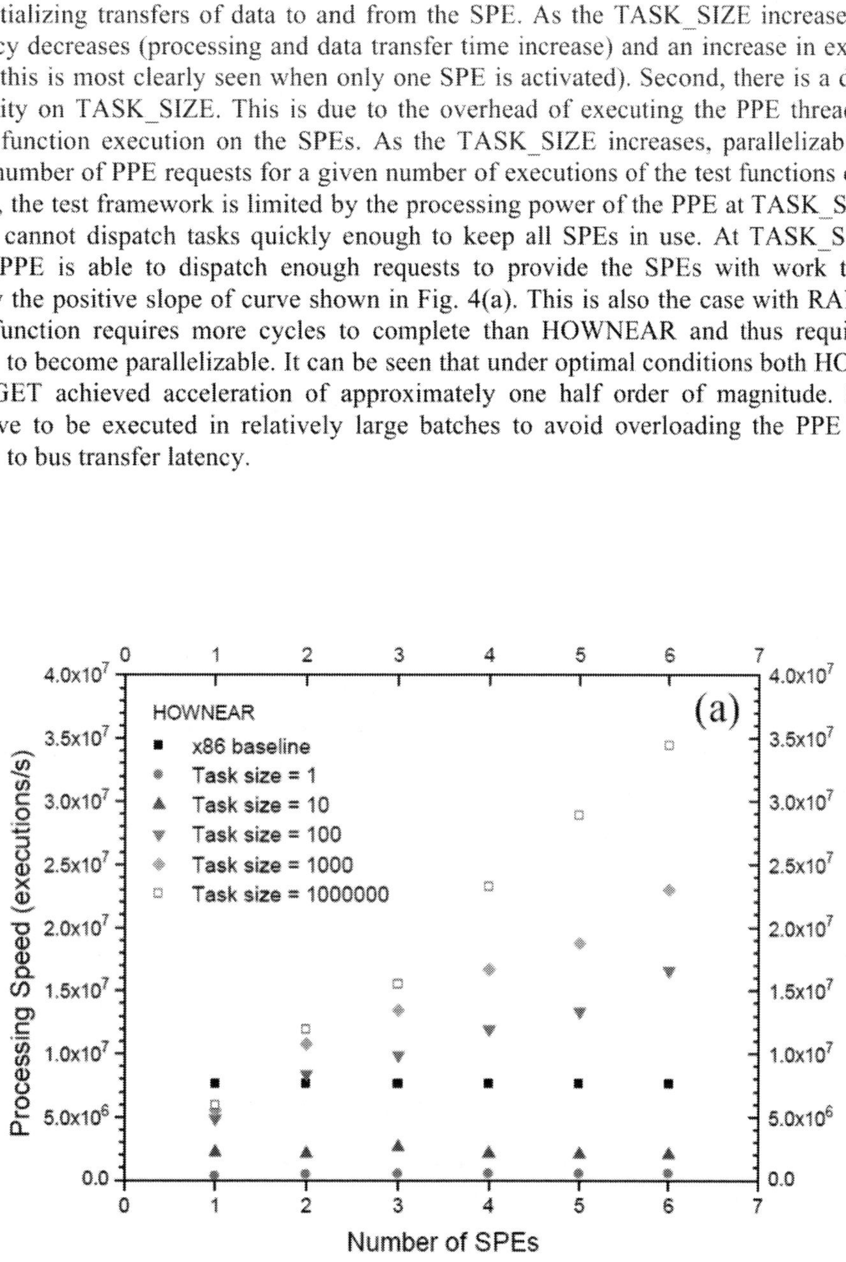

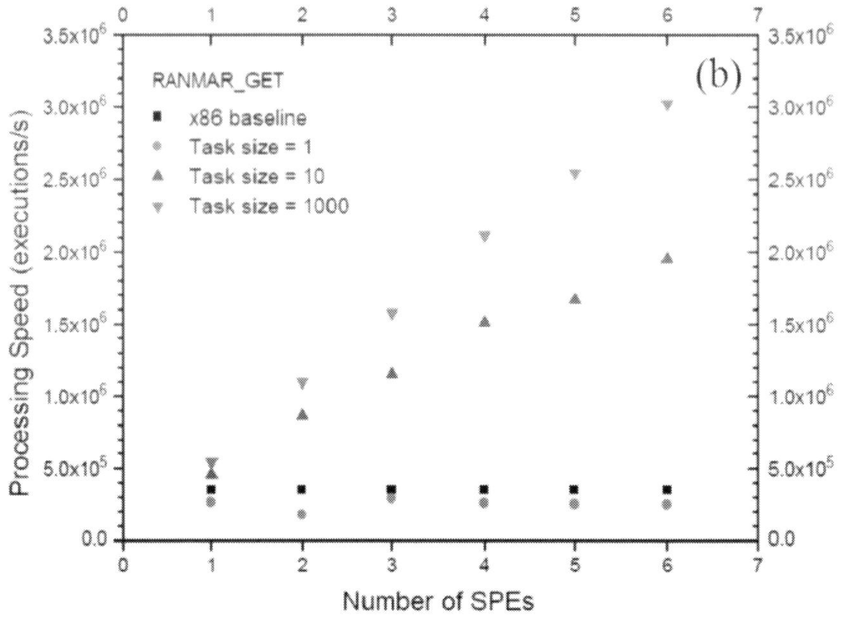

Figure 4. Processing speed vs. number of SPEs for subroutine of (a) HOWNEAR, and (b) RANMAR_GET.

4. Conclusions

We concluded that the algorithms examined in the EGSnrc code are parallelizable on the Cell platform. To implement this, we suggested changing the architecture of the EGSnrc such that EGSnrc is initialized on the PPE, and execution of the particle simulation loop within the program be split between PPE and SPE. The SPE would only perform sections of the code that are appropriate. To sidestep the problem of the tasks being too small (as seen in the test framework for TASK_SIZE < 10), SPE operations would be aggregated before being dispatched to an SPE for calculation. To implement this, we suggest pursuing the architecture that instantiates many simultaneous particle simulations (since each particle is statistically independent) and aggregates small operations across threads for calculation on the SPEs. It is concluded that EGSnrc performance is currently limited on the PS3 by the PPE. For future work, a platform involving a normal PC coupled with a GPGPU may provide a more viable avenue for acceleration. This is due to three factors. First, PC CPUs are much more powerful than the Cell PPE. EGSnrc contains a significant amount of framework code surrounding the actual processing. This may become a bottleneck if many concurrent particle simulations must be maintained at once - on the PS3, the PPE may be too slow to provide work for the SPEs to do. Second, PCs generally have a much larger available memory space than the PS3s 256MB; currently, it is not uncommon for workstations to have more than 8 GB of available memory. Under the proposed architecture of maintaining multiple simultaneous particle simulations, limited memory may quickly become a performance bottleneck. Finally, CPU and GPGPU technology are being improved much more intensively than the Cell architecture. Hence, gains made by reengineering the EGSnrc for the GPGPU architecture would benefit more from technology upgrades than from reengineering made for the Cell.

References

[1] Fabritiis G De. 2007 Performance of the cell processor for biomolecular simulations, Computer Physics Communications **176**, 660-664.

[2] Kurzak J, Buttari A and Dongarra J. 2008 Solving systems of linear equations on the cell processor using Cholesky factorization, IEEE Transactions on Parallel and Distributed Systems **19**(9), 1175-1186.

[3] Williams S, Shalf J, Oliker L, Kamil S, Husbands P and Yelick K. 2006 The potential of the cell processor for scientific computing, in: Proceedings of the 3rd conference on Computing frontiers, pp. 9-20.

[4] Williams S, Shalf J, Oliker L, Kamil S, Husbands P and Yelick K. 2007 Scientific computing kernels on the cell processor, International Journal of Parallel Programming **35**(3) 263-298.

[5] Kawrakow I and Rogers D W O 2000 The EGSnrc code system: Monte Carlo simulation of electron and photon transport, NRCC Report PIPRS-701, National Research Council of Canada, Ottawa.

[6] Nelson W R, Hirayama H, and Rogers D W O. 1965 The EGS4 code system, SLAC Report 265, Stanford Linear Accelerator Center, Stanford, CA.

[7] Chibani O and Li X A. 2002 Monte Carlo dose calculations in homogeneous media and at interfaces: A comparison between GEPTS, EGSnrc, MCNP, and measurements, Medical Physics **29**(5), 835-847.

[8] Chow J C L and Grigorov G N. 2008 Monte Carlo simulation of backscatter form lead for clinical electron beams using EGSnrc, Medical Physics **35**(4), 1241-1250.

[9] Verhaegen F. 2002 Evaluation of the EGSnrc Monte Carlo code for interface dosimetry near high-Z media exposed to kilovolt and ^{60}C photons, Physics in Medicine and Biology **47**, 1691-1705.

[10] Wang R and Li X A. 2001 Monte Carlo dose calculations of beta-emitting sources for intravascular brachytherapy: A comparison between EGS4, EGSnrc and MCNP, Medical Physics **28**(2) 134-141.

[11] Rogers D W O. 2006 Fifty years of Monte Carlo simulations for medical physics, Physics in Medicine and Biology **51**, R287-R301.

[12] Buttari A, Luszczek P, Kurzak J, Dongarra J and Bosilca G. 2007 A rough guide to scientific computing on the Playstation 3, Technical Report UT-CS-07-595.

[13] Uhrig J L. 1978 System requirements specification for real-time systems, in: Computer software and applications conference, 241-245.

[14] Amdahl G. 1967 Validity of the single processor approach to achieving large-scale computing capabilities, *AFIPS* conference proceedings **30**, 483-485.

Journal of Physics: Conference Series **341** (2012) 012029 doi:10.1088/1742-6596/341/1/012029

Optimum off-line trace synchronization of computer clusters

Masoume Jabbarifar, Michel Dagenais, Robert Roy and Alireza Shameli Sendi
Department of Computer and Software Engineering,
École Polytechnique de Montréal, Montréal, Canada

E-mails: {masoume.jabbarifar, michel.dagenais, robert.roy, alireza.shameli-sendi}@polymtl.ca

Abstract. A tracing and monitoring framework produces detailed execution trace files for a system. Each trace file contains events with associated timestamps based on the local clock of their respective system, which are not perfectly synchronized. To monitor all behavior in multi-core distributed systems, a global time reference is required, thus the need for traces synchronization techniques. The synchronization is time consuming when there is a cluster of many computers. In this paper we propose an optimized technique to reduce the total synchronization time. Compared with related techniques that have been used on kernel level traces, this method improves the performance while maintaining a high accuracy. It uses the packet rate and the hop count as two major criteria to focus the computation on more accurate network links during synchronization. These criteria, tested in real-word experiments, were identified as most important features of a network. Furthermore, we present numerical and analytical evaluation results, and compare these with previous methods demonstrating the accuracy and the performance of the method.

1. Introduction

The appearance of multi-core processors in computer clusters represents an evolutionary change in conventional computing to obtain high performance computing [13]. However, these systems may exhibit coherency problems when parallel programs access shared resources, thus creating hard to debug timing related problems. It is therefore crucial to have proper tools to monitor, trace and analyse system execution, in order to identify functional and performance problems. A trace facility aims to keep track of functional flow and report relevant changes at certain times. Global trace analysis, however, faces the problem that the cores of each node of the cluster have their own clock not synchronized with all the others [10]. Dealing with this problem is even more complicated in multi-level tracing (tracing in virtual machines, middleware and applications layer).

The Linux Trace Toolkit next generation (LTTng) [2], developed at École Polytechnique de Montréal, provides a detailed execution trace of the Linux operating system with low overhead. Moreover, LTTng is capable of handling huge traces of several gigabytes or more [16]. However, a new method is required to handle huge traces while allowing the collection of traces from multiple systems and embedded devices, for both on-line and a posteriori off-line analysis and viewing. Furthermore, the user of LTTng expects to see the output of real-time analysis to diagnose possible problems properly. Therefore, LTTng should be able to visualize traces from several distributed systems on a common reference time base.

In a computer cluster, multiple nodes produce separate trace streams independently. There are timestamps on the events of the trace. Since timestamps are recorded based on a logical time that runs natively on each node, the logical order of events cannot be guaranteed [6]. Trace synchronization with due attention to high precision and low intrusiveness has been achieved recently. It complements trace analysis with a post-processing step called offline synchronization [1]. The objective of this paper is to optimize the total trace synchronization time for huge computer clusters. The ideal situation is to improve the total synchronization time while preserving the synchronization accuracy.

This paper is organized as follows: first, we investigate earlier work and several existing synchronization algorithms. Then, we present and discuss our model in Section 3. In Section 4, experimental results are presented and results are compared with existing methods. Finally, we conclude our work by reviewing the significant points of this paper and suggest future work in Section 5.

2. Related Work

This paper focuses on offline synchronization. Several papers have proposed algorithms in this area [3, 4, 5, 6, 8, 9, 10, 11, 12]. When packets are exchanged between a pair of nodes, the receiving and sending times will not be shown properly because the clocks of the two nodes are not synchronized and the time offset should be calculated [9]. Since the packet is necessarily sent before it can be received, this relationship can be used to help synchronize two nodes. Thus, a conversion function is needed to convert these two times to a common time.

The first offline time synchronization method was proposed by Duda et al. [3]. He proposed two synchronization algorithms, *Linear Regression* and *Convex-Hull*. They estimate a conversion function between a pair of clocks. In both algorithms, the conversion function appears as a line. Drift and offset between the two clocks are extracted from this linear model. In a two dimensional space, based on times of the nodes A and B, the Linear Regression algorithm tries to map all points on a line; thus, each point will affect the position of the line. In reality, network latency and such problems between two nodes cause problematic outlying points which ideally should not affect the delay and offset computed from the linear regression and should be ignored to increase the accuracy [3, 8]. Convex Hull is an accurate algorithm that assumes minimum sending times and maximum receiving time; in this way, it finds the area that has minimum latency and ignores outlying points. Therefore, the estimated line is more accurate than with Linear Regression.

Khlifi et al. [4] proposed two algorithms named *average* and *direct skew removal* techniques. The second technique is a brute force algorithm which iteratively divides the duration of the trace into next intervals. It considers the effect of the clock resolution on the accuracy of delay measurements and provides more accurate synchronization compared with other techniques.

Poirier et al. [1] presented an accurate method for synchronizing distributed traces. This method is applied to traces that are recorded at the kernel level with low intrusiveness. They apply Convex-Hull as conversion function to the clocks of traced nodes.

Clement et al. [5] evaluated the system characteristics impact on the trace synchronization accuracy. Firstly, they studied the tracing duration impact. They proposed to divide long duration traces into 30 minute segments since the error in the clock drift calculation started to increase significantly after approximately 45 minutes of tracing, while it was almost stable during the first 30 minutes. The error increases because of the variation of the clock drift with time due to the Allan deviation [14] and environmental effects on the clock circuit frequency such as temperature and supply voltage variations. The second parameter studied is the system load impact, when there is a heavy load on the major subsystems, CPU, memory, and disk. They found that the transmission time and response time measurement variations, caused by interrupt latency in a loaded system, influence the clock drift computation directly and subsequently the time synchronization accuracy [5]. The third parameter studied is hop impact, where there are more than one path between two nodes. In that case, the offset between the two indirectly linked nodes may be computed by adding the offsets along a path, from one intermediate hop to the next. The path that has fewer hops generally provides a higher accuracy. In the case where we have a direct path, it is normally better to choose it to synchronize two nodes [5].

3. Architecture of Proposed Model

Figure 1 illustrates the architecture of the proposed model. It consists in four connected modules. Each module receives input from one or more modules and sends outputs to other modules. The input of this architecture is fed by LTTng [2] and consists of two or more unsynchronized trace files. The output is two or more synchronized traces. The output format is compatible with LTTV (Linux Trace Toolkit Viewer) [2] which is able to show synchronized traces. The following modules are present in this architecture:

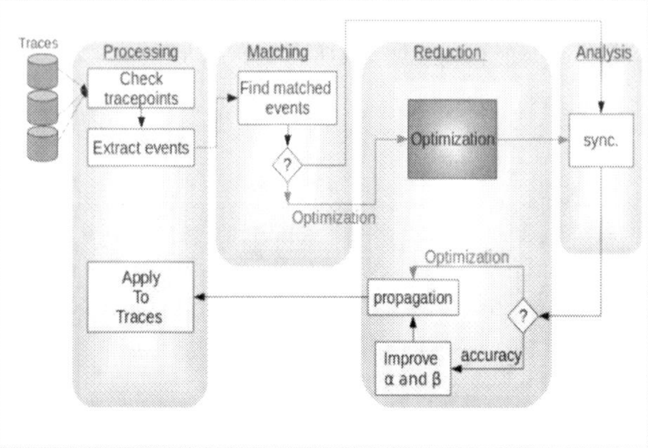

Figure 1. Architecture of proposed model.

- *Processing module:* traces are gathered from all of nodes distributed in a computer cluster and we have to analyze them. In order to synchronize two nodes, we need network traffic exchanges between them. We have to extract packet exchange events and dispatch those events to the next module. Thus, this module captures network traffic and computer activity and extracts the necessary information for the matching module.

- *Matching module:* event processing feeds the events one by one. However, the Analysis module works on groups of events. Thus, the Matching module is responsible for forming these groups. The relations between the packets are of different types ("one to one", "one to many", or a mix) and this will influence the overall behavior of the module for TCP, UDP or MPI. This module must recognize related sent and received packets and group them. For example, for linear regression, we need the round trip time (RTT), so this module makes a group after finding an acknowledgment packet, and the acknowledge time will be assumed as reference time.

- *Analysis module:* as mentioned, there are two methods to synchronize time: *Linear Regression* and *Convex-hull*. In this module, the user can chose any of these methods to synchronize traces. As discussed, the Convex-hull method can synchronize traces with better accuracy so it is chosen as the default option.

- *Reduction module:* the Reduction module consists of two methods. The first method is the previous work [1] where all node pairs are examined and no accuracy is possibly traded off for performance: "Accuracy method" [1]. The second method is our proposed method based on time reduction. We explain the proposed method in the next Section. To sum up, the previous method sends matched packets to the Analysis module without any processing and then synchronizes each node pair based on synchronization algorithm. Then, the reference time is specified and all nodes can be synchronized with this reference time in Reduction module. We therefore have to choose a node as reference node which has the most accurate links. Since each node pair in a computer cluster had to be synchronized in the previous method, there may be indirect paths between two nodes with better accuracy (less drift and offset). To find which links (and drift/offset) to use for synchronizing two nodes, a Minimum Spanning tree (Min-ST) based on drifts and offsets is computed since it gives best accuracy to compare any two nodes in the cluster. The last part in this module is propagation. All nodes should be synchronized based on the reference node and drift and offset factors (α and β) propagated through the routed path to the reference node.

3.1. Main challenge

As we discussed, previous work tried to obtain the best accuracy for network synchronization [1]. In real setups, a scalability problem arises. The total synchronization time increases rapidly with respect to the network size. For example, in experiments generated with the NS2 network simulator, where there are 21 connected computers, the simulated time was seven seconds and during this period, each computer sent and received around 1000 to 4000 packets. The timestamp of each packet helps us to synchronize computers as we have seen before. The output of this simulation was sent to the previous accurate synchronization algorithm [1]. The total synchronization time for such an experiment, which is a realistic setup, is about 27 minutes. Obviously, it is impractical to wait 27 minutes to synchronize 21 computers that have been traced for just seven seconds. This scalability problem grows when adding more computers to trace and analyze simultaneously.

4. Methodology

To tackle the discussed challenges in Section 3.1, we ignore and neglect inaccurate links in the network. Let's take a look at a simple example. Figure 2 shows four connected computers, A, B, C and D. In Figure 2.a, each number labelling an edge represents the number of exchanged packets in the link between that pair of nodes. In previous work, each pair of connected computers were synchronized. This meant computing the drift, offset and accuracy for each pair of connected nodes, A-B, A-D, A-C , B-C and D-C. The direct connection is considered and other indirect connections are considered as a combination of some direct connections. Figure 2.b displays the accuracy on each edge.

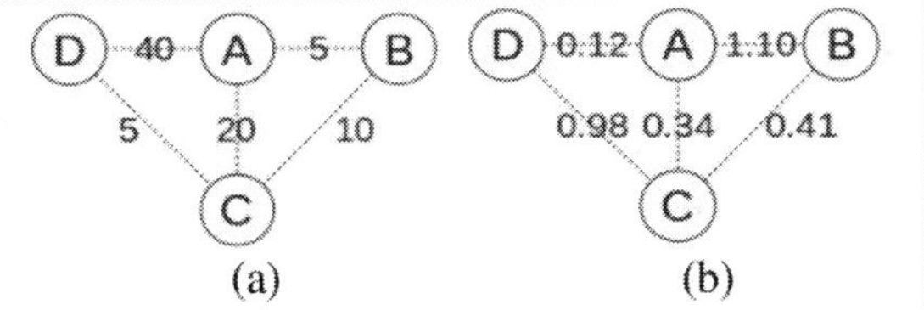

Figure 2. Main graph with 4 nodes.

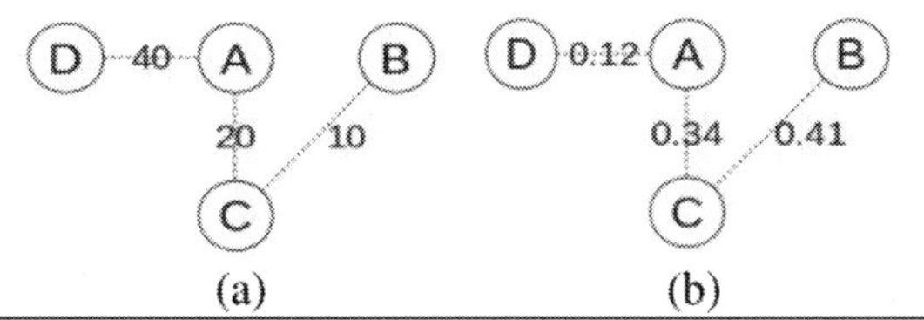

Figure 3. Max / Min-ST.

Then, one of the nodes is selected to be the reference node in the cluster by the Reduction module and its time is assumed as reference time. Adding up the factors of the connected nodes gives total accuracy for each node. The node that has the smallest total bound on accuracy has the smallest error, normally associated with lower latency links to the other nodes [1]. Therefore, the synchronization between this node and others will be achieved with higher accuracy (here we have A: $1.10 + 0.12 + 0.34 = 1.56$). Thus, Accuracy method selects node "A" as reference node. All nodes in the cluster should be synchronized with the reference node. However, there are usually some nodes that do not have a direct connection to the reference node in the network. Furthermore, some nodes may have more than one indirect path to the reference node. Thus, a unique routed path from each node to the reference should be found. To this aim, a Degree-Constrained Spanning Tree [15] is sufficient. This tree connects all nodes together and shuns loop or redundant paths. In Accuracy method, the Reduction module finds Min-ST based on synchronization values [1]. In this example, ignored links are two links; A-B and D-C. A manifest disadvantage of the method is that it has synchronized these links already and now they are ignored and useless after this. The idea is "not to lose time finding these useless factors". Let us now discuss the main features of our methodology:

4.1. Hop count

A criteria that leads us to find accurate paths is the number of hops. Since network delays cause a major part of errors, the accuracy degrades with the number of links from the source node to the destination node. The source node has hop count equal to zero and every node that has a direct link to it has hop count equal to one. Other nodes can count their hop count as the number of links from each one to the source node by incrementing the previous node's hop count with one.

This property is one of the operations used to select not only the reference node but also the routed path to propagate synchronization factors. We compose this criteria with another one, as explained in the next Section.

4.2. Number of exchanged packet

The tests results of the previous method and studies about factors impacting on the accuracy of connection synchronization show that the number of packets exchanged has a most significant impact on accuracy. Thus, we exploit this property to analyse links before synchronizing.

4.3. Reference node selection

Since a cluster includes many connected computers, the first step of optimization is to recognize the best criteria to choose the reference node. The reference node time is considered as reference time and drift and offset times are then calculated to other computers. We propose to select the reference node based on two criteria: 1) the fewest hop count to all nodes and 2) the maximum packet traffic. The first criterion is considered more effective and important. We select a set of nodes having the fewest hop count in the network. In the case where there is a single member in the set, we have found the reference node. Otherwise, we use the second criterion in order to have the minimum latency and delay. Thus, we select from the set the nodes with the highest packet count. Most of the time, these two conditions result in a single reference node candidate. In the case where more than one node qualifies, one is selected at random among these.

4.4.Optimized synchronization time

In the next step, after finding the reference node, we restrict the synchronization computations to the accurate paths to optimize the synchronization time. We obtain a Maximum Spanning Tree (Max-ST) based on the number of exchanged packets to all connected nodes, thus removing any redundant link. In this method, this Max-ST is sent to the Analyse module to synchronize the connected computers. Then, the synchronization factors must be propagated through the Max-ST to synchronize the time of all computers in the network cluster. Let's take a look at the previous example and go through these steps. In the proposed method, Max-ST for each link is used. The specific feature of this Max-ST is that it starts to grow with the Reference as root to the Max-ST. In the optimization step, as we discussed beforehand, the number of exchanged packets is counted and reference node (A) and Max-ST are selected. Figure 3 shows the Max/Min-ST that are exactly the same as with the previous method. The important difference between the two methods is the required computation time. Figure 1 illustrates the optimization variant in the Reduction module. With the proposed method, Max-ST is sent to the Analysis module to synchronize the nodes. The optimized method produces a significant reduction in computation time, while often producing the same or very similar synchronization accuracy results.

5. Real world and Experiment Results

5.1. Network Simulation

Since a large cluster is not always available for experiments, we used a powerful simulator to test more configurations and eventually validate our simulation results using Mammouth as a real world experiment. Many simulator, such as Wireshark, Clownix, and etc., were studied and finally NS2 was selected. NS2 provides substantial support for most network features on both wired and wireless types. It supports TCP, multi-cast, and routing protocols. NS2 simulates not only network latencies but also the distances and many other features.

Table 1. Timing improvement

No. of Nodes	Total No. of Packets	Previous Sync. Time	Optimized Sync. Time	Saved Time	Percentage
4	1437	8 s	5.5 s	2.5 s	29%
5	2098	13 s	7.5 s	5.5 s	41%
6	13044	80 s	69.5 s	10.5 s	13%
21	173985	1157 s	922 s	235s	20%

The simulator has a clock and all node time sources are based on it. However, for synchronization studies, we need to run each computer node with its own independent clock, something none of the simulators offer. Therefore, the core of NS2 has been changed and an independent clock has been provided for each computer. Each computer sends and receives packets recording the events with its own clock and the results can be used as the input for synchronization algorithms.

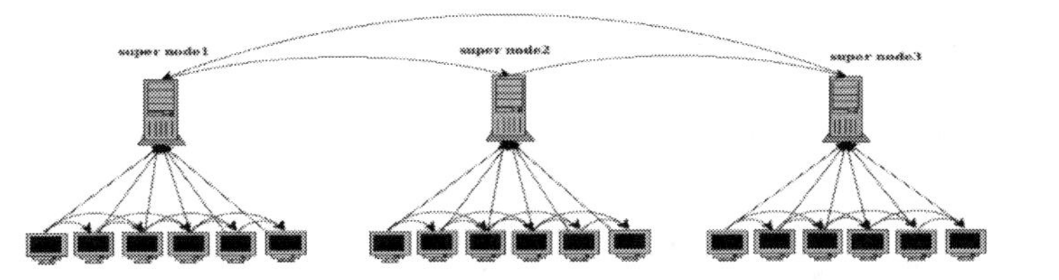

Figure 4. Experiment cluster

The first network that we have simulated had four computers connected with a router. Each link has its own distance and delay. Also, nodes send packets with specified sizes at specified intervals through the links. Then, we expanded our network to 21 nodes and got the same accuracy as the Accuracy method in less time. Table 1 shows the timing improvement with the new method for our experiments.

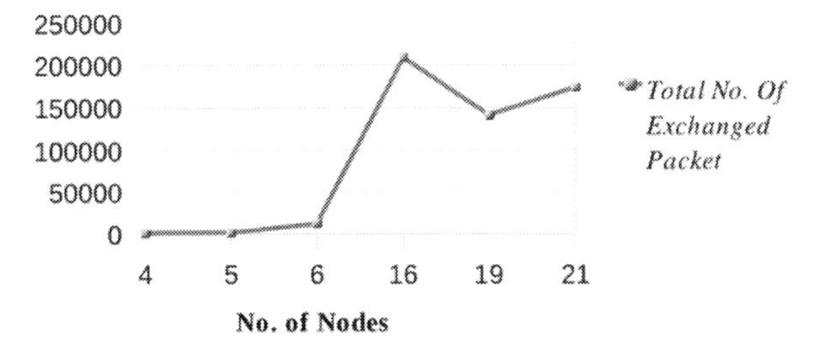

Figure 5. Experimented network traffic

The variation of the percentage mainly depends on the number of ignored links in the optimization step, the more packets are exchanged in the ignored links, the more time is saved. Figure 4 illustrates the network connections of the large experimental cluster with 21 nodes.

In this experiment, the number of packets exchanged between each pair of nodes varies from 500 to 4000, which is more realistic. The optimized method still achieves the highest desired accuracy while improving the total execution time. The result is shown in Table 2. This experiment has been repeated for many models and we obtained improvement for all of them.

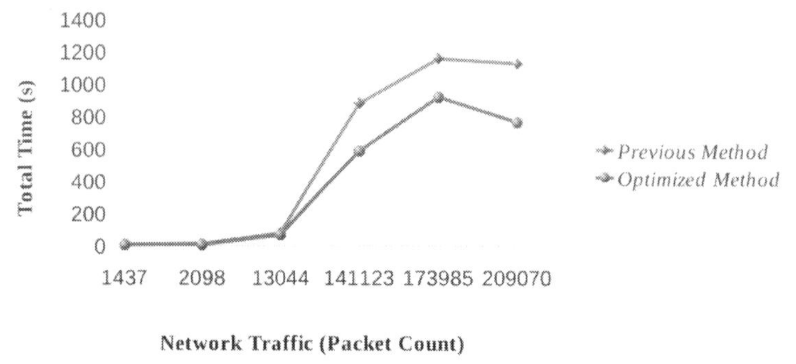

Figure 6. Comparison between optimized and previous methods

Figure 5 illustrates the fact that the traffic of a network is variable. The experiments were performed in different sizes of networks and different intensity of traffic to measure their impact on performance. For example, there is a network with 16 nodes that has higher traffic compared to another network with 19 or 21 nodes.

Figure 6 illustrates the difference between two synchronization method compared for several networks. When there is a network with high traffic, synchronization takes more time. For example, in the network with 141123 packet exchanges, the synchronization methods take much more time than when there is less traffic.

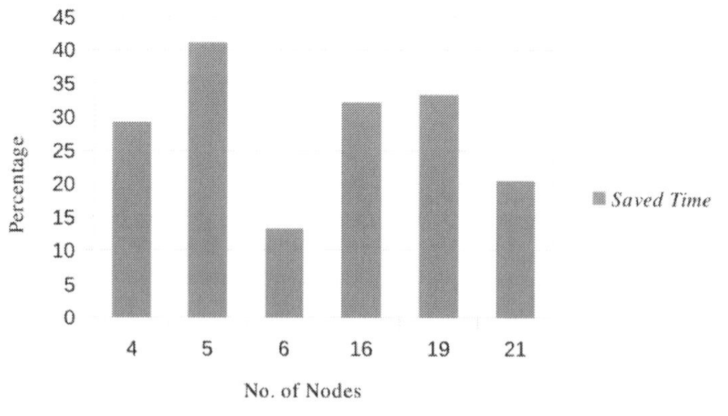

Figure 7. Saved time percentage

Furthermore, it shows the time difference between the previous and optimized methods. As expected, the saved time depends not only on the network traffic but also on the network topology. The synchronization takes more time for a network with more nodes. For example, the total synchronization time for a network with 21 nodes, even with fewer packet exchanges (173985 packets) is higher than for a network with 16 nodes with more packets (209070 packets). Thus, the more nodes to analyze, the more time to synchronize the network. The saved time also depends on the features of the network. If a significant amount of network traffic consists in client to client communications (for example in distributed systems and high performance computing system), not only the number of ignored links is considerable but also ignored links sustain a high traffic. In that case, the saved time is particularly interesting (more than 50%). On the other hand, in typical networks with servers sharing data for numerous clients (the type of the experimented network), the optimized method could on average save 30% of the total synchronization time. Figure 7 shows the percentage of saved time for several networks. The variation of the percentage mainly depends on the number of ignored links in the optimization step, the more packet exchanges in the ignored links, the more time saved.

5.2. Real-world experiment results

One particularly interesting experiment was run on the Mammouth cluster with 55 connected computers. Mammouth is one of the largest Linux clusters in Canada, and is located at the Centre de Calcul Scientifique in Sherbrooke University, funded by the RQCHP. It is divided into two partitions: *Serial* and *Parallel*. Serial consists in Pentium 4 computers connected by an Ethernet Gigabit network. Parallel consists in Opteron computers linked by an infiniband network. The parallel partition includes 1120 Intel Xeon 64bit, 3.6 GHz CPUs in 560 Dell SC1425 compute nodes. Each node has 8 GB RAM (Total 4.5 TB) and 160 GB disk space. Infiniband network non-blocking (800 MB/sec) SDR Cisco-Topspin are used to connect nodes which run Linux CentOS 5 [7].

Table 2. Results of Accuracy for 21 nodes.

Trace	Drift	Offset (* E+009)	Start time	Trace	Drift	Offset (*E+009)	Start time
0	1.000010	8.46	7605.594802	11	1.000020	5.15	7606.28
1	1.000000	8.92	7606.686668	12	1.000020	5.07	7606.286858
2	1.000080	3.57	7607.793202	13	0.999998	5.02	7606.291110
3	1.000040	4.67	7606.285003	14	1.000030	4.98	7606.289300
4	1.000030	4.86	7606.288089	15	1.000040	8.38	7606.294055
5	0.999996	1.45	7606.294936	16	1.000010	5.26	7606.293697
6	0.999998	5.06	7606.284456	17	1.000030	4.95	7606.284242
7	1.000020	8.81	7606.294419	18	1.000020	5.12	7606.292262
8	1.000030	4.97	7606.292909	19	1	4.83	7606.285830
9	1.000030	5.65	7606.290214	20	1.000030	8.56	7606.28
10	1.000030	4.80	7606.285457				

We experimented with many types of network topologies, however we describe here one of the most typical setups to support our claim. Figure 8 illustrates our experimental cluster with 55 nodes. This graph is one of the most popular topologies in networks. Usually, computers are divided into local area networks with nodes having connections to each other in the group. There is also a master node, or gateway, in the group that connects the group to other groups. In this example, different numbers of computers has been assumed in each group in the cluster. The first group consists of 19 connected computers, the second group consists of 21 nodes and third group consists of 15 nodes. A script has been run in each node to generate packet exchanges between each node and its two next neighbor nodes as well as to its master node.

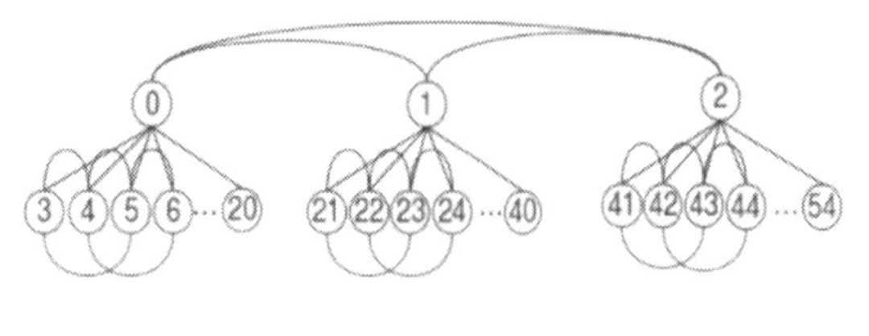

Figure 8. The graph of Main Network

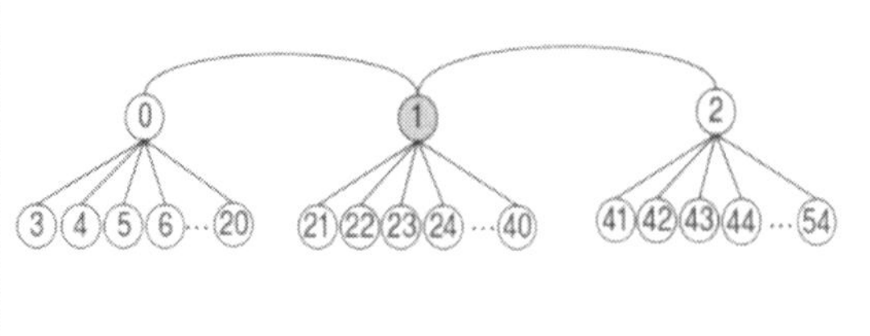

Figure 9. Accurate Graph

It is obvious that there are more connections to the master nodes. In our experiments, each connection is a *"ssh connection"* and generates from 43 to 48 packets. The previous method finds the best links, with the best time synchronization accuracy, then synchronization factors are propagated through these links and all nodes are synchronized based on the reference node's time. Figure 9 displays the final accurate graph in the experimental cluster. Node 1 is selected as the reference node and all 20 nodes of the second group can be synchronized directly with it. Also, each of the other group members can be connected to the reference node through an indirect path through their own master node. Both methods synchronize the nodes through this accurate graph. However, the new proposed method is able to reduce the computation time. Performance improvement depends on the number of ignored links. Thus experiments with real connections and numerous messages are useful to compare the performances of these methods.

6. Conclusion and future work

In this paper, we have discussed offline trace synchronization for computer clusters with large number of communications. We have proposed a method to optimize the synchronization performance. The presented method is applicable when event data is distributed over a computer cluster, and exchanged packets are used for time synchronization. It has been integrated in the Linux Trace Toolkit next generation (LTTng), developed at École Polytechnique de Montréal, and used both for performance debugging and security and performance monitoring.

As future work, we plan to extend the method to synchronize online traces while receiving streaming data from LTTng. Since, in a streaming trace, nodes are discovered as they become active, and may cease operation or activity at any time, nodes joining and leaving at any time during the streaming trace cause alterations to the network topology and consequently complexify the analysis.

7. References

[1] B. Poirier, R. Roy and M. Dagenais, "Accurate Offline Synchronization of Distributed Traces Using Kernel-level Events," Operating Systems Review, vol. 44, 2010, pp. 75-87.

[2] M. Desnoyers, "Low-Impact Operating System Tracing," PhD thesis, École Polytechnique de Montréal, 2009.

[3] A. Duda, G. Harrus, Y. Haddad and G. Bernard, "Estimation global time in distributed system," In proceeding 7th Int. Conf. on Distributed Computing Systems, pp. 299-306, Berlin, 1987.

[4] H. Khlifi and J. C. Gregorie, "Low-complexity offline and online clock skew estimation and removal," The International Journal of Computer and Telecommunications Networking, vol. 50, no. 11, pp. 1872-1884, 2006.

[5] E. Clement and M. Dagenais, "Trace synchronization in distributed networks," Journal of computer system, Network, and Communication, 2009.

[6] A. D. Ksehmkalyani and M. Singhal, "Logical time," in Distributed Computing: Principles, Algorithms, and Systems, 1st ed., USA: Cambridge University Press, 2008, pp. 50-84.

[7] Mammouth project available at "https://rqchp.ca/?mod=cms&pageId=566&lang=EN," Sep. 2010.

[8] P. Ashton, "Algorithms for off-line clock synchronization," Technical report, University of Canterbury, Department of Computer Science, Dec. 1995.

[9] R. Sirdey and F. Maurice, “A linear programming approach to highly precise clock synchronization over a packet network,” 4OR: A Quarterly Journal of Operations Research, vol. 6, no. 4, 2008, pp. 393-401.

[10] B. Scheuermann, W. Kiess, M.Roos, F. Jarre and M. Mauve, “On the time synchronization of distributed log files in networks with local broadcast media,” IEEE/ACM Transactions on Networking, vol. 17 no. 2, 2009, pp. 431-444.

[11] R. Sirdey and F. Maurice, “A linear programming approach to highly precise clock synchronization over a packet network,” 4OR: A Quarterly Journal of Operations Research, 6(4):393–401, 2008.

[12] H. Marouani and M.R. Dagenais, “Internal Clock Drift Estimation in Computer Clusters,” Journal of Computer Systems, Networks, and Communications, vol. 2008, no. 1, 2008, pp. 1-7.

[13] E. Betti, M. Cesati, R Gioiosa and F. Piermaria, “A global operating system for HPC clusters,” IEEE International Conference on Cluster Computing and Workshops, 2009.

[14] NIST Time and frequency from A to Z., February 2011. http://tf.nist.gov/general/glossary.htm.

[15] Raidl, G.R., Julstrom, B.A., "Edge sets: an effective evolutionary coding of spanning trees," Evolutionary Computation, IEEE Transactions on, On page(s): 225 - 239, Volume: 7 Issue: 3, June 2003

[16] J. H. Deschenes, M. Desnoyers and M. Dagenais. “Tracing Time Operating System State Determination,” The Open Software Engineering Journal, vol. 2, 2008, pp. 40-44.

A software product line approach to enhance a meta-scheduler middleware

Rafael F Scheidt, Katreen Schmidt, Gabriel M Pessoa, Matheus A Viera and Mario Dantas
University of Santa Catarina (UFSC), Department of Informatics and Statistic (INE), 88040-900, Florianopolis - SC, Brazil

E-mail: {rfscheidt, katreenschmidt, gabrielm, matviera, mario}@inf.ufsc.br

Abstract. Software Projects in general tend to get more software reuse and componentization in order to reduce time, cost and new products resources. The need for techniques and tools to organize projects of higher quality in less time is one of the greatest challenges of Software Engineering. The Software Product Line is proposed to organize and systematically assist the development of new products in series at the same domain. In this context, this paper is proposed to apply the Software Product Line approach in Distributed Computing Environments. In projects that involve Distributed Environments, each version of the same product can generate repeatedly the same artifacts in a product that evolves its characteristics; however there is a principal architecture with variations of components. The goal of the proposed approach is to analyze the actual process and propose a new approach to develop new projects reusing the whole architecture, components and documents, starting with a solid base and creating new products focusing in new functionalities. We expect that with the application of this approach give support to the development of projects in Distributed Computing Environment.

1. Introduction

A Software Product Line represents a set of systems sharing common functionalities and characteristics that satisfies the needs of a particular market or specific segment [1]. Such a system set also can be named as Product Family [2]. The members of the product family are specific products systematically developed by the Software Product Line also classified as core assets. The core assets are represented by a variable features set which indicate a late decision of design of project [1]. The assets choice and configuration composes a specific product [3].

The Computational Grid approach in general presents a complex and distributed structure involving hardware and software that allows access with low costs to the high performance computing resources [4]. In the context of this approach, could find several elements like graphic interfaces to access the environment, middleware, meta-scheduler and so on, in consequence involve several knowledge areas that can derive other set of elements.

In this sense it becomes a challenge to systematically organize through the Software Product Line, such elements and architecture, aimed at planning systematic reuse and fast customization for products development that complete other needs and have the same structure.

This paper is organized as follows: Section II represents the value of concept Software Product Line. Section III describes the Distributed Computing Environments. Section IV illustrates and focuses on the proposal and related works. Section V discusses the facts and problems found at the moment. Section VI provides comments e directions about future works.

2. Software Product Line

Software Product Line respects to systematical planning and strategic reuse productivity, by exploiting similar features of a product. Its main goal is to achieve significant reduction in organizations in terms of development, cost reduction in maintenance, reduce time-to-market, improve quality productivity and customer satisfaction as well to anticipate the problems encountered [1].

The main difference between conventional Software Engineering and Software Product Line is the variation presence in some or all the software requirements and the focus. In the conventional systems the focuses is a unique goal, in other words, produce a specific product. While the conventional systems development works ad-hoc, in other words, contract oriented, the Product Line development has a strategic vision of the niche market [5].

The implementation of Software Product Line approach requires efforts on the major artifacts in the development cycle. Table 1 shows the main characteristics of the essential artifacts for Software Product Line.

Table 1. Implementation costs of a Software Product Line [6].

Core Artifacts	Implementation cost
Architecture	Must support variation inherited of Product Line
Software Components	It should be design to be general, without loss of performance, must have support of variation points
Test Plans, Test Cases and Test Data	Should be considered variation points and multiple instances of Product Line
Business Case and Market Analysis	Should approach a family of software products, not just a product
Project Plans	It should be generic or extensible to accommodate variations of the product
Tools and Processes	Should be more robust
People, Skills and Training	Should involve training and experience about the artifacts and procedures associated with the Product Line

There are essential activities for Software Product Line. Such activities are Domain Engineering, which aims to develop the core of artifacts and Application Engineering, which will be instantiated in the core of artifacts to generate different products also known as Product Development [1].

Core assets are a set of assets ready to be reused in new product development. The core assets can be software components, project patterns, documents used in development, architecture, schedule and other artifacts that will serve as building blocks in the Product Line. The core assets are present in every product line. The architecture is one of these core assets and carries with it the possibilities of line variability. Architecture core asset, in particular, is seen as a key point of a product line and if it is projected badly can derail the entire project [2].

The main objective of Product Development is the generation of products based on core assets of a Product Line. It corresponds to the instantiation of applications and since found new requirements that had not been specified, it starts a feedback between this activity and the activity of core assets (Domain Engineering), in cycles in an evolutionary way.

A very important concept intrinsic to a Software Product Line is the variability. To form a software family through Software Product Line it is necessary identify possible changes that may occur in the artifacts produced. The products generated during products development phase can be differentiated in terms of behavior, quality attributes, physical configurations, scaling factors, among others [2]. Variations are the tangible differences between the products of Product Line in any device such as architecture, components, interfaces between components and components connection. The variations can be identified in any development phase of Products Line [7].

To represent the possible variations identified in a particular domain becomes necessary an approach of variability, which until now, there is no consensus about how to identify and represent variability in Product Line because it is a recent issue [8]. One possible approach available and based on UML was developed and named SMarty (Stereotype-based Management of Variability) [8]. The SMarty approach was developed based on the activities and on the concepts of variability. A SMarty approach was created to manage the variability consisting of a UML profile called SMartyProfile and a process named SMartyProcess. The main goal of SMarty approach is to allow the variability of a Product Line to be managed effectively supporting UML models.

3. Distributed Computing

Grid computing is characterized by making a variety of distributed resources, including services, devices and applications, available to a wide range of users [9]. Various organizations, both real and virtual (VO), can make different types of resources available under dynamically changing availability constraints and with varying policies for access and use of these resources [10].

As a result of their nature, this variety of services and resources can be incorporated for achieving computational tasks. Specifically, users can access resources, applications and services, submit jobs for execution either via queues or by advance reservation, create combination processes in workflows, and verify the status of jobs or systems. Also, there has been a movement towards integrating grid computing environments with mobile computing [11] [12], creating an interface for users access the resources and services of a grid from anywhere, at anytime.

One the other hand, clusters environments are common configurations used in many organizations to reach high performance computing (HPC) for specific local applications. Multi-clusters, if well orchestrate as a grid environment, can represent a differential computational power for a global execution of several class of applications, including also parallel jobs. A grid environment, composed by multi-clusters configurations, can be considered in a private or in multi-domain organizations. In other words, this metacomputing [13] environment can be employed as a HPC facility inside a specific organization, or among different organizations, creating in both cases the concept of virtual organization (VO).

One important challenge is how to reach an efficient re-utilization of this environment and coordinate all resources and services.

4. Proposal and Related Works

With a focus on Distributed Computing Environment the present study aims to systematically organize the elements and architecture, aiming the systematic reuse planning and rapid customization for new product development to complete other requirements and to have the same structure. To achieve this objective, this work applies the SMarty process specifically in the project being developed by the laboratory LaPesD at the Federal University of Santa Catarina UFSC, contributing to the development of techniques to systematize a process for reuse of the artifacts, architecture and components of Software Products Line in Distributed Computing Environment.

Among the existing approaches of Software Product Line and researches, SMarty approach was chosen due to its ability of modeling artifacts and ability of integrate with the UML.

Currently in the literature are few works in the area of Distributed Environments. One of the few existing, but with excellent quality [14] [15] proposes to build an approach of Software Product Line for Grid-Oriented Middleware. The work developed a Software Product Line Architecture to facilitate the development of Grid-Oriented Middleware Systems. In order to validate the proposed PLA, it is instantiated in the construction of a middleware for grid computing.

Another approach is named GISPL (Grid Infrastructure Software Product Line) adopts SPL concepts in the design, optimization and implementation of Grid-Oriented Middleware. Initially, the requirements are defined through ontology and a specific language. Next, service oriented architecture is defined that incorporates software and hardware artifacts needed to the functioning of VO (Virtual Organization). The architecture is modeled in System Modeling Language (SysML), which allows specifying the infrastructure using specific models derived from the ontology defined in the previous step. Finally, the infrastructure models are prototyped and optimized [16] [15].

Unlike the presented approaches, the purpose of this study is to apply in the Distributed Computing Environment as a whole, covering all the part of domain and instantiating complete applications distributed at the stage of product development.

5. Discussion

The initial motivation to use a Software Product Line approach in the distributed environment, which is often a complex architecture, was based on problems encountered, such as: architecture elements are deployed and are removed a few times, because the client does not know exactly what you want about a project. Such problem without using Software Product Line causes a massive rework.

In the proposed architecture, each element presented (GUI Interface, Meta-scheduler, schedulers) can be componentized as part of a common architecture called Domain Engineering (DE), and then every new product would made an analyze to an instance of this Architecture of Application Engineering (AE) and specialization of the parties not common.

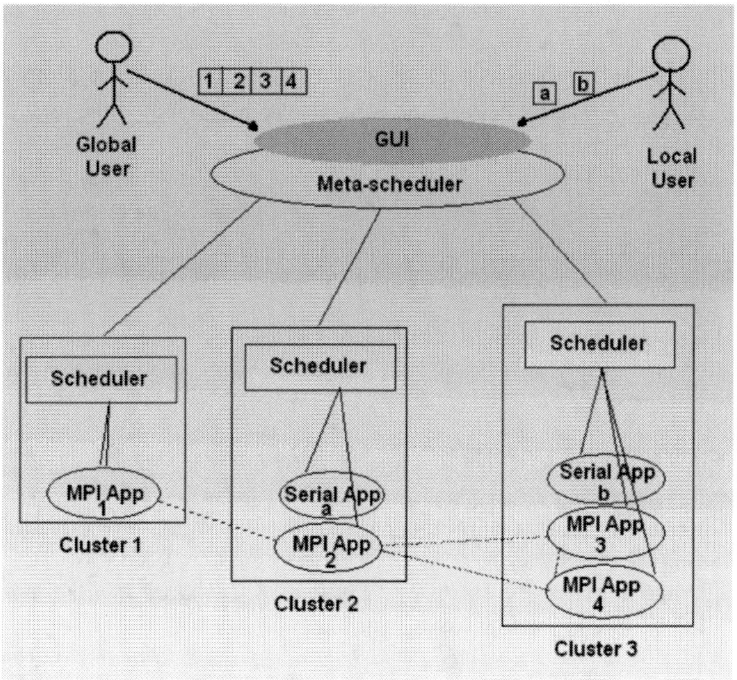

Figure 1. Domain Engineering in the Distributed Computing Environment.

For example it can apply ontologies aimed to describe the existing settings in the available network resources specializing only one of the architecture components. It can also add new elements or components in the architecture, such as an upper layer to replace the GUI for a mobile application with an ontology itself (where this component would be part of the Domain and later reused in other products).

All these elements presented in the architecture and in any layer can be planned carefully developed thinking of reuse and variations points (mandatory, optional and alternative) aimed at creating a Software Product Line in the Distributed Computing Environment.

Study the possibility of using OSGI [17] on the GUI layer and on the cellular phone. This technology increases the level of componentization lowering the granularity of each module.

6. Conclusion and Future Works

This paper presented an approach to Software Product Line architecture specially designed for distributed computing environments. The principal relevance for this approach can be defined as follow.

Increasingly through the Software Engineering seeks to reuse elements for reuse in project from the same domain. This practice is directly related to improvement in the cost of projects, time and productivity [6]. The Software Product Line contributes with all these features and to increase the quality of final product, productivity and so on.

The adoption of Software Products Line concepts provides a way for better integration of environment components quickly and securely, and also supports an approach of mass customization.

The reuse of the artifacts and final product decreases the number of errors, increases compliance of the produced software, eases maintenance and reduces development time.

We are now applying the approach in the projects and documenting information about this. With the application of the approach, we are able to solve practical problems related to the user's indecision as to what elements in its architecture the user would like to have and quickly add and remove elements without major problems.

7. References

[1] Pohl K, Böckle G and van der Linden F J 2005 *Software Product Line Engineering: Foundations, Principles, and Techniques* (Secaucus, NJ: Springer-Verlag)

[2] Clements P C and Northrop L M 2002 *Software Product Lines: Practices and Patterns* (Boston, MA: Addison-Wesley)

[3] Junior E A O, Gimenes I M S and Maldonado J C 2011 A meta-process to support trade-off analysis in software product line architecture *Proc. Int. Conf. on Software Engineering and Knowledge Engineering (Miami)* vol 1 (Skokie, IL: Knowledge Systems Institute) pp 687-692

[4] Silva A P C and Dantas M A R 2007 A selector of grid resources based on the semantic integration of multiple ontologies *Proc. 19th Int. Symp. on Computer Architecture and High Performance Computing (Gramado)* vol 1 (Washington, DC: IEEE Computer Society) pp 143-150

[5] van der Linden F J, Schmid K and Rommes E 2007 *Software Product Lines in Action: The Best Industrial Practice in Product Line Engineering* (Secaucus, NJ: Springer-Verlag)

[6] Software Engineering Institute [homepage on the Internet] Pennsylvania: Carnegie Mellon University 2011 Available from: http://www.sei.cmu.edu/

[7] Gimenes I M and Travassos G H 2002 *O Enfoque de Linha de Produto para Desenvolvimento de Software* (Florianópolis, SC: SBC) pp 1-32

[8] Junior E A O *SystEM-PLA: um método sistemático para avaliaçao de arquitetura de linha de produto de software baseada em UML* PhD [dissertation] São Carlos: Universidade Federal de São Paulo 2010 Avaiable from: Biblioteca Digital de Teses e Dissertações da USP

[9] Foster I 2002 What is the grid? a three point checklist *GridToday* **1** 22-25

[10] Foster I, Kesselman C and Tuecke S 2001 The anatomy of the grid: enabling scalable virtual organizations *In. J. of High Per. Comp. App.* **15** 200-222

[11] Chu D C and Humphrey M 2004 Mobile ogsi. net: grid computing on mobile devices *Proc. 5th IEEE/ACM Int. Work. on Grid Computing (Pittsburgh)* vol 1 (Washington, DC: IEEE Computer Society) pp 182-191

[12] Gomes A T A, Ziviani A, Lima L S and Endler M 2007 DICHOTOMY: a resource discovery and scheduling protocol for multihop ad hoc mobile grids *Proc. of the 7th IEEE Int. Symp. on Cluster Computing and the Grid (Rio de Janeiro)* vol 1 (Washington, DC: IEEE Computer Society) pp 719-724

[13] Czajkowski K, Foster I, Karonis N, Kesselman C, Martin S, Smith W and Tuecke S 1998 A resource management architecture for metacomputing systems *Proc. of the 4th Work. on Job Scheduling Strategies for Parallel Processing (Orlando)* vol 1 (Secaucus, NJ: Springer-Verlag) pp 62-82

[14] Oliveira D and Rosa N 2009 Ubá: a software product line architecture for grid-oriented middleware *Proc. of the 2009 33rd Annual IEEE Int. Computer Software and Applications Conf. (Seattle)* vol 2 (Washington, DC: IEEE Computer Society) pp 160-165

[15] de Oliveira D J S and Rosa N S 2010 Evaluating product line architecture for grid computing middleware systems: ubá experience *Proc. of the 2010 IEEE 24th Int. Conf. on Advanced Information Networking and Applications Work. (Perth)* vol 1 (Washington, DC: IEEE Computer Society) pp 257-262

[16] Goasguen S and McGregor J D 2007 Poster: grid infrastructure software product line *Proc. of the 8th IEEE/ACM Int. Conf. on Grid Computing (Austin)* vol 1 (Washington, DC: IEEE Computer Society)

[17] Open Grid Services Infrastructure [homepage on the Internet] California: OSGi Alliance 2011 Available from: http://www.osgi.org/

Journal of Physics: Conference Series **341** (2012) 012031 doi:10.1088/1742-6596/341/1/012031

Determinant Computation on the GPU using the Condensation Method

Sardar Anisul Haque[1], Marc Moreno Maza[2]
University of Western Ontario, London N6A 1M8, Canada
E-mail: [1] `shaque4@csd.uwo.ca`, [2] `moreno@csd.uwo.ca`

Abstract. We report on a GPU implementation of the condensation method designed by Abdelmalek Salem and Kouachi Said for computing the determinant of a matrix. We consider two types of coefficients: modular integers and floating point numbers. We evaluate the performance of our code by measuring its effective bandwidth and argue that it is numerical stable in the floating point number case. In addition, we compare our code with serial implementation of determinant computation from well-known mathematical packages. Our results suggest that a GPU implementation of the condensation method has a large potential for improving those packages in terms of running time and numerical stability.

1. Introduction

The celebrated algorithm of Charles Lutwidge Dodgson [1] (also known as Lewis Carroll) for computing the determinant of a square matrix $A = (a_{i,j} \mid 0 \leq i, j \leq n-1)$ of order n is a popular trick among students. It is, indeed, much easier to perform by hand on paper than the other classical methods, such as those based on minor expansion or Gaussian elimination. This is due to its amazing data traversal pattern. Each transformation step, from one array to the next one, is a streaming process, called a *condensation*. Dodgson's Algorithm can be executed as a stencil computation: the input data array is transformed into its determinant through $n-1$ successive data arrays. This method suffers, however, from a serious algebraic limitation: it may fail to compute the targeted determinant. Indeed, after each condensation, the newly generated matrix should have no zero elements in its interior [1] for the next condensation step to take place. The interior of A is the submatrix $\text{int}(A) = (a_{i,j} \mid 0 < i, j < n-1)$. One can sometimes reduce to this case by combining rows or columns. When this is not possible, the algorithm terminates without producing any answers. In [2], Abdelmalek Salem and Kouachi Said have solved this difficulty by introducing another type of condensation.

One can easily realize that the condensation method (Dodgson's original method and the improved one by Salem and Said) can be executed in parallel. Moreover, we argue in this paper that its data traversal pattern makes it a good candidate for an implementation within a concurrency platform based on data-parallelism such as CUDA [5].

We report on an implementation of the algorithm described in [2] on GPU using CUDA. We consider two types of coefficients: modular integers and floating point numbers. In the first case, our contribution is to show that the condensation method can be implemented efficiently in terms of memory bandwidth, leading to a very competitive code with respect to popular software packages for computing determinant over finite fields (i.e. with modular integer coefficients).

In the floating point case, our contribution is to show that the condensation method can be implemented efficiently in terms of numerical stability. We observe that the condensation method computes, in some sense, a factorization of the determinant. To take advantage of this fact, we use a new algorithm to compute the product of those factors such that, if overflow/underflow can be avoided then computations are ordered in a way that overflow/underflow is indeed avoided. The challenge is to keep the intermediate products within the range of machine floats; our solution is described in Section 4.

The organization of the paper is as follows. We describe the condensation method in Section 2. Its GPU implementation is presented in Section 3 and 4 for the finite field and floating point case respectively. Both of these two sections contain the corresponding experimental results. Concluding remarks are in Section 5.

2. The condensation method

In this section, we first review the condensation method described in [2]. We will then analyze the algebraic complexity and cache complexity of this condensation method.

2.1. The formula of Salem and Said

As mentioned in the introduction, the authors of [2] have solved the algebraic limitation of Dodgson's Algorithm by introducing another type of condensation, which we summarize below. The input is a square matrix A of order $n > 2$. If the first row of $A = (a_{i,j} \mid 0 \leq i, j \leq n-1)$ is the null vector then the determinant of A is zero and the process terminates. Otherwise, let ℓ be the smallest column index of a non-zero element in the first row of A. The *condensation step* produces a matrix $B = (b_{i,j})$ of order $n-1$ defined by:

$$b_{i,j} = \begin{vmatrix} a_{0,\ell} & a_{0,j+1} \\ a_{i+1,\ell} & a_{i+1,j+1} \end{vmatrix}$$

for $j \geq \ell$ and by $b_{i,j} = -a_{i+1,j}a_{0,\ell}$ for $j < \ell$. The key relation between A and B is the following:

$$\det(A) = \det(B)/(a_{0,\ell})^{n-2} \tag{1}$$

We call $a_{0,\ell}$ the *pivot* of the condensation step. Formula (1) implies that the condensation method of Salem and Said computes $\det(A)$ as a product of powers of the inverse of the successive pivots. In the finite field case, this product can be accumulated in a variable, that is, this variable is updated after each condensation step. In the floating point number case, these powers can be accumulated in a list so that their product can be performed in a way that overflow/underflow is avoided, if possible, as we shall see in Section 4.2.

2.2. The algebraic complexity of the condensation method

Algebraic complexity estimates are given for the RAM model with memory holding a finite number of s-bit words, for a fixed s [9]. Each condensation step involves two matrices: A and B of order n and $n-1$, respectively. Computing B from A requires $2(n-1)^2$ multiplications and $(n-1)^2$ subtractions considering that ℓ refers to the first column (which is the worst case for computing B). The best case happens when ℓ is the last column. When this happens each condensation requires $(n-1)^2$ multiplications. The number of operations involved in finding ℓ is linear in n. The whole algorithm takes at most $n-2$ condensation steps before terminating. So, the total cost for computing the determinant is bounded by $O(n^3)$ arithmetic operations. Moreover, in the worst case, a precise account is $n^3 - 3/2n^2 + 1/2n - 3$, which is comparable to the worst case of Gaussian Elimination.

Journal of Physics: Conference Series **341** (2012) 012031 doi:10.1088/1742-6596/341/1/012031

2.3. *The cache complexity of condensation method*

Cache complexity estimates are given for the ideal cache model. The ideal cache model [10] is a fully associative cache. Its cache replacement policy is optimal in that the cache line to be evicted is one which will be required again furthest in future.

Before estimating the cache complexity of a condensation step, we need to describe the data structures used to represent a square matrix of order n in our implementation. We represent A by a one-dimensional array $\alpha[0, 1, \ldots, n^2 - 1]$ of size n^2. We use *column major layout*, that is, the sub-array $\alpha[i, i+1, \ldots, i+n-1]$, for $i = 0, n, 2n, \ldots (n-1)n$ represents the i-th column. In particular the element $a_{i,j}$ is stored in $\alpha[i + j * n]$.

Consider an ideal cache of Z words, with cache line of L words. To make the analysis simple, we assume that n is large enough such that one column of A does not fit into the cache. Let α and β be two one-dimensional arrays of size n^2 and $(n-1)^2$, representing the input matrix A and output matrix B of the condensation method, respectively. In each condensation step, matrix B is created from A. Assume that, our algorithm will compute B sequentially column-wise. This involves the following data traversals:

- Each element of B is visited only once.
- Each element of A (if it is neither in first row nor in ℓ-th column) is visited only once.
- The ℓ-th column of A is scanned $n-1$ times.
- Each element of the first row of A is visited $n-1$ times consecutively.

It follows that one condensation step incurs $2(n-1)^2/L + n/L + 3$ for α and $(n-1)^2/L + 1$ for β, thus $3(n-1)^2/L + n/L + 4$ cache misses in total. Summing over the condensation steps for $k = Z + 1 \cdots n$ (that is, those for which one column does not fit it in cache) we obtain

$$\frac{(n-Z)\left(n^2 - n + Z^2 - Z + Zn + 1 + 4\,L\right)}{L} \tag{2}$$

Therefore, asymptotically, the ratio between the algebraic complexity and the cache complexity is L. This is similar to Gaussian Elimination. However, the condensation method works in a more data-oblivious way: at each condensation step, apart from the search of the pivot, the same operations are performed independently of the data. This regularity pattern facilitates scheduling, in particular hardware scheduling as it is the case on a GPU. Gaussian Elimination does not have this feature. Indeed, permutations of rows and columns may be needed before proceeding to the next step.

3. GPU implementation: the finite field case

As mentioned in the introduction, the GPU implementation reported in this paper handles two types of coefficients, namely modular integers and floating point numbers. In both cases, each coefficient is stored in a fixed number of machine words and hardware arithmetic is used as much as possible for efficiency consideration. This latter property is more difficult to achieve in the case of modular integers and we discuss it in Section 3.2 Numerical stability is the challenge of floating point number arithmetic and we address it in Sections 4.1 and 4.2. Other types of coefficients, such as multi-precision integers, could be also considered and we leave it for future work.

Before describing the issues specific to the finite field case (in other words to modular integers) we present the part of our implementation which is common to both scenarios. More precisely, we discuss in Section 3.1 our GPU implementation strategy, in particular the mapping between thread blocks and data. In section 3.3, we report on our experimentation with the condensation method for matrices over finite fields. This, of course, is primarily dedicated to evaluate the performance of our GPU implementation, but also to compare it with serial implementations of determinant computation available in computer algebra packages One of our goals is to understand to which extent GPU implementation could improve those packages.

3.1. Data mapping

Each condensation step is performed by one CUDA kernel call. The matrices α and β, introduced in Section 2.3, are stored in the global memory of GPU. After each condensation step, instead of copying β back to CPU main memory, we simply "swap the pointers" to these arrays.

In practice, we find that the index ℓ is small. So we dedicate one kernel call, with a single thread in a single block, to find ℓ. Once we get ℓ, we compute the $(n-2)$-th power of the inverse of $\alpha[\ell * n]$. We will call it *pivot*. We also store the product of the successive pivots in this kernel call.

The kernel performing a condensation step uses one-dimensional blocks and threads. Let T be the number of threads in a block. Each thread is responsible to compute t elements of the array β (representing B). So the total number of blocks required to compute β is $\lceil (n-1)^2/(Tt) \rceil$. Consider thread i is in block j. Then this thread is responsible for computing $\beta[Ttj + it, Ttj + it + 1, \dots Ttj + it + t - 1]$.

3.2. Finite field arithmetic

Multiplying two elements a, b modulo a prime number p is obviously a key routine. Unlike the case of single and double precision floating point arithmetic, the operation $(a, b, p) \longmapsto (ab) \bmod p$, for $a, b, p \in \mathbb{Z}$, is not provided directly by the hardware. This operation is thus an efficiency-critical low-level software routine that the programmer must supply. When p is a machine word size prime, which is the assumption in this paper, two techniques are popular in the symbolic computation community.

The first one takes advantage of hardware floating point arithmetic. We call `double_mul_mod` our implementation of this technique, for which our CUDA code is shown below. The fourth argument `pinv` is the inverse of `p` which is precomputed in floating point.

```
__device__ int double_mul_mod(int a, int b, int p, double pinv) {
   int q  = (int) ((((double) a) * ((double) b)) * pinv);
   int res = a * b - q * p;
   return (res < 0) ? (-res) : res;
}
```

In our implementation, double precision floating point numbers are encoded on 64 bits and make this technique work correctly for primes p up to 30 bits.

The second technique, called the *Montgomery reduction* relies only on hardware integer arithmetic. We refer to Montgomery's paper [11] for details. We have experimented both approaches in [12]. Our CUDA implementation favors the `double_mul_mod` trick.

3.3. Experimental results

We generate random integer matrices modulo a prime number p of machine word size. The order of our test matrices varies from 10 to 4000. We conduct all our experiments on a GPU NVIDIA Tesla 2050 C.

We use *effective memory bandwidth* to evaluate our GPU code. The effective memory bandwidth (measured in GB/seconds) of a kernel run is, by definition,

- the amount of data traversed in the global memory of the GPU during the kernel run,
- divided by the running time of the kernel.

Following a principle proposed by Greg Ruetsch and Paulius Micikevicius in [7], we compared the effective memory bandwidth of our kernel to that of a *copy kernel*, that is, a kernel that simply performs one memory copy from one place to another place in the global area of GPU. Such benchmark kernel can be regarded as a good practical measure of the maximum memory bandwidth of a kernel.

For matrix of order of 3000, the effective memory bandwidth of the copy kernel and our condensation method (with modular integer coefficients) on our card are 96 GB/s and 18.5 GB/s respectively.

Our effective memory bandwidth results show that our code is reasonably efficient considering the following two facts:

- the index calculation in our code is not straightforward and
- finite field arithmetic (see Section 3.2) is time consuming.

Figure 1 reports on the memory bandwidth of our CUDA code for different matrix orders.

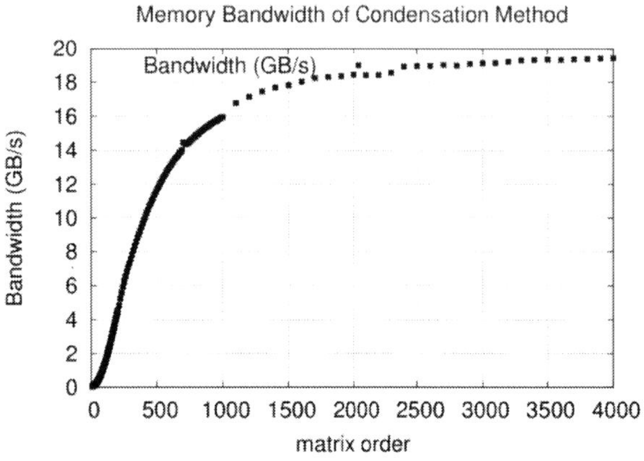

Figure 1. Effective memory bandwidth of condensation method.

To conclude this section, we compare our CUDA code for computing determinants over finite fields with its counterpart in two popular mathematical software packages, namely MAPLE [4] and NTL [3].

Figure 2 compares the computing time between our CUDA code for the condensation method and the NTL determinant code, both with modular integer coefficients. It shows clearly that the condensation method in CUDA outperforms NTL code for computing determinant over finite fields. For example, when $n = 4000$, the condensation method in CUDA takes about 9 seconds, while NTL code takes about 850 seconds.

Figure 3 compares the computing time between our CUDA code for the condensation method and MAPLE's determinant command over finite fields. It shows clearly that the condensation method in CUDA outperforms MAPLE code for computing determinant over finite fields. For example, when $n = 4000$, the condensation method in CUDA takes about 9 seconds, whereas MAPLE code takes about 90 seconds.

4. GPU implementation: the floating point case

In this section, we consider the case of matrices with floating point number coefficients. We adapt our CUDA code described in Section 3 to this new scenario. The modifications are described in Section 4.1. One potential challenge that we found is to multiply the successive pivots. Mathematically, the problem is to multiply a sequence of floating values where the intermediate results might not be in the range of the floating point number data type while the final results might be. We will state the problem and our solution in Section 4.2. We conclude the section by providing experimental results.

4.1. Finding the pivots

Instead of taking the first nonzero from the left in the first row of A, we choose the nonzero element of the first row that is closest to value 1.0; let use call $p = a_{0,\ell}$ this element. We have verified that this modification of the original algorithm in [2] does not invalidate the expected result, namely the determinant. For simplicity, we are describe the procedure for matrices A and B instead of the arrays α and β. Once p is chosen, all elements in the ℓ-th column are divided by p including $a_{0,\ell}$. Thus we modify Formula (1) as follows:

$$\det(A) = \det(B) * p$$

We will call p the pivot for this floating point number implementation.

The benefits of the above transformation are as follows.

- $a_{0,\ell}$ becomes 1.0. So we need neither computing the $(n-2)$-th power of it nor performing any division at the end.
- By choosing an element that is the closest to 1.0, we are expecting to reduce the potential of overflow/underflow.

4.2. Multiplication of the successive pivots

We first state the problem that we wish to address. Consider an array $x[0, 1, \ldots, k-1]$ of k floating point numbers, encoded by a floating point number data type of fixed precision. Then the problem is to write an algorithm for computing the product $X = \prod_{i=0}^{k-1} x[i]$, assuming that X fits within the range of the given floating point number data type. Our solution to this problem is stated as Algorithm 1 hereafter.

We give a sketch of the proof of Algorithm 1. We observe that multiplications occur at Lines 6, 17 and 25. The multiplication at Line 6 cannot lead to overflow/underflow since $|q1| \leq 1 \leq |q2|$ holds. If Lines 17 or 25 would lead to overflow/underflow, this would bring a contradiction to our hypothesis.

We estimate the running time of Algorithm 1. The first *while* loop runs in linear time with the number of elements in x. The second *while* loop runs $m-1$ times considering there exists m elements in R. Each of the iteration takes $O(m)$ time. So the time complexity of the second *while* loop is $O(m^2)$. Considering the inequality $k \geq m$, the time complexity of the Algorithm is $O(k^2)$.

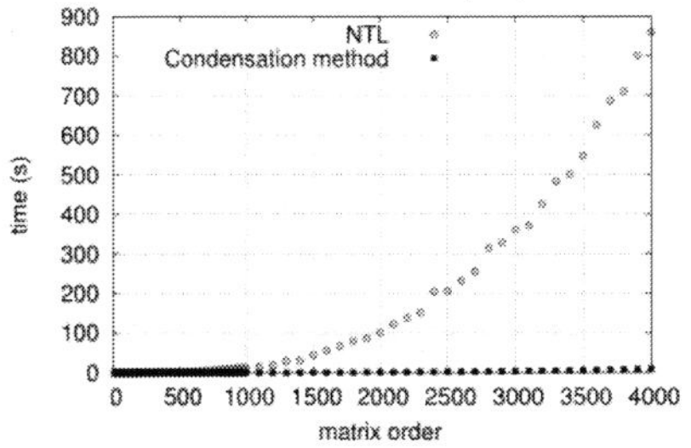

Figure 2. CUDA code for condensation method and determinant on NTL over finite field.

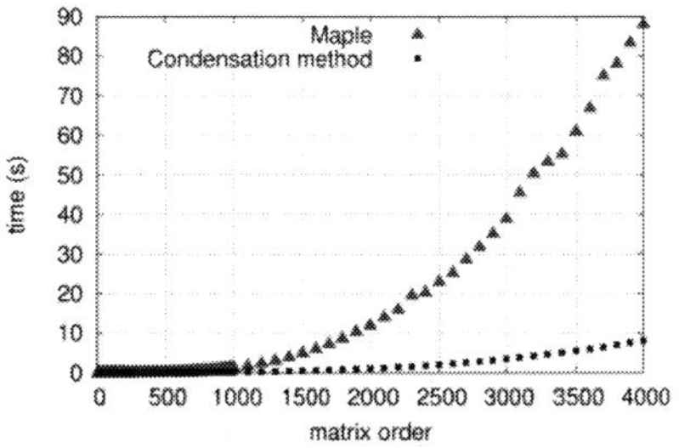

Figure 3. CUDA code for condensation method and determinant on MAPLE over finite field.

Algorithm 1 MulSuccPivot (x)

```
Create a stack S1 of elements of x in [-1.0, 1.0].
Create a stack S2 of the other elements of x not in S1.
while both S1 and S2 are nonempty do
    q1 = pop(S1)
    q2 = pop(S2)
    q = q1 * q2
    if q is in [-1.0, 1.0] then
        push(q, S1)
    else
        push(q, S2)
    end if
end while
if stack S1 is not empty then
    make a list R with the elements in S1
    while R has more than one element do
        select r1 and r2 in R such that r1 and r2 are closest to 0 and |1.0| respectively.
        r = r1 * r2
        delete r1 and r2 from R
        insert r into R
    end while
else
    make a list R with the elements in S2
    while R has more than one element do
        select any r1 and r2 in R.
        r = r1 * r2
        delete r1 and r2 from R
        insert r into R
    end while
end if
return R
```

4.3. Experimentation

For the experimentation in the case of floating point coefficients matrices with MAPLE, we use the *Determinant* command of the *LinearAlgebra* package. In this case, MAPLE may not have the best possible implementation, since MAPLE's primary purpose is symbolic computation. However, MATLAB has certainly a competitive implementation for floating point coefficients. Indeed, efficiently supporting numerical linear algebra is the primary goal for this cutting-edge software.

For small finite field coefficients, the best serial algorithm is simply Gaussian elimination, which is what MAPLE is using. Therefore, in the case of modular integers, our comparison reported in Section 3.3 is also meaningful.

In order to investigate the numerical stability of our GPU implementation of the condensation method, we use the infamous Hilbert matrix $H_{ij} = \frac{1}{i+j-1}$, which is a canonical example of ill-conditioned matrix. This matrix is non-singular, for each value of n. However, for n large enough, any determinant computation of this matrix using a fixed precision floating point number arithmetic will return zero.

$$\begin{bmatrix} 1 & \frac{1}{2} & \frac{1}{3} & \frac{1}{4} & \frac{1}{5} \\ \frac{1}{2} & \frac{1}{3} & \frac{1}{4} & \frac{1}{5} & \frac{1}{6} \\ \frac{1}{3} & \frac{1}{4} & \frac{1}{5} & \frac{1}{6} & \frac{1}{7} \\ \frac{1}{4} & \frac{1}{5} & \frac{1}{6} & \frac{1}{7} & \frac{1}{8} \\ \frac{1}{5} & \frac{1}{6} & \frac{1}{7} & \frac{1}{8} & \frac{1}{9} \end{bmatrix}$$

In the tables below, we compare determinant computation of the Hilbert matrix with

- MAPLE using multi-precision floating point number arithmetic (thus software floating point number),
- MATLAB using double-precision floating point number arithmetic,
- our CUDA implementation using double-precision floating point number arithmetic.

We observe that:

(i) despite of the use of multi-precision floating point, MAPLE is less accurate than MATLAB and our CUDA implementation (this was checked by computing the exact value of the determinant using rational number arithmetic),

(ii) our CUDA implementation of the condensation method can compute determinants of much larger order than MATLAB,

(iii) our CUDA implementation is also competitive with MATLAB in terms of running time.

5. Conclusion

MAPLE and MATLAB commands for computing matrix determinants combine many different state-of-the-art algorithms. On a given input, MAPLE and MATLAB determinant commands choose one of these algorithms by considering the types of the coefficients and the combinatorial properties (size, sparsity) of the input matrix. These choices are heavily tuned since linear algebra is, in the case of MAPLE, at the core of its symbolic routines while it is, in the case of MATLAB, at the core of the whole system. Therefore, comparing our code against those systems is meaningful. From our experimental results, it is clear that the condensation method implemented on the GPU is a promising candidate for computing determinants of matrices with both modular integer coefficients and floating point number coefficients.

Though it seems unfair in the first place that we compare our parallel code with serial codes in mathematical packages, our primary objective is to propose parallel algorithms for computing determinants within mathematical software packages, such as MAPLE and MATLAB. Actually, these two systems are already able today to take advantage of multicore processors and GPUs for certain types of computations. Therefore our objectives are meaningful and motivated by our active cooperation with the Maplesoft company developing MAPLE. We believe that a GPU implementation of the condensation method can be used to improve the efficiency, in terms of running time and numerical stability, of existing mathematical software packages.

Acknowledgments

We like to thank Dr. Wei Pan (Intel Corporation) for discussions around our CUDA implementation in the finite field case. We also like to thank Dr. Jürgen Gerhard (Maplesoft) for his suggestions in the floating point number case.

Matrix order	MAPLE	MATLAB	Condensation Method on GPU
5	0.3239712e-11	3.749295e-12	3.74967e-12
6	-0.1037653175e-16	5.367300e-18	5.36556e-18
7	-0.2940657217e-22	4.835803e-25	4.44292e-25
8	-0.2156380381e-28	2.737050e-33	-3.92813e-33
9	-0.1692148341e-35	9.720265e-43	-2.79235e-41
10	0.4704819751e-42	2.164405e-53	-4.44342e-50
15	0.1386122551e-74	-2.190300e-120	-9.47742e-103
20	0.4711757502e-106	-1.100433e-195	3.81829e-164
25	-0.4092672466-139	5.482309e-274	-3.82134e-239
30	-0.2087134536-174	0	-2.50914e-319
35	0.6863051439e-205	-	3.50293e-398
40	0.3354475665e-237	-	-7.42227e-479
70	-0.1605231989e-443	-	-1.42973e-961
100	-0.1344119185e-667	-	1.96009e-1467
200	-0.1635472167e-1423	-	9.43651e-3169
295	-0.1313897019e-2117	-	3.27673e-4811
300	0.4832058492e-2154	-	-1.95564e-4897
320	0.1012376674e-2298	-	7.2904e-4951
340	0.3198288621e-2442	-	-8.67557e-4949*2.08848e-644
360	0.6712616355e-2593	-	9.84118e-4938*8.32678e-1006
380	-0.1532669346e2736	-	-3.28068e-4950*-6.51644e-1341
400	-0.4230797452e-2881	-	-6.19676e-4945*6.56337e-1696
500	-0.1956609252e-3608	-	1.40177e-4939*-2.22223e-3444
600	-0.4139972675e-4335	-	-2.55164e-4950*1.99856e-4945*6.19736e-232
800	0.4570493645e -5853	-	4.25009e-4940*-2.21715e-4940*-2.17891e-3739

Table 1. Determinant of Hilbert Matrix by MAPLE, MATLAB, and condensation method on both CPU and GPU.

References

[1] C. L. Dodgson. Condensation of Determinants, Proceedings of the Royal Society of London, 15(1866), 150-155.
[2] Abdelmalek Salem, and Kouachi Said. Condensation of Determinants, `http://arxiv.org/abs/0712.0822`.
[3] NTL: A library for doing number theory. `http://www.shoup.net/ntl`.
[4] Maple: The essential tool for mathematics and modeling. `http://www.maplesoft.com`.
[5] NVIDIA developer zone. `http://developer.nvidia.com`.
[6] Matlab- The language of technical computing. `http://www.mathworks.com`.
[7] G. Ruetsch and P. Micikevicius. Optimizing MatrixTranspose in CUDA, NVIDIA Corporation, 2009.
[8] W. Pan. Algorithmic Contributions to the Theory ofRegular Chains, PhD Thesis, The University of Western Ontario, Canada, 2011.
[9] J. E. Savage. Models of Computation, Addison-Wesley Longman, Boston, MA, USA, 1998.
[10] M. Frigo, C. E. Leiserson, H. Prokop and S. Ramachandran. Cache-Oblivious algorithms. FOCS'99: Proc. of the 40th Annual Symp. on Foundations of Comp. Sc., 1999
[11] P. L. Montgomery. Modular multiplication without trial division. *Mathematics of Computation*, 44(170):519–521, 1985
[12] X. Li, M. Moreno Maza, and W. Pan. Computations modulo regular chains. In *Proc. ISSAC'09*, pages 239–246, New York, NY, USA, 2009. ACM Press.

Matrix order	MAPLE	MATLAB	Condensation Method on GPU
5	0.004	0	0.000530
6	0.008	0	0.000570
7	0.012	0	0.000595
8	0.008	0	0.000631
9	0.012	0	0.000741
10	0.012	0	0.000447
15	0.016	0	0.000964
20	0.016	0	0.001078
25	0.020	0	0.001271
30	0.024	-	0.001460
35	0.044	-	0.001671
40	0.036	-	0.001896
70	0.188	-	0.003083
100	0.588	-	0.005145
200	5.988	-	0.012488
295	20.733	-	0.023402
300	21.661	-	0.023759
320	26.741	-	0.026633
340	31.677	-	0.029433
360	38.150	-	0.032401
380	46.146	-	0.035940
400	54.099	-	0.038955
500	104.334	-	0.058193
600	187.151	-	0.081969
800	467.541	-	0.147037

Table 2. Time(s) Required to compute determinant of Hilbert Matrix by MAPLE, MATLAB, and condensation method on both CPU and GPU.

Integration of a neuroimaging processing pipeline into a pan-canadian computing grid

S. Lavoie-Courchesne[1,2,3], P. Rioux[3], F. Chouinard-Decorte[1,3], T. Sherif[3], M.-E. Rousseau[3], S. Das[3], R. Adalat[3], J. Doyon[1,2], C. Craddock[4,8], D. Margulies[5,8], C. Chu[6,8], O. Lyttelton[7,8], A. C. Evans[3], P. Bellec[1,2,8]

[1]Centre de recherche de l'institut de gériatrie de Montréal, Montréal, CA
[2]Département d'informatique et de recherche opérationnelle, Université de Montréal, Montréal, CA
[3]McConnell Brain Imaging Centre, Montreal Neurological Institute, McGill University, Montréal, CA
[4]Virginia Tech Carilion Research Institute, Roanoke, VA, USA
[5]Max Planck Institute for Human Cognitive and Brain Sciences, Leipzig, DE.
[6]Section of Functional Imaging Methods, Laboratory of Brain and Cognition, NIMH, NIH, Bethesda, USA
[7]Mind, Brain Imaging and Neuroethics Unit, Institute of Mental Health Research, Royal Ottawa Health Care Group, University of Ottawa, Ottawa, CA
[8]The Neurobureau Research Institute, Chicago, USA

E-mail: {bellecp@iro.umontreal.ca, lavoiecs@iro.umontreal.ca}

Abstract. The ethos of the neuroimaging field is quickly moving towards the open sharing of resources, including both imaging databases and processing tools. As a neuroimaging database represents a large volume of datasets and as neuroimaging processing pipelines are composed of heterogeneous, computationally intensive tools, such open sharing raises specific computational challenges. This motivates the design of novel dedicated computing infrastructures. This paper describes an interface between PSOM, a code-oriented pipeline development framework, and CBRAIN, a web-oriented platform for grid computing. This interface was used to integrate a PSOM-compliant pipeline for preprocessing of structural and functional magnetic resonance imaging into CBRAIN. We further tested the capacity of our infrastructure to handle a real large-scale project. A neuroimaging database including close to 1000 subjects was preprocessed using our interface and publicly released to help the participants of the ADHD-200 international competition. This successful experiment demonstrated that our integrated grid-computing platform is a powerful solution for high-throughput pipeline analysis in the field of neuroimaging.

1. Introduction

With most modern scientific projects relying heavily on computer programs, the dissemination of research software prototypes is of central importance at the very least to enable the reproduction of published results by independent research groups [1]. In the neuroimaging field, such dissemination raises a number of domain-specific computational challenges:

(i) **Complex multi-stage processing.** Neuroimaging datasets need to go through a large number of preprocessing steps before a scientifically meaningful outcome can be inferred.

These steps can include such operations as non-linear realignment of the individual brain anatomy into a common stereotaxic space [2], the correction for motion of the participant during the data acquisition [3] or the correction of physiological noise [4], amongst many others. Even for a common type of analysis such as the multi-level general linear model [5], it is only recently that fully automated scripts, which include all analysis steps going from raw data to the final results, have been made widely available [6].

(ii) **Computational challenge.** A neuroimaging database includes a collection of multiple brain volumes for anywhere between tens to thousands of subjects, and typically representing at least several gigabytes of raw data. Some of the processing steps mentioned above can take up to several hours of computation for a single subject on a single machine, such as the extraction of the cortical surface [7]. Processing a full database can thus range from tens of hours for a small study to possibly thousands of hours for a large database or a complex analysis, e.g. [8]. Rather than a very large computational challenge faced by a single research group, hundreds of research groups in the neuroimaging community are independently facing moderately large computational tasks.

(iii) **Diversity of production environments.** The type of computational resources that are being used to process neuroimaging databases varies widely in the neuroimaging community, ranging from a personal workstation to cloud or grid computing [9] and including multi-core workstations as well as computer clusters. To reach a wide audience, a software package should ideally be able to take advantage of any of these computing environments.

The most common approach to high-performance computing proceeds by parallelizing computations using a message-passing interface (MPI) [10]. The processing tools used in neuroimaging are generally not available in such a form, and re-implementing them with MPI is not feasible given the volume and heterogeneity of the existing code base and the fact that new algorithms are constantly being released by the community in a non-parallel format. The processing of a neuroimaging database is however inherently parallel, as multiple subjects are often treated independently. A solution to the issue of the computational load is thus to formally combine tools into a so-called pipeline, which represents the sequential application of a given series of tools to a particular dataset, and then to take advantage of the parallel nature of the pipeline flowchart to use distributed computation resources. If necessary, jobs simply interact through a file system which is shared by all the processing units. Rather than strict high-performance computing, this type of approach belongs to the emerging family of many-tasks computing [11]. A number of independent frameworks have been proposed in the neuroimaging community to design pipelines composed of several heterogeneous, non-parallel tools and then execute them in a distributed computing environment, e.g. [12]. The LONI pipeline[1] [13] offers a graphical interface to design pipelines, as well as a web-based system to execute them on a dedicated supercomputer, but it is not easy to deploy that system on any given computing facility. Some libraries have recently been implemented in high-level scientific programing languages to design and execute pipelines : the pipeline system for Octave[14] and Matlab (PSOM[2]), Nipype[3] [15] and SOMA workflow[4] for Python [16]. These code-based frameworks are powerful for the implementation of a fully automated prototype pipeline, and can be deployed in a variety of computing environments, including supercomputers, but not on a grid of heterogeneous supercomputers. The CBRAIN[5] initiative addresses this last issue by providing a flexible platform to run pipelines on a grid of supercomputers; however, CBRAIN

[1] `http://pipeline.loni.ucla.edu/`
[2] `http://code.google.com/p/psom/`
[3] `http://nipy.sourceforge.net/nipype/`
[4] `http://brainvisa.info/doc/soma-workflow`
[5] `http://cbrain.mcgill.ca/`

was designed to run existing packages rather than build pipelines from scratch.

In this paper, we propose an integrated solution for the development and deployment of neuroimaging pipelines by building an interface between PSOM and CBRAIN. Since the proposed interface is generic, it is very easy to take any PSOM pipeline and make it publicly accessible via CBRAIN. We estimate that the integration of a new pipeline in CBRAIN could take as little as one day of work. The PSOM/CBRAIN integration addresses the issues (i-iii) we initially listed, as it implements fully automated multi-step analysis that can be deployed in virtually any computing environment, from a single workstation with multiple CPUs to a grid of supercomputers. To demonstrate the feasibility of our approach, we have integrated a pipeline for the preprocessing of functional magnetic resonance imaging datasets into CBRAIN. This pipeline is available as part of the "Neuroimaging Analysis Kit" (NIAK)[6] [17], and was developed using the PSOM framework. We demonstrate the benefits of this integration on a large database, the ADHD-200 sample[7].

2. PSOM and NIAK

PSOM is a framework to implement pipelines in a high-level programming open source GNU[8]-compatible language called Octave, and is also compatible with the proprietary language Matlab[9]. The core idea of PSOM is to build a representation of all the steps of the pipeline using a specific native Octave structure. Once this structure is defined, a generic tool called the pipeline manager can execute the pipeline in a number of computational environments, i.e. a local computer or a cluster of computers. A number of coding guidelines have also been established to help the developers keep the code concise, and ensure that the modules used in pipelines as well as the pipelines themselves can easily be combined, while keeping access to, and track of, all pipeline parameters. The description of a pipeline involves a list of jobs which are defined through five attributes :

- The **command** executed by the job. This is a (short) piece of Octave/Matlab code.
- A list of **input files** necessary to run the job.
- A list of **output files** generated by the job.
- A list of **clean-up files** that will be deleted by the job, in order to limit the amount of intermediate outputs of the pipeline.
- A list of **options** (or parameters).

These attributes are coded in a so-called Octave/Matlab structure, where each field represents a job, and each job has up to five fields coding for its attributes. An example of a simple pipeline follows:

```
% Job "sample" :    No input, generate a random vector a
command = 'a = randn([opt.nb_samps 1]); save(files_out,''a'')';
pipeline.sample.command      = command;
pipeline.sample.files_out    = 'sample.mat';
pipeline.sample.opt.nb_samps = 10;
% Job "quadratic" : Compute a.^2 and save the results
command = 'load(files_in); b = a.^2; save(files_out,''b'')';
pipeline.quadratic.command   = command;
pipeline.quadratic.files_in  = pipeline.sample.files_out;
pipeline.quadratic.files_out = 'quadratic.mat';
```

[6] www.nitrc.org/projects/niak
[7] fcon_1000.projects.nitrc.org/indi/adhd200
[8] http://www.gnu.org/copyleft/gpl.html
[9] http://www.mathworks.com/

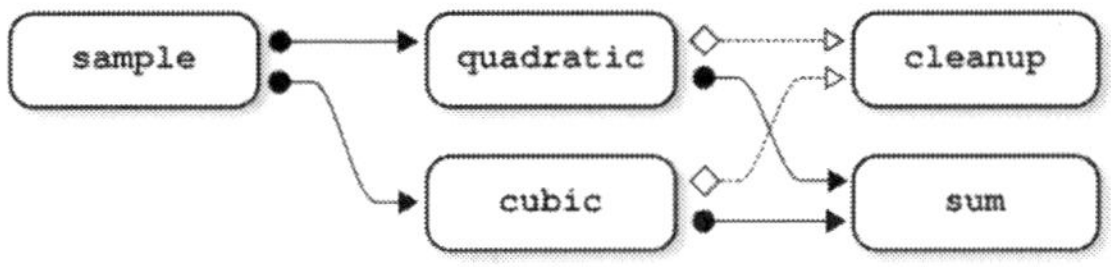

Figure 1. Dependency graph of a toy pipeline. Each job (i.e. running a specific tool on a specific dataset) is represented by a box. A plain arrow from box `A` to box `B` means that the job `B` uses one or multiple file(s) generated by job `A`. A dotted arrow from box `A` to box `B` means that the job `A` is using one or multiple file(s) deleted by job `B`.

```
% Adding a job "cleanup" : delete the output of "sample"
pipeline.sample.command       = 'delete(files_clean)';
pipeline.sample.files_clean   = pipeline.sample.files_out;
% Adding a job "cubic" : Compute a.^3 and save the results
command = 'load(files_in); c = a.^3; save(files_out,''c'')';
pipeline.cubic.command        = command;
pipeline.cubic.files_in       = pipeline.sample.files_out;
pipeline.cubic.files_out      = 'cubic.mat';
% Adding a job "sum" : Compute a.^2+a.^3 and save the results
command = 'load(files_in{1}); load(files_in{2}); d = b+c, save(files_out,''d'')';
pipeline.sum.command        = command;
pipeline.sum.files_in{1}    = pipeline.quadratic.files_out;
pipeline.sum.files_in{2}    = pipeline.cubic.files_out;
pipeline.sum.files_out      = 'sum.mat';
```

The order in which the jobs are specified is meaningless. There are however some constraints on the order of execution for the jobs:

- If job `A` uses `file` as input, and `file` is an output of job `B`, then the execution of `B` must be complete before `A` is started.
- If job `A` uses `file` as input, and `file` is deleted by job `B`, then the execution of `A` must be complete before `B` is started.

These constraints define a dependency graph. For a pipeline to be solvable, the dependencies must form a directed acyclic graph. The dependency graph for the toy pipeline above is presented in Figure 1.

The basic approach to integrating a PSOM pipeline into the CBRAIN platform was to generate the pipeline structure and dependencies, and then convert it into an XML structure that could be handled by the CBRAIN server. The PSOM pipeline manager is not used at all in this context. The feasibility of this approach was tested with a particular pipeline, designed to preprocess a neuroimaging database including structural magnetic resonance imaging (MRI) as well as functional magnetic resonance imaging (fMRI). This pipeline along with all the necessary tools is distributed publicly as part of the neuroimaging analysis kit (NIAK)[10] [17]. The outline of the processing steps for one subject was as follows. The three first volumes of each run were suppressed to allow the magnetisation to reach equilibrium. Each dataset was corrected of inter-slice difference in acquisition time, rigid body motion, slow time drifts (high-pass filter with a 0.01 Hz cut-off) and physiological noise [4]. Slow time drifts and physiological noise correction were implemented in an attempt to reduce the spatially correlated noise present in the fMRI time series, which may induce spurious spatial correlations unrelated to neural activity. For each subject, the mean motion-corrected volume of all the datasets was coregistered with

[10] www.nitrc.org/projects/niak

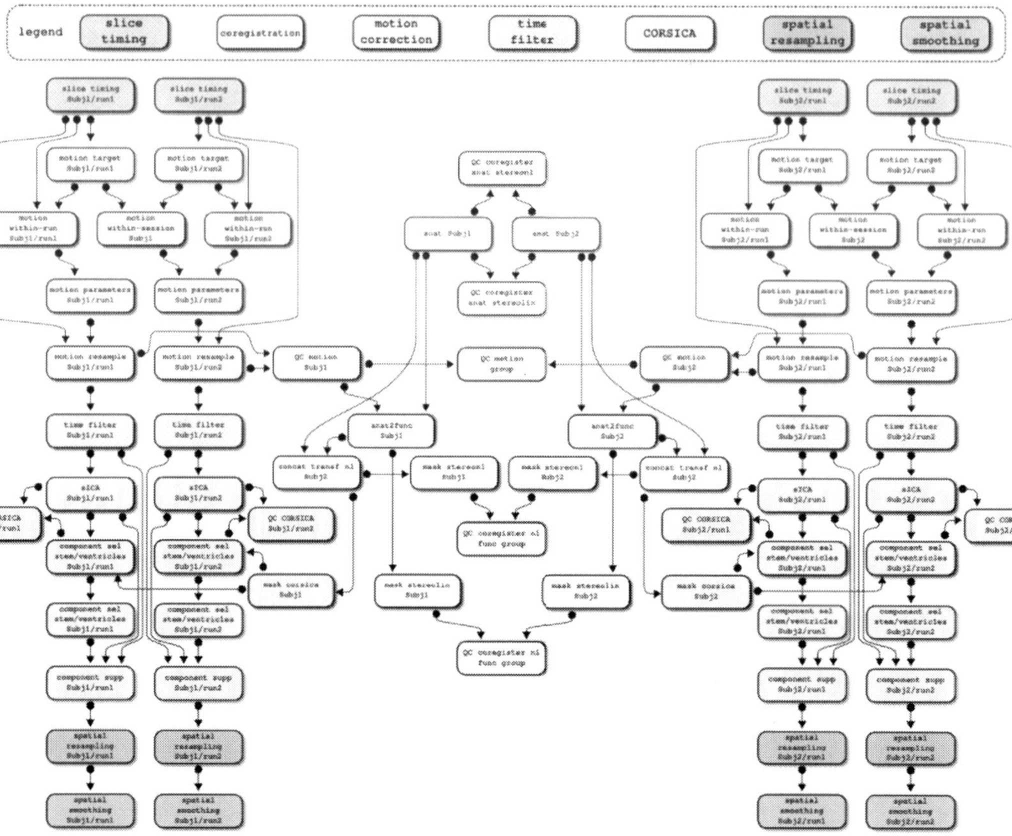

Figure 2. An example of dependency graph for the NIAK fMRI preprocessing pipeline. This example includes two subjects with two fMRI datasets each. The pipeline includes close to 100 jobs, and clean-up jobs have been removed to simplify the representation. Colors have been used to code the main subpipelines in the analysis: correction for slice timing and motion, coregistration, temporal and spatial filtering, spatial resampling and correction of physiological noise (CORSICA).

a T1 individual scan using minctracc [18], which was itself non-linearly transformed to the Montreal Neurological Institute (MNI) non-linear template [2] using the CIVET pipeline [7]. The functional volumes were resampled in the MNI space at a 3 mm isotropic resolution and spatially smoothed with a 6 mm isotropic Gaussian kernel. The spatial smoothing was implemented in an attempt to minimize the residual variability in anatomy and functional organization of individual brain in stereotaxic space. These operations were implemented through generic medical image processing modules, the so-called MINC tools[11]. Some of the operations (motion correction, correction of physiological noise, CIVET) are themselves pipelines, involving a number of steps. A flowchart for the pipeline is represented in Figure 2.

3. CBRAIN

CBRAIN's primary goal is to render the computational aspects of neurological analyses as simple as possible. Researchers interact with the platform through an intuitive web interface. Complex

[11] `http://en.wikibooks.org/wiki/MINC`

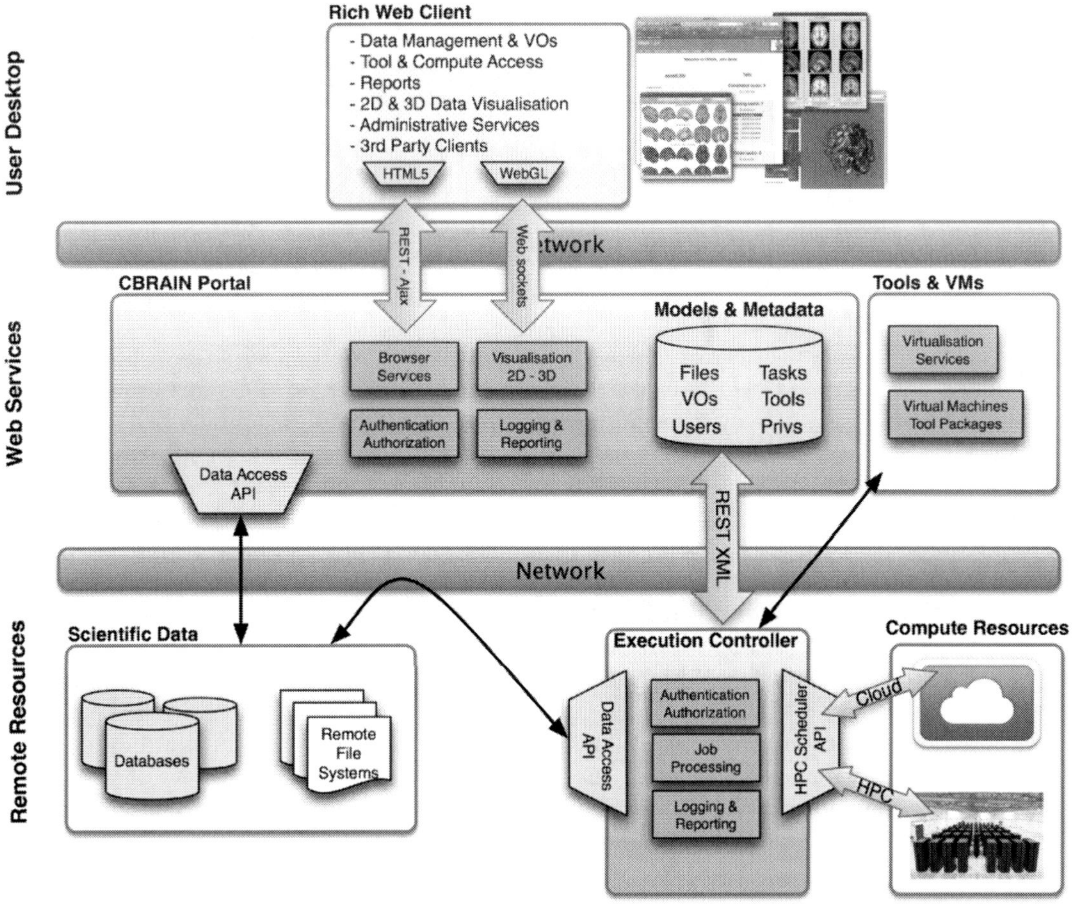

Figure 3. CBRAIN Platform Architecture Layers. The CBRAIN platform is a fully distributed grid of data and compute resources implemented using the Ruby on Rails Framework and common UNIX tools and libraries. It consists of 3 main layers: 1- The user layer (top), which represents the Rich Web Client. An authenticated user has full control of his/her data, tasks, tools or resources through a standard web browser. 2- The service layer (middle) contains all web services, presentation layers, models, logic, authentication/authorization, reporting and platform metadata. This layer does not contain scientific data, nor does it perform heavy computation. 3- The resource layer (bottom) represents all registered data storage (file systems or databases) and compute resources. A low-footprint Execution Controller (EC) coordinates and reports all activities on each HPC resource. Note that the Data Access API on the EC accesses data directly from the data resource without creating a bottleneck in the service layer, similar to every other grid architecture.

command-line tools are wrapped in simple web forms. On the back end, interactions with remote data and High Performance Computing centers (HPCs) are handled as transparently as possible. CBRAIN connects to HPCs and data around the world, giving users access to processing power previously unavailable to most research groups, while abstracting away all the complexities of handling the notoriously heterogeneous environments of distributed resources. CBRAIN's fundamental architecture is modular and generic, meaning that its applicability is by no means restricted to neuroimaging.

CBRAIN is currently connected to 9 HPCs across Canada and Europe, for a total of around 80 000 cores. Given the heterogeneous nature of HPC setups and policies around the country and the world in general, it was necessary to create an API for CBRAIN to interact with various cluster management programs simply and effectively. Although some active projects, such as the DRMAA working group [19], have been carefully examining scheduler integration issues, none of them handled the sheer variety of schedulers, queuing setups and policies they encountered in a satisfactory way. Issues vary greatly, from unsupported scheduler versions, core scheduling versus full node scheduling, multi-level queuing, queuing limits to completely home-brewed schedulers (i.e. using UNIX scripts or Perl). To avoid imposing some queuing requirements on every compute center, the CBRAIN design aims at keeping any remote footprint as lightweight as possible. CBRAIN's Simple Cluster Interface in Ruby (SCIR) library does just that and can currently interact with HPCs using various flavors and setups of SGE, Torque, PBS, Moab, Maui as well as a variety of custom setups.

Data in CBRAIN is handled through a flexible data provider API, allowing for the integration of any network-enabled storage or database into the platform. All registered data sources in CBRAIN are transparently presented as if they were from a single system. A project may contain data from multiple sites shared under the "virtual organization" model, thereby restricting access on an individual or group basis. By default, all data or generated results can only be accessed by the user whom registered them in the system. The user can then share these data with other registered members of CBRAIN. Collaborative projects usually share common data and, if necessary, roles can be established so that not all project members have read and write access.

4. Interface between PSOM and CBRAIN

At a user level, the execution of the pipeline starts with the selection of a file collection, see Figure 8 (Annex). The file collection has to follow a rigorous organization, making it possible to infer the identification labels as well as the neuroimaging datasets available for each subject. In this case, the canonical file organization followed the one adopted by the ADHD-200 consortium (see next Section 5 and Figure 4 for an outline). Adapting the pipeline to other types of file organization involves minimal coding, but it is one of the few development steps that is not automated at this point. The user then specifies the parameters of the pipeline in a dedicated web form coded in Ruby on Rails, see Figure 9 (Annex). Once the parameters of the pipeline have been specified, the interactions between CBRAIN and PSOM are as follows (see Figure 5):

(1) CBRAIN saves the pipeline parameters in an XML file, along with a list of all the files that are going to be used in the pipeline execution.

(2) A generic tool available as part of PSOM (`psom_read_xml`) is used to read the XML parameters file, parse it and convert it into an Octave/Matlab structure.

(3) A dedicated tool (`niak_pipeline_fmri_preprocess`) is used to generate an fMRI preprocessing pipeline in NIAK based on the Octave/Matlab structure. At this stage, no dataset is actually processed but all steps of the analysis are described in an Octave/Matlab structure, as described in Section 2.

(4) A generic tool available in PSOM (`psom_write_pipeline2xml`) is used to convert the pipeline into a series of XML files containing the job name, the input and output file names,

```
ADHD-200 file collection
|__ sites ................................................... kki
|   |__ subjects ......................................... 1018959
|   |   |__ sessions .................................... session_1
|   |   |   |__ anats ...................................... anat_1
|   |   |   |   |__          mprage_noface.nii.gz
|   |   |   |__ anats ...................................... anat_2
|   |   |   |   |__          mprage_noface.nii.gz
|   |   |   |__                    ...
|   |   |   |__ rests ...................................... rest_1
|   |   |   |   |__              rest.nii.gz
|   |   |   |__ rests ...................................... rest_2
|   |   |   |   |__              rest.nii.gz
|   |   |   |__                    ...
|   |   |__ sessions .................................... session_2
|   |   |   |__                    ...
|   |   |__                        ...
|   |__ subjects ......................................... 1019436
|   |   |__                        ...
|   |__                            ...
|__ sites ................................................... nyu
|   |__                            ...
|__                                ...
```

Figure 4. File organization of the ADHD-200 database. The files are organized with one folder per site contributing neuroimaging datasets (e.g. Kennedy Krieger Institute[13]). The site-specific folder contains a file with comma-separated values coding for the phenotypic information for all participants, as well as one folder per subject named with the unique subject ID. The folder of each subject contains a number of subfolders with arbitrary name, each one corresponding to an imaging session (visit of the participant to the scanning site). For each imaging session, there can be one or multiple folders `anat_k`, where `k` is an positive integer, each containing a structural MRI volume of the subject, called `mprage_noface.nii.gz`. There are also one or multiple folders `rest_k`, where `k` is an positive integer, each containing a resting-state fMRI acquisition of the subject, called `rest.nii.gz`.

the dependencies between jobs, as well as a comprehensive description of the attributes of the job in Octave/Matlab format. This last description can be used to execute single jobs using a dedicated command in PSOM.

(5) The execution manager in CBRAIN finally reads the series of XML file, groups some jobs to increase efficiency and executes each defined job (using `psom_run_job`) on the specified execution server while respecting the graph of dependencies. The pipeline execution terminates when all jobs have completed or an error has occured. The user monitors the execution of the pipeline in real time on a dedicated web page, see Figure 10 (Annex).

Even though the feasibility of the integration of PSOM in CBRAIN was demonstrated using the NIAK fMRI preprocessing pipeline, the interface described above can be applied to any PSOM pipeline. The creation of the interface for a new pipeline can be achieved by creating a new parameter page on CBRAIN and replacing the call to the pipeline generation used in step (3). Since the steps (1), (2), (4) and (5) are completely generic, the generation of a new interface can be done rapidly. The time of development for such new interface is estimated to

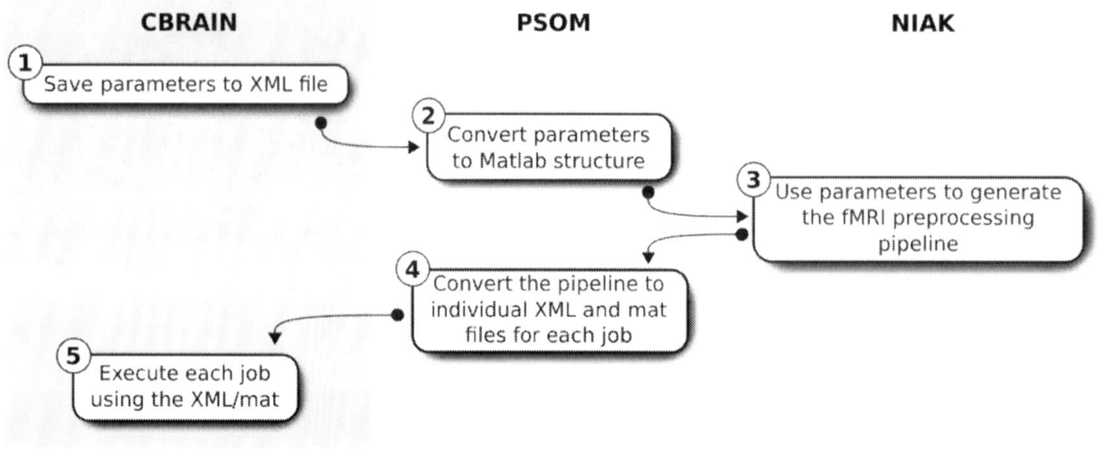

Figure 5. Workflow of the interface between CBRAIN and the NIAK fMRI preprocessing pipeline. The simple application program interface is based on generic tools implemented in PSOM. The main pipeline-specific stage is number 3, which only involves a few line of pipeline-specific code.

a few hours on a test execution server. The main work for integrating new tools in CBRAIN using the PSOM framework is to install all the necessary libraries in all the execution servers available to CBRAIN, which already adds up to 9 supercomputers and may require the limited involvement of a system administrator at each site.

5. The ADHD-200 global competition

Overview of the competition. The ADHD-200[14] sample is a data-sharing initiative dedicated to accelerating the scientific community's understanding of the neural basis of attention deficit hyperactivity disorder (ADHD) in human development. The project was coordinated by Michael Milham at the Child Mind Institute, NY, and involved 8 sites which contributed neuroimaging datasets as well as various phenotypic informations and clinical evaluation. One of the immediate goals of the ADHD-200 release was to organize a global competition for the image analysis pipeline that provided the best predictor (biomarker) of ADHD and ADHD subtype. The first (training) database comprised 776 children, with and without ADHD, and was released in April 2011. This database featured all phenotypic information, including ADHD diagnosis, and was provided to researchers in order to design an effective method. The competition will last until August 30th 2011, at which time competing researchers will submit an neuroimaging-based prediction of ADHD diagnosis and subtype for the unlabeled subjects. This is an unusually large volume of datasets, as a typical sample for an fMRI study was about 20 subjects before the advent of open data-sharing efforts such as the 1000-functional connectomes project [20]. Moreover, the training and test databases have to be preprocessed under very stringent deadlines.

The NIAK preprocessing. We decided to use the ADHD-200 global competition to demonstrate that our pipeline analysis tools can be utilized in a real large-scale project. Specifically, we preprocessed, quality-controled and publicly released the training database as well as the test database. Table 1 shows a summary of the computational aspects of the preprocessing of the training ADHD-200 database. The initial size of the database was about 70 Gb, and over

[14] `http://fcon_1000.projects.nitrc.org/indi/adhd200/`

Site	n	iSize	fSize	%size	jobs	eTime	pTime	accel
Peking University	195	9.1	27	296.70%	5951	155:32:07	60:21:30	257.69%
Kennedy Krieger Institute	84	4.9	15	306.12%	2551	66:36:17	22:00:22	302.66%
neuroIMAGE	49	6.8	20	294.12%	1477	47:35:30	16:22:14	290.72%
New-York University Child Study Center	223	42	125	297.62%	9574	239:11:14	106:52:49	223.79%
Oregon Health and Science University	80	3.5	10	285.71%	6487	56:46:19	32:47:32	173.13%
University of Pittsburgh	90	3.4	10	294.12%	2738	64:58:40	23:31:16	276.25%
Washington University in St-Louis	61	3.1	9	290.32%	3311	44:08:53	23:49:23	185.32%
Total	**782**	**72.8**	**216**	**296.70%**	**32089**	**674:49:00**	**285:45:06**	**236.15%**

Table 1. Computational metrics for the preprocessing of the ADHD-200 training sample. The columns respectively code for: scanning **site**; the number **n** of participants; the size **iSize** of the datasets before preprocessing (in Gb); the size **fSize** of the datasets after preprocessing (in Gb); the ratio **fSize/iSize**, called **%size**; the number of **jobs** composing the NIAK pipeline; the total time **eTime** spent running jobs; the effective computation time **pTime** using parallelization; the ratio **eTime/pTime**, called **accel**.

200 Gb of final results were generated. This does not include the intermediate results that were deleted as soon as possible in the pipeline. The NIAK fMRI preprocessing pipelines combined across all the sites comprised over 30000 jobs, which took a total of about 675 hours to execute. A factor of 2 speedup was achieved using parallel computing on a machine with 20 computing cores available for the pipeline, which represents a low degree of parallelization efficiency. Computational efficiency was not a primary objective of this proof of concept. We believe that the low parallelization efficiency is related to the mechanisms of job submission and can be dramatically improved in the future. Possible remedies to that issue are presented in the discussion section. As an indication of a minimal level of achievable efficieny, the parallelization efficiency observed with the PSOM pipeline manager in a similar computing environment and the same pipeline is about 67%, i.e. the computation time was divided by 67% times the number of cores [17]. When the tools will be deployed in a large supercomputer, we expect that the whole analysis will run in less than a day.

Quality control. A common concern expressed in the neuroimaging community regarding the pipeline processing of large volumes of data is the lack of control over the quality of individual steps, which is perceived as greater than with tools that are less automatic. This question is of central importance, because it undermines the scientific credibility of pipeline analysis, despite obvious advantages in terms of computational power and reproducibility. The results of the preprocessing were carefully yet rapidly quality controled using summary indices derived at the group level. These indices notably included the spatial (Pearson's) correlation coefficient between each individual brain volume and the average of the population, after non-linear coregistration to stereotaxic space was performed, which was done for both structural and functional MRI. For example, at the KKI site, there were 4 subjects for which the coregistration between the structural and the functional volumes clearly failed, see Figure 6. This was further confirmed by the visual examination of individual datasets. A battery of summary brain maps were also generated, such as the average and standard deviation (std) of all brain volumes after non-linear coregistration, both for anatomical and functional scans, as well as the average of brain masks extracted in both modalities to identify variations in the field of view across subjects. For the KKI site, the std images clearly revealed substandard coregistration, with significant signal found outside of the brain in the stereotaxic space, see Figure 7. Those abnormalities were created by the four subjects where the coregistration between the structural and functional volumes catastrophically failed. Some group-level summary metrics were also generated to summarize

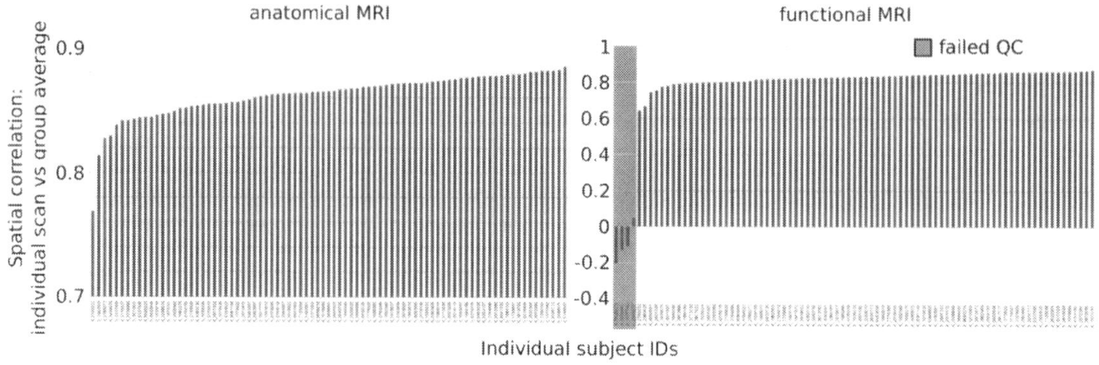

Figure 6. Group-level summary metrics of coregistration procedures.

the maximal transition in motion parameters, translation and rotation, for each subject (results not shown). This allowed us to tag subjects with excessive, moderate and negligible levels of motion during the acquisition of the fMRI datasets. A detailed report of the results of our quality control procedure was released publicly on the internet[15]. Note that the results of the subjects where coregistration problems were identified could all be fixed by adjusting the parameters of the pipeline.

Scientific significance. The ultimate proof of the scientific value of a tool is that it is used by the community and significantly contributes to scientific discoveries. All of the preprocessed datasets have been publicly released as part of a community initiative organized by the Neurobureau[16], which includes a variety of tools and analysis besides the results of the NIAK fMRI preprocessing pipeline. In less than 6 months, this public sharing has generated over 5500 downloads on the NITRC website[17]. Note that this number includes all downloads of individual preprocessing packages, yet we estimate based on our analytics software that at least 134 different researchers have accessed these resources, 30 of which have accessed the NIAK release. The winners of the ADHD-200 competition [18], a group from Johns Hopkins University, was made up of statisticians who used the Neurobureau release. Processing rapidly a large volume of neuroimaging datasets is a real challenge, especially for researchers who may lack some of the domain-specific knowledge. The NIAK pipeline as integrated in the CBRAIN platform has thus proven to be a powerful solution for large-scale processing tasks. The NIAK pipeline will be made accessible soon and without restrictions to the whole CBRAIN users community.

6. Discussion

In this paper, we presented a general strategy to implement complex pipelines composed of heterogeneous, non-parallel tools, and deploy them in a grid-computing platform. The core of the work consisted of writing an interface between two systems: PSOM, a code-oriented pipeline development framework, and CBRAIN, a web-oriented platform for grid computing. We demonstrated the feasibility of this approach by integrating a pipeline for preprocessing structural and functional magnetic resonance imaging into CBRAIN. We also showed the capacity of this approach to handle real large-scale databases by processing close to 1000 subjects

[15] `http://www.nitrc.org/plugins/mwiki/index.php/neurobureau:NIAKPipeline`
[16] `http://www.theneurobureau.org`
[17] `http://www.nitrc.org/projects/neurobureau`
[18] `http://fcon_1000.projects.nitrc.org/indi/adhd200/results.html`

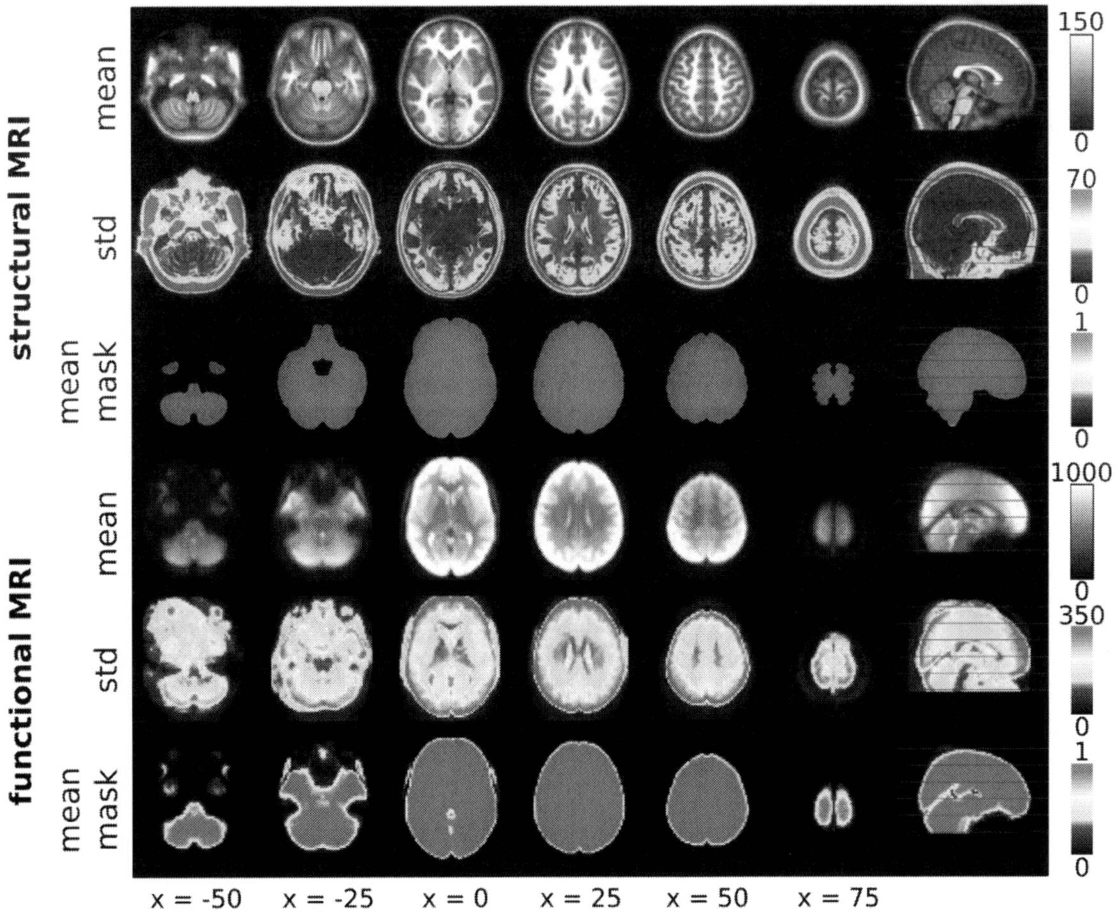

Figure 7. Group-level summary maps of coregistration procedures. The first three rows are derived from anatomical MRI (mean across subjects, standard deviation across subjects, mean of the individual brain masks). The last three rows are derived from the individual average of all functional MRI (mean across subjects, standard deviation across subjects, mean of the individual brain masks).

released as part of the ADHD-200 competition. Despite the limited computational resources available to us in our test environment, we were able to process this massive dataset in a timely fashion, using semi-automated metrics to perform a quality control of the results of the pipeline. We publicly released the preprocessed datasets which have been used by competing teams around the world. We believe that this experiment makes a strong case for the feasibility and usefulness of grid-computing in neuroimaging.

The major limitation we have experienced thus far using the proposed approached is computational resources. We notably deployed the pipeline on a test execution server featuring only 20 computing cores, because we did not have time to deploy the software packages in a supercomputing environment for this pilot project due to the tight release schedule of the ADHD-200 competition. There is however a more fundamental issue: the fMRI preprocessing pipeline has been designed for PSOM which can submit at least three jobs per second to a queuing system, and much more for multi-core processing on a single machine. The pipeline was thus implemented at a very high level of granularity. There are notably hundreds of cleaning jobs which only delete files on the system and usually take a few hundreds of milliseconds to run. This becomes a problem with a grid-level job submission system like CBRAIN, which has an overhead of several seconds to submit jobs. In addition, due to the nature of queuing in public resources, pipelines featuring large number of jobs are more exposed to long queue wait times. Another potential issue is that many HPC centers allocate a full multi-CPU node for a single task, so using a single core for a task would represent a considerable waste of resources. The CBRAIN framework addresses these issue by meta-scheduling: several PSOM jobs are regrouped under the control of serializers and parallelizers, and these meta-jobs are actually submitted to the cluster. Serializers combine sets of PSOM jobs that are directly dependent on each other in the graph of dependencies. Parallelizers are used to run several serialized and/or independent PSOM jobs in parallel. The end result is that any single cluster's job will typically be responsible for running about 20 PSOM jobs from different parts of the pipeline, and all of this in the proper order such that dependencies are respected. Achieving an optimal grouping of tasks however requires precise estimates for the execution time of the jobs composing the pipeline, and we are currently working on a generic approach to optimize that process.

Another future improvement of the framework is to modularize the pipeline parameters, and give some memory to the CBRAIN pipeline manager. Currently, the same parameters apply to all subjects in a database. To reprocess a subset of a database with different parameters, the only solution is to copy this subset and then treat it as an independent database. In addition, if a parameter of the processing changes, CBRAIN is not able to identify which jobs of the pipeline actually changed and which did not. As a consequence, the whole pipeline has to be restarted if any change is made to the pipeline parameters. This is quite computationally inefficient, because the datasets that do not pass quality control are usually a small subset of subjects and the issues can usually be fixed easily by tweaking parameters and restarting a very limited number of stages. PSOM features these capabilities, and we will extend them to the CBRAIN pipeline manager in the near future.

The interface between PSOM and CBRAIN represents a very powerful approach for integrating neuroimaging pipelines in a web-based grid computing environment that can easily be shared with a large community of people with very limited information technology skills. We see this as a fantastic opportunity to connect a new community with the world of high-performance computing. We also believe that such progress would dramatically help a large number of neuroscientists perform the research they want at a higher pace or at a scale they would not currently envision. One priority is thus to port existing popular tools such as AFNI, FSL and SPM as pipelines conforming to the PSOM framework, such that they can be easily deployed into CBRAIN. Of important note is that a parallel effort to PSOM/NIAK, the so-called Nipype/Nipy projects [15], is being developed in Python by a very active community and already includes a number of pipelines coded under a format very similar to the one used in PSOM. We are hoping that creating an interface between Nipype/Nipy and CBRAIN is going to be as straightforward as interfacing PSOM/NIAK in CBRAIN. This will dramatically accelerate the integration of new tools into that platform.

While the early days of neuroimaging have been dominated by small studies collected by independent researchers who kept their datasets private, there are now substantial efforts toward the systematic sharing of large databases. This will open new research avenues for large-scale data processing and data mining in the field [21]. The pilot project reported here demonstrate that CBRAIN has the potential to stand as a key computing infrastructure to face this upcoming challenge.

Acknowledgements

The authors are grateful to the members of the ADHD-200 consortium for publicly releasing the eponym data sample. SLC and PB were partly supported by a pilot project grant of the Quebec bioimaging network (QBIN), "fonds de recherche en santé du Québec" (FRSQ). The CBRAIN project is funded by Canada's advanced research and innovation network (CANARIE).

Bibliography

[1] J. P. Mesirov, "Accessible reproducible research," *Science (New York, N.Y.)*, vol. 327, no. 5964, pp. 415–416, Jan. 2010. [Online]. Available: http://dx.doi.org/10.1126/science.1179653

[2] V. Fonov, A. C. Evans, K. Botteron, C. R. Almli, R. C. McKinstry, and D. L. Collins, "Unbiased average age-appropriate atlases for pediatric studies," *NeuroImage*, vol. 54, no. 1, pp. 313–327, Jan. 2011. [Online]. Available: http://dx.doi.org/10.1016/j.neuroimage.2010.07.033

[3] D. R. Pickens, Y. Li, V. L. Morgan, and B. M. Dawant, "Development of computer-generated phantoms for FMRI software evaluation," *Magnetic Resonance Imaging*, vol. 23, no. 5, pp. 653–663, Jun. 2005. [Online]. Available: http://dx.doi.org/10.1016/j.mri.2005.04.007

[4] V. Perlbarg, P. Bellec, J.-L. Anton, M. Pelegrini-Issac, J. Doyon, and H. Benali, "CORSICA: correction of structured noise in fMRI by automatic identification of ICA components," *Magnetic Resonance Imaging*, vol. 25, no. 1, pp. 35–46, Jan. 2007. [Online]. Available: http://dx.doi.org/10.1016/j.mri.2006.09.042

[5] K. Worsley, "A General Statistical Analysis for fMRI Data," *NeuroImage*, vol. 15, no. 1, pp. 1–15, Jan. 2002. [Online]. Available: http://dx.doi.org/10.1006/nimg.2001.0933

[6] Y. Chao-Gan and Z. Yu-Feng, "DPARSF: A MATLAB Toolbox for "pipeline" data analysis of resting-state fMRI," *Frontiers in systems neuroscience*, vol. 4, 2010. [Online]. Available: http://dx.doi.org/10.3389/fnsys.2010.00013

[7] A. P. Zijdenbos, R. Forghani, and A. C. Evans, "Automatic "pipeline" analysis of 3-D MRI data for clinical trials: application to multiple sclerosis," *IEEE Transactions on Medical Imaging*, vol. 21, no. 10, pp. 1280–1291, Oct. 2002. [Online]. Available: http://dx.doi.org/10.1109/TMI.2002.806283

[8] P. Bellec, P. Rosa-Neto, O. C. Lyttelton, H. Benali, and A. C. Evans, "Multi-level bootstrap analysis of stable clusters in resting-state fMRI," *NeuroImage*, vol. 51, no. 3, pp. 1126–1139, Jul. 2010. [Online]. Available: http://dx.doi.org/10.1016/j.neuroimage.2010.02.082

[9] M. Hanke and Y. O. Halchenko, "Neuroscience runs on GNU/Linux," *Frontiers in Neuroinformatics*, vol. 5, 2011.

[10] M. Snir, J. Dongarra, J. S. Kowalik, S. Huss-Lederman, S. W. Otto, and D. W. Walker, *MPI: The Complete Reference (Vol. 1) - 2nd Edition, Vol. 1 - The MPI Core*, second edition ed. The MIT Press, Sep. 1998. [Online]. Available: http://www.amazon.com/exec/obidos/redirect?tag=citeulike07-20&path=ASIN/0262692155

[11] I. Raicu, I. T. Foster, and Y. Zhao, "Many-task computing for grids and supercomputers," *Many-Task Computing on Grids and Supercomputers (MTAGS08), Proceedings of the IEEE Workshop on*, pp. 1–11, Nov. 2008. [Online]. Available: http://dx.doi.org/10.1109/MTAGS.2008.4777912

[12] J. D. D. Van Horn and A. W. Toga, "Neuroimaging workflow design and data-mining: a frontiers in neuroinformatics special issue," *Frontiers in neuroinformatics*, vol. 3, 2009. [Online]. Available: http://dx.doi.org/10.3389/neuro.11.031.2009

[13] I. D. Dinov, J. D. Van Horn, K. M. Lozev, R. Magsipoc, P. Petrosyan, Z. Liu, A. Mackenzie-Graham, P. Eggert, D. S. Parker, and A. W. Toga, "Efficient, distributed and interactive neuroimaging data analysis using the LONI pipeline," *Frontiers in neuroinformatics*, vol. 3, 2009. [Online]. Available: http://dx.doi.org/10.3389/neuro.11.022.2009

[14] J. W. Eaton, D. Bateman, and S. Hauberg, *GNU Octave Manual Version 3*. Network Theory Ltd., Oct. 2008. [Online]. Available: http://www.amazon.com/exec/obidos/redirect?tag=citeulike07-20&path=ASIN/095461206X

[15] K. Gorgolewski, C. D. Burns, C. Madison, D. Clark, Y. O., Halchenko, M. L. Waskom and S. S. Ghosh, D. S."Nipype: A flexible, lightweight and extensible neuroimaging data processing framework in python." *Frontiers in neuroinformatics*, vol. 5, 2011. [Online]. Available: http://dx.doi.org/10.3389/fninf.2011.00013

[16] W. F. Punch and R. Enbody, *The Practice of Computing using Python*, 1st ed. Addison Wesley, Mar. 2010. [Online]. Available: http://www.amazon.com/exec/obidos/redirect?tag=citeulike07-20&path=ASIN/0136110673

[17] P. Bellec, F. M. Carbonell, V. Perlbarg, C. Lepage, O. Lyttelton, V. Fonov, A. Janke, J. Tohka, and A. C. Evans, "A neuroimaging analysis kit for Matlab and Octave," in *Proceedings of the 17th International Conference on Functional Mapping of the Human Brain*, 2011.

[18] D. L. Collins, P. Neelin, T. M. Peters, and A. C. Evans, "Automatic 3D intersubject registration of MR volumetric data in standardized Talairach space," *Journal of computer assisted tomography*, vol. 18, no. 2, pp. 192–205, 1994. [Online]. Available: http://view.ncbi.nlm.nih.gov/pubmed/8126267

[19] P. Troger, H. Rajic, A. Haas, and P. Domagalski, "Standardization of an API for distributed resource

management systems," *Cluster Computing and the Grid, IEEE International Symposium on*, vol. 0, pp. 619–626, 2007. [Online]. Available: http://dx.doi.org/10.1109/CCGRID.2007.109

[20] B. B. Biswal, M. Mennes, X.-N. Zuo, S. Gohel, C. Kelly, S. M. Smith, C. F. Beckmann, J. S. Adelstein, R. L. Buckner, S. Colcombe, A.-M. Dogonowski, M. Ernst, D. Fair, M. Hampson, M. J. Hoptman, J. S. Hyde, V. J. Kiviniemi, R. Kötter, S.-J. Li, C.-P. Lin, M. J. Lowe, C. Mackay, D. J. Madden, K. H. Madsen, D. S. Margulies, H. S. Mayberg, K. McMahon, C. S. Monk, S. H. Mostofsky, B. J. Nagel, J. J. Pekar, S. J. Peltier, S. E. Petersen, V. Riedl, S. A. R. B. Rombouts, B. Rypma, B. L. Schlaggar, S. Schmidt, R. D. Seidler, G. J. Siegle, C. Sorg, G.-J. Teng, J. Veijola, A. Villringer, M. Walter, L. Wang, X.-C. Weng, S. Whitfield-Gabrieli, P. Williamson, C. Windischberger, Y.-F. Zang, H.-Y. Zhang, F. X. Castellanos, and M. P. Milham, "Toward discovery science of human brain function," *Proceedings of the National Academy of Sciences*, vol. 107, no. 10, pp. 4734–4739, Mar. 2010. [Online]. Available: http://dx.doi.org/10.1073/pnas.0911855107

[21] D. S. Marcus, J. Harwell, T. Olsen, M. Hodge, M. F. Glasser, F. Prior, M. Jenkinson, T. Laumann, S. W. Curtiss, and D. C. Van Essen, "Informatics and data mining tools and strategies for the human connectome project," *Frontiers in neuroinformatics*, vol. 5, 2011. [Online]. Available: http://dx.doi.org/10.3389/fninf.2011.00004

Annex

Figure 8. Caption of the CBRAIN interface : file collections. The CBRAIN users can upload and download file collections through the web, and transfer these between data repositories. A secured sftp mechanism is running in the background, but this process is transparent for registered users.

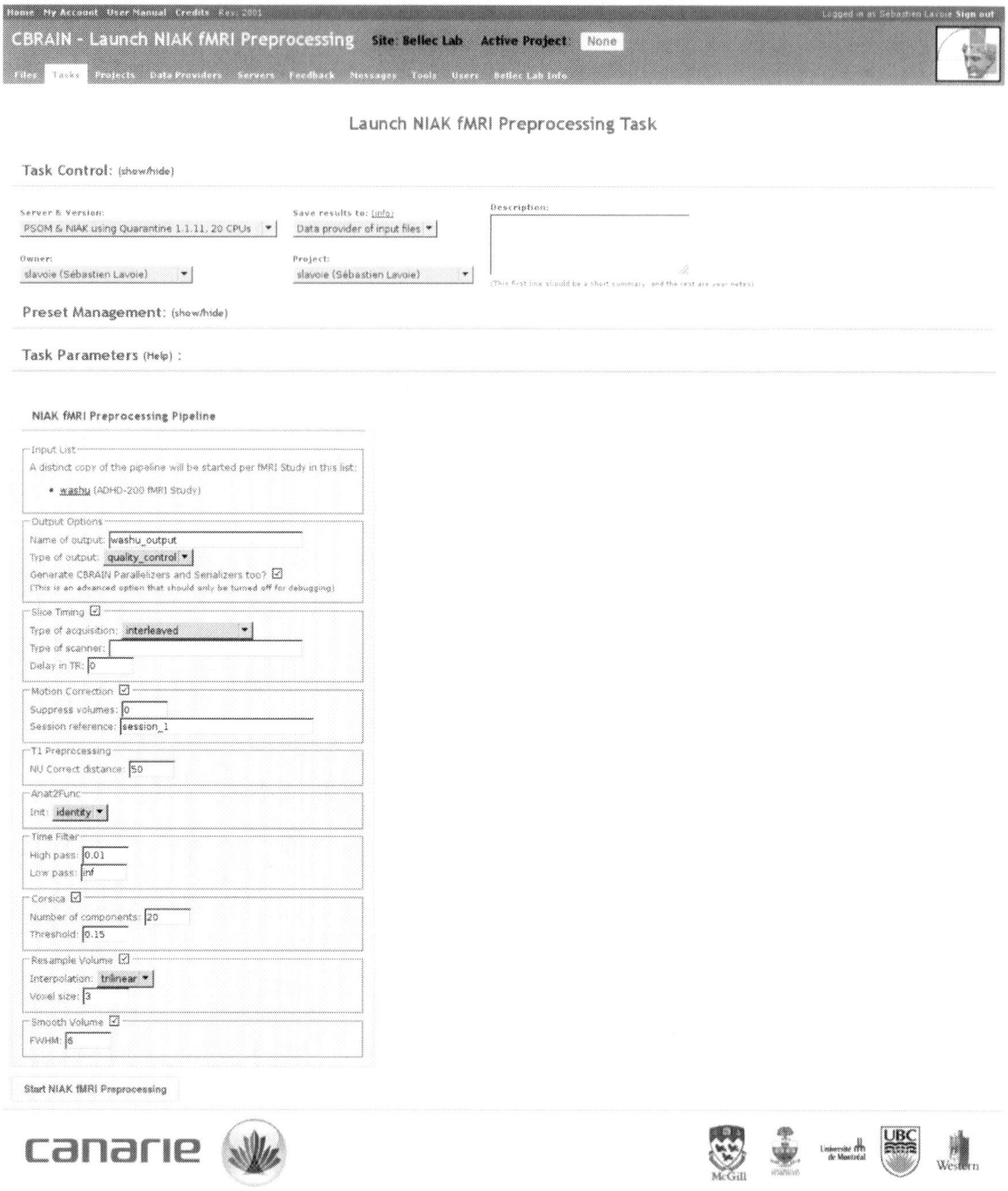

Figure 9. Caption of the CBRAIN interface : running NIAK. A simple web form is available to set the parameters of the NIAK fMRI preprocessing pipeline.

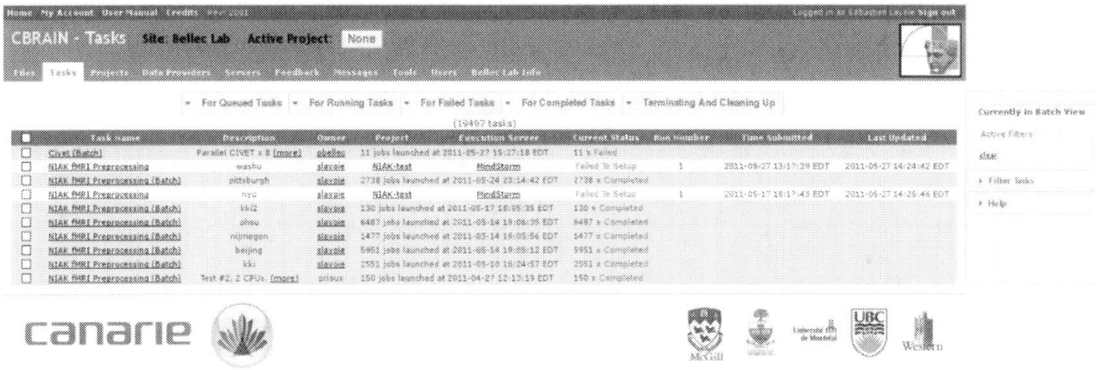

Figure 10. Caption of the CBRAIN interface : execution servers. The status of jobs running an any execution server can be monitored in real-time on a dedicated web page.

Journal of Physics: Conference Series **341** (2012) 012033 doi:10.1088/1742-6596/341/1/012033

Domain decomposition method of stochastic PDEs: a two-level scalable preconditioner

Waad Subber[1] **and Abhijit Sarkar**[2]

Department of Civil and Environmental Engineering, Carleton University, Ottawa, Ontario K1S5B6, Canada

E-mail: [1]`wsubber@connect.carleton.ca`, [2]`abhijit_sarkar@carleton.ca`

Abstract.
For uncertainty quantification in many practical engineering problems, the stochastic finite element method (SFEM) may be computationally challenging. In SFEM, the size of the algebraic linear system grows rapidly with the spatial mesh resolution and the order of the stochastic dimension. In this paper, we describe a non-overlapping domain decomposition method, namely the iterative substructuring method to tackle the large-scale linear system arising in the SFEM. The SFEM is based on domain decomposition in the geometric space and a polynomial chaos expansion in the probabilistic space. In particular, a two-level scalable preconditioner is proposed for the iterative solver of the interface problem for the stochastic systems. The preconditioner is equipped with a coarse problem which globally connects the subdomains both in the geometric and probabilistic spaces via their corner nodes. This coarse problem propagates the information quickly across the subdomains leading to a scalable preconditioner. For numerical illustrations, a two-dimensional stochastic elliptic partial differential equation (SPDE) with spatially varying non-Gaussian random coefficients is considered. The numerical scalability of the the preconditioner is investigated with respect to the mesh size, subdomain size, fixed problem size per subdomain and order of polynomial chaos expansion. The numerical experiments are performed on a Linux cluster using MPI and PETSc parallel libraries.

1. Introduction

In modeling numerous engineering and physical systems, it is necessary to consider the random heterogeneity of the model parameters for more realistic computer simulation. When sufficient statistical information is available, the probability theory offers a rich mathematical framework to represent uncertainty in terms of random variables and stochastic processes(e.g. [1, 2]). The spectral stochastic finite element method [1] is a popular computational tool for uncertainty quantification of SPDEs. For large-scale engineering problems of practical interest, the application of SFEM becomes computationally challenging as the size of the algebraic linear system grows rapidly with the spatial mesh resolution and the order of the stochastic dimension. To address this issue, a non-overlapping domain decomposition approach (namely a *substructuring* method) of SPDEs is introduced in [2] in the framework of SFEM. The methodology is based on Schur complement based geometric decomposition and an orthogonal decomposition and projection of the stochastic processes using the polynomial chaos expansion [1]. To tackle the interface problem of the stochastic system using the substructuring method, a parallel preconditioned conjugate gradient method (PCGM) adopted in [3] using a lumped preconditioner.

Despite being superior to the direct solver (in terms of memory requirement and floating point operations), the PCGM based iterative solvers equipped with the lumped preconditioner exhibit performance degradation as the number of subdomains increases. Consequently, a one-level Neumann-Neumann domain decomposition preconditioner is introduced in [4] which demonstrates a fairly good scalability for moderate number of subdomains. As already demonstrated for deterministic PDEs, the one-level preconditioners do not scale well for large number of subdomains as their convergence rates depend on the subdomain size and thus a number of two-level preconditioners containing *coarse* grids are proposed (see, e.g. [5, 6, 7, 8, 9]). A two-level preconditioner consists of local (fine) and global (coarse) components leading to the convergence rate of the iterative solver independent of the problem size, subdomain size and fixed problem size per subdomain(e.g. [5, 6, 7, 8, 9]). In this paper, a two-level scalable preconditioner is proposed for stochastic PDEs.

Previously, the authors have reported a two-level preconditioner for SPDEs[10] based on the domain decomposition method [2]. As a continuation of our previous work, we propose a new two-level preconditioner for iterative substructuring methods for SPDEs. In the context of SPDEs, the proposed preconditioner may be viewed as an extension of the Balancing Domain Decomposition by Constraints (BDDC) [9, 7] in the context of stochastic PDEs. The preconditioner involves construction of a stochastic coarse problem that enforces the global exchange of information across the subdomains . At each iteration of the PCGM solver, the local problems are solved on each subdomain in parallel which collectively construct the fine level of the preconditioner. Additionally the coarse grid accomplishes the global exchange of information which makes the preconditioner scalable. The parallel performance of the algorithm is studied for two-dimensional stochastic elliptic PDE with spatially varying non-Gaussian random coefficients. PETSc [11] and MPI [12] parallel libraries are used for efficient parallel implementation of the proposed algorithm. The graph partitioning tool METIS [13] is used for optimal decomposition of the finite element mesh. The numerical experiments are performed on a Linux cluster consists of 22 nodes with an InfiniBand interconnect (2 Quad-Core 3.0 GHz Intel Xeon processors and 32 GB of memory per node).

2. Uncertainty representation by stochastic processes

This section provides a brief review of the theories of stochastic processes relevant to the subsequent developments of the paper [1, 2]. We assume the data induces a representation of the model parameters as random variables and processes which span the Hilbert space $\mathcal{H}_G$. A set of basis functions $\{\xi_i\}$ is identified to characterize this space using the Karhunen-Loeve expansion. The state of the system resides in the Hilbert space $\mathcal{H}_L$ with basis functions $\{\Psi_i\}$ being identified with the Polynomial Chaos expansion (PC). The Karhunen-Loeve expansion of a stochastic process $\alpha(x,\theta)$ is based on the spectral expansion of its covariance function $R_{\alpha\alpha}(x,y)$. The expansion takes the following form

$$\alpha(x,\theta) = \bar{\alpha}(x) + \sum_{i=1}^{\infty} \sqrt{\lambda_i}\xi_i(\theta)\phi_i(x), \tag{1}$$

where $\bar{\alpha}(x)$ is the mean of the stochastic process, θ represents the random dimension, and $\xi_i(\theta)$ is a set of uncorrelated (but generally not independent) random variables, $\phi_i(x)$ are the eigenfunctions and λ_i are the eigenvalues of the covariance kernel which can be obtained as the solution to the following integral equation

$$\int_{\mathcal{D}} R_{\alpha\alpha}(x,y)\phi_i(y)dy = \lambda_i\phi_i(x). \tag{2}$$

where $\mathcal{D}$ denotes the spatial dimension over which the process is defined. The covariance function of the solution process is not known *a priori*. Hence the Karhunen-Loeve expansion cannot be used to represent it. Therefore, a generic basis, that is complete in the space of all second-order random variables will be identified and used in the approximation process. Since the solution process is a function of the material properties, nodal solution variables, denoted by $u(\theta)$ can be formally expressed as some nonlinear functional of the set $\xi_i(\theta)$ used to represent the material stochasticity. It has been shown that this functional dependence can be expanded in terms of polynomials in Gaussian random variables, namely Polynomial Chaos [1, 14] as

$$\begin{aligned} u(\theta) &= a_0\Gamma_0 + \sum_{i_1=1}^{\infty} u_{i_1}\Gamma_1\left(\xi_{i_1}(\theta)\right) \\ &+ \sum_{i_1=1}^{\infty}\sum_{i_2=1}^{i_1} u_{i_1 i_2}\Gamma_2\left(\xi_{i_1}(\theta), \xi_{i_2}(\theta)\right) + \ldots . \end{aligned} \tag{3}$$

In this equation, the symbol $\Gamma_n(\xi_{i_1}, \ldots, \xi_{i_n})$ denotes the polynomial chaos ([15, 1]) of order n in the variables $(\xi_{i_1}, \ldots, \xi_{i_n})$. Introducing a one-to-one mapping to a set with ordered indices denoted by $\{\Psi_i(\theta)\}$ and truncating the polynomial chaos expansion after the Nth term, Eq.(3) can be rewritten as,

$$\mathbf{u}(\theta) = \sum_{j=0}^{N} \Psi_j(\theta)\mathbf{u}_j. \tag{4}$$

These polynomials are orthogonal in the sense that their inner product $\langle \Psi_j \Psi_k \rangle$, defined by the statistical average of their product, is equal to zero for $j \neq k$.

3. Review of Schur complement based domain decomposition method of SPDEs

In this section, we provide a brief review of the domain decomposition method for SPDEs based on [2, 3, 4, 10]. For an elliptic SPDE defined on a domain Ω with a prescribed boundary conditions on $\partial\Omega$, the finite element discretization leads to the following linear system

$$\mathbf{A}(\theta)\mathbf{u}(\theta) = \mathbf{f}, \tag{5}$$

where $\mathbf{A}(\theta)$ is the stiffness matrix with random coefficients, $\mathbf{u}(\theta)$ is the random vector representing the response vector and $\mathbf{f}$ is the external force. For large-scale system, Eq.(5) can be solved efficiently using domain decomposition methods [2, 3, 4, 10].

In the non-overlapping domain decomposition method (for example, see [5, 6, 7, 8, 9]), the spatial domain Ω is partitioned into n_s non-overlapping subdomains $\{\Omega_s, 1 \leqslant s \leqslant n_s\}$ such that $\Omega = \bigcup_{s=1}^{n_s}\Omega_s$, $\Omega_s \bigcap \Omega_r = 0$, $s \neq r$ and $\Gamma = \bigcup_{s=1}\Gamma_s$ where $\Gamma_s = \partial\Omega_s \backslash \partial\Omega$. For a typical subdomain Ω_s the nodal vector $\mathbf{u}^s(\theta)$ is decomposed into the interior unknowns $\mathbf{u}_I^s(\theta)$ associated with the nodes in the interior of Ω_s and the interface unknowns $\mathbf{u}_\Gamma^s(\theta)$ corresponding to the nodes shared among two or more subdomains. This decomposition leads to the following equilibrium equation for the subdomain

$$\begin{bmatrix} \mathbf{A}_{II}^s(\theta) & \mathbf{A}_{I\Gamma}^s(\theta) \\ \mathbf{A}_{\Gamma I}^s(\theta) & \mathbf{A}_{\Gamma\Gamma}^s(\theta) \end{bmatrix} \begin{Bmatrix} \mathbf{u}_I^s(\theta) \\ \mathbf{u}_\Gamma^s(\theta) \end{Bmatrix} = \begin{Bmatrix} \mathbf{f}_I^s \\ \mathbf{f}_\Gamma^s \end{Bmatrix}. \tag{6}$$

Enforcing the transmission conditions (compatibility and equilibrium) along the subdomain interfaces and expanding the solution vector by the polynomial chaos (as in Eq.(4)) and then performing Galerkin projection, the following block linear systems of equations express the global equilibrium equation of the stochastic system ([2, 3, 4, 10]):

$$
\langle \sum_{i=0}^{L} \Psi_i(\theta) \begin{bmatrix} \mathbf{A}_{II,i}^1 & \dots & 0 & \mathbf{A}_{I\Gamma,i}^1 \mathbf{R}_1 \\ \vdots & \ddots & \vdots & \vdots \\ 0 & \dots & \mathbf{A}_{II,i}^{n_s} & \mathbf{A}_{I\Gamma,i}^{n_s} \mathbf{R}_{n_s} \\ \mathbf{R}_1^T \mathbf{A}_{\Gamma I,i}^1 & \dots & \mathbf{R}_{n_s}^T \mathbf{A}_{\Gamma I,i}^{n_s} & \sum_{s=1}^{n_s} \mathbf{R}_s^T \mathbf{A}_{\Gamma\Gamma,i}^s \mathbf{R}_s \end{bmatrix} \sum_{j=0}^{N} \Psi_j(\theta) \begin{Bmatrix} \mathbf{u}_{I,j}^1 \\ \vdots \\ \mathbf{u}_{I,j}^{n_s} \\ \mathbf{u}_{\Gamma,j} \end{Bmatrix} \Psi_k(\theta) \rangle
$$

$$
= \langle \begin{Bmatrix} \mathbf{f}_I^1 \\ \vdots \\ \mathbf{f}_I^{n_s} \\ \sum_{s=1}^{n_s} \mathbf{R}_s^T \mathbf{f}_\Gamma^s \end{Bmatrix} \Psi_k(\theta) \rangle, \quad k = 0, ..., N. \tag{7}
$$

The restriction operator $\mathbf{R}_s$ is a Boolean rectangular matrix of size $(n_\Gamma^s \times n_\Gamma)$ that maps the global interface vector $\mathbf{u}_\Gamma(\theta)$ to the local interface vector $\mathbf{u}_\Gamma^s(\theta)$ as $\mathbf{u}_\Gamma^s(\theta) = \mathbf{R}_s \mathbf{u}_\Gamma(\theta)$. In parallel implementation, $\mathbf{R}_s$ represents a *scatter* operator while $\mathbf{R}_s^T$ acts as a *gather* operator. Compactly, Eq.(7) can be expressed as

$$
\begin{bmatrix} \mathcal{A}_{II}^1 & \dots & 0 & \mathcal{A}_{I\Gamma}^1 \mathcal{R}_1 \\ \vdots & \ddots & \vdots & \vdots \\ 0 & \dots & \mathcal{A}_{II}^{n_s} & \mathcal{A}_{I\Gamma}^{n_s} \mathcal{R}_{n_s} \\ \mathcal{R}_1^T \mathcal{A}_{\Gamma I}^1 & \dots & \mathcal{R}_{n_s}^T \mathcal{A}_{\Gamma I}^{n_s} & \sum_{s=1}^{n_s} \mathcal{R}_s^T \mathcal{A}_{\Gamma\Gamma}^s \mathcal{R}_s \end{bmatrix} \begin{Bmatrix} \mathcal{U}_I^1 \\ \vdots \\ \mathcal{U}_I^{n_s} \\ \mathcal{U}_\Gamma \end{Bmatrix} = \begin{Bmatrix} \mathcal{F}_I^1 \\ \vdots \\ \mathcal{F}_I^{n_s} \\ \sum_{s=1}^{n_s} \mathcal{R}_s^T \mathcal{F}_\Gamma^s \end{Bmatrix}, \tag{8}
$$

where

$$
[\mathcal{A}_{\alpha\beta}^s]_{jk} = \sum_{i=0}^{L} \langle \Psi_i \Psi_j \Psi_k \rangle \mathbf{A}_{\alpha\beta,i}^s \ , \quad \mathcal{F}_{\alpha,k}^s = \langle \Psi_k \mathbf{f}_\alpha^s \rangle,
$$
$$
\mathcal{U}_I^m = (\mathbf{u}_{I,0}^m, \dots, \mathbf{u}_{I,N}^m)^T, \ \ \mathcal{U}_\Gamma = (\mathbf{u}_{\Gamma,0}, \dots, \mathbf{u}_{\Gamma,N})^T.
$$

The subscripts α and β represent the index I and Γ. The restriction operator $\mathcal{R}_s$ for the stochastic problem in Eq.(8) takes the following form: $\mathcal{R}_s = blockdiag(\mathbf{R}_s^0, \dots, \mathbf{R}_s^N)$, where $\{\mathbf{R}_s^j, \ j = 1, .., N\}$ are the restriction operators that map the polynomial coefficients of the global interface vector to those of the local interface vector. Performing a Gaussian elimination in Eq.(8), we obtain the global *extended* Schur complement system for the interface variable $\mathcal{U}_\Gamma$ as

$$
\mathcal{S}\, \mathcal{U}_\Gamma = \mathcal{G}_\Gamma, \tag{9}
$$

where the global extended Schur complement matrix $\mathcal{S}$ is given by

$$
\mathcal{S} = \sum_{s=1}^{n_s} \mathcal{R}_s^T [\mathcal{A}_{\Gamma\Gamma}^s - \mathcal{A}_{\Gamma I}^s (\mathcal{A}_{II}^s)^{-1} \mathcal{A}_{I\Gamma}^s] \mathcal{R}_s, \tag{10}
$$

and the corresponding right hand side vector $\mathcal{G}_\Gamma$ is defined by

$$\mathcal{G}_\Gamma = \sum_{s=1}^{n_s} \mathcal{R}_s^T[\mathcal{F}_\Gamma^s - \mathcal{A}_{\Gamma I}^s(\mathcal{A}_{II}^s)^{-1}\mathcal{F}_I^s]. \tag{11}$$

For clarity, the global interface solution $\mathcal{U}_\Gamma$ of Eq.(9) relates to the interface nodes shown schematically in Fig(1).

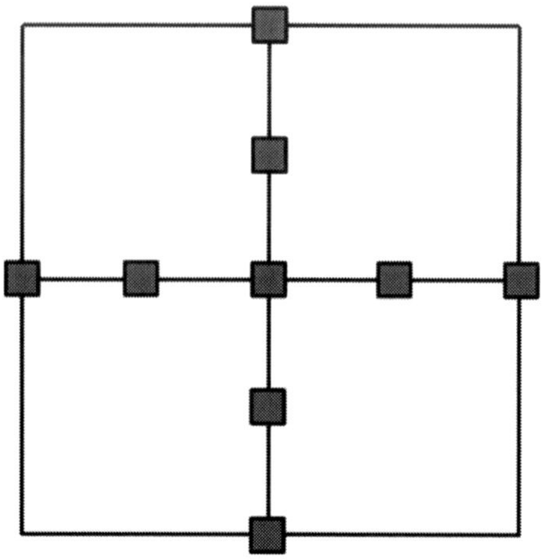

Figure 1. The interface boundary (■).

4. Iterative Solutions of the interface problem

For a large-scale system, the solution of the interface problem in Eq.(9) using direct solvers becomes impractical due to overwhleming memory requirements and poor scalability to large number of processors and thus iterative solution techniques are generally adopted [16]. In domain decomposition approach the Schur complement system is typically solved using preconditioned Krylov subspace method such as preconditioned conjugate gradient method (PCGM). The preconditioned Krylov subspace methods circumvents the need to construct the Schur complement matrix explicitly. Instead only the action of a vector on the Schur complement matrix is needed while using preconditioned Krylov subspace methods. Such matrix vector product can be obtained by solving Dirichlet problem in each subdomain and gathering the resulting vectors at each iteration (e.g. [5, 6]). Non-overlapping domain decomposition method or *iterative substructuring* can be viewed as a preconditioned iterative method to solve the Schur complement system with a parallel preconditioner[5, 6]

$$\mathcal{S}\,\mathcal{U}_\Gamma = \mathcal{G}_\Gamma. \tag{12}$$

For symmetric positive-definite system such as Schur complement system, the Conjugate Gradient Method (CGM) is generally used. The performance of CGM mainly depends on the spectrum of the coefficient matrix. However, the rate of convergence of the iterative method can generally be improved by transforming the original system into an equivalent system that has better spectral properties (i.e. lower condition number $\kappa(\mathcal{S})$) of the coefficient matrix. This transformation is called preconditioning and the matrix used in the transformation is called the preconditioner. In other words, the transformed linear system becomes:

$$\mathcal{M}^{-1}\mathcal{S}\,\mathcal{U}_\Gamma = \mathcal{M}^{-1}\mathcal{G}_\Gamma, \tag{13}$$

where $\mathcal{M}^{-1}$, the preconditioner, is a symmetric and invertible matrix that approximates $\mathcal{S}^{-1}$, the coefficient matrix, in the sense that the condition number of the preconditioned system $\kappa(\mathcal{M}^{-1}\mathcal{S})$ is much smaller than that of the original system $\kappa(\mathcal{S})$ and the eigenvalues of the preconditioned system $\mathcal{M}^{-1}\mathcal{S}$ are clustered near one. This procedure known as Preconditioned Conjugate Gradient Method (PCGM). In practice, the explicit construction of $\mathcal{M}^{-1}$ is not needed. Instead, for a given vector r_Γ, a system of the the following form is solved

$$\mathcal{M}\mathcal{Z} = r_\Gamma. \tag{14}$$

The combination of a good parallel preconditioning technique and the CGM solver makes the iterative solver well-suited to tackle large-scale systems in parallel computers [16].

5. Preconditioners For Stochastic PDEs

As mentioned previously, an efficient scalable preconditioner is required to enhance the convergence rate, reliability and performance of the PCGM iterative solver used to tackle the stochastic Schur complement system. Next we delineate the procedure to construct such preconditioners.

5.1. One-Level Stochastic Neumann-Neumann Preconditioners

A one-level Neumann-Neumann domain decomposition preconditioner is presented in [4] in the context of stochastic PDEs. This preconditioner approximates the inverse of the global extended Schur complement matrix by a weighted sum of the inverse of the local extended Schur complement matrices [5, 17, 6]. The implementation of the algorithm requires a local solve of a stochastic Dirichlet problem followed by a local solve of a stochastic Neumann problem in each iteration of the PCGM solver. The stochastic one-level Neumann-Neumann preconditioner formally takes the following form

$$\mathcal{M}_{NN}^{-1} = \sum_{s=1}^{n_s} \mathcal{R}_s^T \mathcal{D}_s^T [\mathcal{S}^s]^{-1} \mathcal{D}_s \mathcal{R}_s, \tag{15}$$

where

$$\mathcal{D}_s = blockdiag(\mathbf{D}_s^0, \cdots, \mathbf{D}_s^N), \tag{16}$$

and $\mathbf{D}_s^j$ represents a diagonal scaling matrix that derives from the partition of unity as

$$\sum_{s=1}^{n_s} \mathbf{R}_s^T \mathbf{D}_s^j \mathbf{R}_s = \mathbf{I}. \tag{17}$$

The diagonal entries of $\mathbf{D}_s^j$ are the reciprocal of the number of subdomains that share the boundary nodes [5, 17].

In practice, the local extended Schur complement matrix $\mathcal{S}^s$ is not constructed explicitly, and therefore the inverse of the matrix $\mathcal{S}^s$ is not available. Instead the effect of the inverse of the Schur complement matrix on the residual vector is computed by solving the following subdomain level stochastic Neumann problem

$$\begin{bmatrix} \mathcal{A}_{II}^s & \mathcal{A}_{I\Gamma}^s \\ \mathcal{A}_{\Gamma I}^s & \mathcal{A}_{\Gamma\Gamma}^s \end{bmatrix} \begin{Bmatrix} \mathcal{X}^s \\ \mathcal{U}_\Gamma^s \end{Bmatrix} = \begin{Bmatrix} \mathbf{0} \\ \mathbf{r}_\Gamma^s \end{Bmatrix}. \tag{18}$$

For the case of floating subdomains (subdomains without enough essential boundary conditions), the solution to the local stochastic Neumann problem defined in Eq.(18) exists only if the right hand side vector satisfies the compatibility condition (i.e. the right hand side vector is orthogonal to the null space of the coefficient matrix). A computationally expensive Moore-Penrose pseudo-inverse can be used to obtain a solution to this singular system. Alternatively a small value can be added to the diagonal entries of the local stochastic stiffness matrix to ensure the solvability of this singular system. Such regularization does not change the problem, but only alters the preconditioner.

5.2. *Two-Level Stochastic Neumann-Neumann Preconditioner*

For a moderate range of subdomains the stochastic Neumann-Neumann preconditioner demonstrates a fairly good scalability [4]. However, the convergence rate of the one-level Neumann-Neumann preconditioner degrades with increasing the number of subdomains. This performance degradation can be attributed to the lack of global exchange of information. For the one-level preconditioner, the information is transferred only among the neighboring subdomains. In the two-level preconditioner, the coarse problem provides a mechanism for global propagation of information across the subdomains to effectively enhance the convergence rate of the PCGM solver.

In this section, we formulate a novel two-level preconditioner for stochastic PDEs. The preconditioner is equipped with a coarse problem constructed using a collection of corner nodes. The coarse problem connects the subdomains globally via the corner nodes and thus provides a mechanism to propagate the information quickly across the subdomains leading to a scalable preconditioner. Following a version of the Balancing Domain Decomposition by Constraints (BDDC) [9, 7], we formulate a two-level preconditioner for stochastic PDEs. The methodology is detailed next.

In the one-level Neumann-Neumann preconditioner in Eq.(15), a stochastic Neumann problem (see Eq.(18)) is solved to calculate the preconditioned residual for each iteration of PCGM solver. To ensure the solvability of the Neumann problem (defined in Eq.(18)), the artificial Dirichlet boundary conditions are imposed to remove the singularity from the local stiffness matrix of the floating subdomains.

Similar to the dual-primal domain decomposition methods [9, 8, 10], the boundary nodes can be partitioned into two subsets: remaining nodes (boundary nodes shared only between two adjacent subdomains) and corner nodes (boundary nodes shared between more than two subdomains plus the nodes on the ends of interface edges). The corner nodes serve the following purposes: (a) they impose artificial Dirichlet boundary conditions to remove the singularity from the local Schur complement matrix, (b) they introduce a coarse grid that provides a mechanism to propagate information globally.

For a typical subdomain Ω_s, the local unknown vector consists of the interior $\mathcal{X}_i^s$, remaining (fine) $\mathcal{U}_r^s$ and corner (coarse) $\mathcal{U}_c^s$ unknown vectors as schematically shown in Fig.(2). Therefore the local interface unknown vector $\mathcal{U}_\Gamma^s$ is given by

$$\mathcal{U}_\Gamma^s = \left\{ \begin{array}{c} \mathcal{U}_r^s \\ \mathcal{U}_c^s \end{array} \right\}, \tag{19}$$

where

$$\mathcal{U}_r^s = \mathcal{R}_s^r \mathcal{U}_\Gamma^s, \tag{20}$$
$$\mathcal{U}_c^s = \mathcal{R}_s^c \mathcal{U}_\Gamma^s. \tag{21}$$

The stochastic Boolean splitting operators $\mathcal{R}_s^r$ and $\mathcal{R}_s^c$ are defined by

$$\mathcal{R}_s^r = blockdiag(\mathbf{R}_s^{r,0}, \ldots, \mathbf{R}_s^{r,N}),$$
$$\mathcal{R}_s^c = blockdiag(\mathbf{R}_s^{c,0}, \ldots, \mathbf{R}_s^{c,N}),$$

where N is the number of the polynomial chaos basis, and $\mathbf{R}_s^{r,j}$ and $\mathbf{R}_s^{c,j}$ are defined by

$$\mathbf{u}_{r,j}^s = \mathbf{R}_s^{r,j}\mathbf{u}_{\Gamma,j}^s,$$
$$\mathbf{u}_{c,j}^s = \mathbf{R}_s^{c,j}\mathbf{u}_{\Gamma,j}^s.$$

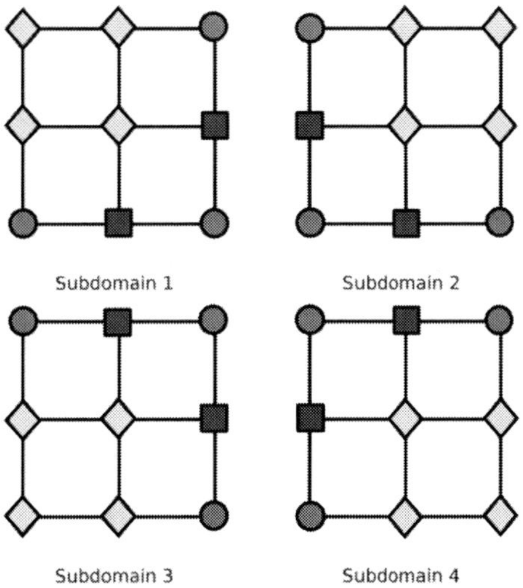

Figure 2. Partitioning the subdomain nodes into: interior (◆), remaining (■) and corner(●)

Performing partial assembly, the following linear system is obtained

$$\begin{bmatrix} \mathcal{A}_{ii}^s & \mathcal{A}_{ir}^s & \mathcal{A}_{ic}^s\mathcal{B}_c^s \\ \mathcal{A}_{ri}^s & \mathcal{A}_{rr}^s & \mathcal{A}_{rc}^s\mathcal{B}_c^s \\ \sum_{s=1}^{n_s}\mathcal{B}_c^{sT}\mathcal{A}_{ci}^s & \sum_{s=1}^{n_s}\mathcal{B}_c^{sT}\mathcal{A}_{cr}^s & \sum_{s=1}^{n_s}\mathcal{B}_c^{sT}\mathcal{A}_{cc}^s\mathcal{B}_c^s \end{bmatrix} \begin{Bmatrix} \mathcal{X}_i^s \\ \mathcal{U}_r^s \\ \mathcal{U}_c \end{Bmatrix} = \begin{Bmatrix} \mathbf{0} \\ \mathcal{F}_r^s \\ \sum_{s=1}^{n_s}\mathcal{B}_c^{sT}\mathcal{F}_c^s \end{Bmatrix}, \quad (22)$$

where

$$\mathcal{F}_r^s = \mathcal{R}_s^r r_\Gamma^s,$$
$$\mathcal{F}_c^s = \mathcal{R}_s^c r_\Gamma^s.$$

The restriction operator $\mathcal{B}_c^s$ in Eq.(22) is a Boolean rectangular matrix that maps the global coarse (corner) unknown $\mathcal{U}_c$ into the local coarse unknown $\mathcal{U}_c^s$ as

$$\mathcal{U}_c^s = \mathcal{B}_c^s\mathcal{U}_c,$$

where

$$\mathcal{B}_c^s = blockdiag(\mathbf{B}_{c,0}^s, \ldots, \mathbf{B}_{c,N}^s),$$

and $\mathbf{B}_{c,j}^s$ is defined by

$$\mathbf{u}_{c,j}^s = \mathbf{B}_{c,j}^s \mathbf{u}_{c,j}.$$

Consequently we obtain the following extended Schur complement system

$$\begin{bmatrix} \mathcal{S}_{rr}^s & \mathcal{S}_{rc}^s \mathcal{B}_c^s \\ \sum_{s=1}^{n_s} \mathcal{B}_c^{sT} \mathcal{S}_{cr}^s & \sum_{s=1}^{n_s} \mathcal{B}_c^{sT} \mathcal{S}_{cc}^s \mathcal{B}_c^s \end{bmatrix} \left\{ \begin{array}{c} \mathcal{U}_r^s \\ \mathcal{U}_c \end{array} \right\} = \left\{ \begin{array}{c} \mathcal{F}_r^s \\ \sum_{s=1}^{n_s} \mathcal{B}_c^{sT} \mathcal{F}_c^s \end{array} \right\}, \tag{23}$$

where

$$\mathcal{S}_{\alpha\beta}^s = \mathcal{A}_{\alpha\beta}^s - \mathcal{A}_{\alpha i}^s [\mathcal{A}_{ii}^s]^{-1} \mathcal{A}_{i\beta}^s,$$

the subscripts α and β represent the index r and c.

Due to the artificial Dirichlet boundary conditions imposed by corner nodes, the local Schur complement matrix $\mathcal{S}_{rr}^s$ in Eq.(23) is positive-definite and invertible for the floating subdomains. One could concurrently obtain $\mathcal{U}_r^s$ for each subdomain as

$$\mathcal{S}_{rr}^s \mathcal{U}_r^s = \mathcal{F}_r^s - \mathcal{S}_{rc}^s \mathcal{B}_c^s \mathcal{U}_c. \tag{24}$$

For the global corner node unknown $\mathcal{U}_c$, the following extended Schur complement system is obtained

$$F_{cc} \mathcal{U}_c = d_c, \tag{25}$$

where

$$F_{cc} = \sum_{s=1}^{n_s} \mathcal{B}_c^{sT} (\mathcal{S}_{cc}^s - \mathcal{S}_{cr}^s [\mathcal{S}_{rr}^s]^{-1} \mathcal{S}_{rc}^s) \mathcal{B}_c^s,$$
$$d_c = \sum_{s=1}^{n_s} \mathcal{B}_c^{sT} (\mathcal{F}_c^s - \mathcal{S}_{cr}^s [\mathcal{S}_{rr}^s]^{-1} \mathcal{F}_r^s).$$

The coarse problem defined in Eq.(25) couples the subdomains by the corner nodes in order to transfer information globally. Although the size of the coarse problem is relatively small compared to that of the fine problem, it is still computationally expensive to construct the coarse extended Schur complement system explicitly. Therefore we solve the coarse problem iteratively in parallel without constructing the coarse extended Schur complement matrix explicitly.

Next, the local interface unknowns can be recovered as

$$\mathcal{U}_\Gamma^s = \mathcal{R}_s^{r^T} \mathcal{U}_r^s + \mathcal{R}_s^{c^T} \mathcal{U}_c^s. \tag{26}$$

After some algebraic manipulations, the two-level Neumann-Neumann preconditioner for the stochastic system can be obtained as

$$\mathcal{M}_{CNN}^{-1} = \sum_{s=1}^{n_s} \mathcal{R}_s^T \mathcal{D}_s^T (\mathcal{R}_s^{r^T} [\mathcal{S}_{rr}^s]^{-1} \mathcal{R}_s^r) \mathcal{D}_s \mathcal{R}_s + R_0^T [F_{cc}]^{-1} R_0, \tag{27}$$

where the fine to coarse space restriction operator is defined by

$$R_0 = \sum_{s=1}^{n_s} \mathcal{B}_c^{s^T} (\mathcal{R}_s^c - \mathcal{S}_{cr}^s [\mathcal{S}_{rr}^s]^{-1} \mathcal{R}_s^r) \mathcal{D}_s \mathcal{R}_s. \tag{28}$$

Clearly, the two-level stochastic Neumann-Neumann preconditioner expressed in Eq.(27) consists of local problems $[\mathcal{S}_{rr}^s]^{-1}$ in order to construct a subdomain level preconditioner, and a coarse problem $[F_{cc}]^{-1}$ for global information propagation. This global exchange of information leads to a scalable preconditioner.

5.3. Parallel Implementation

The parallel PCGM iterative solver is used to tackle the extended Schur complement system defined by

$$\mathcal{M}_{CNN}^{-1} \mathcal{S}\, \mathcal{U}_\Gamma = \mathcal{M}_{CNN}^{-1} \mathcal{G}_\Gamma. \tag{29}$$

The procedure is listed in Algorithm(1) [10].

Algorithm 1: The Parallel PCGM Procedure

input : $(\mathcal{U}_{\Gamma_0})$

`compute` : $\mathbf{r}_{\Gamma_0} = \mathcal{G}_\Gamma - \sum_{s=1}^{s=n_s} \mathcal{R}_s^T \mathcal{S}_s \mathcal{R}_s \mathcal{U}_{\Gamma_0}$

`precondition` : $\mathcal{Z}_0 = \mathcal{M}_{CNN}^{-1} \mathbf{r}_{\Gamma_0}$

`compute` : $\mathcal{P}_0 = \mathcal{Z}_0$

`compute` : $\rho_0 = (\mathbf{r}_{\Gamma_0}, \mathcal{Z}_0)$

for $j = 1, 2, \cdots$, **until convergence**:

do

  `compute` : $\mathcal{Q}_j = \sum_{s=1}^{s=n_s} \mathcal{R}_s^T \mathcal{S}_s \mathcal{R}_s \mathcal{P}_j$

  `compute` : $\rho_{tmp_j} = (\mathcal{Q}_j, \mathcal{P}_j)$

  `compute` : $\alpha_j = \rho_j / \rho_{tmp_j}$

  `update` : $\mathcal{U}_{\Gamma_{j+1}} = \mathcal{U}_{\Gamma_j} + \alpha_j \mathcal{P}_j$

  `update` : $\mathbf{r}_{\Gamma_{j+1}} = \mathbf{r}_{\Gamma_j} - \alpha_j \mathcal{Q}_j$

  `precondition` : $\mathcal{Z}_{j+1} = \mathcal{M}_{CNN}^{-1} \mathbf{r}_{\Gamma_{j+1}}$

  `compute` : $\rho_{j+1} = (r_{\Gamma_{j+1}}, \mathcal{Z}_{j+1})$

  `compute` : $\beta_j = \rho_{j+1} / \rho_j$

  `update` : $\mathcal{P}_{j+1} = \mathcal{Z}_{j+1} + \beta_j \mathcal{P}_j$

output : $(\mathcal{U}_\Gamma)$

The vector $\mathcal{Q}_j$ in steps 1 and 7 of Algorithm (1), which represents the action of the extended Schur complement matrix on a vector, is computed in parallel by three matrix vector products and one direct solve of a local Dirichlet problem (with non-homogeneous boundary condition) on each subdomain [10] as shown in Algorithm (2).

Algorithm 2: Dirichlet-Solver Procedure

---

input : $(\mathcal{P})$

for $s = 1, 2, \cdots, n_s$ **in parallel**:

do

- `scatter` : $\mathcal{P}^s = \mathcal{R}_s \mathcal{P}$
- `compute` : $v_1 = \mathcal{A}^s_{I\Gamma} \mathcal{P}^s$
- `solve` : $\mathcal{A}^s_{II} v_2 = v_1$
- `compute` : $v_3 = \mathcal{A}^s_{\Gamma I} v_2$
- `compute` : $v_4 = \mathcal{A}^s_{\Gamma\Gamma} \mathcal{P}^s$
- `compute` : $\mathcal{Q}^s = v_4 - v_3$
- `gather` : $\mathcal{Q} = \sum_{s=1}^{n_s} \mathcal{R}_s^T \mathcal{Q}^s$

output : $(\mathcal{Q})$

The two-level preconditioner effect on the residual vector in steps 2 and 12 of Algorithm (1) defined by

$$\mathcal{Z} = \mathcal{M}^{-1}_{CNN} \mathbf{r}_\Gamma, \tag{30}$$

can be implemented in parallel as outlined in Algorithm(3)

Algorithm 3: Two-Level Neumann-Neumann Preconditioner Effect Procedure

input : $(\mathbf{r}_\Gamma)$
1 **for** $s = 1, 2, \cdots, n_s$ **in parallel**:
2 **do**
3 `scatter` : $\mathbf{r}_\Gamma^s = \mathcal{D}_s \mathcal{R}_s \mathbf{r}_\Gamma$
4 `compute` : $\mathcal{F}_r^s = \mathcal{R}_s^r \mathbf{r}_\Gamma^s$
5 `compute` : $\mathcal{F}_c^s = \mathcal{R}_s^c \mathbf{r}_\Gamma^s$
6 `solve` : $\mathcal{S}_{rr}^s v_1^s = \mathcal{F}_r^s$
7 `update` : $d_c^s = \mathcal{F}_{\Gamma_c}^s - \mathcal{S}_{cr}^s v_1^s$
8 `gather` : $d_c = \sum_{s=1}^{n_s} \mathcal{B}_c^{sT} d_c^s$
9 `solve` : $F_{cc} \mathcal{Z}_c = d_c$
10 **for** $s = 1, 2, \cdots, n_s$ **in parallel**:
11 **do**
12 `scatter` : $\mathcal{Z}_c^s = \mathcal{B}_c^s \mathcal{Z}_c$
13 `update` : $v_2^s = \mathcal{F}_{\Gamma_r}^s - \mathcal{S}_{rc}^s \mathcal{Z}_c^s$
14 `solve` : $\mathcal{S}_{rr}^s \mathcal{Z}_f^s = v_2^s$
15 `update` : $\mathcal{Z}^s = \mathcal{R}_s^{r^T} \mathcal{Z}_f^s + \mathcal{R}_s^{c^T} \mathcal{Z}_c^s$
16 `gather` : $\mathcal{Z} = \sum_{s=1}^{n_s} \mathcal{R}_s^T \mathcal{D}_s \mathcal{Z}^s$

output : $(\mathcal{Z})$

The local solve in steps 6 and 14 of Algorithm (3) constitutes a subdomain level stochastic Neumann problem which can be tackled using Algorithm (4) as follows

Algorithm 4: Neumann-Solver Procedure

input : $(\mathcal{F}_r^s)$
1 `solve` : $\begin{bmatrix} \mathcal{A}_{ii}^s & \mathcal{A}_{ir}^s \\ \mathcal{A}_{ri}^s & \mathcal{A}_{rr}^s \end{bmatrix} \begin{Bmatrix} \mathcal{X}_i^s \\ \mathcal{U}_r^s \end{Bmatrix} = \begin{Bmatrix} 0 \\ \mathcal{F}_r^s \end{Bmatrix}$

output : $(\mathcal{U}_r^s)$

The coarse problem in step 9 of of Algorithm (3) is solved iteratively in parallel using PCGM solver equipped with a parallel lumped preconditioner as

$$\mathcal{M}_c^{-1} F_{cc} \mathcal{Z}_c = \mathcal{M}_c^{-1} d_c, \tag{31}$$

where

$$\mathcal{M}_c^{-1} = \sum_{s=1}^{n_s} \mathcal{B}_c^{sT} \mathcal{A}^s{}_{\Gamma\Gamma} \mathcal{B}_c^s. \tag{32}$$

6. Numerical Results

For numerical illustration, we consider a two dimensional Poisson PDE with randomly heterogeneous permeability coefficient defined by

$$\nabla \cdot (\kappa(\mathbf{x},\theta)\nabla u(\mathbf{x},\theta)) = f(\mathbf{x}), \quad \mathbf{x} \in \Omega, \tag{33}$$

$$u(\mathbf{x},\theta)) = 0, \quad \mathbf{x} \in \partial\Omega. \tag{34}$$

The coefficient $\kappa(\mathbf{x},\theta)$ is modeled as a lognormal stochastic process obtained from the underlying Gaussian process with an exponential covariance function given by

$$C_{\alpha\alpha}(\mathbf{x},\mathbf{y}) = \sigma^2 \exp\left(-\frac{|x_1 - y_1|}{b_1} - \frac{|x_2 - y_2|}{b_2}\right). \tag{35}$$

The lognormal process is approximated using the four-dimensional third order polynomial chaos expansion ($L = 34$). The maximum size of the finite element mesh considered consists of $494,322$ elements and $247,159$ nodes. The non-Gaussian response is expressed using the third order polynomial chaos expansion ($N = 34$) leading to a linear system of order $8,650,565$.

In PCGM implementation, the forcing term is taken to be the initial residual. The iterations are terminated when the ratio of L_2 norms of the current and the initial residual is less than 10^{-6} defined by

$$\frac{\|\mathcal{G}_\Gamma^k - \mathcal{S}\mathcal{U}_\Gamma^k\|_2}{\|\mathcal{G}_\Gamma^0\|_2} \leqslant 10^{-6}. \tag{36}$$

Numerical experiments are performed in a Linux cluster having 22 nodes (2 Quad-Core 3.0 GHz Intel Xeon processors and 32 GB of memory per node) with InfiniBand interconnect using Message Passing Interface (MPI) [12] and PETSc [11] parallel libraries.

6.1. *Stochastic features*

Fig.(3-a) shows the physical domain and Fig.(3-b) represents a typical partitioned mesh using METIS graph partitioner [13]. METIS performs mesh decomposition using the criteria of load balancing among the CPUs and minimum interface boundary among subdomains.

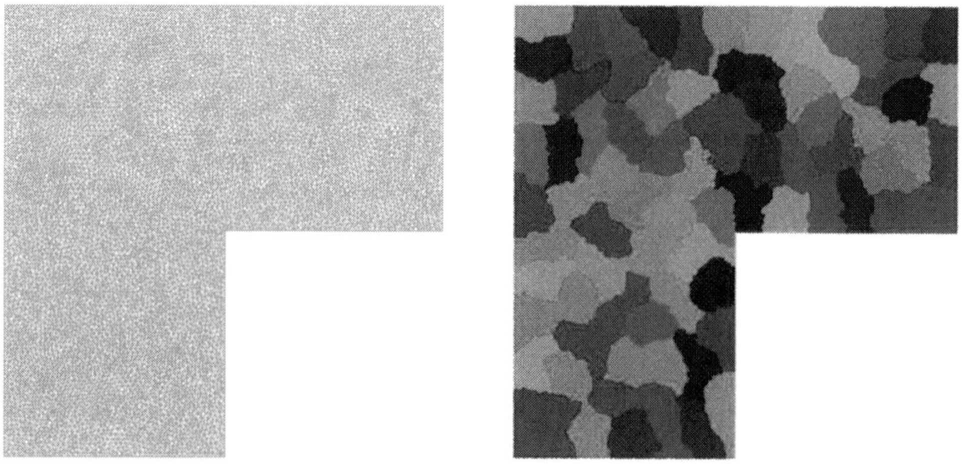

(a) FEM mesh (b) Decomposed mesh

Figure 3. FEM mesh and typical mesh decomposition

The mean and standard deviation of the non-Gaussian solution process are shown in Fig.(4-a) and Fig.(4-b) respectively. The maximum value of the coefficient of variation (CoV) of the solution field is about 21% highlighting the effect of uncertainty.

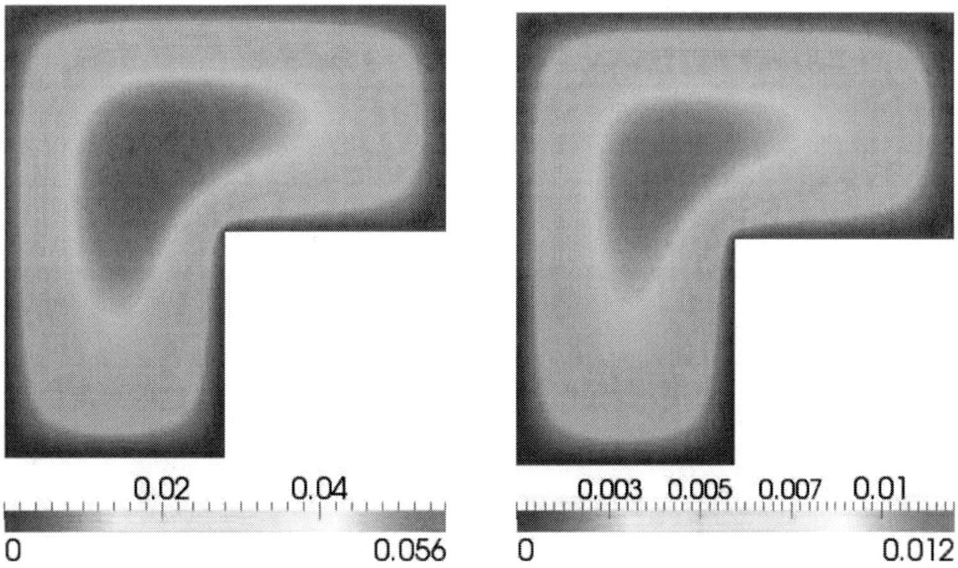

(a) Mean (b) Standard deviation

Figure 4. The mean and standard deviation of the solution process

6.2. Scalability study

Next we investigate the numerical scalability of the two-level Neumann-Neumann preconditioner (CNN) with respect to mesh size (h), subdomain size (H) and fixed problem size per subdomain (h/H) as discussed in the following subsections. In order to demonstrate the usefulness of the coarse grid, we contrast the performance of the one-level (NN) and two-level Neumann-Neumann (CNN) preconditioner.

6.2.1. Scalability with respect to mesh size: For this scalability test we fix the number of subdomains to 128 while increasing the mesh resolution in the geometric dimension. Fig.(5) shows the results for both NN and CNN preconditioners for the first, second and third polynomial chaos (PC) expansions while Fig.(6) shows the same results for NN and CNN preconditioners separately for various order of PC expansions. Unlike NN preconditioner, increasing mesh resolution does not deteriorate the performance of the CNN preconditioner as the iteration counts of PCGM solver remains nearly constant. For NN preconditioner the iteration counts increases as the problem size grows indicating the dependency of the NN preconditioner on the problem size. While increasing the order of PC expansions does not affect the performance of CNN preconditioner, it degrades the performance of NN preconditioner. For a given spatial problem size (n), the first and third order PC expansions lead to a total problem size of ($5 \times n$) and ($35 \times n$) respectively. These performance results suggest that the CNN preconditioner is numerically scalable with respect to the problem size and order of PC expansion.

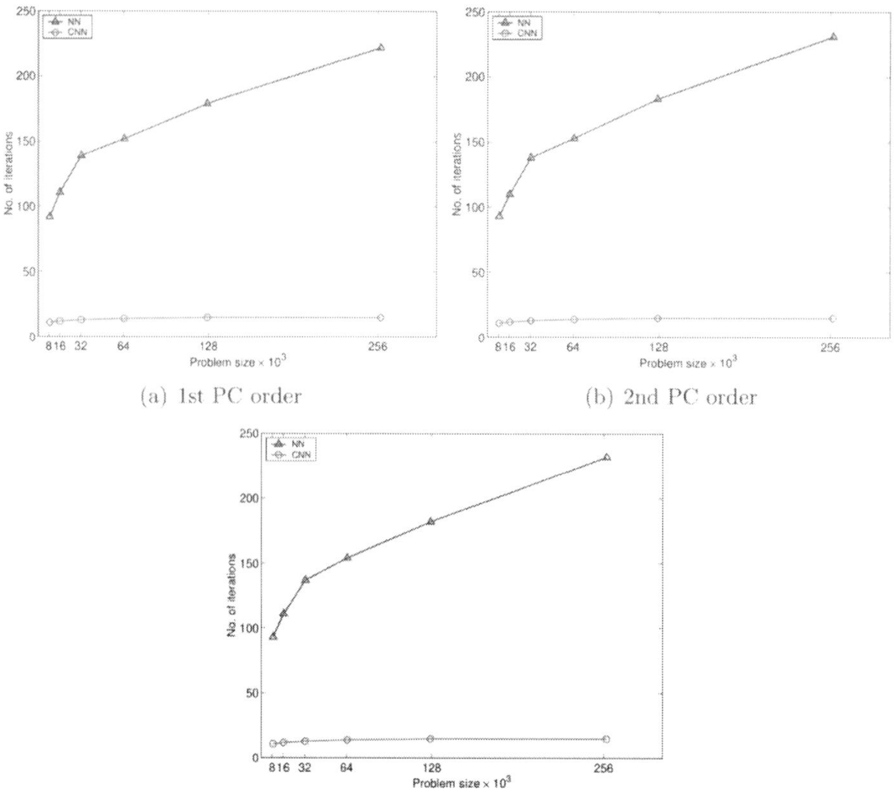

(a) 1st PC order (b) 2nd PC order

(c) 3rd PC order

Figure 5. Iteration counts for fixed number of subdomains

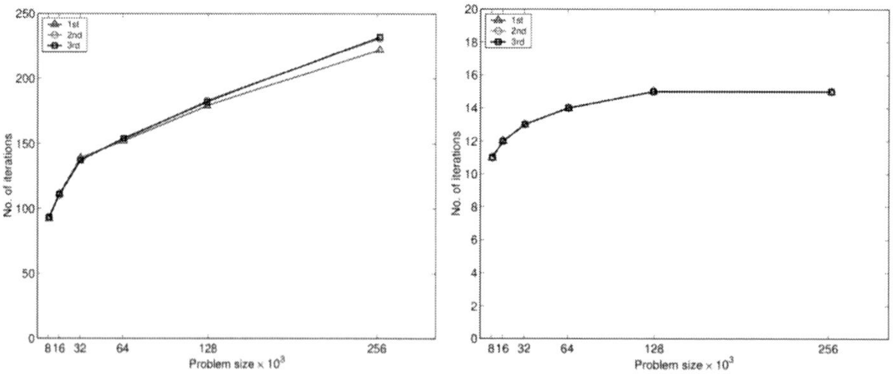

(a) NN preconditioner (b) CNN preconditioner

Figure 6. Iteration counts for fixed number of subdomains

6.2.2. Scalability with respect to subdomain size: Next we fix the problem size in the spatial dimension to 151, 179 and increase number of subdomains to solve the problem. In Fig.(7) and Fig.(8), the results are reported, respectively, for the first, second and third order PC expansions for both NN and CNN preconditioners. These performance results suggest that the

CNN preconditioner is numerically scalable with respect to number of subdomains. Clearly, NN preconditioner requires much more number of iterations to converge compared to CNN preconditioner.

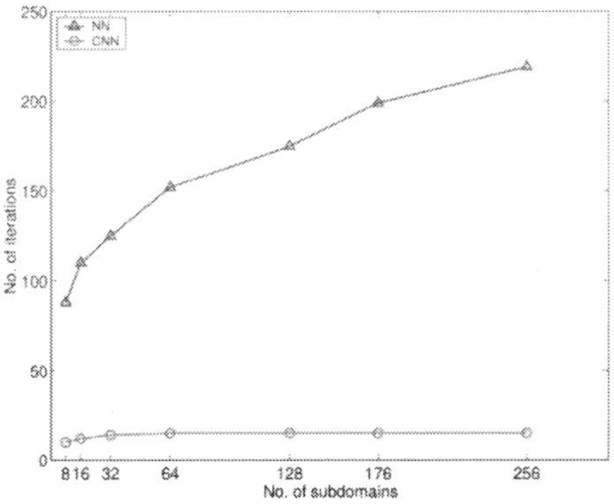

(a) 1st PC order

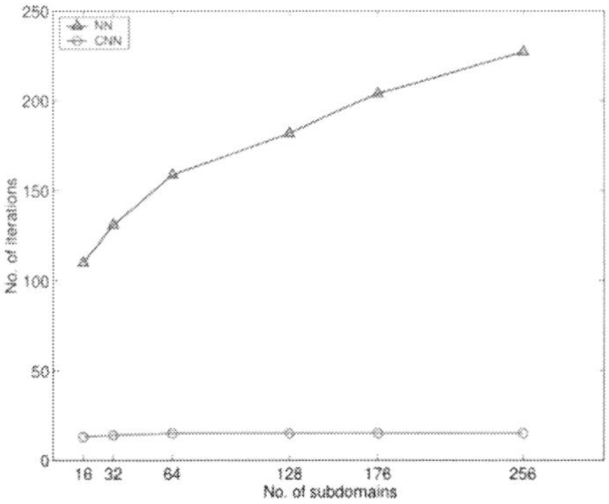

(b) 2nd PC order

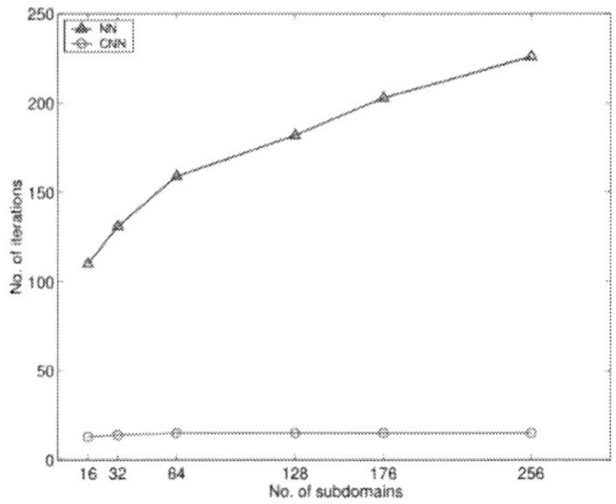

(c) 3rd PC order

Figure 7. Iteration counts for fixed problem size

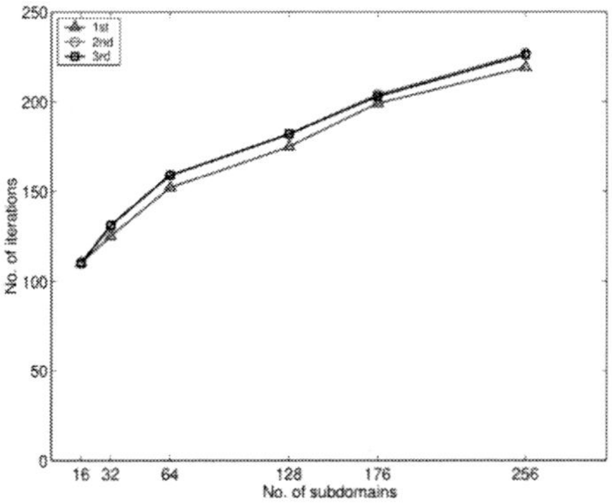

(a) NN preconditioner

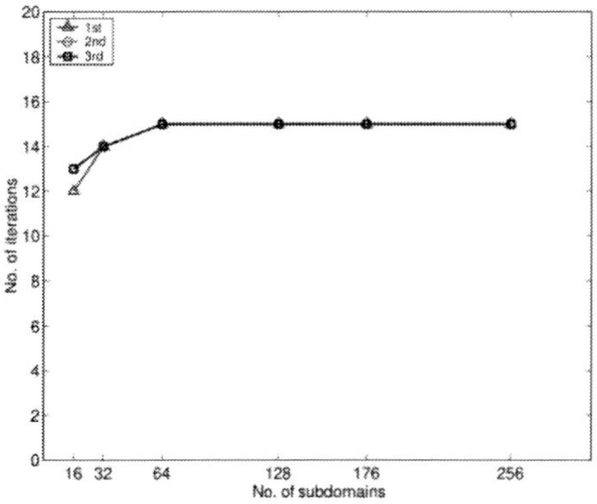

(b) CNN preconditioner

Figure 8. Iteration counts for fixed problem size

Journal of Physics: Conference Series **341** (2012) 012033 doi:10.1088/1742-6596/341/1/012033

6.2.3. Scalability with respect to the fixed problem size per subdomain: Fig.(9) and Fig.(10) demonstrate the performance of NN and CNN preconditioners for the fixed problem size per subdomain while increasing the overall problem size by adding more subdomains. The results are shown for the first, second and third order PC expansions for both NN and CNN preconditioners. In contrast to NN preconditioner, these performance results suggest that CNN preconditioner is scalable with respect to the fixed problem size per subdomain.

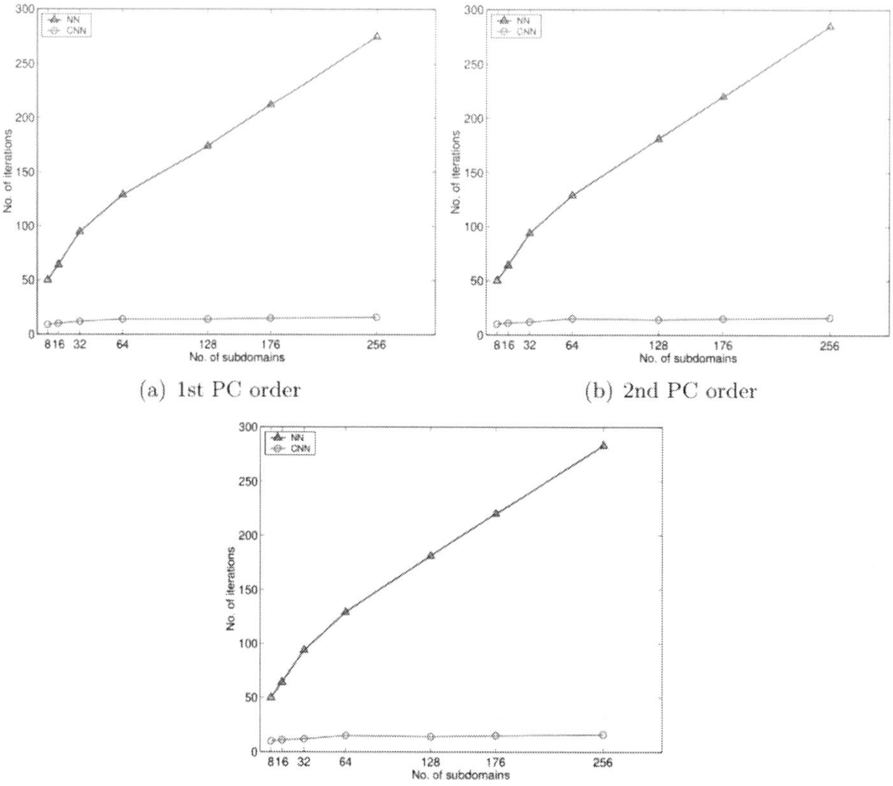

(a) 1st PC order

(b) 2nd PC order

(c) 3rd PC order

Figure 9. Iteration counts for fixed problem size per subdomain

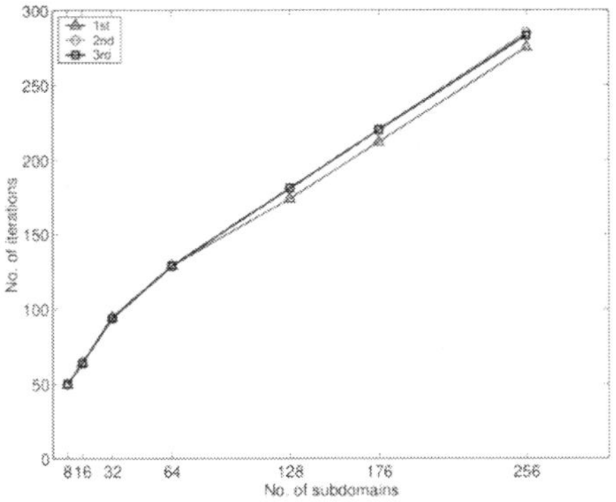

(a) NN preconditioner

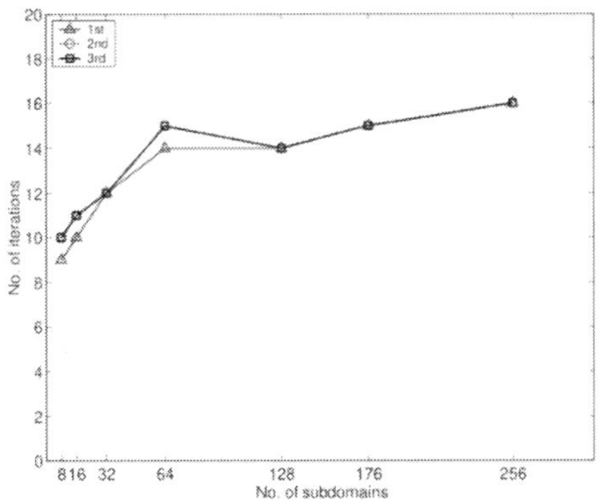

(b) CNN preconditioner

Figure 10. Iteration counts for fixed problem size per subdomain

7. Conclusions

A novel two-level scalable domain decomposition preconditioner is described for the iterative solutions of large-scale linear systems arising from the stochastic finite element method. The preconditioner is equipped with a coarse grid to propagate information globally across the subdomains both in geometric and stochastic dimensions. The proposed preconditioner may be viewed as an extension of the Balancing Domain Decomposition by Constraints (BDDC) method for stochastic PDEs. For the specific illustrations considered in the paper, the numerical experiment demonstrates that the preconditioner is numerically scalable with respect to the problem size, subdomain size and fixed problem size per subdomain. Furthermore, the preconditioner shows a convergence rate nearly independent of the order of the polynomial chaos expansion (namely the order of stochastic dimension).

References

[1] Ghanem R and Spanos P 1991 *Stochastic Finite Element: A Spectral Approach* (New York: Springer-Verlag)
[2] Sarkar A, Benabbou N and Ghanem R 2009 *International Journal for Numerical Methods in Engineering* **77** 689–701
[3] Subber W, Monajemi H, Khalil M and Sarkar A 2008 *International Symposium on Uncertainties in Hydrologic and Hydraulic*
[4] Subber W and Sarkar A 2010 *Proceddings of High Performance Computing Systems and Applications* **5976** 251–268
[5] Toselli A and Widlund O 2005 *Domain Decomposition Methods - Algorithms and Theory* (*Springer Series in Computational Mathematics* vol 34) (Berlin: Springer)
[6] Mathew T 2008 *Domain Decomposition Methods for the Numerical Solution of Partial Differential Equations* (*Lecture Notes in Computational Science and Engineering* no 61) (Berlin: Springer)
[7] Sousedik B and Mandel J 2008 *Electronic Transactions in Numerical Analysis* **31** 384–402
[8] Farhat C, Lesoinne M and Pierson K 2000 *Numerical Linear Algebra with Applications* **7**(7-8) 687–714
[9] Cros J 2003 *Fourteenth International Conference on Domain Decomposition Methods* 373–380
[10] Subber W and Sarkar A 2010 *Journal of Physics: Conference Series* **256**
[11] Balay S, Buschelman K, Gropp W, Kaushik D, Knepley M, McInnes L, Smith B and Zhang H 2009 PETSc Web page http://www.mcs.anl.gov/petsc
[12] Message passing interface forum http://www.mpi-forum.org
[13] Karypis G and Kumar V 1995 METIS unstructured graph partitioning and sparse matrix ordering system
[14] Cameron R H and Martin W T 1947 *Ann. Math* **48** 385 – 392
[15] Wiener N 1938 *Amer. J. Math.* **60** 897 – 936
[16] Saad Y 2003 *Iterative methods for sparse linear systems* 2nd ed (Philadelphia: SIAM)
[17] Smith B, Bjorstad P and Gropp W 1996 *Domain Decomposition: Parallel Multilevel Methods for Elliptic Partial Differential Equations* (New York: Cambridge University Press)

High Performance Computing Symposium 2011 IOP Publishing
Journal of Physics: Conference Series **341** (2012) 012034 doi:10.1088/1742-6596/341/1/012034

Bioinformatics algorithm based on a parallel implementation of a machine learning approach using transducers.

Abiel Roche-Lima[1]

Department of Computer Science, E2-445 EITC. University of Manitoba, Winnipeg, Manitoba R3T 2N2, Canada

E-mail: aroche@cs.umanitoba.ca

Ruppa K. Thulasiram

Department of Computer Science, E2-445 EITC. University of Manitoba, Winnipeg, Manitoba R3T 2N2, Canada

E-mail: tulsi@cs.umanitoba.ca

Abstract. Finite automata, in which each transition is augmented with an output label in addition to the familiar input label, are considered finite-state transducers. Transducers have been used to analyze some fundamental issues in bioinformatics. Weighted finite-state transducers have been proposed to pairwise alignments of DNA and protein sequences; as well as to develop kernels for computational biology. Machine learning algorithms for conditional transducers have been implemented and used for DNA sequence analysis. Transducer learning algorithms are based on conditional probability computation. It is calculated by using techniques, such as pair-database creation, normalization (with Maximum-Likelihood normalization) and parameters optimization (with Expectation-Maximization - EM). These techniques are intrinsically costly for computation, even worse when are applied to bioinformatics, because the databases sizes are large. In this work, we describe a parallel implementation of an algorithm to learn conditional transducers using these techniques. The algorithm is oriented to bioinformatics applications, such as alignments, phylogenetic trees, and other genome evolution studies. Indeed, several experiences were developed using the parallel and sequential algorithm on Westgrid (specifically, on the Breeze cluster). As results, we obtain that our parallel algorithm is scalable, because execution times are reduced considerably when the data size parameter is increased. Another experience is developed by changing precision parameter. In this case, we obtain smaller execution times using the parallel algorithm. Finally, number of threads used to execute the parallel algorithm on the Breezy cluster is changed. In this last experience, we obtain as result that speedup is considerably increased when more threads are used; however there is a convergence for number of threads equal to or greater than 16.

[1] To whom any correspondence should be addressed.

1. Introduction and Motivation

Machine learning is concerned with the design and development of algorithms and techniques that allow computers to "learn". Many applications of Machine Learning methods have been developed for use in Bioinformatics. Some of the most important approaches applied to this field are Neural Network, Support Vector Machine, Bayesian Networks and Hidden Markov Model [1-4].

On the other hand, finite automata, in which each transition is augmented with an output label in addition to the familiar input label, are considered *finite-state transducers* [5]. Transducer algorithms have been successfully used in a variety of applications such as speech recognition [6-8], optical character recognition [9], machine translation, a variety of other natural language processing tasks including parsing and language modeling, and image processing [10]. Indeed, transducers have been used to solve some problems in bioinformatics; weighted finite-state transducers have been proposed to pairwise alignments of DNA and protein sequences; as well as to develop kernels for computational biology [2, 11, 12]. Also, machine learning algorithms for conditional transducers have been implemented and used for DNA sequence analysis [13].

Transducer learning algorithms are based on the conditional probability computation. It is calculated using techniques, such as the creation of a pair database, normalization (with Maximum-Likelihood normalization) and parameters optimization (with Expectation-Maximization - EM) [14, 15]. All these techniques are intrinsically costly for computation when they are implemented; and when they are applied to bioinformatics the computational cost further increases. That means, it is almost unviable to use these techniques with serial implementation.

Therefore, the first objective of this work is to describe the implementation of these techniques in a bioinformatics application using parallel technologies to be utilized for conditional transducers learning. Secondly, we discuss the results when this application was executed on a grid to evaluate execution times.

WestGrid is a grid that operates across western Canada offering high performance computing, collaboration and visualization infrastructure [16]. The system software that handles the batch jobs consists of two pieces: a resource manager (TORQUE) and a scheduler (Moab). Together, TORQUE and Moab provide a suite of commands for submitting jobs, altering some of the properties of waiting jobs (such as reordering or deleting them), monitoring their progress and killing ones that are having problems or are no longer needed. In this work Westgrid is used as a baseline to test our algorithm.

The remainder of this paper is organized as follows: section 2, background and related works are described; section 3 states the specific problem; solution strategy and implementation are detailed in section 4; section 5 describes the experimental framework; finally results are illustrate in section 6, and conclusion and future works presented in section 7.

2. Background and Related Work

There have been several efforts to improve the speed and scalability of bioinformatics applications using parallel techniques. De-Cypher [17] is such a product that utilizes reconfigurable hardware to accelerate sequence analysis. Other strategies use multicore network processors for bioinformatics processing [18], as well as GPU techniques. For example, ClawHMMER [19] is a GPU-based streaming Viterbi algorithm that scales nearly linear within a cluster of Radeon 9800 Pro GPUs. Another example is P7Viterbi algorithm described in Lindahl [20]. They used Altivec's, including Single Instruction Multiple Data (SIMD) instructions to achieve speedup beyond Intel Pentium.

Nevertheless, there are not too many parallel algorithm related to bioinformatics applications using machine learning approaches. There are some efforts in specific implementations related to learning transducers (e.g. Maximum-Likelihood normalization, Expectation-Maximization, etc.) that has been taken into account in this work.

On the other hand, there are some bioinformatics applications running on grids. For example, on Breezy cluster (i.e. a cluster on Westgrid) there is available BLAST (Basic Local Alignment Search Tool), which is a suite of tools for assessing the similarity of a given sequence of proteins or nucleotides with a database of sequences [21].

Next section describes in details the problem we studying.

3. Specific Problem Statement

One of the main uses of the learning conditional transducer algorithm applied to bioinformatics is for classification. For example, based on a dataset with DNA sequences, classified following a criteria (i.e. ...ACTGCTACTAGGGGCCTTTA... – Methanogenic; ...CAGCTAAGAGCTTCTCTTA... – Protelitic; ...TAGACTACTAGGGGCCTTTA.. – Cellulolitic; .TGCAGCTAAGAGCTTCTCTTA... – Protelitic; ..CAACTGCTACTAGGGGCCTTTA... – Methanogenic; etc.) a transducer can be learned from this dataset in order to classify future sequences. The classification could be made by calculating the stochastic edit distance based on the learned transducers.

3.1 Learning transducer algorithm

This algorithm is divided in two parts and describes in Table 1.

Table 1. General idea of the Learning Conditional Transducer Algorithm

Part I. Learning a stochastic transducer
Step 1: The dataset will be considered the Learning Set (L_i), i.e. each sequence is classified under some consideration (i.e. type 0 - "methanogenics ", type 1 - "protelitics", type 2 - "celulolitic", etc.).
Step 2: from each sequence in L_i, a set of string pairs (pair dataset) P_i will be built in the form (x, y), where $y = NN(x)=argmin_{y \in Li} d_E(x,y)$ ($y=NN(x)$ is the Nearest Neighbor of x, calculated as the minimum value (*argmin*) of the d_E (the classic edit distance) between x and all other values in the same dataset ($y_i \in L_i$)).
Step 3: a unique conditional transducer t will be learned from all pair dataset ($\cup_i P_i$)
Part II. Classification process
For a new given sequence z,
Step 4: z will be classified as the same type as y, where $y \in L_i$ maximizing $p(y\|z)$, where $p(y\|z)$ *is the* ***conditional probability calculated using the learned transducer*** t.

As can be seen in Table 1, the algorithm is based on a dataset (L_i) of classified sequences under some conditions, e.g. the sequences are classified by their functionalities as type 0 - methanogenics, type 1 - protelitics, type 2 - celulolitic, etc. A new dataset, defined as pair-database (P_i), is created as $(x, NN(x))$, by calling the subalgorithm "pair-database creation", where each sequence x is going to be paired with its nearest neighbor $y = NN(x)$, defined as the smallest edit distance between x and y ($d_E(x,y)$).

The conditional transducer (t) is a matrix with the probabilities of the edit costs (insertions, deletion and substitutions) obtained by learning these values from the pair-database (P_i). The learning process is implemented using EM algorithm to estimate the parameters until a threshold precision is achieved.

Finally, a new sequence (z) can be classified comparing it with all elements in the original dataset (L_i) and calculating the stochastic distance using the transducer t. Thus, the sequence (y) where the stochastic distance is smallest (highest conditional probability - $p(y|z)$) will be chosen to classify z (i.e. z will be classified as y is in L_i: type 0 - methanogenics, type 1 - protelitics, type 2 - celulolitic, etc.

This algorithm is intrinsically high time-consuming. The "pair-database creation" algorithm has to create a new dataset using each element from the original dataset (L_i) and look for its nearest neighbor on the same dataset. Indeed, EM algorithm uses whole pair dataset (P_i) and calculates the probability to obtain a new transducer t, as many times as threshold precision is achieved.

When this algorithm is used in bioinformatics dataset with several thousands of sequences, high execution times are expected. Based on our experiences, 95% of total execution time is related to "pair-database creation" and EM algorithms (approximately, "pair-database creation" 55% and EM 40%).

In the next section, we describe the solution for parallel approach and the implementation.

4. Solution Strategy and Implementation

Based on the available parallel techniques, a previous study was made to decide what method to use to tackle this problem. Initially, dependencies along the variables and data were taken into account. In the case of the "pair-database creation" algorithm, there were not considerable dependencies between the *i*-iteration and the *(i-1)*- iteration. However, EM algorithm has remarkable dependencies for the different stages in the computation process.

Other considered issues to decide what parallel technique should be used were volume and communication data. In the "pair-database creation" algorithm the database size might be large, as well as communication among processors (if a distributed memory technique is chosen). In the EM algorithm will be the same, actually even worse, considering that the pair database (P_i) size is twice the original dataset (L_i) size. Taking these issues into consideration, a shared memory machine technique for parallel implementation was finally selected, using OpenMP (a C based parallel programming language) and C++ language.

In order to programming the general algorithm in parallel, these two high-consuming time sub-algorithms ("Pair-database creation" and EM) were re-implemented using OpenMP instructions. "Pair-database creation" algorithm uses as an input a database of sequences (L_i). For each sequence x a method is called to calculate its Nearest Neighbor ($NN(x)$) in the same dataset (this algorithm order is $O(n^2)$). When the sequential implementation of this algorithm runs with a dataset of few thousands of sequences, the execution time is around one hour (depending of the hardware configuration). During the implementation of the parallel version, "*pragma omp parallel for*" instruction of OpenMP language was used, in the outer *for loop* of this algorithm. Initially, we got some errors regarding "*Fault Segmentation*", due to "*race condition*". A race condition exists when two unsynchronized threads access the shared memory to try to modify a variable at the same time. In our algorithm, it occurred when the variable used to create the pair database was accessed by different threads at the same time. This error was avoided in the implementation by including "*#pragma critical*" instruction, just before this variable assignation. On that way, we give exclusive right on the address space to one processor at the same time, avoiding race conditions.

The EM algorithm was also implemented in its parallelized version, using the "*#pragma omp parallel for*" instruction in outer *for loop*. In this case, dependencies are in the internal *for loop*, being each thread responsible for the execution of the *Forward, Backward* and *Expectation* methods involved in this algorithm. Finally, the probability is computed by summing the obtained partial sum values for each iteration. Thus, a "*reduction sum*" instruction was used in order to obtain the final sum of each thread.

Considering our possibility to access to Westgrid infrastructure, the parallel and sequential versions of this algorithm was executed, changing some parameters. In the next section, we describe the experiments that have been developed to evaluate this implementation.

5. Experimental Framework

Two different parameters have been changed in the algorithm to develop the experiments to compare the parallel and sequential implementation performances. These parameters are: data size and precision. Indeed, we have considered one more parameter related to the grid, i.e. the number of threads to be used during the execution of the parallel implementation.

5.1. Data size

The sequential and parallel implementations were executed for different values of the database (L_i) size (number of sequences). In this case, values have been increased by:

Database size	=	405	810	1620	3240	6480

These algorithms were executed for these different values. Precision parameter was used as a constant (1E-03) as well as thread numbers (8).

5.2. Precision

By the contrary, in other set of experiments, precision values were changed to analyze execution times of these implementations. The precision parameter has an important role in the algorithm to define how many iterations have to be executed to reach the "best" transducer (*t*) and, consequently, a more precise classification. Precision parameter is used to stop the algorithm when the difference between the calculate transducer (*t*) and the one obtained in the previous iteration (*t-1*) is smaller than the precision value. In this case the parameter database size was kept as a constant (1620) as well as thread numbers (8), changing the precision values as follow:

Precision parameter	=	1E-01	1E-02	1E-03	1E-04	1E-05

5.3. Numbers of Threads

Finally, a parameter related to the grid configuration was changed. It was the numbers of threads. Particularly, in this set of experiments database size was kept constant (1620), as well as precision parameters (1E-03). Thread numbers were:

Number of threads	=	2	4	8	16	24

When this research was started, different resources (clusters) on the Westgrid were interested to use that permits execution of OpenMP programs on a grid. Unfortunately, no other resources were open on Westgrid that allowed us to compare them.

In the next section, the main results are presented.

6. Results and Discussions

The experimental testbed, as well as the main results are described here.

6.1. Experimental testbed and environment

The experiments were developed on Breezy, a cluster of the Westgrid [16] as we have mentioned above. It is a Linux AMD stanbul cluster with large-memory nodes connected by Infiniband. Breezy is intended for jobs that need a big amount of memory on a single node, such as for large OpenMP-based parallel jobs. The Breezy specifications are summarized in Table 2.

Table 2. Specifications of the Breezy cluster on the Westgrid

Processors:
▪ Quad-socket, 6-core AMD Istanbul processors (24 cores @ 2.4 GHz) per node ▪ 256 GB memory per compute node (64 GB on login node) ▪ 16 nodes conected with 4X DDR InfiniBand at 20 Gbits/s and Gigabit Ethernet
***Interconnect*:**
▪ 6 nodes conected with 4X DDR InfiniBand at 20 Gbits/s and Gigabit Ethernet
***Storage*:**
▪ 10 TB total for /global/scratch ▪ 10 TB total for /home

6.2. Evaluation

On this configuration, the sequential and parallel implementations were executed, following the designed experiments.

6.2.1. Data size

Regarding the different sizes of the database, important improvement were obtained by the parallel implementation. These results can be seen in table 3 and figure 1.

Table 3. Results of the sequential and parallel implementation for different data sizes

Database size (# sequences)	**Execution Time (sec)**		**Speedup** $S = T_s/T_p$	**Reduction in execution time (%) when parallelized**
	Parallel Implementation	Sequential Implementation		
405	17.8	66.9	3.8	73.7
810	34.0	188.6	5.6	82.1
1620	90.0	567.3	6.3	84.1
3240	290.2	1857.1	6.4	84.4
6480	816.4	5307.0	6.5	84.6

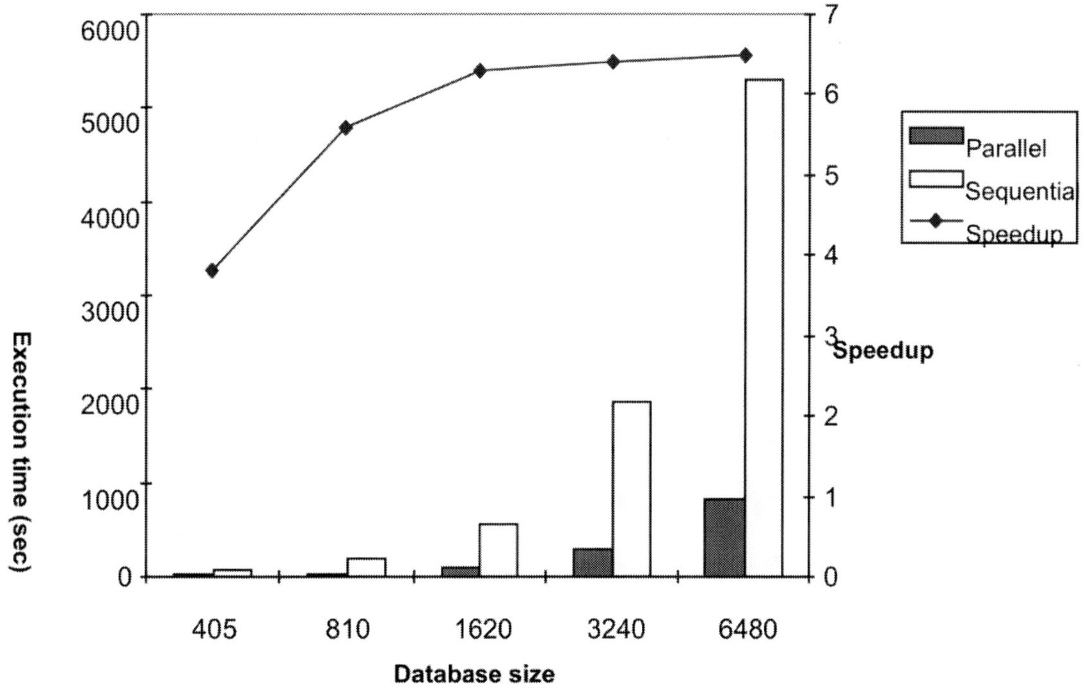

Figure 1. Results for different values of database size for sequential and parallel implementation

The execution times and speedups were obtained for several database (L_i) sizes. Table 3 shows these values. They are plotted in figure 1. As can be seen in these results, when database size increases, execution times of the sequential implementation are exponentially increased. However, in the parallel implementation, execution times were lower and for larger database sizes, these values have been increased close to linear. We have also used to show the performance, the speedup metric. Speedup is defined as the ratio between the sequential and parallel execution times (i.e. $S = T_s/T_p$). This is a better measure that the actual execution times to set a better picture of the parallel implementation.

6.2.1. Precision

As it was mentioned above, other experiments were related to change precision parameter to obtain the learned transducer. These results can be seen in table 4 and figure 2.

Table 4. Results of the sequential and parallel implementation for different precision values

Precision	**Execution Time (sec)**		**Speedup**	**Reduction in execution time (%) when parallelized**
	Parallel Implementation	Sequential Implementation		
1E-01	14.9	68.0	4.5	77.7
1E-02	17.3	97.2	5.6	82.1
1E-03	18.4	118.0	6.3	84.1
1E-04	18.8	116.8	6.2	83.8
1E-05	19.3	121.6	6.3	84.1

As can be seen in table 4, data are collected in relation to execution times for different precision values. Indeed, speedups and reduction in execution times are computed. As it was mentioned in section 5.2. precision is used to stop the loop when the difference between *t* and *t-1*(obtained in the last iteration) transducers is smaller than the precision parameter. For example, after a number of repetitions (e.g. 100000) the difference between transducer *t* and transducer *t-1* is 0.000001. In this example the threshold value which the algorithm converges is 100000 (i.e. if the difference between *t-1* and *t-2* (iteration 99999) is 0.01, and the difference between *t* and *t-1* is 0.000001, execution times will be same value for precision parameter = 1E-03, 1E-04, and 1E-05). Thus, for precision smaller than a certain value, the number of loops will be similar, because the algorithm converges.

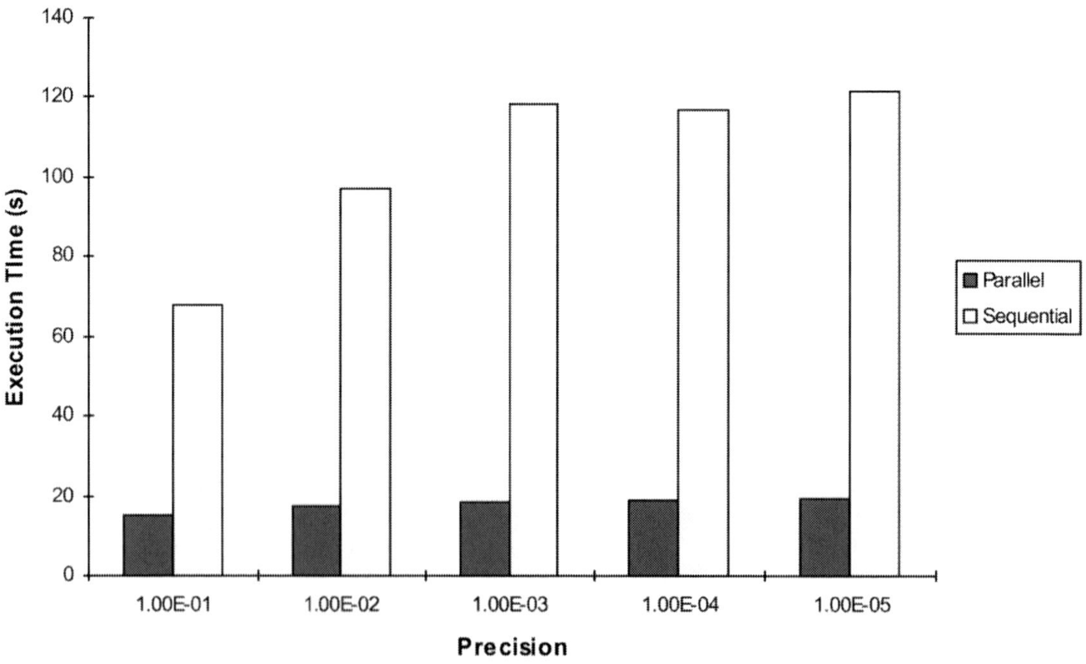

Figure 2. Results for different values of precision for sequential and parallel implementation

In figure 2 it is clear to see how the parallel algorithm starts to converge almost from the first precision value (1.00E-1). However the sequential implementation converges around the third computed precision value (1.00E-3). This tendency is due to the threshold value of the parallel implementation is reached faster, allowing the algorithm to converge from the first precision values. In other words, better solutions are achieved at much overhead implementing the algorithm in a parallel environment.

6.2.2. Numbers of Threads
Another changed parameter was the number of threads, a grid related parameter. Results are shown in figure 3.

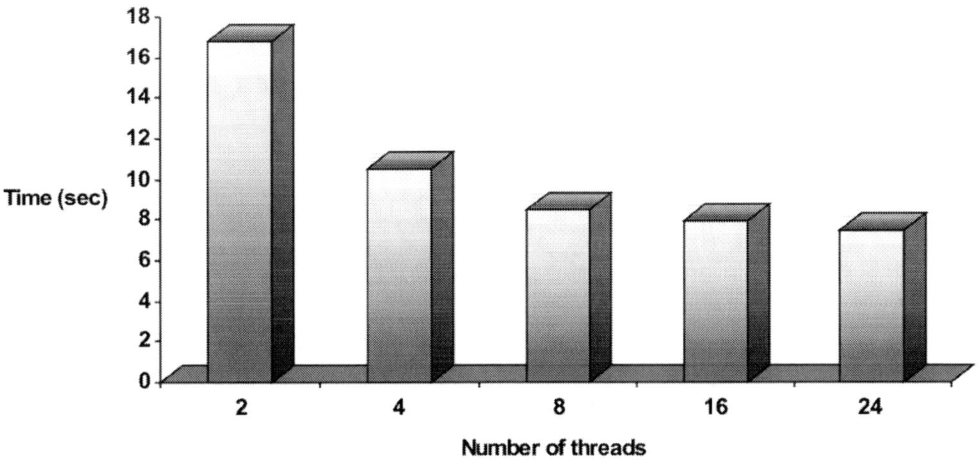

Figure 3. Results for different values of numbers of threads

As can be seen in this figure, data are collected in relation to execution times for different number of threads. Execution times decrease when thread numbers are added in the system, i.e. speedup is increased when the number of threads is larger. However, these nodes on Breezy cluster have 24 cores per node (see section 6.1 above), executing one thread per core. When the algorithm is executed using more than 24 threads, no result is obtained.

Indeed, for threads number equal to 16 and 24, the results are almost the same. This is because using these values, the threshold is achieved relatively close (.i.e. almost at the same time). In other words, with number of thread greater than 16, the algorithm converges around nine seconds.

7. Conclusions and Future Work
In this paper, we have described the problem, approach and evaluation of a parallelized implementation of the conditional transducer learning algorithm applied to bioinformatics.

The most time consuming sub-algorithms (95% of the total time) were parallelized using OpenMP. These are "pair-database creation" and EM. Several experiments were designed to evaluate our implementation, using as a baseline Breezy cluster, in Westgrid grid. Experiments were grouped in three different goals: (1) increasing database size, (2) increasing precision, and (3) changing the numbers of threads.

Obtained results have permitted to analyze this application on the grid. When database sizes were increased, execution time values of the sequential algorithm rose in an exponential trend. However, these values for parallel implementation were increased close to linear.

On the other hand, important results were obtained when parallel implementation was executed on the grid changing the precision parameter. Execution time values of the parallel algorithm were lower than sequential implementation. Indeed, the algorithm converged for the first precision values in the parallel proposal, while it was achieved later (for larger precision values) in the sequential approach.

Another important result was related to the number of threads used to execute the parallel implementation of the algorithm on the Westgrid. Considering the configuration of this cluster, the maximum number of threads available for nodes is 24. When the number of threads was increased, the speedup rose. However, there is a convergence in the execution times when the number of threads is equal to or greater than 16.

Concluding, the parallel implementation was able to guarantee considerably low execution times when it was executed on the Breezy cluster. It permits to speed up the performance of the algorithm.

In future work, some changes will be implemented in the algorithm to use larger sequences. When these modifications are made, a new parallelized implementation will be programmed and evaluated.

References

[1] S. J. Hua, and Z. R. Sun, "A novel method of protein secondary structure prediction with high segment overlap measure: Support vector machine approach," *Journal of Molecular Biology,* vol. 308, no. 2, pp. 397-407, 2001.

[2] R. Durbin, S. R. Eddy, A. Krogh *et al.*, "Biological sequence analysis: Probalistic Models of Proteins and Nucleic Acids.," Cambridge University Press, 1998.

[3] J. Soding, "Protein homology detection by HMM-HMM comparison (vol 21, pg 951, 2005)," *Bioinformatics,* vol. 21, no. 9, pp. 2144-2144, 2005.

[4] G. Pollastri, D. Przybylski, B. Rost *et al.*, "Improving the prediction of protein secondary structure in three and eight classes using recurrent neural networks and profiles," *Proteins-Structure Function and Genetics,* vol. 47, no. 2, pp. 228-235, 2002.

[5] J. Berstel, "Transductions and Context-Free Languages.," Teubner Studienbucher:Stuttgart, 1979.

[6] C. Allauzen, and M. Mohri, "Linear-Space Computation of the Edit-Distance between a String and a Finite Automaton," College Publications, 2009.

[7] M. Mohri, *Statistical natural language processing.*: Cambridge University Press, 2005.

[8] M. Mohri, F. C. N. Pereira, and M. Riley, "Speech recognition with weighted finite-state transducers.," In Larry Rabiner and Fred Juang, editors, Handbook on speech processing and speech communication, Part E: Speech recognition. Springer-Verlag, Heidelberg, Germany, , 2008.

[9] T. M. Breuel, "The OCRopus open source OCR system," *Document Recognition and Retrieval Xv,* vol. 6815, 2008.

[10] J. Albert, and J. Kari, "Digital image compression," In Manfred Droste, Werner Kuich, and Heiko Vogler, editors, Handbook of weighted automata, EATCS Monographs on Theoretical Computer Science, 2009.

[11] C. Allauzen, M. Mohri, and A. Talwalkar, "Sequence kernels for predicting protein essentiality."

[12] P. A. Pevzner, "Computational Molecular Biology: an Algorithmic Approach.," MIT Press, 2000.

[13] A. Roche-Lima, R. A. Grave de Peralta, M. A. Cora *et al.*, "Bioinformatics applied to genetic study of rumen microorganism," 2007.

[14] J. Oncina, and M. Sebban, "Learning stochastic edit distance: Application in handwritten character recognition," *Pattern Recognition,* vol. 39, no. 9, pp. 1575-1587, 2006.

[15] M. Bernard, J. C. Janodet, and M. Sebban, "A discriminative model of stochastic edit distance in the form of a conditional transducer," *Grammatical Inference: Algorithms and Applications, Proceedings,* vol. 4201, pp. 240-252, 2006.

[16] "WestGrid | Western Canada Research Grid," http://www.westgrid.ca/.

[17] http://www.timelogic.com/, "TimeLogic biocomputing products from Active Motif, Inc.," 23-12-2010.

[18] B. Wun, J. Buhler, and P. Crowley, "Exploiting coarsegrained parallelism to accelerate protein motif finding with a network processor," In PACT '05: Proceedings of the 2005 International Conference on Parallel Architectures and Compilation Techniques, 2005.

[19] D. R. Horn, M. Houston, and P. Hanrahan, "Clawhmmer: A streaming hmmer-search implementation," SC '05: The International Conference on High Performance Computing, Networking and Storage, 2005.

[20] E. Lindahl, "Altivec-accelerated HMM algorithms," http: //lindahl.sbc.su.se/, 2008.

[21] "BLAST - Basic Local Alignment Search Tool | WestGrid," http://www.westgrid.ca/support/software/blast.

AUTHOR INDEX

AUTHOR INDEX